North Shore

NORTH
SHORE

A NATURAL HISTORY OF
MINNESOTA'S
SUPERIOR COAST

CHEL ANDERSON

AND

ADELHEID FISCHER

UNIVERSITY OF MINNESOTA PRESS
MINNEAPOLIS · LONDON

Published by the University of Minnesota Press
111 Third Avenue South, Suite 290
Minneapolis, MN 55401-2520
http://www.upress.umn.edu

Design and production by Mighty Media, Inc.
Text design by Chris Long and Anders Hanson
Page layout and production by Chris Long
Image processing and color correction by Kelly Doudna
Copyediting by Mary Keirstead
Project management by Daniel Ochsner

Unless otherwise credited, illustrations are by Patti Isaacs
Shaded relief base maps copyright Michael Schmeling, www.aridocean.com

Library of Congress Cataloging-in-Publication Data
Anderson, Chel.
North shore : a natural history of Minnesota's Superior coast / Chel Anderson and Adelheid Fischer.
Includes bibliographical references and index.
ISBN 978-0-8166-3232-9 (hc : alk. paper)
1. Coastal ecology—Superior, Lake, region. 2. Ecology—Minnesota. 3. Lake ecology— Superior, Lake, region. 4. Lake ecology—Minnesota. 5. Natural history—Superior, Lake, region. 6. Natural history—Minnesota. 7. Superior, Lake. I. Fischer, Adelheid. II. Title.
QH104.5.S85A53 2015
577.5'1097749—dc23

2014039243

Printed in China on acid-free paper

The University of Minnesota is an equal-opportunity educator and employer.

22 21 20 19 18 17 16 15   10 9 8 7 6 5 4 3 2 1

To wonder, curiosity, inquiry, humility, imagination, reciprocity, and reverence in the pursuit of belonging.

—Chel Anderson

In memory of Paul "Lizard" Rothstein, who was there from the beginning and is here at the end.

—Adelheid Fischer

The University of Minnesota Press gratefully acknowledges the generous support provided for this publication by the Hamilton P. Traub University Press Fund and from the following institutions and individuals:

Elmer L. and Eleanor J. Andersen Foundation
Philip J. and Karna Anderson
Mary Lee Dayton
The Duff Family: In Memory of Nick Duff
Richard and Carol Flint
Glenn and Chelly Gilyard
John C. and Janet C. Green
Martin and Esther Kellogg
Kathy and Allen Lenzmeier
Stephanie Hemphill and William Miller
David and Karen Nasby
Ann Possis
Walter and Harriet Pratt
Eleanor and Frederick Winston

Contents

Abbreviations

BNL	benthic nepheloid layer
BP	before present
BWCAW	Boundary Waters Canoe Area Wilderness
DNR	Minnesota Department of Natural Resources
EDCS	endocrine-disrupting compounds
EPA	U.S. Environmental Protection Agency
GCM	global circulation model
GIS	geographic information system
GLFC	Great Lakes Fishery Commission
IJC	International Joint Commission
LGM	Last Glacial Maximum
LLO	Large Lakes Observatory
LMF	Laurentian Mixed Forest
MBS	Minnesota Biological Survey
MBWSR	Minnesota Board of Water and Soil Resources
MDH	Minnesota Department of Health
MNDOT	Minnesota Department of Transportation
MPCA	Minnesota Pollution Control Agency
NPS	National Park Service
NRRI	Natural Resources Research Institute
PCBS	polychlorinated biphenyls
SML	surface microlayer
SNF	Superior National Forest
UCS	Union of Concerned Scientists
USFS	U.S. Forest Service
USFWS	U.S. Fish and Wildlife Service
USGS	U.S. Geological Survey

A Gathering of Waters

It takes much less knowledge to exploit an ecosystem
than to live as part of one.

Kai Lee (1999)

Pay attention to the mystery. Apprentice to the best apprentices.
Rediscover in nature your own biology. Write and speak with
appreciation for all you have been gifted. Recognize that a politics
with no biology, or a politics without field biology, or a political platform
in which human biological requirements form but one plank,
is a vision of the gates of Hell.

Barry Lopez, "The Naturalist" (2001)

THE MINNESOTA NORTH SHORE HAS BEEN A MAGNET FOR LIFE FOR TWELVE thousand years or more. Some organisms arrived by chance; others came with intention. Some were propelled by wings, fins, feet, or the wind. Others navigated their way by the stars, or from behind a ship's wheel or the steering wheel of an automobile. Some have been transients, occupying the region's land and waters for mere days; others have put down deep roots, persisting beyond precise reckoning.

Today, these living beings form communities of great diversity in a complex mosaic of land and water. Here, longtime residents and long-distance migrants can take wildly different forms and exhibit a dramatic range of behaviors from the microscopic to the massive, the bold to the elusive, the fragile to the seemingly indestructible. Together with the shore's ancient mountains, azure lakes, and free-flowing rivers, they weave a living fabric of sublime beauty.

Documented within the pages of our book are the natural histories of these extraordinary people, plants, and animals. They celebrate what is distinctive and glorious about the North Shore's ecology and its inhabitants. They enable readers not only to locate and name the other members of their North Shore community but also to better understand how their neighbors live and what they need to thrive.

To help in this endeavor, we focus on the complex interplay among landforms, biotic systems, and organisms, including people. And we do so in considerable depth— for several reasons. Scientific investigation plays an essential role in our culture; yet the resulting knowledge and its real-world applications too often are communicated to the public only in sound bites or are framed in sensationalist terms—if at all. Much research is simply ignored because it is deemed too complex, too difficult to interpret, or too difficult to access since it is written in the technical language of science journals and public agency reports. Most troubling of all, scientific research has become a target of contention in our culture wars and increasingly is dismissed as a factor for consideration in public decision making. Our intent is to counter these tendencies and present science in a manner that is accurate, relevant, compelling, and accessible to nonscientists.

Our goals are premised on the belief that an advanced ecological education is important not simply for scientists but also for the general public. There is a critical need for each of us to thoughtfully examine how we live in our own habitats and to ensure that our lifeways further the welfare of the larger whole. As we explore the environmental legacy of those who came before us, we also pose the questions, What will be the environmental legacy of our own era? How can we live in a manner that sustains this home so that all its inhabitants might flourish?

This kind of reflection has become ever more urgent along the North Shore as tensions surrounding land management grow. How, for example, can we protect water quality as lakeshore development intensifies, conserve migratory bird corridors while pursuing alternative energy sources such as wind power, or sustain forest habitats and natural processes in the face of demands for forest products, mineral extraction, recreational access, and energy? How do we pay for services such as wildfire protection and road construction as homes and resorts are built in ever more remote corners of the North Shore watershed? How do we ensure that these roads do not become death traps for small, slow-moving animals, such as salamanders in their seasonal walkabouts, or that houses do not stand in the way of cleansing, renewing wildfires? How do we make land-use and management decisions in a way that does not foreclose the opportunity for these inhabitants to meet their needs in the future?

Whether we are residents or visitors, sustaining the living splendor of this place for generations to come requires a change in our way of responding to these environmental challenges. In this, holistic knowledge is key. We believe that "instead of learning more and more about less and less, we must learn more and more about the whole biotic landscape," as Aldo Leopold posited in *Round River,* particularly since humans now are exerting unprecedented influence on the biotic landscape to which they belong. Few books have attempted to knit this information into a coherent whole that might prove useful to North Shore residents and visitors. And yet it is precisely this kind of deeper understanding that could help ensure that people, as well as bears, butterworts, and ovenbirds, continue to flourish in the region.

We are ably guided in this task by the work of a devoted and gifted cadre of scientists. While researching our book, we were continually impressed by the remarkable scientific study in the Lake Superior region. We stand in awe of the dedicated people who carry out this research and their love for the organisms and natural systems they study. Among the heroes of science is a biologist who has spent summer after summer hunched over Isle Royale's chilly shore-edge pools in an effort to carefully unravel the life drama of chorus frogs. There is a physical oceanographer who has painstakingly teased apart the mystery of the currents that course beneath her boat in Lake Superior. There is a forest ecologist who has scoured the floor of northern hardwood forests to document the activities of exotic earthworms and their devastating effects on its inhabitants including many of our beloved spring wildflowers, plants that winter-weary northerners have come to depend on to herald the arrival of spring. There is a neurobiologist, a MacArthur Fellow, who has peered into the cryptic lives of star-nosed moles and revealed their extraordinary sensory gifts. There is a fisheries biologist who has trudged the length of North Shore streams in the dead of winter to locate the clear, cold springs that will become nurseries for new generations of endangered coaster brook trout. There is a wildlife biologist who has walked in the footsteps of black bears, recording their dietary habits berry by berry, nut by nut.

The list of the scientists we have come to admire is long. Yet, despite their engaged and passionate commitment, enormous gaps remain in our knowledge of some of the most basic dynamics of northern forests and aquatic ecosystems. But the missing pages in the book of North Shore research should not keep us from acting on the knowledge we already have and, in doing so, proceeding with caution and humility. The livelihoods of many species are threatened by toxic pollution, exotic species introductions, habitat destruction and fragmentation, and, perhaps the most worrisome and sweeping of all, rising global temperatures. Nonetheless, much of the region has survived the massive changes brought about by the economic activities of Euro-Americans since the mid-eighteenth century. Unlike in many other parts of the world, we have a chance to conserve much of the shore's original complement of plants and animals—and the ecosystem processes they need to thrive.

So, of all the things we discovered in our long learning journey of writing this book, what are the general themes we want to sound?

Well, for starters, we want to emphasize that the North Shore is old, really old. The shore-edge rocks that are so beloved by picnickers and sunset watchers are among the oldest exposed bedrock on the planet, shaped by ancient forces that date to Earth's infancy. About 2.5 billion years old, these rocks are part of North America's great geologic basement known as the Canadian Shield. To stand on the shore edge, then, is to step back into ancient time.

Melted and molded, squeezed and folded, this volcanic rock has also been whittled and chiseled over the past two million years by a series of twenty glaciations reaching down from the north to sculpt what became Lake Superior and its watershed within Minnesota. Together, volcanic fire and glacial ice drafted a script over the past two billion years for the course of both human economies and nature's ecologies. These ancient forces determined where water collects and flows, where soils will build or be stripped away, where snowfall can accumulate in deep, insulating drifts or be scoured by

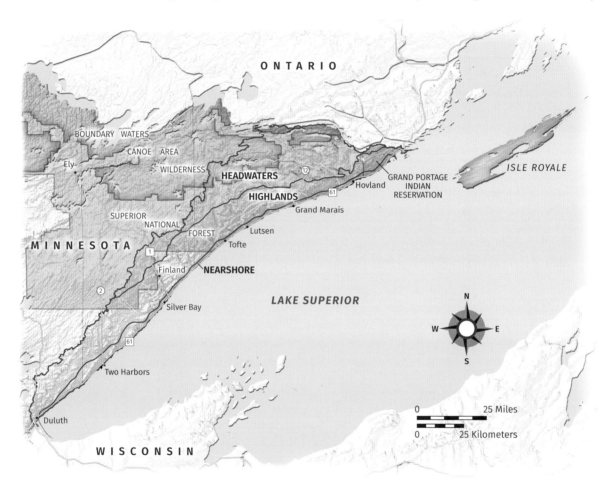

The concept of the watershed is the organizing structure of *North Shore*. The mainland landscape is divided into three sections *(outlined in red)*: the farthest inland reach encompasses the land and waters that form the headwaters for the North Shore's rivers and streams, the middle zone is defined by the Highland Moraine and a series of bedrock ridges that parallel Lake Superior, and the nearshore terrestrial ecosystem refers to the coastal terrace and shore edge along Lake Superior. Each division has its own set of ecological conditions and processes, resulting in the formation of distinctive forest communities across the landscape *(opposite)*. Lake Superior and selected islands that closely share the shore's geologic and botanical heritage complete the watershed picture.

the winter blasts off the lake or down from the Arctic. From maple syrup production to the timber, tourism, and commercial fishing industries, the shore's history of fire and ice continues to influence virtually every aspect of its economic past and present.

In the midst of such great antiquity, the North Shore has retained functioning wildness that has long been planted to cornfields and housing developments in most other parts of the Midwest. The region's forests still roll uninterrupted from the shoreline of Lake Superior into the far northern reaches of Canada. Their dense tangle of tree limbs and shadows, and lacework of bogs and open lakes end only where trees toe the line of Arctic tundra. On backcountry roads, black bears briefly appear and then vanish into

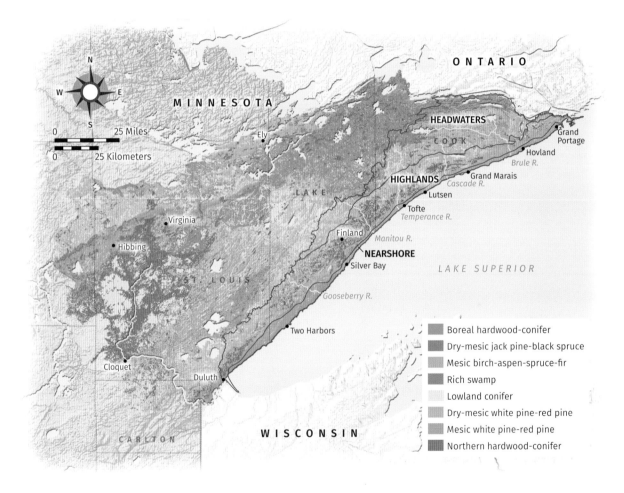

the forest, as if they had slipped through a stage-set curtain painted with trees. Otters drag their thick tails through the snow, leaving long grooves along the riverbanks. Moose posthole trails that are cruised by cross-country skiers. Ravens croak from the ragged treetops of black spruce. Timber wolves—the largest population in the lower forty-eight states—still go about their daily and seasonal rounds—denning and resting, feeding and breeding—without once crossing a Walmart parking lot or the concrete trough of an interstate highway. And this place lies not in a remote wilderness of the American West but east of the Mississippi, in one of the most heavily settled regions of the country.

The third overarching message that we would like to convey to readers is this: If we are going to build durable, robust human economies and sustain the highly prized quality of life on the North Shore, then we need to understand and preserve the ecological underpinnings that allow the flourishing of all the shore's inhabitants, both human and wild. And we must prioritize this ecological knowledge in our decision making—both in our individual households and in our communities—so that our best hopes and dreams for the North Shore's future can be realized.

Human destinies have been shaped by the shore's natural systems and biotic communities, and we, in turn, have altered the course of their futures. How we will rectify our missteps and take more ecologically informed actions in the future to benefit the diversity of life, including humans, is up to us. May our book help in these endeavors.

A Few Notes about the Book's Structure

North Shore is divided into five sections: Headwaters, Highlands, Nearshore, Lake Superior, and Islands. Together they cover a geographic scope that ranges from Two Harbors to the Canadian border, the full expanse of Lake Superior, and selected islands that closely share the shore's geologic and botanical heritage, most notably Isle Royale and the Susie Islands.

The concept of the watershed is the organizing structure for our discussion of the terrestrial portion of the Minnesota North Shore. We divide the North Shore watershed into three sections: the farthest inland reach encompasses the land and waters that form the headwaters for the North Shore's rivers and streams, the middle zone is defined by the Highland Moraine and a series of bedrock ridges that parallel the lake, and the nearshore terrestrial ecosystem refers to the coastal terrace and shore edge along Lake Superior. None of these ecozones is scribed by hard boundaries, whether they be ecological or cultural. Although each division presents a distinctive set of ecological conditions and processes, they share a geologic history and an associated complement of flora and fauna.

They also share a long history of human occupation and are influenced by the ongoing evolution of cultural attitudes. In this discussion, we emphasize Euro-American culture largely because of the comparative abundance of historic reference material available and the fact that it is our own cultural milieu. Nonetheless, we reference historic and contemporary Native scholarship to illustrate indigenous people's long and rich cultural relationship with the land and water, the differences and similarities between indigenous and settlement cultures, the influence of Anishinabe culture on Euro-Americans who settled in their ancestral grounds, and current tribal initiatives that have arisen from Native people's continued connection to this place. We hope that our comparatively limited coverage of indigenous history and culture will motivate curious readers to further explore these subjects in the work of authentic Native voices, such as Theresa M. Schenck, Carl Gawboy, and Emily Faries.

Each of the five sections of our book begins with a general overview of its distinctive ecological history, attributes, relationships, and habitats. These introductions are followed by more detailed and in-depth profiles of representative organisms and ecosystem processes, discussions of environmental issues that are of special concern to each ecozone, and explorations of the legacy of historic land uses and management practices. These profiles provide ports of entry into the larger North Shore narrative. Depending on readers' particular interests or time constraints, they may begin at the scale of an individual plant or animal (such as bears, diatoms, pitcher plants, salamanders, galls) or choose a narrative that focuses on the effects of people on larger ecosystems (the reintroduction of coaster brook trout, wildcrafting, exploding gull populations, septic-system pollution, proposed diversions of Lake Superior's waters). Other readers may choose to do the reverse and begin with the longer ecological overview for each section before drilling down to the smaller scale of individual organisms and environmental issues. Having provided a beginning here, our hope is that readers will pick up the threads of knowledge within each entry and continue to weave them into an understanding of the complex relationships that support the health and wholeness of the North Shore's great tapestry of life. ❧

HEADWATERS

I F YOU HAVE EVER PAUSED IN THE SILENCE OF A FROZEN CLEARING TO WATCH THE northern lights strobe overhead, then you know that the interior of the Minnesota North Shore is a special place. But to understand just how special, you have to take a step back—way back—and examine a nighttime view of the continental United States from deep space. Satellite pictures show that throughout much of the American West, the land is like a moonless sky, so dark that you can pick out pinpricks of settlement scattered like constellations. East of the Great Plains, however, the country is a smear of electricity. Thick impastos of light cover the heavily settled Northeast, fading into a thinner, more even wash in the Midwest. Here and there cities such as Minneapolis, Kansas City, and Atlanta explode on the surface in ragged flashes. The nation falls neatly into halves, the yin and yang of dark, open space and the glare of city streets.

Follow the continent north to the mouth of the St. Louis River, past the glowing bulb of Duluth, and head east along the Minnesota shoreline of Lake Superior. Behind the steep Highland Moraine and bedrock ridges that parallel the lake lies a wedge of land as dark as any you will find in the far expanses of the American West. This is the North Shore headwaters, a country still wild and whole enough to support the largest

The headwaters landscape is characterized by a complex patchwork of upland forests, wetlands, and thousands of lakes and ponds connected by a vast network of streams and rivers. This portion of the watershed is primarily federal and state land, a key characteristic affecting land uses and watershed health.

population of timber wolves in the lower forty-eight states. Here, moose sink into cool beds of sphagnum moss to escape the summer heat. Nesting boreal owls peer from cavities in aged aspens. And out of the shoreline rocks of inland lakes grow gnarled cedars that tendered their first shoots toward the light during the fall of the Roman Empire.

In the jostling to settle the vast American continent, few places in the eastern half of the United States have escaped the carving knife of subdivisions, shopping malls, and cornfields. The headwaters is such a place. Most of the land that makes up its reaches falls within the boundaries of county, state, and federal forest including the Superior National Forest (SNF). At three million acres, it is the largest national forest in the eastern United States. Along its northern border lie portions of the Boundary Waters Canoe Area Wilderness (BWCAW). Covering an area one and a half times the size of Rhode Island, the BWCAW is the largest wilderness east of the Rockies.

Although dramatically reshaped by logging and radically changed fire regimes, the North Shore's inland landscape remains as forbidding and astonishingly beautiful to today's visitors as it did to the U.S. survey teams who charted this backcountry in the nineteenth and early twentieth centuries. In a 1903 report published by the U.S. Geological Survey (USGS), J. Morgan Clements hints at the heart-sinking frustration experienced by some of his colleagues when navigating the vast stretches of dense brush, wind-fallen trees, and swamps. "One inexperienced in a country of this character," he wrote, "would feel that the task were well-nigh hopeless were he called upon to leave the canoe routes and beaten trails and explore this wilderness. It sometimes requires two hours to advance a mile, and to run a line 5 miles in length and explore the area for a few hundred yards on both sides is a good day's work."

In 1906 Charles Armstrong, a surveyor employed by the Government Land Office for the U.S. Public Land Survey, stated his exasperation far more succinctly. He documented a series of townships (a township is a designated area of thirty-six square miles) just northeast of present-day Isabella in Lake County. His notes describe the landscape as the "most forsaken country it has ever been my misfortune to encounter. There is apparently nothing within the boundaries of the four townships that would induce a sane person to enter within the unsacred domain of moose, wolves, bears, snow, rain, mosquitoes, flies, rocks, swamps, brush and rapids."

But not all of these early visitors concurred with such bleak assessments. William A. Kindred and Charles E. Thurston painted an almost idyllic picture of the terrain they surveyed in the Pigeon River valley near the Canadian border. "The lakes in this township," they wrote in their field notes of 1876, "are beautiful, water very clear and in many places very deep, enabling one to see boulders on the bottom at a depth of from 50 to 60 feet. They abound in trout, pickerel and other tribes of fish. Wild fruits, such as currants, raspberries and blueberries, grow in great abundance."

Whether they regarded the land with delight or disgust, the surveyors seemed to share one common response: awe in the presence of a vast forest that was sparsely occupied by humans. Although they occasionally stumbled across the trails, sugar camps, burial grounds, and abandoned villages of the region's native people, they had few face-to-face encounters. More often, their field notes hinted at an almost sublime and terrible isolation, feelings made all the more resonant by the brevity of their entries. "I have not been able to find any recent trace of humankind within this township," noted one surveyor. Wrote another: "If any whiteman ever traversed this township, he was careful to leave no trails."

Infinite Variety: The Ancient Legacy of Water and Ice

The southern boundary of the North Shore headwaters skirts the leeward base of the Highland Moraine, Sawtooth Mountains, and other bedrock ridges paralleling the Lake Superior coast. Its northern border lies deep in the BWCAW. The region is so named because most of the twenty-two major rivers that empty into Lake Superior originate here. On average, their headwaters—that is, the place where the rivers' waters first gather and begin their journey to Superior—lie within fifteen miles of the big lake.

Together, the watersheds of these rivers contain an astounding diversity of vegetation, topography, soils, and water bodies. The field notes of nineteenth-century surveyor Charles Davis sum up the great natural variety that can be found throughout the headwaters as a whole. Visiting an area near Crescent Lake, Davis observed that the "surface of this township is hilly and in places mountainous. The soil is second rate and rocky. . . . Fire swept through portions of the township destroying a large area of valuable timber . . . but there are considerable bodies of Birch, Spruce and Pine in the east half . . . while the burnt district is covered with a dense growth of young Birch, Aspen, Fir, Spruce and Cedar. There are numerous swamps in the township, all heavily timbered with principally spruce and cedar. There are also numerous lakes some of considerable sizes and all of deep, clear water."

Much of the credit for this great medley of natural features goes to the glaciers, the most recent of which, the Wisconsin ice sheet, retreated from the area about ten thousand years ago. Serving as rock crusher, backhoe, and dump truck, the restless mountains of advancing and retreating ice rearranged everything in their path. As they spilled across the landscape, the glaciers scraped and scoured the ancient bedrock while obliterating the formations of previous glaciations. Their handiwork created clay as fine as flour, as well as gravel, palm-sized cobbles, and boulders.

This jumbled mix of rocky debris, known as glacial till, became incorporated into the ice sheets. As the glaciers melted, they redeposited these materials in landforms, such as gently undulating blankets of sediments known as ground moraines.

In other cases, the stalled ice sheets shed till along their margins, producing what are called end moraines. These are linear ridges mounded with rolling hills. For example, abutting the southern border of the headwaters region is a massive end moraine known as the Highland Moraine. This twelve-mile-wide band of irregular ridges, which runs parallel to the Lake Superior shoreline, was created when the ice sheet temporarily stopped in its tracks. Even though the glacier stagnated, it continued to drop copious amounts of debris, which accumulated in great piles along the ice front.

In addition to the shifting ice, the land was shaped by meltwaters that streamed from the glaciers. These watercourses cut drainages into the land as they followed its contours, shaping and depositing glacial debris along the way. Some of these streams formed paths through which rivers course to this day. Others have disappeared, but they have left behind a series of landforms like topographical calling cards. Surrounding Plouff Creek on the Sawbill Trail, for example, are sinuous, steep-sided landforms known as eskers, which from the air resemble the outlines of snakes wriggling under a floor rug. An esker begins to form when the pressurized flow of a torrential river carves a tunnel beneath the surface of the ice sheet. In time, the river's flow slackens and drops sand, gravel, and boulders along the course of the tunnel. The water and glacial ice eventually vanish, but these deposits remain in place.

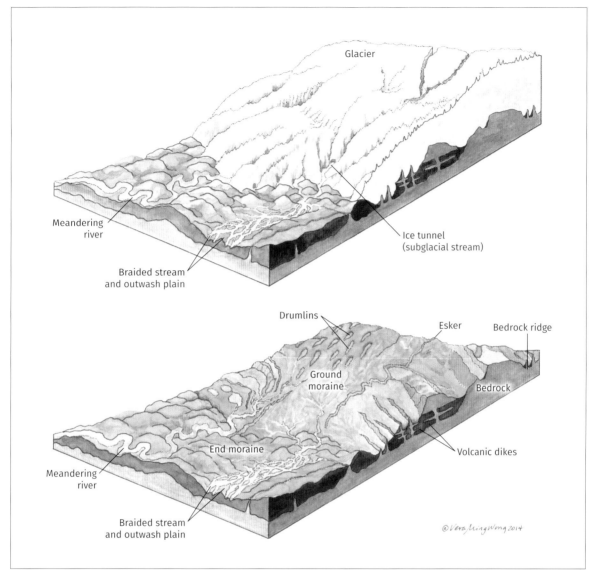

The Wisconsin glacier created a variety of landforms whose distinctive moisture, soil, and nutrient conditions support today's great diversity of plant communities. Illustration by Vera Ming Wong.

But not all glacial features are so dramatic. Portions of the headwaters of the Temperance and Poplar Rivers, by contrast, occupy a comparatively flat terrain known as an outwash plain. Once located at the foot of the ice margin, outwash plains are accumulations of sand, gravel, and cobbles that were deposited by braided streams of meltwater.

In northeastern Minnesota, the glaciers added a distinctive touch to the land's extraordinary composition—they laid bare numerous outcrops of volcanic bedrock some 2.5 billion years old. Part of the Canadian Shield—Earth's great backbone of rock—they are among the oldest exposed bedrock on the planet. Unlike northwestern Minnesota, where glacial action buried the bedrock under a more evenly distributed blanket of sediments as much as 490 feet deep, glaciers in the northeastern part of the state either scraped away surface deposits or lightly mantled the bedrock in 50 feet or less of glacial debris. Only in occasional moraines are glacial materials found in any depth. Thanks to the ice sheets, say Richard W. Ojakangas and Charles L. Matsch, authors of *Minnesota's Geology,* northeastern Minnesota has more bedrock visible on the surface than anywhere else in the state.

The Character of the Headwaters: An Interplay of Land and Water

If you peeled away the headwaters' forests and meadows, you would discover that the underlying bedrock has been ice-gouged with numerous basins. So pitted is the landscape that the headwaters region hosts nearly 11 percent of the state's sixteen thousand lakes, along with thousands of wetlands and connecting rivers and streams.

The headwaters' great wealth of water is due to a hydrologic regime that is fundamentally different from the one that operates throughout the rest of the state. Instead of having large, deep aquifers that absorb water like a porous underground sponge, much of the headwaters is characterized by shallow soils and exposed bedrock that have little or no capacity for storing water. Add to this the headwaters' hilly topography, and you can see why a very large portion of its ample rainfall and melting snowpack quickly courses off the land into lakes, rivers, and ponds. Because precipitation exceeds evapotranspiration (the combined loss of water through evaporation and transpiration by plants), for much of the year the headwaters is a soggy place scored with numerous small creeks, streams, and flowages among lakes.

Many of these waterways are subject to boom-and-bust hydrologic cycles. As a result, water levels can fluctuate wildly. Torrential flows that roar through streambeds in spring can wither to a trickle in August. Only rivers such as the Brule, Cascade, and Temperance, which are fed by sizable watersheds and large headwaters lakes, have significant year-round flows.

The headwaters' thin soils and poor water-holding capacity largely prevented the wholesale conversion of the land to agriculture that occurred in other parts of Minnesota. Even today, approximately 85 percent of the headwaters remains covered by forest. Within this landscape, thanks to the roving Wisconsin ice sheet, myriad combinations of different soil types, landforms, and microclimates have allowed a great diversity of plant communities to take hold and persist. For example, although the headwaters is home to only 24 tree species throughout its entire reach (a tropical rainforest, for example, can harbor 750 different kinds of trees in a mere twenty-five acres), the region supports more than eight hundred species of vascular plants and hundreds of species of mosses and lichens. Together they form dozens of plant communities across the landscape.

This vegetational collage was the very hallmark of Minnesota's forests, wrote Horace Beemer Ayres of the U.S. Forestry Division in his 1900 report *Timber Conditions in the Pine Region of Minnesota.* Indeed, "the so-called original forest," he wrote, "was a complicated patchwork of kinds and conditions due to a great variety of surface and soil." Formally documenting the species composition of the headwaters forest on paper, Ayres added, was challenging work for surveyors on the ground. "The trees have their preferences as to soil, subsoil, and exposure," he wrote, "but there is so little difference in large areas and so much variety on almost every 40-acre tract that, excepting the larger tracts of sandy lands and muskegs, the classes are so intermingled that they can not be differentiated on a map."

Rivers in the headwaters follow a broad, meandering route in contrast to the swift, waterfall-studded chutes of their lower reaches near Lake Superior. These waters, such as Tait Creek, a tributary of the Poplar River, provide unique habitats for fish and other wildlife. Photograph by Bob Firth/Firth Photo-Bank.

Wetlands in the headwaters, such as this conifer swamp and fen, form an important transition between upland forests and slow-moving rivers and shallow lakes. Photograph by Chel Anderson.

Black spruce swamps, a common wetland plant community in the headwaters, host a variety of mosses that serve as a kind of sponge through which water is captured, filtered, and slowly released to surrounding lands and waters. Copyright Minnesota Department of Natural Resources–Lynden B. Gerdes.

Jack pine woodland communities are common in the headwaters, thriving in the warm but droughty soils of eskers and outwash plains and in the thin soils over bedrock. Photograph by Chel Anderson.

Add to this mix the great number and variety of water bodies, and the result is an ecological complexity that defines the headwaters region. Take, for example, the Temperance River. One of the North Shore's longest waterways, it originates in a series of small beaver-ponded streams and in Brule Lake about twenty-five miles from the Superior shore. Most people know the river only in its final form as it crashes through a series of narrow chutes and torrential waterfalls into Lake Superior. Farther inland, however, the Temperance "takes its waking slow," to paraphrase a line by the poet Theodore Roethke. Following a meandering, circuitous route, the river assumes many different personalities. As it crosses the Sawbill Trail and the Grade, for example, the Temperance is a clear, rippling, musical stream fringed by a dark forest of pine, cedar, and spruce. Behind boulders in the riverbed, native brook trout escape the fast current in pockets of calm. Not far upstream, however, the river widens to form a series of broad, shallow lakes where animals such as ducks and moose find sustenance in expansive beds of pondweed, water lilies, horsetail, and sedges that favor such wandering, sluggish waters.

Follow any backcountry road near the Temperance headwaters, and the ecological mosaic gets even more intricate. A small low-lying glen at the bottom of a small hillock, for example, might collect enough groundwater for a poor conifer swamp of black spruce to take hold. On the forest floor, sphagnum mosses cover a messy jumble of deadfall trees. In this acidic environment, downed wood, mosses, and other plants decompose at an infinitesimally slow rate, adding mere fractions of an inch each year to a ground layer of peat that may be one thousand years old but only three feet deep. The darkness cast by the overstory of black spruce branches together with the limited nutrients prevent most tall plants and shrubs from taking hold, enabling visitors to peer deeply into the swamp's dim, tangled interior. To the uninitiated, it might appear ragged and lifeless like the nightmare forest of legends and fairytales. In fact, the woodland floor is alive with small plants that have adapted to the spruce swamp's low light levels and nutrient-poor conditions. From spring to early summer, some of the northland's most delicate flowers blossom from cool, mossy mounds: goldthread, starflower, Canada mayflower, sweet coltsfoot, three-leaved false Solomon's seal, and pink moccasin flower.

Just across the road, on the other hand, jack pines may have colonized a sunny slope, taking advantage of the warm, droughty soils. Not far from this spacious, open forest—so dry that the soil crunches like cinders and tufts of reindeer lichen

Deep peat formed by sphagnum and feathermosses accumulates at an infinitesimally slow rate of mere fractions of an inch per year. Photograph by Craig Blacklock.

crackle underfoot—you gradually descend through a thicket of young black spruce and balsam fir and suddenly pass into the great clearing of a northern bog. As if held back by an invisible barrier, trees give way to low-growing shrubs, such as leatherleaf, bog rosemary, and bog laurel, and the ground becomes carpeted by soft, multicolored mounds of sphagnum moss. The few trees that manage to survive here cluster on scattered islands, an indication of higher and drier spots in a place where each footstep fills with water. Walk farther into the bog, and even the shrubs become scarce. Here, cottongrass, pitcher plants, and small bog cranberry emerge from magenta and emerald-green mounds of

Lichens, such as these star-tipped and green reindeer lichens, thrive in woodlands, along with such low-growing shrubs as bearberry and blueberry. Together they protect soil from erosion by slowing the movement of water off the surface of the land. Copyright Minnesota Department of Natural Resources–Lynden B. Gerdes.

sphagnum. The terrain becomes lumpier, harder to navigate, what some have described as trying to traverse a leaky water bed filled with beach balls.

Near the edge, where the floating mat of vegetation meets the open water of an oxbow lake, the dominant moss species change, and the hummocky terrain levels out into what is known as a poor fen—a less acidic and wetter cousin of the bog. Resembling an unmown lawn, the narrow margin of the fen features a community of fine-leaved sedges such as fen wiregrass sedge, bog wiregrass sedge, candle-lantern sedge, and white beak-rush. Together their stems wave in the breeze over rosettes of the carnivorous sundew plant. With each step you slowly sink up to your ankles in water as the mat of vegetation under your feet gently ripples and sways, barely keeping you afloat.

If you can steady your balance long enough to look up, you might glimpse a nearby hill whose mass of bedrock withstood the grinding of the glaciers. Its flanks and crown support yet another series of plant communities that are typically found in the headwaters. On talus slopes and exposed cliffs, lichens thrive in pioneer conditions that are little changed since the passing of the last glacier. Rare species such as encrusted saxifrage, maidenhair spleenwort, Rocky Mountain woodsia, and rock fir moss perch on ledges and sprout from cracks. Free of competition from trees, they occupy these vertical, rocky habitats with more common species such as harebell, fragile fern, and rough bentgrass.

In the shallow soils and bare rock outcrops around the cliff top, dry red and white pine woodlands have taken hold. Nearby, where soils are deeper and moister, forests of mesic red and white pines flourish. Also thriving on these soils are other mesic mixed forests dominated by early-successional hardwood species, the most common forest in the headwaters today. They include such species as birch, aspen, white spruce, white cedar, and balsam fir.

Poor fens often float along the edges of ponds, shallow bays, and streams. Here, a waterlogged band of sphagnum supports a "lawn" of fine-leaved sedges adapted to the saturated conditions. Copyright Minnesota Department of Natural Resources–Lynden B. Gerdes.

Bog laurel and Labrador tea are joined by the showy white tassels of cottongrass on the sphagnum hummocks of this shrubby poor fen. Poor fens are peat-forming wetlands that, unlike bogs, receive nutrients from sources other than precipitation, usually from groundwater and the runoff from the mineral soils of surrounding uplands. Copyright Minnesota Department of Natural Resources–Lynden B. Gerdes.

Harebells frequently nod from bedrock cervices in cliffs. Photograph by Chel Anderson.

The crevices and ledges of rocky cliffs are home to plant species that can tolerate their harsh conditions as well as benefit from the diminished competition with other plants. Photograph by Gary Alan Nelson.

Dynamic Equilibrium and the Forces of Disturbance

The diversity of communities in the headwaters, together with those of the neighboring North Shore Highlands to the south, are part of a broad zone known as the Laurentian Mixed Forest (LMF) Province. According to a 1993 ecological classification system developed by the Minnesota Department of Natural Resources (DNR), University of Minnesota, and U.S. Forest Service (USFS), the LMF covers the northeastern third of the state, cutting a path through northern Minnesota and southern Canada before veering up into northern New England. Characterized by a largely homogenous climate and similar mix of broadleaf and needle-leaf trees, the LMF forms a continental-scale ecotone between two major biomes: the eastern deciduous forest and the boreal forest. University of Minnesota ecologist Lee Frelich points out that ecologists often refer to the forests of the headwaters as "near boreal" forests "because they are literally near the true boreal forest that extends across Ontario about 50 to 100 miles north of the Minnesota-Canada border."

Both the headwaters and highlands landscapes are lumped into this broad classification because they share many of the same plant species, including a suite of needle-leaf trees such as black spruce, white cedar, white spruce, balsam fir, tamarack, and red, white, and jack pines. These conifers are joined by a deciduous component of boreal hardwoods such as trembling aspen, paper birch, black ash, and balsam poplar. The two landscapes also feature a similar list of common of herbaceous plants, mosses, and lichens.

Despite this shared botanical endowment, the headwaters and highlands landscapes are distinctly different from one another. You can think of each landscape as a vast patchwork quilt composed of patterned building blocks that represent various plant communities. Although the individual blocks are the same, their arrangement in larger, repeating patterns across the landscape defines the distinctive character of the headwaters and highlands ecoregions, making them as different from one another as the Jacob's Ladder pattern is from the Dutchman's Puzzle in a quilt.

But unlike quilts crafted by humans, nature's handiwork exists in a constant state of flux. This shuffling of vegetational building blocks is due to natural disturbances that occur on multiple temporal and spatial scales. Like obsessive quilt crafters, agents of change, such as fire, disease, insect infestations, windthrow, and flooding, visit

Dry, woodland communities, such as this red and white pine woodland, favor the warm, well-drained conditions found on cliff tops, rock outcrops, and the shallow soils of bedrock ridges. Copyright Minnesota Department of Natural Resources–Lynden B. Gerdes.

headwaters ecosystems again and again. Within the time span of a century or more, small parts of the quilt may be removed or rearranged. In other cases, cataclysmic fire or windstorms can take apart and re-create huge swaths of the quilt, literally overnight. Ecologists use the word *succession* to refer to the process of change over time in the composition and structure of vegetation within an ecosystem.

In the headwaters' forests, insects can cause radical changes that trip the process of succession. John Pastor, a University of Minnesota forest ecologist, has studied the intricate twists and turns of one such scenario in a mixed forest of aspen, balsam fir, and spruce—with fascinating results. At one stage of forest development, for example, aspens of mostly similar mature age dominate the forest canopy, keeping shade-tolerant fir and spruce in check. From time to time, however, the aspens are beset by major outbreaks of forest tent caterpillars, whose numbers swell to such proportions that the forest crackles with the sounds of their munching as they feed on leaves. The defoliated branches allow more sunlight to filter to the forest floor, spurring the growth of young spruce and fir. As they age, the aspens are felled by disease and damaging winds and ice, releasing this new generation of conifers that have been biding their time in the ground layer and understory. The conifers eventually replace aspen as the dominant trees in the canopy.

But the story does not end there. The life spans of spruce and fir too can be shortened by disease or insects, among them spruce budworm, a common moth larva that, despite its name, prefers balsam fir as a host. In early summer, adult moths deposit their eggs on balsam needles. Within two weeks, the larvae hatch and begin feeding. The caterpillars overwinter in spun shelters. By spring, they are ready to resume their voracious feeding habits, transform into adult moths, and repeat the cycle all over again. If they continue over several years, severe infestations can rampage through large areas, killing balsam and white spruce in both the canopy and understory.

In the upland forests of the headwaters, where droughty conditions are common, these forests of dead trees stoke fires that reshuffle the pieces and patterns of the landscape quilt. As they consume snags and downed timber, fires open up the forest floor to the sun, stimulating fast-growing and sun-loving species such as the fire-adapted aspen. Flourishing in these favorable circumstances, aspen eventually dominate the forest again. But within thirty years, balsam fir and spruce will have recolonized the forest floor. It is only a matter of time before both populations of tree and insect species resume their complicated, but rhythmic, dance of give and take.

Wind also plays a role in maintaining heterogeneity within both individual forest communities and across the landscape. On average, storms strong enough to topple a few localized trees occur every ten years. Gusts that blow down 30 to 40 percent of a forest's canopy trees happen at three-hundred-year intervals. Winds that completely level the canopy occur only once every two thousand years.

Rarely are such major blowdowns caused by tornadoes, however. The devastating winds that accompany big storm systems in the headwaters region are likely to be microbursts. These winds are created when rain or snow in storm clouds evaporates, cooling the air. Since cool air is denser—and heavier—than warm air, these parcels sink, causing a downdraft of fast-moving air to form. Downdrafts can hit the ground at speeds in excess of 100 miles per hour, where they flatten trees in a telltale radial pattern like that from a bomb blast. Fast-moving storm fronts can produce multiple microbursts, some measuring up to 2.5 miles in diameter.

Wind plays an important role in maintaining the forest mosaic at both the plant community and landscape scale. On July 4–5, 1999, a major windstorm swept through nearly one million acres in northern Minnesota and Canada, toppling trees in a highly variable pattern across the forest. Courtesy of Superior National Forest.

Northern Minnesota and southern Canada experienced such a major blowdown on July 4–5, 1999. According to Frelich, the storm was the equivalent of a category three or four hurricane, packing as much of a wallop as Hurricanes Hugo and Andrew. And its damage was just as extensive. Slamming into the ground at ten- to fifteen-minute intervals, the microbursts created localized and highly variable gaps in the forest that together added up to some 478,000 acres in northern Minnesota and at least that many in Canada.

Up until about eighty years ago, however, when fire suppression became a forest-management priority, by far the most common and pervasive agent of change in the headwaters was fire. Compared with other parts of the North Shore watershed, fires in the headwaters maintained their own timing and character. Just south of the head-waters, for example, along the Superior lakeshore and in the North Shore Highlands, catastrophic fires (severe fires that killed overstory and understory trees as well as the seedbed in the soil) were relatively rare, occurring on average every 250 to 1,000 years.

Behind the highlands ridge, however, the wildfire regimes were markedly different. According to Frelich, the "mixture of conifers and birch and aspen that covered much of northern Minnesota had severe canopy-killing fires with average fire intervals ranging from 50–200 years," depending on the ecosystem type. These forests were vulnerable to scorching blazes at every stage of growth whether they were dominated by tender new seedlings recovering from recent firestorms or mature forests that had escaped strafing for decades. In extreme drought years, these consuming fires were intense and widespread. Frelich points out that "crown fires with flame lengths of 50–100 feet swept across large tracts of forest ranging from a few thousand to several hundred thousand acres in size."

These conflagrations made a deep impression on the region's earliest visitors. The Jesuit missionary Fr. Jean-Pierre Aulneau in a letter of 1735, for example, described how he and his companions traveled from Lake Superior to Lake of the Woods through continuous smoke-choked forests without "even once catching a glimpse of the sun." Some of these fires appear to have been deliberately set by Native people and "assisted to spread, to kill the timber, and so give better feeding ground for the moose and deer," wrote O. E. Garrison, a member of the state's Geological and Natural History Survey, in 1881.

Catastrophic blazes left a unique imprint on the landscape, burning so hot that they incinerated the organic material blanketing the soil, thereby exposing the mineral bed laid down by glacial action thousands of years before. These fires favored the regeneration of species such as jack pines. Because the low moisture content of their leaves and thin bark leave them very vulnerable to fire, individual jack pines were incinerated in hot blazes. Their tight, gnarly cones, on the other hand, are serotinous, which means that they open when exposed to searing heat. When temperatures rise above 116 degrees Fahrenheit, the resin that seals each seed under a thick scale melts, releasing as many as fifteen to seventy-five seeds per cone, or up to two million per acre in a mature forest.

This adaptation to fire enables jack pines to seed in thickly, forming dense forests. The following year after a severe fire, Frelich points out, it is not uncommon for jack pine seedlings to sprout in densities of from three hundred thousand to four hundred thousand new individuals per acre. In a mere five years, the survivors are able to produce seeds of their own. One of the secrets of their success is that jack pine seeds not only tolerate—but prefer—dry sites with poor soils, inhospitable conditions that would kill most other seedlings. As a result, they are often found in shallow soils or well-drained, nutrient-poor sites such as sandy plains, ridges, slopes, and bedrock outcrops. Not surprisingly, these forest communities, as well as uplands dominated by black spruce, which exhibit similar characteristics, were visited by severe, stand-consuming fires at least every two hundred years.

The concentration of dead trees in these forests triggered burgeoning populations of wood-boring insects—a feast for black-backed woodpeckers. And by opening the forest canopy, fire also stimulated the growth of disturbance-loving shrubs such as raspberry, blueberries, serviceberries, chokecherry, and pin cherry, which provide forage for birds and mice as well as for such larger mammals as black bears and humans.

Although widespread, major conflagrations were uneven in their effects, skipping over some stands while charring others. Even in fire-strafed forests of jack pine and black spruce, Frelich observes, the flames spared some stands—sometimes for periods of up to 250 years. During fires or in their aftermath, animals found a haven within these localized patches and an array of amenities that were scarce in a postfire landscape: shade; coarse woody debris whose cool, damp undersides shelter insects, amphibians, reptiles, and small mammals; sturdy trees for roosting, foraging, and nesting; and a forest layered with a variety of trees and shrubs that increased feeding, breeding, nesting, and resting opportunities for specialized species.

Unburned or lightly burned areas also served as reservoirs for plants associated with the forests' older growth stages and as a seed source for other species that were wiped out in surrounding areas. Together, they helped to ensure that the burned forest community could begin the process of re-creation with the potential for a full complement of species and their ecosystem relationships.

But not all fires burned with fierce intensity. Ayres was among the first to recognize that the mosaic of the northern forest was maintained by, among other forces,

Wildfires, even cata-strophic ones such as the 2006 Cavity Lake Fire near the Gunflint Trail, are uneven in their effects, skipping over some areas while charring others. Photograph by Dennis Neitzke.

"ever-varying fires." Extensive wet lowlands or lakes with their rocky shorelines and higher humidity acted like firebreaks, sometimes transforming hot crown fires into relatively "cool" surface fires. A change to cooler, wetter weather could dampen some blazes, while others died down into a smolder after nightfall or when they encountered plant communities with limited fuels. These low-level blazes often meandered along the forest floor, consuming forest litter, dead woody debris, and understory vegetation.

Such fires, which prevented the buildup of materials that fueled major conflagrations, enabled some of the headwaters' most spectacular residents—red and white pines—to reach old ages of three to four hundred years. In these forests, fire could race through the highly flammable litter without destroying the old trees. (A mature white pine, for example, can withstand surface fires for up to ten minutes before the heat penetrates its thick bark and injures the tree.) These surface fires typically traveled through the understory of mature pine forests every twenty-five to one hundred years, killing fire-sensitive invaders such as spruce, balsam fir, and red maple and consuming dead woody debris. Left unchecked, these species could become lethal ladders that carried fires into the crowns of mature pine trees or elevated fire temperatures long enough to kill them.

These cool fires not only performed light housekeeping duties in mature pine forests, but they also fed their occupants. Unlike deciduous trees, which shed their leaves annually, conifers retain their needles, replacing them individually on an as-needed basis. Like people who make do with last year's fashions rather than buy a new wardrobe with each season, they avoid making a costly investment in the annual overhaul of their leaves. Such frugality enables conifers to occupy sites where soil nutrients are scarce.

Retaining their needles also allows them to get a jump start on photosynthesizing as soon as water is available in the spring. This nimbleness is critical since the trees live in a virtual desert for much of the winter when moisture is locked up in the form of ice. To help stem water losses during the winter "drought," conifers sport thin, needle-shaped leaves that minimize the amount of leaf surface exposed to the drying sun and wind. High concentrations of lignin in the leaves' thick outer layers also help retard moisture loss.

The very chemical and physical characteristics that confer such environmental benefits make conifer leaves highly resistant to decay on the forest floor. Thus, much of the conifer forest's source of nutrients is bound up in the needle litter. Decomposition of the litterfall slowly releases nitrogen in compounds such as nitrate or ammonia. But fire speeds the breakdown of the highly flammable conifer needles, enriching the forest with life-giving nutrients.

Because white pines evolved multiple strategies for fostering longevity, unseating forests of veteran trees was not easy. Even after great conflagrations, some mature trees invariably survived to reseed in the burned patches. Among the old-growth forests Ayres visited, he noted, for example, what he called "fire scattered" pine forests in which a few old trees presided over a progeny of young and middle-aged pines.

In time, however, they too would succumb to disturbances such as windfall, insect infestations, disease, and old age. The toppling of these trees would clear the way for the emergence of a new growth stage of the existing forest or a changing of the guard in the canopy to other species such as balsam fir, white cedar, white spruce, white pine, and paper birch.

Beavers: The Ecosystem Engineers

By analyzing tree-ring records, pollen cores, fire scars, and surveyors' notes, researchers have been able to cobble together a fairly accurate history of many disturbances in the headwaters forest. Far more difficult to tease out of the ecological record, however, is the effect of beavers on the landscape's changing mosaic. Nonetheless, many scientists suspect that their influence—then as now—was substantial before their numbers were decimated in the fur trade.

Wetlands created by beaver dams were oases of life in the vast forest. Beaver dams captured the free-flowing, nutrient-poor waters of headwaters streams and transformed them into warm, plankton-rich pools. As the pools grew, the waters flooded large areas of forest, killing the trees by starving their roots of oxygen. Even black spruce and tamarack, the conifers most tolerant of saturated soils, succumbed and were replaced by lowland shrubs or grasses and sedges. "The frequency with which beaver killed lowland conifer forest is not known precisely," Frelich observes, "but it is likely that certain tracts of forest situated at the edge of the peatlands where the beaver had access to aspen were killed repeatedly."

The expanded food base and creation of new backwater habitats multiplied the number of plants and animals that could take root and flourish. Beaver-dammed pools also served to cleanse the water. According to Alice Outwater in *Water: A Natural History,*

The European demand for hats made of beaver pelts drove the animals to near extinction across North America by the late nineteenth century. Beaver populations have rebounded, and today they are common residents in the headwaters. Photograph copyright D. W. Ross (CC BY 2.0).

Beavers help maintain a diversity of habitats in the headwaters. By damming the running waters of streams and flooding surrounding forests, they kill trees and in their place create warm, plankton-rich pools and open wetlands. Photograph copyright Jerry Hiam (CC BY 2.0).

"Water detained in the wetlands behind a beaver dam is more likely to percolate down to the groundwater, raising the water table and creating springs and freshets throughout the watershed. A land with hundreds of millions of beavers is truly a rich land, and the wetlands associated with beaver dams made the New World's water plentiful and clear as the dew."

Beavers were roving agents of change across the landscape. Whole families colonized a wetland. Eventually they succumbed to predators or disease or simply exhausted the available food and building materials for their lodges—aspen, alder, and willow—and moved on. In time, their dams collapsed. Water levels dropped, and dry land—and a new plant community—emerged.

The European Economy Discovers the Headwaters

Change is a precondition of life for organisms in the headwaters region, from glaciers periodically resculpting the landscape to fires shifting the mosaic of forest communities across large areas. But the region as a whole had reached a kind of equilibrium in presettlement times, Frelich says. Even though at any given location one tree species was replacing another species, one forest community or growth stage succeeding another, the proportion of forest types had largely stabilized across the entire landscape. These forest types had arranged themselves in predictable patterns based on the response of vegetation to stable abiotic conditions, such as landforms, soils, and the hydrologic cycle, as well as unstable ones, including climate and natural disturbances.

But that changed when Euro-Americans entered the forest. Although no tree species has become extinct, the composition and landscape pattern of today's forest communities are unlike anything that existed in the ecological record in the three thousand years prior to Euro-American settlement.

These dramatic changes can be traced to the industrial-scale exploitation of the region that began in the seventeenth century. The French were the first Europeans driven by dreams of material riches to set foot in the Lake Superior region. Writing about this early history in his book *Superior,* Arno Karlen observes: "Quakes and glaciers left behind a rocky bowl of water. The explorers, Jesuits and trappers of New France heard of it and thought it was the Northwest Passage to Asia. They took along silk robes, in case the Great Khan should greet them at Duluth."

Europeans did not find the hoped-for shortcut to the Orient in the Great Lakes. Instead, they discovered a land of great natural resources that would yield even greater wealth. In exploiting these resources, they forever altered rhythms of change that had evolved in the region over thousands of years.

One of the first and most sought-after riches of the New World was the beaver. What made beavers so commercially valuable was their fur. The guard hairs in beaver pelts could be plucked from the hide, leaving an underlying wool that was carded, boiled, and pressed to make felt. Hats made of beaver fur not only retained their shape better than those made of most other materials, but were lightweight and cool.

And Europeans were wild for them. "Hats. Hats was the name of the game," writes historian Fred Gowans of the mania for beaver skins, which lasted until the 1830s when fashion tastes switched to silk. "The carriage trade was the thing that set the standard of fashions. The carriage trade was nothing more than the rich driving around in their carriages in Boston or London or Paris . . . and whatever they wore, everybody else wanted to wear. Well, they were wearing beaver hats. And thus it made a demand for beaver."

To the Europeans, the abundance of beaver in the New World must have seemed nothing less than divinely ordained. By the mid-1500s, beavers had disappeared from everywhere but the most remote reaches of the European continent. Beaver hats were so scarce that even the wardrobes of kings had to do without great numbers of them. Passed down from generation to generation, they were such a valuable commodity that in 1659 Thomas Mayhew sold most of Nantucket for thirty pounds and two beaver hats, one for himself and one for his wife.

Estimated at a population of two hundred million animals before the advent of the fur trade, the beaver was all but eliminated east of the Appalachians by 1700 and nearly extinct in the once fur-rich Great Lakes region by the mid-1800s. Undeterred, the traders pressed on to the Rocky Mountains and into the Pacific Northwest. Ironically, by the time trappers hit a dead end on the West Coast, silk top hats replaced those made of beaver fur. It was just as well. There were few beavers left to kill.

No one knows to what extent the loss of the beavers' vast network of reservoirs altered the lives of the plants and animals that had come to depend on its lifesome waters. But their disappearance undoubtedly had an enormous effect on both terrestrial and aquatic ecosystems.

A Young Nation Hungry for Timber

As the fledgling nation grew, it continued to look to the hinterlands to finance its economic growth. And in the very center of the nation's industrial crosshairs were the rich copper troves of the Lake Superior region. In 1856, when the Methodist minister James Peet arrived on the North Shore near the Knife River, he noted in his diary that for thirty miles along the coastline "nearly all the land is claimed and a shanty built every half mile."

The shacks were built by mineral seekers hoping to strike it rich in copper as many of their fellow prospectors had done along the lake's southern shore. But the dreams of great mineral wealth that drew newcomers to the North Shore in the latter part of the nineteenth century never came to pass. According to the writers of the *WPA Guide to the Minnesota Arrowhead Country,* "The copper boom was shortlived. No paying ore having been found, and with the panic of 1857 imminent, the prospectors, who a few short months before could not get in fast enough, now stampeded out by ox team over rude trails, by rowboat, or on foot with a pack and blankets."

But the great industrial machine that powered the country required another commodity that the northland possessed in abundance: lumber. The first sawmills on the North Shore were established in the 1850s largely to serve the needs of area settlements. After the Civil War, however, the nationwide consumption of timber skyrocketed from eight billion board feet in the mid-1860s to forty-five billion board feet by the turn of the nineteenth century. The country's northern forest, which stretched from New England to Minnesota, supplied most of this new lumber, and, according to forest ecologist Pastor, "arguably experienced the most rapid and destructive exploitation of any forest region on the continent." (For a more detailed discussion of North Shore logging, see "Nearshore.")

Prized above all was white pine. According to historian Agnes M. Larson:

No other soft wood had so many desirable qualities. It was strong, slow to decay, light in weight, odorless, and easy to cut, thus yielding readily to pattern work. It seasoned well and it had a strong resistance to weather and time.

No wood has served more usefully in America than has the white pine. Its abundance, cheapness, and varied usefulness made it an important factor in the westward movement. It furnished the early settler with shelter. The logs for the cabin, the shakes for the roof, and the puncheons for the floor all came from the white pine. It also provided the incoming settler with implements, furniture, fences—all of them necessities in a new country.

Forest historian John Fritzen points out that by the late 1800s timber companies had largely exhausted the great pineries of Michigan and Wisconsin. They turned their sights northward to Minnesota, where the infrastructure already was in place to accommodate industrial-scale logging. In 1854 the Ojibwe formally ceded most of the territory in the Arrowhead region to the federal government. The following year the river channel at Sault Ste. Marie was deepened to accommodate heavier commercial hauls of lumber and other commodities to eastern ports. More important, a rail line connecting St. Paul to Duluth was completed in 1870, which established Duluth as an important portal for the transfer of commodities to and from the prairie and forest interiors.

On August 1, 1882, President Chester A. Arthur announced that two million acres of Minnesota pinelands stretching from Cook County to Itasca County would be put on the auction block that December. The sale attracted prominent lumbermen from as far away as Michigan, Wisconsin, and the Twin Cities. Among them were some of the most influential business and political leaders in their communities, including two future governors of the state of Minnesota.

Such mass transfers of public land via auctions favored wealthy individuals and corporations. To ensure a more equitable distribution of land, these auctions would be outlawed ten years later by the Land Act of 1891. As a result, lumbermen would be forced to gain title to land by less honest means. Under the Preemption and Homestead Acts that followed, individual settlers could purchase land for a nominal fee provided they themselves occupied their lands for a designated period of time and made certain improvements on it. But even though they swore an oath that the property was for their exclusive use and benefit, many claimants in the Arrowhead region had no intention of inhabiting the lands on which they staked their claims. Vast numbers of them were on the payroll of timber companies. No sooner would they file a claim than they would hand over their acres to their corporate patrons, who sheared the lands of their valuable timber. Indeed, government officials investigating legally mandated improvements to these properties routinely discovered sham dwellings made of saplings, some without roofs, floors, windows, and doors. In other instances, writes SNF historian J. Wesley White, the deception took an even more preposterous turn. "Since the law specified that a homestead of 160 acres would be given to anyone who would build a residence of at least 10 × 12 × 6 on the land," White writes, "unscrupulous operators would build a house of 10 × 12 × 6 *inches* to fulfill the requirements, would cut the timber, and then abandon the land." So successful were lumbermen at consolidating timber lands by fraudulent means that by 1913, Larson points out, six holders were in possession of 54 percent of the white and red pine forests in Minnesota.

And they cut trees as if there were no tomorrow—literally. The amount of timber extracted from the North Shore forests seems almost incomprehensible to today's visitor. In June 1899 the tugboat *Gettysburg* towed a raft of logs cut from the Pigeon River watershed to the Alger-Smith Company's sawmill in Duluth. Measuring sixty acres, the raft was so enormous that two days after its launch near Hovland the people there could still see it on the lake.

And lumberjacks could not cut timber fast enough to meet the demand. Take, for example, the timber needs in the city of Duluth alone. According to Larson, within a mere six months in 1869, Duluth's population soared from fourteen families to thirty-five hundred individuals. So acute was the lumber shortage that most of them weathered the winter in tents. From 1890 to 1895 the city saw an 80 percent increase in population, more than any other municipality in the state.

But by far the biggest consumers of timber were commercial enterprises. By the late 1890s Duluth had become a major port for the shipping of wheat, flour, iron ore, and lumber. Larson observes that in 1892 alone an estimated seventy-five million to one hundred million board feet of lumber—enough to build the equivalent today of nearly nine thousand two-bedroom homes—were taken up in the construction of grain elevators and docks, each structure consuming between four million and six million board feet of wood. To meet this demand, the city was forced to import raw materials from outside the region.

Just as voracious in their appetite for wood were the railroads. According to USFS historian Douglas W. MacCleery, between 1850 and 1910, the number of railroad miles jumped from less than 10,000 miles to more than 350,000 miles. By the turn of the nineteenth century, railroads made up 20 to 25 percent of the nation's timber consumption. Railroads, MacCleery points out, were "made of wood: cars were wood, ties were wood, the fuel was wood, the bridges and trestles were wood, and station houses, fences, and telegraph poles were wood."

To meet the demand, timber companies clear-cut vast acres of Michigan and Wisconsin, turning their sights to Minnesota's woodlands by the end of the nineteenth century. But far from learning from their mistakes, logging outfits simply repeated their cutting practices in Minnesota—with the same catastrophic results. In 1900, for example, the John Schroeder Lumber Company began white pine harvesting in the Cross River watershed near the town that today bears the company's name. Within five years, the company had "high-graded," that is, selectively cut, the white pine from a thirty-six-square-mile area that extended inland from the edge of the lake some twenty miles. When the harvest was suspended in 1905, the company had stripped the forest of an estimated twenty million board feet of white pine.

Early logging operations left virtually no ecosystem untouched. For the Cross River, one of the few North Shore streams big enough to drive logs, the impacts were devastating. For example, to ensure that logs would float unimpeded down the Cross River to Lake Superior, the company removed all the vegetation in and near the river. To provide an adequate flow of water, seven dams were built on the upper reaches of the Cross River and its tributary lakes and streams. The largest dam alone created a lake more than one mile long. During a log drive, the sluice gates of the dams would be opened in succession, beginning with those farthest upstream. Torrents of water sent logs careening down the river channel, where they tore up the banks and scoured streambeds. So rough and tumble was the ride to the lake that many logs arrived at the mouth of the river in splinters.

To head off some of this damage to their logs, company engineers used dynamite to reshape the river channel and tame the wild switchbacks of chutes and gorges and the free fall of cataracts. Historian Willis H. Raff in his book *Pioneers in the Wilderness* describes the regret that many North Shore settlers felt about the Cross River's destruction.

> Pioneers told [Schroeder resident] Horace Stickney of the unspoiled natural beauty of the Cross River gorge, with stands of majestic white pine; in "improving" the river for

Early logging operations used the larger North Shore rivers to float trees from the headwaters to Lake Superior. The clearing of riparian vegetation, construction of dams, and blasting of streambeds for log drives significantly altered these river ecosystems. Photograph by Crandall & Fletcher, circa 1899. Courtesy of the Kathryn A. Martin Library, University of Minnesota Duluth, Archives and Special Collections.

driving logs, and dynamiting the three-tiered, sheer-drop falls (the tumbled remnants are still visible, a few feet north of Highway 61), great amounts of gravel and huge cracked boulders eroded and tumbled down to fill and choke the deep, lovely bay and harbor; and a new sandbar was created at the river's mouth! In addition, obviously, the plunging logs and springtime floods added up-country mud and debris erosion to the mess in the lower reaches of the river.

Even though Stickney himself later built tourist cabins on the cleared lake-front flats on the west bank of the river, he nonetheless regrets that the thick stand of white pine that had grown there was logged off by the Company. The aged merchant has strenuously denied that there ever were economic benefits to the area that could justify such drastic loss of beauty and esthetic qualities.

According to historian Fritzen, by the turn of the century most of the pine near "drivable" rivers and Lake Superior had been cut. To reach the stands that grew in the headwaters, logging outfits penetrated the interior with series of railway spurs and branches. Tiny settlements sprouted along the tracks, such as Finland, which became a station of the Alger Line, a 153-mile railway network built by the big logging outfit Alger, Smith and Company. In time, even the pine in remote parts of the region disappeared.

As Minnesota pine became increasingly scarce and expensive to harvest and transport, logging companies turned their attention to the untapped abundance of newly opened forests in the American West. But the logging era was far from over in northeastern Minnesota. Species once considered second-rate became desirable as feed stock to meet a new demand—the growing pulp trade. The state's first paper mill was built in Cloquet in 1898. Shortly thereafter, in 1906, Fritzen observes, the region saw its first

Early in the twentieth century, rail lines were built deep into the headwaters, giving sawmill operations access to once-remote stands of white pine. Photograph circa 1890. Courtesy of the Minnesota Historical Society.

timber sale that included species other than just pine. Timber claims began to routinely include spruce, tamarack, and balsam for pulpwood, and cedar for poles, railroad ties, and posts. "As fast as the pine logger was done with a tract of timber," Fritzen says, "the pulp and cedar men moved in."

In 1910 a train with sixty-six cars of lumber from the Gooseberry River watershed pulled into the Duluth mill of the Virginia and Rainy Lake Company. Bearing 42.6 million board feet of timber—enough to build 4,100 two-bedroom homes today— the haul was said to be a train-load record. But there would be a grim irony in the record-setting cut. The year 1910 would mark the beginning of what Fritzen calls a "rapid decline" of the lumber industry in Duluth and along the North Shore. The math was simple: forests were being cut faster than they could regenerate themselves.

Other forces would take their toll as well. The faster and more furious the pace of logging, the greater the strafing of the land by wildfires. Logging operations routinely left behind great slash piles—treetops, trimmed branches, brush, and unwanted trees. These leftovers, together with natural windfalls, created debris piles some twelve to fifteen feet tall, says fire historian Stephen Pyne. The massive fuel loads were ripe for conflagrations, which could be ignited by lightning, sparks from the smokestacks and ash pans of passing locomotives, or the runaway fires of settlers, who routinely torched meadows, brush piles, and tree stumps. According to Fritzen, "During the period of 1900 to 1909 there were many destructive forest fires in northern Minnesota, Wisconsin, and Michigan, and several of them reached disastrous proportions." The year 1910 was the worst on record in terms of acres burned. Uncontrolled wildfires swept the North Shore beginning in May. Following a brief respite in June, the blazes continued until the snow fell in late October. "Large areas on the North Shore had been burned all the way from Duluth to the Canadian border at Lakewood, Knife River, Two Harbors, Silver Creek, Gooseberry River, Split Rock River, Beaver River, Tofte, and Grand Marais. One observer stated that everything appeared to be black all the way along the North Shore as far back as he could see," Fritzen writes.

According to Ayres, 90 percent of the state's cutover lands were visited by postlogging fires. In the early days of industrial logging, voices of protest over such careless and extravagant waste were largely drowned out by the stampede to put farmers on cutover acres. In fact, as historian Pyne points out, under the guiding philosophy of settlement, which posited that the plow followed the ax, postlogging fires were seen as a labor-saving force rather than as agents of human and ecological tragedy. In 1881, the *Detroit Post* warbled that "where the fires have raged the forests have been killed, the underbrush burned and the ground pretty effectively cleared. There are square miles and whole townships where the earth is bare of everything except a light covering of ashes; and other square miles where all that is needed to complete the clearing is to gather up a few scattered chunks per acre and finish burning them."

And lumber companies urged an even greater pace of forest clearing in order to safeguard communities from the threat of future fires as well as to salvage timber before it went up in smoke. "It is a question whether this valuable timber shall be saved to be used for the convenience of human beings, or be wasted by destructive forest fires," declared the *Lumberman's Gazette* in 1881. "If it is to be saved, it must be cut as fast as possible. It cannot be husbanded and preserved for the future."

Ironically, the greater the clearing the more violent the flames. From survivors'

White pine forests were intensely harvested, leaving behind large amounts of slash, which quickly became highly flammable fuel for fires. Sparks from the stacks of logging railroads were a common source of ignition. Photograph by W. E. LaFountain, 1904. Courtesy of the Minnesota Historical Society.

accounts, Pyne has compiled a description of the horror of these conflagrations whose "peculiar physics," he writes in *Fire in America,* released energy that was "equivalent to the chain reaction of a thermonuclear bomb":

> The first sign of impending doom for most communities was a preternatural darkness. . . .
>
> Out of the darkness came "currents of air on fire," a "sirocco," a withering blast of heat. In areas otherwise untouched by the fire itself, heat and heated gasses claimed lives. So fierce was the heat that it alone drove hundreds to shelter in root cellars, wells, and stone buildings. The vast majority of those who fled to cellars perished from asphyxiation; those in wells, from asphyxiation and incineration; and those in buildings, from suffocation and flame. . . . Heat was the apparent cause; the landscape for miles ahead of the flames was violently preheated and desiccated.
>
> The fire itself was preceded by a barrage of firebrands and a thunderous roar. A "storm of fire brands, cinders, and ashes" showered the landscape with "fire flakes." Other people spoke of "balls of fire" and "fire balloons," probably bubbles of heated gasses distended from the main burn itself and ignited by radiant heat. Many of the "balloons" exploded above ground, blasting fire like shrapnel. An island half a mile out in Lake Michigan erupted into fire. One terrified eyewitness described the fires as a "veritable cyclone of flames. There came, as it seemed to me, great balls of fire from the sky, and when they were within 20 feet of the ground, they burst, sending down a heavy rain of flashing sparks, like a mighty sky rocket exploding with a brilliant display of flashing light."
>
> Overpowering everything was the roar. It sounded like thunder, the pummeling of a dozen cataracts, the pounding of heavy freight trains, and "all the hounds of hell. . . ."

When the fire arrived, it came not as a wave or a surge of flame but as though suddenly dropped from the sky. The landscape was instantly enveloped in a "tornado of flame," a "hurricane of fire." Firewhirls traveled 6–10 mph, "the pine tree tops were twisted off and set on fire, and the burning debris on the ground was caught up and whirled through the air in a literal column of fire." One witness exclaimed, "it was a waterspout of fire." Its winds, like its heat, sometimes preceded, sometimes coincided with the advent of the main fire and reached staggering velocities. Surface winds were rarely excessive, frequently 15–40 mph, but the turbulence from the violent convection was awesome. Winds of 60–80 mph uprooted trees like match sticks; a 1,000-pound wagon was tossed like a tumbleweed. Papers were lofted by the winds from Michigan across Lake Huron to Canada.

Just such a catastrophic fire flattened the Hinckley–Sandstone area on September 1, 1894. In an account of the fire written by the Reverend William Wilkinson, survivors described heat so intense that "kegs of nails that had been for sale in the hardware stores were found one melted mass," and the tracks of the railway "ran liquid steel." The cyclone of fire incinerated 160,000 acres and left 418 people dead in its wake, including twelve children who sought refuge in the town well, and scores of "rabbits, deer and birds." The bodies of most of the casualties were so badly disfigured they could only be identified by small scraps of clues such as a label or a button on a shirt, a locket, a wristwatch, a jackknife, a shoe. Others, such as Hinckley resident Thomas Sanderstrom, age fifty-six, simply died in heartbreaking anonymity in the flames, the codicil to the death list noting that "nothing certain of his being identified; buried in Hinckley."

A catastrophic wildfire in 1894 leveled the town of Hinckley, Minnesota. Courtesy of the Minnesota Historical Society.

Perhaps the most chilling account of the fire came from Two Harbors, located some ninety-five miles to the northeast and far removed from the horrors of the actual conflagration. At three o'clock on the afternoon of the fire, the skies over this tiny Lake Superior settlement took on a "sulphurous hue, suggesting an approaching storm. Two hours later the atmosphere was filled with smoke, cinders and ashes. It had become dark as night and lamps were lighted. A red flush covering the entire heavens gave to it the semblance of the canopy of hades, and the average sinners felt that the forerunner of the wrath to come was with them. Kerosene lamps burned with a flare as blue and clear as an arc light, a singular phenomenon of the atmospheric conditions resultant from the great fires. . . . All night long and part of Sunday, the ashes fell, their silent and peaceful settling back to earth little indicating the roar of the breakers of flame, and tortured shrieks which preceded their long flight."

After the Fires

On the night of September 1, a small group of Hinckley residents emerged from the quarry at the edge of town where they had found refuge from the flames. Dazed and bewildered, they stumbled across a handful of watermelons in a nearby garden that too had miraculously survived the inferno. After eating the fruit, they used the rinds as drinking cups. The fire had stripped them of their belongings, even scorching most of the clothes off their backs.

By three o'clock in the morning relief supplies from Duluth had arrived by train. Two days later, Minnesota governor Knute Nelson established a state commission to collect and dispense private donations to the fire survivors, an amount that totaled $185,000. The funds would pay for supplies such as food, household furnishings (everything from sheets and crockery to salt and pepper shakers), farming implements, grass seed, and seed potatoes.

But as Pyne points out, such aid was motivated as much by economic as humanitarian concerns. "One of the arguments for relief," he writes, "was the fear that farmers would shun the north woods if their plight was ignored, that the tide of progress, capital, and settlement would move elsewhere. Land speculation was big business; boosterism, a reflex arc. The generous outpouring of voluntary contributions was overwhelmingly humanitarian, but its ultimate goal was to keep the farmer where he was, not to move him to a less fire-prone environment."

A flurry of editorial debates in newspapers around the country, however, reflected a vigorous national questioning of the public policy that led to such catastrophes in the first place. The Hinckley fire would bring this deep uneasiness to the surface. The "Hinckley holocaust," writes historian R. Newell Searle, "traced an indelible scar on the public's memory. Everyone demanded action."

Events such as the Hinckley fire helped to expose the greed of private interests at the heart of the government policy to settle farmers on cutover land. Land speculators bought up the shorn acres for as little as twenty-five cents to a half dollar an acre and then resold them to hopeful farmers for five dollars to fifteen dollars an acre. But contrary to their rosy promises, the climate and soils of the north country would prove unsuitable to anything more than subsistence farming. Farmers abandoned hope— and their acres of grubbed tree stumps. In county after county, tax delinquency rates skyrocketed.

Herbert H. Chapman was one of the earliest reformers to decry what he called the "heartless exploitation of would-be settlers by land speculators." His speech on the subject to a meeting of the American Forestry Association in Minneapolis in 1903 was so controversial that he resigned from his post as the superintendent of the University of Minnesota's Agricultural Experiment Station in Grand Rapids shortly after delivering it. To his critics, who feared that Chapman's honest assessments of the north country's agricultural promise would dampen migration to the region, he charged, "I could wish no worse fate than to condemn them to earn a living on the 'farm land' they have sold to poor but honest settlers with families to support. And the federal policy which permits misinformed people to homestead such land, is false and should be rigorously condemned." (For a more detailed discussion of agricultural settlement in the cutover area and its aftermath, see "Highlands.")

At the same time, Chapman offered hope for the economic redemption of the north's blasted acres. "Land which is not fit for farming can still grow trees," argued Chapman, along with a host of other land-use reformers who determined that the highest and best use of north-country lands lay in forest regeneration.

Proponents put forward several arguments to justify their position. They warned that the United States was facing a timber famine as supplies were systematically exhausted from one end of the nation to another. Without forests to ensure a steady supply of wood, nothing less than the country's economic prosperity was at stake, they posited. The problem of wood supply reached crisis proportions as the last pineries east of the Mississippi—those in the American South—too were laid waste. In 1925 USFS Chief W. B. Greeley issued a warning. Sixty percent of the nation's wood grew in forests west of the Great Plains, he wrote in an article in *Economic Geography,* "whereas two-thirds of the population and an even larger proportion of our agriculture and manufactures are east of the Great Plains. . . . Already the unbalanced geographical distribution of this resource is creating well-nigh famine prices in the parts of the United States where forest products are used in the largest quantities."

There was no better illustration of Greeley's concern than Minnesota. A 1929 report published by the Minnesota Forest Service observes that while the state's lumber yards once carried northern pine exclusively, "Eighty-five percent of the stock [is] made up of Douglas fir from the Pacific Coast and yellow pine from the South for which the consumers are paying a freight charge of $18.00 and $11.00 a thousand board feet, respectively. It is only now that we are beginning to realize what the destruction of our forests is costing us and will continue to cost us until we are growing our own timber supply again."

Regenerating forests made good economic sense for other reasons as well. Chapman was among the earliest in a growing chorus of voices to promote green forests teeming with abundant wildlife and clean lakes and streams as the cornerstone of a vibrant tourism economy. "We in northern Minnesota," he pointed out in 1903, "have yet to learn the tremendous volume of the summer tourist business, the underlying sentiment of which goes deep into the heart springs of humanity—the longing for fresh air, freedom, and Nature. This movement continually increases in strength."

Chapman's words were prescient. The demand for recreation would soon intensify. Thanks to Henry Ford, scores of Americans began to climb into their automobiles to satisfy their longings for fresh air, freedom, and nature. According to SNF historian

White, by 1919 nearly thirteen thousand tourists had sought refuge in the new forest reserve, which had been set aside a mere decade before. With few facilities to accommodate them, most recreationists roughed it as campers and canoeists. In 1924–25, however, a newly realigned and upgraded Highway 61, one of the first major highway projects in the Arrowhead region, would help ease entry into Lake Superior country.

But who would replant the cutover acres and protect them from wildfires and a second round of clear-cutting? There were few incentives for timber companies to take on this task since growing trees provided a slow rate of return on investment. Besides, critics of past logging practices charged that the country's forest future could not be returned to the hands of a few reckless and self-interested individuals and private corporations. Instead, they proposed a radically new role for state and federal governments— to become stewards of the nation's great land trust rather than its liquidators. Proponents of public forests advocated that governments set aside large tracts of forest in public ownership with the aim of putting their "most productive use for the permanent good of the whole people, and not for the temporary benefit of individuals or companies," as Secretary of Agriculture James Wilson wrote in 1905.

Federal forest conservation began in 1891 with the passage of the Forest Reserve Act, which gave the president authority to declare forest reserves in public domain lands. Under the leadership of Theodore Roosevelt, the most ardent presidential champion of conservation in U.S. history, and Gifford Pinchot, the first chief of the fledgling Forestry Bureau (later renamed the U.S. Forest Service), the size of the national forests tripled. Included in Roosevelt's conservation legacy is the Superior National Forest, established in northeastern Minnesota on February 13, 1909.

The driving force for the formation of the SNF came from Christopher C. Andrews, Minnesota's first chief fire warden and forestry commissioner. In his first report to the Minnesota legislature as fire warden in 1895, he warned that the rapid cutting of trees would leave the state without a commercial forest in a mere two decades. Andrews advocated the establishment of a series of federal forest reserves. In 1902 he persuaded the General Land Office to withdraw from sale about five hundred thousand acres of public land in Cook and Lake Counties.

Instead of selecting acreage with prime timber or fertile soils for the new SNF, the General Land Office targeted cutover, burned, and exhausted tracts that had fallen into public hands through tax delinquency. Settlers, loggers, and miners had depleted populations of fur-bearing animals, fish, woodland caribou, moose, and deer. Logging and fires had reduced the forest to brush or to then-unmarketable species such as spruce, balsam, jack pine, and aspen. "The lands contain no timber; they are unfit for farming and there is no settlement in them. No one cares much what becomes of them" is how the Forestry Bureau's timber examiner E. A. Braniff in his 1903 boundary report described the forest that became the SNF's fledgling core.

Nonetheless, the forests' despoilers continued to oppose conservation of even the most blighted lands. An exasperated Pinchot condemned the "fierce desire for development which marks the frontier, the hunger for profits of land agents and other speculators in land, and the determination of the lumber men to let no tree escape that would put a dollar in their pockets." Opposition from citizens in northern Minnesota was also fueled by a deep mistrust of government that colluded with mining companies to battle organized labor during the same period.

Many residents viewed the federal action as a land grab that shut out farmers from potential homesteads and robbed the region of future economic prosperity, despite the widespread abandonment of homesteads and soaring levels of tax delinquency that refuted these charges. Surfacing in these early debates was a theme that would be sounded repeatedly in later land battles, such as the especially bitter controversy surrounding the establishment of the BWCAW: that the decisions about the development of the north country were wrested out of local control and placed in the hands of affluent urbanites looking for a forest playground. A circular published by George H. Lommen of the St. Louis County Club during the campaign to expand the SNF in the early 1920s contains many of the themes voiced by opponents. "If the forest is extended," Lommen writes, "then automatically will the doors be closed against the farmer, the man we have prayed would come as the Savior of the North, to give us permanent prosperity and progress. We have but started in the work of selling Northern Minnesota to the thrifty tiller of the soil. We have spent thousands of dollars in advertising to the world the fertile lands and golden opportunities that await the farmer's plow in a virgin field. To the accomplishment of this great enterprise every civic club, Chamber of Commerce, County Club and public-spirited agency have dedicated themselves for the past dozen years, and now, with a falsity that cuts and stings to the bone, the eastern advocates of the Fuller bill declare that the million acres of Minnesota's land proposed to be annexed into the forest 'is valuable ONLY for forestry purposes.'" (See also "Highlands" and "Nearshore.")

But the contrarians did not halt the momentum for conservation. By 1908 two additional tracts of forest on the Minnesota–Ontario border had been added to the SNF to form a core of more than one million acres. Over time, additional lands were added both by presidential proclamation as well as purchase.

To ensure that "the nation could have its forests and use them too," as environmental historian Roderick Nash put it, Pinchot called for a new generation of foresters to oversee their management. They were trained in an agricultural model of forestry known as sustained yield, an approach to timber management that Pinchot had studied in the forests of Germany in the late nineteenth century. Lacking ecological training and reflecting a culture in which exploitable commodities were highly valued, these new foresters viewed trees as an agricultural crop. Their job was to determine the maximum potential of wood production for each site and prescribe a planting and cutting cycle that would ensure continuous supply of the greatest number of desirable trees. Pinchot pioneered the notion of "wise use" of the nation's natural resources. Conservation, he maintained, "stands for development" that looks after the "welfare of this generation first, and afterward the welfare of the generations to follow."

But not everyone viewed forests as little more than a ready source of timber. An 1884 report from the Minnesota State Horticulture Society, for example, declared that unchecked deforestation would "transform our naturally rich lands into desert conditions, by drying up our springs and depleting our lakes and rivers to the injury of navigation and all other industries." Reformers linked forests to soil conservation, flood prevention, protection of water quality and wildlife habitat, and even climate change. And so when a highly publicized series of major forest fires and heavy flooding on the Ohio River claimed property and human lives from 1907 to 1911, the public demanded—and got—passage of the Weeks Act in 1911, which established federal matching funds for state fire-control efforts and authorized the acquisition of public lands to protect the watersheds of navigable streams. (The Clarke–McNary Act of 1924 authorized the federal government to purchase "forested, cut-over or denuded lands within the watersheds of navigable streams" for the production of timber.)

Forest conservation efforts by the state paralleled many of the actions taken on a federal level. Activists such as Andrews and fellow conservationists from groups such as the Minnesota State Forestry Association urged the establishment of state forest reserves, arguing that forest protection "must be a protégé of the state." By 1900, the Minnesota State Legislature had passed bills that delineated a process for donating private lands to state forest reserves. That same year former Minnesota governor John S. Pillsbury was the first to step forward with a donation of a thousand acres of cutover land. A 1914 amendment to the Minnesota constitution authorized the setting aside of state-owned nonagricultural lands for state forests, a move that created 350,000 acres of public forest in St. Louis, Lake, and Cook Counties.

The bill also established a statewide system of fire wardens for the prevention of fires. But several decades would pass before wildfires were brought under control. According to the 1927 report by J. A. Mitchell of the Lakes States Forest Experiment Station, an average of one thousand fires had strafed Minnesota forests each year during the prior decade. With nearly 400,000 acres burned, 432 human lives lost, and $35 million ($454 million in 2013 dollars) in property damage, "It must be conceded that Minnesota has a forest fire problem," Mitchell wrote. The number of acres requiring fire control was staggering. Ninety-nine percent of the total land area of Cook and Lake Counties, for example, was said to require fire protection. In 1927 alone, Mitchell observed, the State of Minnesota expended some $200,000 ($2.6 million in 2013 dollars) on fire protection, with the federal and private sectors contributing an additional $100,000.

What to Do with a Changed Land?

But those seeking to wring new profits from northern forests would be forced to contend with an altogether new set of ecological conditions caused by the cutover fires. In former bogs, Pyne writes, up to twelve feet of organic material had been vaporized, "leaving bare soil and rock as a residue. In many former pine sites only sand remained, a landscape not unlike the glacial outwash left by the retreating ice sheets."

Extreme soil erosion was commonplace. In a 1903 report for the USGS, J. Morgan Clements described how in some of the areas he visited in northern Minnesota these severe fires destroyed the "humus as well as the timber." "As a consequence of the removal of these protections, the major portion of the soil, and even in some cases the subsoil, has been washed into the valleys, and the hills are now practically bare rock," Clements wrote.

The composition of the forests too was dramatically changed. Ayres noted that a "large proportion of the area north of Red Lake and eastward to Lake Superior (several thousand square miles) has been reduced to brush land, and several thousand acres are now bare rock on which dead stubs and partly burned roots show that timber once grew." University of Minnesota foresters Joseph Kittredge and Suren Gevorkiantz observed in a 1929 report that "the most characteristic sight in traveling through northern Minnesota and Wisconsin is the monotonous and scrubby looking growth of small aspen trees. Actual surveys indicate that the aspen, often somewhat mixed with paper birch, is the most abundant and wide-spread type of forest vegetation in the northern Lake States." In St. Louis County alone, scientists from the Minnesota Agricultural Research Station estimated that aspen and birch occupied 70 percent of the cutover land.

At the time, researchers knew little about the forest ecology of northern woodlands, including the growth and reproduction of tree species, the diseases and insect pests that afflicted them, the complex relationships among species, and ecological functions and processes including natural disturbances such as fire and windthrow and patterns of succession. Forest historian Douglas W. MacCleery points out that in 1910 only two schools in the United States—Yale and Cornell—offered forestry curricula. Most scientists, including Pinchot, received their training abroad, primarily in the woodlands of Germany, where the forests' native ecology—as well as many species and communities— were lost due to centuries of intensive use that emphasized wood production. By 1915 the number of schools offering forestry training jumped to thirteen. During that time, the USFS also established its Forest Products Laboratory in Madison, Wisconsin, whose purpose was to find new uses for the kinds of trees that now dominated the Northwoods' second-growth forests.

In the changed ecological conditions that followed the cutover, it is not surprising that large tracts of the burned-over forest were colonized by second-growth aspen and birch. Both species have light seeds that can be blown great distances to recolonize treeless sites. Both species also can regenerate themselves from underground parts that often escape unscathed by the flames. Birches can resprout from the base of their stumps. Aspen trees also take advantage of their ability to regenerate through asexual reproduction. Although individual trees may be killed by a fire, they have a lateral root system loaded with potential budding sites that are protected from all but the most intense conflagrations. When the aboveground part of a tree succumbs to fire, windthrow, or a beaver's bite, these budding sites—a close examination by one scientist revealed six hundred per eighteen inches of root—vigorously sprout suckers known as ramets. In the course of a single summer, these new shoots may grow as much as three feet. At densities of four hundred thousand ramets per acre, aspens can easily crowd out most competitors.

Kittredge and Gevorkiantz argued that the aspen-dominated forests of Michigan, Wisconsin, and Minnesota combined constituted a "major problem of land use." Although, as they pointed out, aspen was utilized in making paper pulp, excelsior boxes, headers, pails, staves, lumber, and fuel, the supply far outstripped the demand for this postfire "weed" species. The researchers praised aspen's virtues—fast-growing, it could produce large volumes of wood in as little as forty to fifty years. "The aspen is therefore an excellent timber crop for an industry that can use it to advantage or for any owner so located that he can sell it readily," they wrote. "Few owners of aspen land, however, are now so situated." The commercial value of vast acres that once had built the fortunes of an entire nation now was nearly worthless. "What then, is to be done with these millions of acres to obtain the best use and highest value from them?" asked Kittredge and Gevorkiantz.

The researchers suggested two alternatives to the aspen "problem." They advocated, on the one hand, that public agencies buy aspen lands and actively work on converting them to more valuable conifer forests. An alternative, on the other hand, was to advertise aspen's abundance and actively promote "its value for present forms of utilization, and study of its properties for the purpose of developing new uses."

Corporate giants in the wood industry had begun their own research into ways of capitalizing on such abundant raw material. Pyne observes that industry magnates Harry Hornby and Rudolph Weyerhaeuser put two of the nation's top-notch wood chemists to work on finding uses for plentiful and fast-growing trees such as aspen, balsam fir, and jack pine. By 1920, the chemists devised a range of uses for these trees, including pulp and paper, as well as new uses for wood fiber and particles, including insulation and synthetic lumber. In 1922, on the site of a burned-out sawmill in Cloquet that once cut stately white pines into planks, Weyerhaeuser built a new plant that churned out new products using the next generation of north-country trees. Nearby, just south of Cloquet, the University of Minnesota established an experimental forest to study the cultivation of this new source of fiber. With its economic engines idling only temporarily, the industry simply switched to new technology "that transformed worthless reproduction into raw materials of real economic value," Pyne writes.

Managing the New Forest

The rhythms of the pre-European-settlement forest were like the workings of a complex clock in which gears of various sizes and shapes whirred at many different speeds. The proponents of sustained yield sought to standardize the components and impose a greater predictability on their functioning. Growing trees like agricultural crops and suppressing wildfires have, however, propelled the northern forest "into states that rarely if ever existed naturally," says forest ecologist Pastor.

But there were other factors that also contributed to these radical shifts. Take, for example, the arrival of white-tailed deer, a nonnative species that invaded from the south and west. Deer found nutritious forage in the leaves, twigs, and bark of shrubs and aspen trees that flourished in the young, postfire forests and abandoned homestead fields. While conditions grew more hospitable for deer, habitat for native species of ungulates, such as woodland caribou, disappeared. These animals were adapted to forests of red, white, and jack pine, where the sparsely vegetated ground layer allowed caribou lichen, their favorite browse, to flourish.

Nearly extirpated too was one of the northland's most majestic animals—moose—which were slaughtered to feed hungry workers in the backcountry logging camps. Survivors found that suitable habitats were hard to come by. Severely diminished were the forests of dense conifers that the animals relied on for thermal cover in the subzero temperatures of winter and for escape from the summer's heat. Many succumbed to pests and diseases, such as ticks and brainworm carried by the burgeoning deer herds.

The proliferation of deer, in turn, had other effects on the future of the forest. In the burned cutover lands, tree species such as white pine failed to recover. For one thing, logging had eliminated most of the trees that served as a seed source. And in the hot, postlogging fires, as Ayres pointed out, "most of the seeds, seedlings, and seed trees are killed. . . . The reappearance of white and Norway [i.e., red] pine on severe burns is rather unusual."

The few seedlings that managed to sprout in the postfire landscape were browsed by the increasing numbers of deer. Also taking a toll were snowshoe hare, whose appetites are particularly destructive when their numbers peak during boom-and-bust population cycles. Some of the greatest damage occurs throughout late fall and spring when

White-pine blister rust, here infecting a white pine sapling, is a fungal disease that was introduced to North America in the early 1900s. This disease, along with animal browsing and loss of seed sources, has prevented the comeback of white pine in some parts of the landscape. Photograph by Chel Anderson.

the animals seek out the succulent needles and branch tips of white pine and cedar trees over a dry, woody diet of dormant deciduous shrubs and trees.

It was clear to even the earliest observers that white pine could not stage a comeback on its own. Beginning in the early 1900s, government agencies and citizen groups mounted massive replanting campaigns using white pine seedlings imported from European nurseries. But the plants accidentally introduced a deadly fungus known as white-pine blister rust. Favorable climatic conditions and changed fire patterns, as well as an abundance of alternative host plants such as gooseberry and currant shrubs, allowed blister rust to run rampant through patches of young pine seedlings.

The loss of a seed source in combination with disease, browsing, and fire suppression have largely prevented the resurgence of these common but majestic forests.

According to Frelich, in pre-European times, pine forests (mostly white pine with a secondary component of red pine) covered some 3.5 million acres statewide. Only a fraction of the original expanse—about 0.5 million acres—exists today, most of it as second-growth forests.

The scarcity of mature white pines has had critical consequences for species that depend on these tall trees for the rearing and protection of their young. Studies conducted in the SNF, for example, have found that 80 percent or more of osprey and bald eagle nests are built in the tops of mature white pines. Boreal owls use the trees' laddered branches as song perches. And black bears search out tall pines as protective roosts for their cubs. Studies conducted by biologists at the North Central Forest Experiment Station found that radio-collared black bears and their cubs made 95 percent of their beds at the base of these trees. Scientists hypothesize that mothers prefer mature white pine since cubs can sink their tiny claws into the fissured bark and scramble more easily to safety.

And the species is as valuable to forest creatures dead as it is alive. Slow to decay and topple, white pine snags provide decades of nesting opportunities for cavity-loving birds and mammals. Once the dead trees crash to the forest floor, they offer an excellent sounding board for drumming grouse. As the tree deteriorates even further into coarse, woody debris, insects and microbes take advantage of the moist, nutrient-rich environment. In the woody remains, rodents and salamanders find food and shelter.

Not only has the abundance of individual tree species such as white pine changed but so too the type, size, age, and shape of forest patches across the landscape. According to Pastor, today's headwaters' forests, like other forests across the upper Great Lakes region, are either being logged (resulting in young stands of clear-cut trees) or are maturing as even-aged forests. It is as if the forests were composed primarily of small, isolated groups of grade schoolers or retirees. Gone, Pastor says, is the "complex patch mosaic of differing forest types and ages. Past exploitation and recent cutting patterns have cumulatively simplified and fragmented the natural landscape by creating more, smaller, and simpler patches than existed originally."

The magnitude of these changes has been quantified in a variety of studies conducted by researchers from the University of Minnesota, the Minnesota DNR, and the Minnesota Forest Resources Council. Using records from the General Land Office Survey, which was conducted in the late 1880s, and the 1990 USFS Inventory and Analysis Survey of the 7.9 million acres that make up the Arrowhead region of northeastern Minnesota, these studies have identified major trends in forest composition and spatial patterns that have occurred from pre-European settlement to modern times.

Research shows that the abundance of the most common species prior to the arrival of Euro-Americans—white pine, spruce, tamarack, and paper birch—has declined significantly. These species now have largely been replaced by aspen. In some cases, changes in the forest over the past century have created fire situations as dangerous as those of the late nineteenth and early twentieth centuries. Reporting at a public meeting in June 1998, Tim Norman, former USFS fire management officer for the Tofte and Gunflint ranger districts, declared that during the drought conditions of 1996, 1997, and 1998, Cook County had the "worst fire conditions in the United States."

Much of the fire danger can be attributed to the highly flammable stands of dead spruce and balsam fir killed by spruce budworm. According to the Minnesota DNR,

Black bear cubs, one of many forest creatures that rely on white pines, can easily climb the tree's deeply fissured bark to protective roosts in the canopy. Courtesy of the Northwoods Research Center, Wildlife Research Institute, Ely, Minnesota.

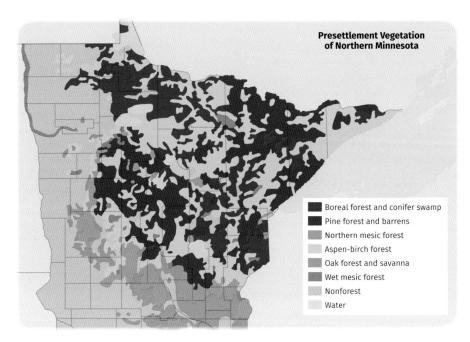

Surveys conducted by the U.S. General Land Office in the mid-1800s were used to create this general map of the presettlement forest in the headwaters. It shows that prior to the arrival of Euro-Americans pine-dominated forests and woodlands and conifer-rich boreal forests were common throughout northern Minnesota.

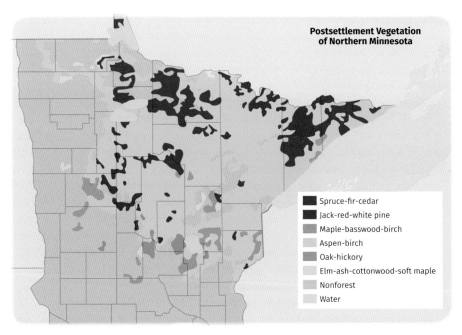

A general map of the northern Minnesota forest, based on a compilation of forest surveys from 1977 to 1983, shows a reduction of pine and a marked increase in aspen and birch as logging activities converted more acres to early-successional forest.

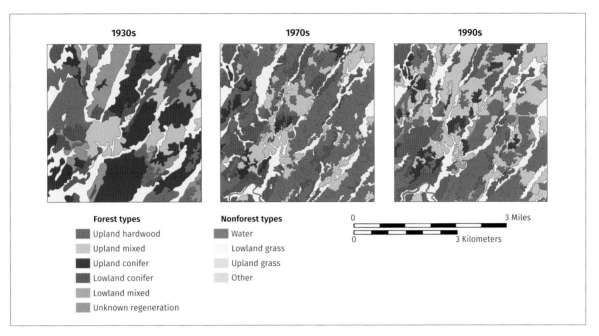

| 1930s | 1970s | 1990s |

Forest types
- Upland hardwood
- Upland mixed
- Upland conifer
- Lowland conifer
- Lowland mixed
- Unknown regeneration

Nonforest types
- Water
- Lowland grass
- Upland grass
- Other

0 ——————————— 3 Miles
0 ——————————— 3 Kilometers

outbreaks by this native insect normally last from eight to ten years, such as the major outbreak that occurred from 1912 into the early 1920s. During these normal episodes, the insect usually kills most of the balsam fir and some white spruce in a given area. Insect populations then subside to barely detectable levels for thirty to sixty years until another generation of balsam fir reaches maturity, allowing budworm numbers to soar once again.

But throughout northeastern Minnesota, an outbreak has persisted since 1954. The altered cycle is consistent with that of other forests across eastern North America, where spruce budworm outbreaks have become longer and more severe and have claimed more acres than in the past three hundred years or more. Studies of Canadian forests from 1900 to 1980 found twenty-one outbreaks during this period compared to only nine in the preceding one hundred years. These modern outbreaks also decimated far greater acreage than previous ones.

Why the change? Forest ecologists point to shifts in fire regimes. Before Euro-American settlement, budworm outbreaks and periodic fires worked hand in hand to largely restrict balsam fir patches to islands within the forested landscape. Budworm outbreaks would primarily target trees weakened by age or disease, making these stands susceptible to fires, which cleared the stage for a vigorous new generation of balsam fir trees as well as created niches for other species such as aspen and pine.

The relative isolation of these patches helped to control the severity, extent, and timing of insect outbreaks. But in many parts of the headwaters landscape, balsam fir forests have expanded, allowing spruce budworm to reach epidemic proportions. Studies by Frelich in the 1990s in the BWCAW, for example, show that a relative lack of fires since 1910 has allowed the amount of spruce-fir-birch forests to double while reducing pine forests to less than a third of their original extent.

In some cases, these forests packed with dead snags pose hazards to human life and

Logging practices, from historic times to the present, have decreased patch size, thereby simplifying and fragmenting the forest landscape.

47

property. In many instances, reducing dangerous fuel loads through prescribed burns in the human-wildland interface is considered too dangerous. Commercial logging is by far the cheapest option for lowering fire hazards, but not all stands contain commercially desirable trees. Furthermore, logging is extremely controversial. Many resort owners regard shorn forests as eyesores that deter tourists.

Instead, USFS personnel have begun crushing the standing dead balsam and live understory trees with heavy machinery, a short-term solution that helps contain potential fires by reducing the number of "ladders" that flames use to climb into the canopy, where they can leap from tree to tree across the forest. But the procedure is extremely expensive, allowing managers to treat only a small fraction of the total number of acres affected.

Evolving concepts of forest-management practices could not only help to eliminate some of these problematic conditions in the future but also have the potential to restore or maintain biodiversity in managed forests. Passage of revolutionary legislation known as the 1976 National Forest Management Act, for example, recognized the need to manage federal forests for more than wood production, water-quality protection, and recreation. The new legislation opened avenues for silvicultural methods that take a more holistic, ecologically based approach to the land instead of simply focusing on regenerating desirable tree species. This forestry philosophy is known under a variety of names, including new forestry, sustainable forestry, and adaptive ecosystem management. Guided by a growing body of ecological knowledge, it aims to preserve the full complement of the forest's native plant and animal life, as well as their related biological and chemical processes.

More recently, the market demand for sustainably grown forest products is driving the expansion of such ecologically informed management on state and private forests. In 2010, for example, the top two forest-certification organizations in North America—the Sustainable Forestry Initiative and the Forest Stewardship Council—certified 4.9 million acres of forests in the state. The designations have made Minnesota one of the largest land bases of sustainably managed forests in the United States. To become certified, forests must undergo a rigorous review by independent auditors, who examine such management practices as reforestation, harvesting methods, and the conservation of water quality, wildlife, and plants other than just tree species.

The toolbox available to sustainable-forest managers includes a suite of techniques that attempts to simulate the effects of natural processes at multiple scales. Such management requires thoughtful consideration, since prescriptions must be appropriately tailored to many different ecological conditions.

Take disturbances such as wildfire. In forest habitats shaped by cool to moderately hot fires, practices such as selective and partial harvests are appropriate. In forests that occupy landforms that were typically scoured by big fires, treatment could include large-scale cuts that mimic the effects of these blazes. But such management practices would not call for an indiscriminate clearing of the land. Studies of a 2,600-square-mile area in the fire-prone boreal forests of Canada show that conflagrations burned some areas to cinder while others stood unscathed. According to Yves Bergeron and his colleagues, up to 80 percent of the fire-strafed area had more living trees than dead ones.

Sustainable forestry is as concerned with mimicking the disturbance patterns of fire on individual sites as much as it is on a landscape scale. As Jerry Franklin, the father of sustainable forestry, observes, "In new forestry, what's left behind on the site is more important than what is taken out," the goal being "continuation not termination" of the forest community. Cutting prescriptions, for example, attempt to reflect natural disturbance in the number, type, and configuration of living trees, which serve as roosts for wildlife, provide seed sources, and jump-start structural diversity in the regenerating stand. Some dead trees are also left standing as hunting perches for hawks and owls; as habitat for cavity nesters such as flying squirrels, flickers, chickadees, nuthatches, creepers, and kestrels; and as a source of insect food for woodpeckers and songbirds.

Even microsites on the forest floor require a range of treatment prescriptions. Downed logs and woody debris are left to rot on the forest floor to serve as nurse logs for some tree species and as food and housing for fungi, insects, amphibians, and vital soil organisms. In other cases, managers carry out more aggressive measures such as scarification. This technique plows up the forest floor to expose the underlying mineral soils that some species require for germination.

And sustainable-forestry techniques would also factor in natural time frames of disturbance. Using a one-size-fits-all approach, conventional management of woodlands that have evolved with fire currently clear-cuts most forests at intervals of thirty to one hundred years. Under sustainable-forestry prescriptions, such aggressive harvests in fire-dependent forests would occur on natural time frames of up to 250 years or more.

Like cyclical fire, the natural disturbance regimes of insect infestations can also be used to structure sustainable-management approaches. Pastor points out that most stands of aspen trees, for example, are currently managed using a single-species approach in which forests of even-aged aspens are clear-cut every thirty to fifty years. Aspen trees sprout vigorously following a disturbance such as clear-cutting, ensuring that a pure stand of similarly aged trees will be available for harvest in a few decades. This prescription keeps aspen forests in a perpetual state of arrested development.

Natural disturbances, even severe ones such as the 2006 Cavity Lake fire, leave behind large amounts of organic matter. Residual snags and downed logs can help feed and shelter organisms for repopulating these disturbed sites. Photograph by Chel Anderson.

In sustainable-forestry management, what is left behind is more important than what is harvested, including large downed logs and woody debris.

Under pre-European-settlement conditions, however, the headwaters landscape would have contained aspen-dominated communities in many different stages of maturity. In older forests, a diverse array of conifers would have grown up alongside aspens, jostling for dominance in alternating intervals. To mimic the cycle of insect infestations that regulate this interplay between aspens and conifers, forest ecologist Pastor recommends a staggered cutting technique known as partial harvest or uneven-aged management. "Maintaining such mixed stands and alternating aspen and conifer dominance," he observes, "results in a reduction of large-scale infestations of pests such as spruce budworm . . . and tent caterpillar . . . and of pathogens, which can be significant problems

As forests age, conifers mingle with early-successional species such as aspen and birch. These mixed-species communities are less vulnerable to large-scale infestations of pests and pathogens, which can be significant problems in the pure stands of aspen, paper birch, and conifers that are intensively managed for pulp and paper production. Photograph by Gary Alan Nelson.

in intensively managed pure stands of both aspen and conifers. . . . Also, because the site is never completely clear-cut, problems of soil erosion and loss of nutrients, soil organic matter, and important soil organisms are minimized, even on difficult sites."

Playing Catch-Up: The Science of Ecosystem-Based Management

Although state and federal legislation mandates an ecosystem-based approach to forest management, change in the field has been slow. Ecologically informed forestry policies routinely hit a wall of resistance from economic interests—both individuals and corporations—whose short-term business models prescribe management regimes that are out of sync with longer-term natural cycles.

But enacting ecosystem-based forestry practices in the field depends as much on expanding the scientific knowledge about the extremely intricate workings of northern forest ecosystems and their species as it does on changing the priorities of public agencies and private players. "What constitutes adequate knowledge is a highly relevant question," say University of Minnesota scientist John Tester and his colleagues in a 1997 paper published in *Biological Conservation*. "The complexity of forest ecosystems allows scientists to possess detailed knowledge of ecosystem processes for only a few locations where case studies have been done. Thus, managers are often faced with making decisions that affect an entire landscape with only rudimentary information on ecosystem parameters for the vast majority of the area involved."

Just how complex is complex? In 1997 University of Minnesota researchers published the results of a study that showed that the microtopography of the forest floor—decaying stumps and downed logs or pits and mounds created by the uprooting of mature trees—may play a critical role in the survival of seedlings for species such as northern white cedar. Scientists refer to these microfeatures as "safe sites." Especially favorable for the germination and survival of white cedar seedlings are decaying logs. Their rough surfaces, largely free of shading from litter, offer hidden crevices for seeds to germinate away from the hungry mouths of birds or rodents. And the decaying wood supplies seedlings with a critical boost of moisture and nutrients. Especially beneficial is the presence of nitrogen-fixing bacteria and the abundant mycorrhizal fungi that colonize plant roots and increase their effectiveness in scavenging nitrogen and phosphorus from the decaying wood.

But upland forests in which white cedar play a dominant role are becoming increasingly rare. Logging reduces the number of safe sites by removing the large canopy trees

Microfeatures on the forest floor, such as decaying wood and the pits and mounds created by the uprooting of mature trees, are essential to the germination and survival of many plant species, including the seedlings of northern white cedar. Photograph by Bob Firth/ Firth PhotoBank.

that, weakened by disease or old age, would otherwise topple to the forest floor and provide shelter for a new generation. Invading these forests are the young of species that are less exacting about their requirements for germination and survival.

What goes on underground may be just as important, however, as what happens on the surface, as scientists are just now discovering. For more than a century, researchers have known about the existence of soil mycorrhizae, thanks to the work of A. B. Frank, a German professor, who in the late 1880s set out to find ways to increase the supply of truffles, a prized fungus that grows on the roots of beech and oak trees. Frank never discovered a recipe for truffle propagation. He did, however, stumble across vast subterranean partnerships between fungi and plant roots, which scientists now can observe in considerable detail thanks to the invention of sophisticated molecular tracking techniques. As a result, "When it comes to ecology and evolutionary biology," says John N. Klironomos, a leading soil ecologist at the University of Guelph in Ontario, Canada, "the [mycorrhizal] field is cracking wide open."

Tiny soil fungi, it turns out, live in association with plant roots in nearly every terrestrial ecosystem, from tropical and boreal forests to grasslands and tundra. Indeed, mycorrhizae (literally translated as "fungus-roots") attach themselves to the feeding roots of plants, covering up to 90 percent of them with fine threads known as mycelia. Klironomos observes that some 352 miles of mycelia can be found in one cubic foot of soil.

Measuring one-sixtieth the diameter of a plant root, the mycelia lace the soil for considerable distances, penetrating tiny pores that are off-limits to plant roots. Their benefits are many. As soil binders, they help prevent erosion and water loss. They also provide a measure of protection against pathogens and toxic wastes. Indeed, mycorrhizae are used to help remediate soils damaged by heavy metals or excessive use of fertilizers.

But of special interest to ecologists has been their food-gathering prowess. Using enzymes to degrade organic matter, mycorrhizae are especially adept at scavenging phosphorus and nitrogen. A portion of the minerals they absorb are "bartered" for energy-rich plant sugars from the host plant, which manufactures them using carbon that is captured from sunlight. So efficient is this system that some 95 percent of plants rely on these subterranean partners for their necessary quotients of phosphorus and nitrogen.

Mycorrhizae are so essential that trees may find it difficult to establish a toehold in their former territories without fungal friends, as biologist John Terwilliger discovered in his study of beaver-created meadows in the SNF. Dam-building activities by beavers routinely flood forests adjacent to streams, killing trees and their root-dwelling fungi, which cannot tolerate the anaerobic, waterlogged condition of the sediments in these impoundments. When beavers abandon their structures, as they often do, the untended dams deteriorate, and the ponds drain. Over time, these wet areas dry out and become meadows. But Terwilliger noticed something curious about these meadows. Even though many of them were surrounded by conifer forests, the trees did not quickly recolonize the new open spaces. The reason, he discovered, was the absence of nutrient-boosting fungi in the soils. It turns out that the feces of red-backed voles are a potential vector for the reintroduction of ectomycorrhizae into these drained meadows. But the forest-dwelling voles rarely ventured beyond the woodland edge into the meadows, thereby limiting the spread of the fungi that would improve the rate of tree colonization and growth. By killing mycorrhizae through their impoundments, beavers add to the complexity of habitats long after their waters have drained away.

Complicating the picture is the recent discovery that these fungal partners benefit not only individual plants but also link with one another to help maintain an equitable distribution of resources throughout entire communities. For example, young fir trees often are found growing in the shade of older birch trees. In 1997 researchers from British Columbia's Ministry of Forests published research showing that these sunlight-starved trees obtained carbon from their better-fed birch neighbors. The mode of transport for the carbon exchange was the great network of ectomycorrhizal fungi. These fungal life-lines may be especially important for the survival of seedlings, which can spend years in the shady understory waiting for a chance to reach the light. Why the fungi give up some of their hard-earned carbon to subsidize the growth of surrounding trees remains a mystery to the team. "Rather than plants competing with each other, fungi are often facilitating one plant getting nutrients from another so they can co-exist," Klironomos observes. "So rather than competition among plants being the center of plant ecology, it could be mutualism, which could change everything."

Even more surprising still, research published in the journal *Nature* in 2001 demonstrates that fungi may seek out live food in addition to dead and decaying material. Common insects known as springtails feast on fungal networks in the soil. Some fungi, such as those of the common species *Laccaria bicolor,* exude what researchers suspect is a toxin that paralyzes the hungry springtails. Mycelia then invade the animal's body in search of nitrogen, some of which they share with the host plant, says Klironomos, one of the article's main authors. Studies in the laboratory of eastern white pine seedlings indicate that the survival of this species may depend as much on these hidden liaisons as on the control of blister rust and browsing by deer and snowshoe hares. Soils seeded with springtails and *L. bicolor,* the scientists discovered, derived 25 percent of their nitrogen from the activities of these fungal predators. "Should this phenomenon prove to be widespread," Klironomos observes, "forest-nutrient cycling may turn out to be more complicated and tightly linked than was previously believed."

The Sociology of Forest Management

The job of carrying out ecologically based forest management is not hampered simply by lack of scientific knowledge. It is shaped as much by cultural values and human perceptions as it is by ecological processes. Take that great American icon Smokey Bear. Polls of the general public have shown that his image is second only to that of Santa Claus in recognizability. Smokey and his fire-suppression mantra, "Only you can prevent forest fires," arose during World War II when the United States feared incendiary attacks on its forests from wartime enemies. With time, Smokey's message—green is good, charred is wasteful—became ingrained in the national consciousness. But many of the forests that this clever ad campaign sought to protect had evolved a dependence on fire to maintain their vitality. As the great fire year of 2000 demonstrated, Smokey Bear has not served the best interests of these ecosystems. Across the nation, decades of fire suppression have created abnormal fuel buildups, increasing the intensity, destructiveness, and danger of wildfires.

Complicating matters is the public's perception of what a healthy forest should look like. Our aesthetic biases can trace their origins to the romantic and transcendentalist movements of the mid-1800s, says Paul Gobster, a research social scientist with the USFS. Artists such as Frederic Church and Thomas Cole of the Hudson River school influenced a whole generation of writers, photographers, and landscape architects. These painters, Gobster says, "often stylized the nature they saw, carefully composing a scene by adapting formal design principles such as balance, proportion, symmetry, order, vividness, unity, and variety in line, form, color, and texture." The public, in turn, learned how to appreciate landscapes by viewing these serene, orderly scenes. As a result, Gobster says, "the 'scenic aesthetic' became the dominant form of landscape appreciation."

Unfortunately, the scenic aesthetic is often at odds with what Gobster calls the "ecological aesthetic," which has its roots in Aldo Leopold's land ethic. Leopold cautioned that our perception of the natural world—and our intervention in it—should be guided by the land's ecological needs and constraints: "A thing is right when it tends to preserve the integrity, stability, and beauty of the biotic community. It is wrong when it tends otherwise."

What is good for the forest, however, may not be pleasing to the eye. Take, for example, the rough-and-tumble world of the forest floor. Scientists have shown that dead and dying trees may be more important to long-term forest health than living trees. Yet, leafless snags and the messy crosshatch of downed trunks, punky limbs, and upturned tree roots are viewed as negatives by many forest visitors. Research on visual perception, Gobster points out, "has shown that, whether it occurs naturally or through timber harvesting, dead and down wood has one of the biggest negative impacts on the perceived visual quality of new-view forest scenes."

As science brings to light new information about the needs of healthy ecosystems, the public will need to adopt a new set of aesthetic criteria. It means a move away from taking pleasure in the mere appearance of a landscape toward taking pleasure in the health of its intricate ecological workings. Making this shift, Gobster says, requires changing the "focus of our relationship with the landscape from a homocentric one toward one that is more biocentric."

Will History Repeat Itself?

As our knowledge of the ecological intricacies of the headwaters forest grows, so too does a whole array of mounting economic pressures that are no less intense than when the first timber cruisers staked out groves of valuable white pine. This time, however, the wood from Minnesota forests is being cut not to build factories in Chicago or homesteads on the treeless prairie but to feed a global demand for wood and paper products.

At the heart of the rhetoric and economic policies of this global trade lies an unsettling echo of the boosterism that fueled the boom-and-bust cycles of the past one hundred years. Take, for example, the go-go decades of the 1980s and 1990s. During this time, according to a 2003 report by the Minnesota Department of Economic Security, per capita wood consumption in the United States rose more than 31 percent. Minnesota paper mills were quick to respond. Between 1986 and 1993 alone, they invested $1.5 billion in new plants and upgrades of older facilities and projected billions more in mill expansions. Regional and state leaders in business and government applauded the rapid escalation of capacity, touting windfalls in new job creation and economic growth.

Critics countered this enthusiasm, warning that the ramped-up consumption would prove destructive to the ecology of Minnesota's forests. Pulp and paper mills rely heavily on aspen trees. The demand stepped up the conversion of native mixed forests to aspen monocultures that were cut at the shortest possible intervals. These expanding acres of single-species forests have resulted in a damaging simplification of the northern forest's native diversity.

Critics also expressed uneasiness about the growing monoculture of the forest products industry in many northern Minnesota communities. Their economies have been especially hard hit by the industry's historic vulnerability to volatile market fluctuations. But the warnings went largely unheeded. The rush to expand continued, and by 2000 capacity had surpassed demand. A recession in 2001–2 brought plant closures, downsizing, production slowdowns, and job losses to many communities.

The economic woes were deepened by a dwindling supply of merchantable Minnesota-grown aspen trees, a shortage that was anticipated by both industry and state regulators as mills expanded in the 1980s and 1990s. The shortfall drove the cost of existing stumpage (merchantable trees in the woods) ever higher, making Minnesota's mills among the least competitive in North America. Some mills made do by substituting other species as well as by importing increasing amounts of aspen from Canada.

By 2006 a precipitous drop in home construction further depressed the demand for a wide array of wood products ranging from oriented strand board to pallets. Unsold products piled up at mills around the state. Those that had stockpiled wood against the predicted aspen shortfall sharply cut the price they paid for wood to independent loggers, the backbone of the forest-products industry. The financial squeeze forced so many of them out of business that calls soon were heard in the halls of the Minnesota legislature for state and federal assistance. As a 2006 story in the *Timberjay,* a northeastern Minnesota weekly, put it, "The current industry upheaval was not just a matter of time; it was a matter of state policy." Wood prices and demand for wood products stubbornly resisted recovery as a severe recession gripped the U.S. economy beginning in 2008. Between 2005 and 2010, timber harvest in the state dropped by one million cords. Although prices had stabilized by 2012, the dip in the forest industry continued with two more mill closures and no significant improvement in sight.

Emerging technologies will continue to challenge the management of the forest for ecological health as well as economic gain. For example, the forest industry has developed techniques for exploiting ever-new types of raw material. Among the most recent is the use of woody biomass (the tops and branches of harvested trees and whole small trees and shrubs) for energy production. Bringing new and existing technologies into a sustainable relationship with the ecology of the northern forest will have a crucial influence on the future of all its inhabitants, including its human communities. As former U.S. senator and longtime conservationist Gaylord Nelson observed, the economy of human communities "is a wholly-owned subsidiary of the environment."

The question is, after a century's worth of hindsight and science, will we serve as better stewards of this rare and beautiful place? ❧

SUGGESTIONS FOR FURTHER READING

Ayres, Horace B. *Timber Conditions in the Pine Region of Minnesota*. Washington, D.C.: U.S. Geological Survey, 1900.

Bergeron, Yves, Alain Leduc, Brian D. Harvey, and Sylvie Gauthier. "Fire Regimes in the Eastern Canadian Boreal Forest: Implications for Sustainable Forest Management at Different Scales." Plenary presentation, Third North American Forest Ecology Workshop, Duluth, Minn., 2001.

Chapman, Herman H. "The Influence of the Chippewa Forest Reserve." *Journal of Forestry* 38, no. 11 (November 1940): 867–70.

Cheyney, E. G., and O. R. Levin. *Forestry in Minnesota*. St. Paul: State of Minnesota Forest Service, 1929.

Cinzano, P., F. Falchi, and C. D. Elvidge. "The First World Atlas of the Artificial Night Sky Brightness." *Monthly Notices of the Royal Astronomical Society* 328 (2001): 689–707.

Clements, J. Morgan. *The Vermilion Iron-Bearing District of Minnesota*. Washington, D.C.: U.S. Geological Survey, 1903.

Committee on Land Utilization. *Land Utilization in Minnesota: A State Program for the Cut-Over Lands*. Minneapolis: University of Minnesota Press, 1934.

Cornett, Meredith W., Peter B. Reich, and Klaus J. Puettmann. "Canopy Feedbacks and Microtopography Regulate Conifer Seedling Distribution in Two Minnesota Conifer-Deciduous Forests." *Ecoscience* 4 (1997): 353–64.

Crocker, Douglas A. *Pulpwood in the Lake States*. New York: American Pulp and Paper Association, 1926.

Friedman, Steven K., and Peter B. Reich. "Regional Legacies of Logging: Departure from Presettlement Forest Conditions in Northern Minnesota." *Ecological Applications* 15 (2005): 726–44.

Fritzen, John. *History of North Shore Lumbering*. Duluth, Minn.: St. Louis County Historical Society, 1968.

Gobster, Paul H. "An Ecological Aesthetic for Forest Landscape Management." *Landscape Journal* 18 (1999): 54–64.

Greeley, W. B. "The Relation of Geography to Timber Supply." *Economic Geography* 1 (1925): 1–14.

Heinselman, Miron. *The Boundary Waters Wilderness Ecosystem*. Minneapolis: University of Minnesota Press, 1996.

Host, George E., and Mark A. White. *Changes in Forest Spatial Patterns from the 1930s to the Present in North Central and Northeastern Minnesota: An Analysis of Historic and Recent Air Photos*. Report LT-1203c. St. Paul: Minnesota Forest Resources Council, 2003.

———. *Contemporary Forest Composition and Spatial Patterns of North Central and Northeastern Minnesota: An Assessment using 1990s LANDSAT Data*. Report LT-1203b. St. Paul: Minnesota Forest Resources Council, 2003.

Kates, James. *Planning a Wilderness: Regenerating the Great Lakes Cutover Region*. Minneapolis: University of Minnesota Press, 2001.

Kittredge, Joseph, Jr., and Suren R. Gevorkiantz. *Forest Possibilities of Aspen Lands in the Lake States*. Technical Bulletin No. 60. St. Paul: University of Minnesota Agricultural Experiment Station, 1929.

Klironomos, John N., and Miranda M. Hart. "Animal Nitrogen Swap for Plant Carbon." *Nature* 410 (2001): 651–52.

Larson, Agnes M. *History of the White Pine Industry in Minnesota*. Minneapolis: University of Minnesota Press, 2007 [1949].

Lommen, George H. "Subject: Shall We Enlarge the Superior National Forest? Puzzle: Find the Nigger in the Woodpile." Biwabik, Minn.: printed by author, circa 1924.

MacCleery, Douglas W. *American Forests: A History of Resiliency and Recovery*. Durham, N.C.: Forest History Society, 1992.

Manolis, Jim. *Project Summary: Results from the Minnesota Forest Spatial Analysis and Modeling Project*. Report LT-1203g. St. Paul: Minnesota Forest Resources Council, 2003.

Marshall, Robert. *The People's Forests*. New York: Harrison Smith and Robert Hass, 1933.

Minnesota Department of Natural Resources. "Spruce Budworm and Balsam Fir: How Much Is Enough?" *Forest Insect & Disease Newsletter,* August 21, 1996.

Mitchell, John A. *Forest Fires in Minnesota.* St. Paul: State of Minnesota Forest Service, 1927.

Mladenoff, David J., and John Pastor. "Sustainable Forest Ecosystems in the Northern Hardwood and Conifer Forest Region: Concepts and Management." In *Defining Sustainable Forestry,* ed. Gregory H. Aplet, Nels Johnson, Jeffrey T. Olson, and V. Alaric Sample. Washington, D.C.: Island Press, 1993.

Morris, R. F., W. F. Cheshire, C. A. Miller, and D. G. Mott. "The Numerical Response of Avian and Mammalian Predators during a Gradation of the Spruce Budworm." *Ecology* 39 (1958): 487–94.

Nute, Grace Lee. *Lake Superior.* New York: Bobbs-Merrill Company, 1944.

Ojakangas, Richard W., and Charles L. Matsch. *Minnesota's Geology.* Minneapolis: University of Minnesota Press, 1982.

Pennisi, Elizabeth. "The Secret Life of Fungi." *Science* 304 (2004): 1620–22.

Pielou, E. C. *The World of Northern Evergreens.* Ithaca, N.Y.: Cornell University Press, 1988.

Pyne, Stephen J. *Fire in America: A Cultural History of Wildland and Rural Fire.* Seattle: University of Washington Press, 1982.

———. *Year of the Fire: The Story of the Great Fires of 1910.* New York: Viking, 2001.

Raff, Willis H. *Pioneers in the Wilderness.* Grand Marais, Minn.: Cook County Historical Society, 1981.

Read, David. "The Ties That Bind." *Nature* 388 (1997): 517–18.

Rogers, Lynn L. "Are White Pines Too Valuable to Cut?" *Minnesota Volunteer,* September/October 1991: 9–21.

Scott, Michael L., and Peter G. Murphy. "Regeneration Patterns of Northern White Cedar, An Old-Growth Forest Dominant." *American Midland Naturalist* 117 (1987): 10–16.

Searle, Newell. "Minnesota Forestry Comes of Age: Christopher C. Andrews, 1895–1911." *Forest History,* July 1973: 15–25.

———. "Minnesota National Forest: The Politics of Compromise, 1898–1908." *Minnesota History* 42 (Fall 1971): 243–57.

———. "Minnesota State Forestry Association: Seedbed of Forest Conservation." *Minnesota History* 44 (Spring 1974): 16–29.

———. *Saving Quetico-Superior: A Land Set Apart.* St. Paul: Minnesota Historical Society Press, 1977.

Simard, Suzanne W., David A. Perry, Melanie D. Jones, David D. Myrold, Daniel M. Durall, and Randy Molina. "Net Transfer of Carbon between Ectomycorrhizal Tree Species in the Field." *Nature* 388 (1997): 579–82.

Terwilliger, John, and John Pastor. "Small Mammals, Ectomycorrhizae, and Conifer Succession in Beaver Meadows." *Oikos* 85 (1999): 83–94.

Tester, John R., Anthony M. Starfield, and Lee E. Frelich. "Modeling for Ecosystem Management in Minnesota Pine Forests." *Biological Conservation* 80 (1997): 313–24.

White, J. Wesley. *Historical Sketches of the Quetico-Superior.* U.S. Forest Service, Superior National Forest, 1974.

White, Mark A., and George E. Host. *Changes in Disturbance Frequency, Age, and Patch Structure from Pre-European Settlement to the Present in North Central and Northeastern Minnesota.* Report LT-1203a. St. Paul: Minnesota Forest Resources Council, 2003.

Williams, Michael. *Americans and Their Forests: A Historical Geography.* New York: Cambridge University Press, 1989.

Wilkinson, William. *Memorials of the Minnesota Forest Fires in the Year 1894 with a Chapter on the Forest Fires in Wisconsin in the Same Year.* Minneapolis: Norman E. Wilkinson, 1895.

Wolff, Julius F., Jr. "Some Vanished Settlements of the Arrowhead Country." *Minnesota History* 34, no. 5 (Spring 1955): 177–84.

Bird Diversity on the North Shore

Once I went into the woods to find an almost unfindable bird,
a blue grosbeak. And I found it: a rough, deep blue, almost black,
with heavy beak; it was plucking one by one the humped,
pale green caterpillars from the leaves of a thick green tree.

Mary Oliver, *Winter Hours* (1999)

ASK THE AVERAGE PERSON TO NAME A HOT SPOT FOR BIOLOGICAL DIVERSITY, and the answer is likely to be tropical rain forests or Caribbean coral reefs. Suggest a less exotic locale, like the Superior National Forest (SNF), and most people, even Minnesotans, are apt to draw a blank. Still, when it comes to birds—warblers, hawks, sparrows, woodpeckers, thrushes, and ducks—northern Minnesota is *the* place to be.

Yet hardly a week seems to pass without a news report chronicling the link between tropical deforestation and the decline of forest birds, especially migrating songbirds. Far less publicized is the importance of healthy *northern* forests to the future of these imperiled populations. In what has been called the "tattered robe" effect, the woodlands that once blanketed much of the eastern half of the United States have been cut into smaller, isolated tracts by housing developments and cornfields, parking lots and roadways. For some species, the fragmentation has been disastrous. Slicing and dicing the forest have left birds that require deep forest interiors for nesting more vulnerable to predators that patrol forest edges, from household cats to fellow birds. Other birds arrive on their customary breeding grounds or migratory stopovers only to find that a Walmart or seaside resort has replaced the habitats they were counting on for foraging or resting.

Only a few large chunks of forest—the Great Smoky Mountains, the Adirondack Mountains, and parts of northern Minnesota and Maine—remain as viable refugia for birds, that is, places with a complex of habitats that are intact enough to support diverse assemblages of birds in the long term. These "large, contiguous forested regions . . . are

Mean number of species per route

- 61 to 67
- 51 to 60
- 41 to 50
- 31 to 40
- 15 to 30

Northeastern Minnesota lies at the heart of a relatively narrow swath of prime bird-breeding territory with the highest bird species richness of any region north of Mexico.

postulated to be the best hope for conservation of both resident and neotropical song-birds," writes Minnesota ornithologist Janet Green in her book *Birds and Forests.*

Among the havens for songbirds is the three-million-acre SNF. It lies at the heart of a relatively narrow swath of prime bird-breeding territory that dips down from the prairie provinces of Canada, stretches across the northern Great Lakes, and then veers up through New England and the Maritime Provinces. Annual surveys of breeding birds have revealed an average of sixty-one to sixty-seven species per surveyed route, giving these regions "the highest bird species richness of any region north of Mexico," Green observes. That is more breeding birds than in Florida or California, more than in Texas or South Carolina. And unlike in many other parts of the country, the populations of most birds in the SNF are stable or increasing.

But bird lovers do not need to look at a map to know that the SNF is home to an astounding variety of birds. In early summer avian songsters stake out breeding terri-tories, broadcasting their melodies as a way to muscle out competitors. For a few brief weeks, the woods of northern Minnesota become an open-air concert hall that swells with music. With their new leaves rustling in the wind, aspen, birch, and maple trees supply a soft background percussion. Against the rise and fall of their pattering are motifs that are as recognizable to bird lovers as the arias of Puccini and Mozart are to operagoers. Take, for example, the tiny winter wren, which attacks his complicated score with speed and reckless abandon, pausing in his runs of impossibly high, sweet notes to trill before zipping up and down the musical scale again. By contrast, the thrushes favor compositions marked by greater gravity and deliberation. Like a clarinet solo in an atonal composition by Bartok, the melancholy hermit thrush sings out the long fading light of a summer night in phrases that seem shy, almost halting, and improvisational.

Some birds include far simpler refrains in their repertoire. The ovenbird, the John Philip Sousa of the avian world, calls out, "teacher, *teacher,* TEACHER," a vocal signature that writer Neltje Blanchan compares to a "series of little explosions, softly at first, then louder and louder and more shrill until the bird that you at first thought far away seems to be shrieking his penetrating crescendo into your very ears."

Bugs, Bugs, Bugs: The Forest Feast

Much of this enthusiastic music making can be attributed to the tiny forest visitors that arrive in waves each spring. Of the 155 species of birds that nest in the SNF, about 43 percent are neotropical migrants, mostly flycatchers, vireos, and warblers that beat their way north from places as far south as Peru. But why would a warbler that could fit in the palm of your hand—with room to spare—leave its Caribbean paradise to risk flying across open water or colliding with radio towers and spring ice storms just to nest in a particular patch of forest in northern Minnesota?

Food—and lots of it. It is no coincidence that songbirds begin raising families when many north-woods hikers reach for their head nets. In late June and early July the parents of nestlings become veritable feeding machines. Particularly prized are the juicy, protein-packed larvae of moths and butterflies. "Insects are what lure these tropical birds away from the rich forests of Central America and the tranquility of the Carib-bean islands," write John J. H. Albright and Richard Podolsky in *From Cape Cod to the Bay of Fundy.* "With the increasing day length, out from the bark of trees and up from

FOLIAGE GLEANERS
Red-eyed vireo
Scarlet tanager
Chickadee, Boreal and
   Black-capped
Brown creeper
Winter wren
Sedge wren
Ruby-crowned kinglet
Magnolia warbler
Nashville warbler
Bay-breasted warbler

GROUND FORAGERS
Ovenbird
Wood thrush
Chipping sparrow
White-throated sparrow

SALLIERS
Eastern wood pewee
Alder flycatcher
Least flycatcher
Eastern kingbird
Eastern phoebe

BARK GLEANERS
White-breasted nuthatch
Black-capped chickadee
Brown creeper
Pine warbler
Black and white warbler

BARK DRILLERS
Hairy woodpecker
Black-backed woodpecker
Northern flicker
Pileated woodpecker

HOVER GLEANERS
Blackburnian warbler
Northern parula
Golden-crowned kinglet

HAWKERS
Tree swallow
Northern goshawk
Merlin

PREDATORS
Barred owl
Broad-winged hawk
Saw-whet owl
Red-shouldered hawk

Birds in the headwaters belong to a wide variety of feeding guilds—groups that exploit resources in similar ways. They include birds that probe bark, snatch insects out of the air, root through debris on the forest floor, or glean the undersides of leaves.

the ground and from fresh meltwaters emerges a cornucopia of grubs, beetles, mayflies, moths, butterflies, blackflies, and mosquitoes. If you enjoy seeing and hearing birds, you are indebted to the insects."

This threesome—birds, plants, and insects—goes back a long way. Many scientists believe that the sudden explosion of new songbird species during the Tertiary period, which began some sixty-five million years ago, was due to the appearance of angiosperms, or flowering plants. Entomologist Gilbert Waldbauer observes that the "rapid evolution among the birds, especially the passerines, or perching birds (also known as the songbirds)—probably [came] in response to the new food resources provided by the rapid proliferation of insects that was, in turn, a response to the appearance and burgeoning of a new and more varied plant resource for insects, the angiosperms. Today most insects depend for their sustenance either directly or indirectly upon these flowering plants."

Not surprisingly, in the course of sixty-five million years of evolution, birds have fine-tuned their relationship to plants and insects in exquisitely detailed ways. For example, a bird's very "architecture"—the shape of its bill or feet—is specially adapted to the manner in which it forages. Birds that exploit resources in similar ways—groups that probe bark, snatch insects out of the air, root through debris on the forest floor, or glean the undersides of leaves—are considered members of what ornithologists call a feeding guild.

But if they are all competing for the same resources—like a surfeit of dentists in a small town—how do members of a guild keep from putting each other out of business?

"Guilds present an interesting problem," says John Pastor, a forest ecologist at the University of Minnesota Duluth. "If species in a guild exploit similar resources, how do they manage to coexist and keep from driving each other into extinction?"

This was the question that ecologist Robert MacArthur posed in the 1950s while observing the community interactions among a guild of wood warblers in the forests of Maine. MacArthur noted that five species of warblers—Cape May, yellow-rumped, black-throated green, blackburnian, and bay-breasted—all preferred nesting in mature spruce-fir forests. All were insect eaters. Their occupation of the same habitat niche in the forest seemed to fly in the face of an ecological principle known as competitive exclusion. Simply put: two species cannot occupy the same niche because sooner or later one will eliminate the other through competition. So how then did *five* species seem to coexist while feeding in the same spruce or fir tree?

In what has become a classic case in ecology studies, MacArthur discovered that the birds divided up the forest structure and became feeding specialists in certain sections of the trees. Some combed the outside of the trees, while others probed more deeply near the trunks. Not only did the birds forage at different depths, but they also staked out feeding territories at different heights. The Cape May warblers, for example, concentrated their foraging on the surface and at the very top of the spruce trees. At the opposite end, the yellow-rumped warblers preferred to feed at the very bottom of the trees and on low shrubs. Because of this vertical and horizontal partitioning, the birds largely avoided competition by eating many different kinds of insects, each of which flourished in a different locale within the forest. Furthermore, because they nested at slightly different times, the peak demands for food by hungry nestlings did not overlap.

The abundance of vegetational niches helps to make the snf a good place for insects—and the birds that prey on them. Characteristic of the snf is its great variety of natural communities. The snf lies at the confluence of three major forest types: the boreal forest of conifers, deciduous aspen, and birch bearing down from the north; the hardwood forest of sugar maple and northern red oak reaching up from the south; and the pine forest stretching from the east. In this transition area, what ecologists call an ecotone, each forest type is at the limits of its geographic range, and together they support a rich commingling of species.

But it is not just the diversity of tree species in a forest that creates such great habitats for insects and birds. Albright and Podolsky point out that "predicting which forests will support the most species of birds is in many ways the same problem as predicting (or measuring) which forests provide insects with standing and moving water for reproduction, a mixture of standing and fallen dead vegetation for overwintering and feeding, and availability of a diversity of flowering plants."

The sculpting action of the glaciers laid the groundwork—literally—for the structural complexity that supports diverse bird life. The glaciers left behind a topography of lakes and pocket swamps, river plains and windy ridges, rocky gorges and mounded hillocks. As a result, the temperature, soils, and hydrology may vary significantly, even among places that are in relative proximity to one another. A low-lying patch of thick, wet clay, for example, might provide home ground for a small forest of northern white cedar, while only a short walk away a ridge rises up out of the land, and in the crevices of its exposed bedrock, enough soil has accumulated to support a cluster of dwarfed fir trees in a mat of caribou lichen. Maintaining this tight mosaic are disturbances—from fires and outbreaks of defoliating insects to windthrow—that periodically reshuffle the locations, composition, and ages of different forest communities on the land.

The complex and ever-shifting mosaic of the forest has upped the reproductive ante for many birds, especially neotropical migrants. Consider their task. After logging thousands of air miles, songbirds get right down to the business of raising a family. In a couple of weeks not only do the birds have to find optimal nesting habitat in the great expanse of forest, but they also have to locate a suitable mate. These choices are critical since most songbirds lay a single clutch of eggs, which means that they have only one chance to reproduce.

Some birds have the advantage of experience. The average life span of a songbird is two to three years, allowing many birds to nest twice before they die. But given the changeability inherent in the northern forest, one season's nesting site may not be the same the following year. To get an idea of what birds are up against, imagine entering your house after a long journey and suddenly discovering that not only has the furniture been completely rearranged, but also essential objects, such as beds, tables, and refrigerators, have been removed!

Remarkably, despite the dramatic natural and human-made changes in the SNF that have occurred since the late nineteenth century, only one species—the passenger pigeon—has become extinct. This high survivorship is largely due to the fact that birds that breed in northern forests have evolved the capacity to thrive in a state of flux. "Flying gives birds a dispersal ability in an ever-changing forest mosaic, and their populations have adapted to variations in forest age and patch heterogeneity," Green observes. "What the biological limits are to that adaptive ability is the crucial question for both researchers and managers."

Pushing the Envelope: Timber Harvesting and Forest Diversity

The timber industry, among the most volatile sectors of state and national economies, can test those limits. To serve the growing national and global demand for paper and wood products, the major pulp, paper, and chipboard mills throughout northern Minnesota embarked on multibillion-dollar expansions beginning in the early 1980s. In 1993, a Generic Environmental Impact Study (GEIS), a study of statewide timber harvesting, was completed. The study projected that five million cords of wood could be sustainably harvested from Minnesota's forests. During the 1990s, Minnesota boasted the fastest-growing timber economy in the nation. By 2005 loggers were harvesting nearly four million cords of wood each year from Minnesota forests, an increase of some 60 percent from the preexpansion years in the late 1970s. (Four million cords is equivalent to a stack of wood measuring four feet wide, three feet high, and six thousand miles long. The amount of wood logged each year was enough to build two three-foot-by-three-foot stacks from coast to coast across the continental United States.)

Many scientists worried that the volume and pace of logging could eliminate the very features that have made the SNF so hospitable to birds. Forest ecologists point out that structural complexity, both within a single stand of trees and across the larger landscape, is essential to healthy forest habitat for birds. "Structural diversity, the size, shape, and number of layers (or physiognomy) of the forest . . . provides more microniches and offers the opportunity for many more insects to distribute themselves spatially

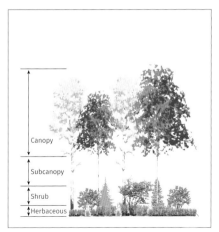

A forest managed to sustain its multilayered structure, shown here in cross section, provides animals with more opportunities for breeding and feeding, nesting and resting. In even-aged stands simplified by management *(right)*, the number of niches is diminished or missing.

throughout the forest," say Albright and Podolsky. "All else being equal, a large, multi-layered forest with a well-defined groundcover layer, a subcanopy, and a large, expansive canopy will support far more species of birds than a small forest with only a single layer. Physical diversity within a forest begets species diversity among the birds, small mammals, insects, and amphibians, because the physical spaces in a forest translate into greater ecological opportunities (or niches) for breeding and feeding. . . . Thus, maximizing landscapes for forest birds translates into optimizing insect abundance and physical structure in forest lands." (See also "The Secret Life of Salamanders.")

Such diversity is important both within an individual forest community as well as across the larger landscape. Experience has shown, however, that a wood-products industry dominated by pulp and paper companies often drives the forest into a more simplified ecological condition by focusing on aspen production. (Remember the six-thousand-mile-long woodpile harvested each year from Minnesota forests in 2005? One-half of this harvested wood was composed of aspen trees.) Valued by the industry for their dense fibers, aspen trees also have other desirable characteristics. In Minnesota they are early-successional, or pioneer, species; that is, they have evolved to take quick advantage of large disturbances such as fire. They sprout quickly after fires or clear-cutting, and the new shoots thrive, forming a dense cover of similar age. These stands can be clear-cut in short rotations, that is, every forty to fifty years. Managing aspen stands in this even-aged condition allows loggers to harvest stands with great efficiency and maximum fiber output. "When loggers enter a stand, 90 percent of the time they clear-cut," says forest ecologist Pastor. "It doesn't take a rocket scientist to understand that if you log a stand every fifty years, and you do that everywhere across the landscape, you don't have a lot of diversity left."

New technologies have enabled the industry to exploit aspen and other species at ever-younger ages, a trend that Andrew Pollack noted in a 1991 *New York Times* article. The forest-products industry, he observed, "is undergoing a major shift from an era in which it took big pieces of wood and cut them into smaller products to one in which it takes small pieces and builds them into bigger products." He points to the success of a product pioneered by Louisiana Pacific, known as oriented strand board (OSB), in which wood flakes from young trees such as aspen are glued and pressed together to form sheets that can be substituted for plywood, a product normally made of strips peeled from large trees. Since the early 1980s, OSB manufacturers in Minnesota have shelled out more than $3 billion constructing new factories or upgrading existing ones. (Recall again the six-thousand-mile-long wood pile. Some two thousand miles of it are devoted to OSB production alone.)

Especially damaging to birds in the conversion of the forest to aspen trees, says Gerald Niemi, former director of the Center for Water and the Environment at the Natural Resources Research Institute (NRRI), Duluth, is the continued loss of the forest's conifer component. "Forest birds can generally be divided into two groups: those that like coniferous trees and those that like deciduous trees," Niemi points out. "The biggest change in the landscape of the region over the past 100 to 150 years has been the reduction of long-lived conifer cover in favor of aspen-birch-fir forests. We have had a drastic reduction in white pine and white spruce in the past 150 years, and the bird species that like these conifer trees have certainly declined." (See also "Highlands.")

Niemi and his colleagues at the University of Minnesota have developed computer models of forest management based on natural disturbance regimes. The scientists are trying to develop a picture of what the forest looked like before Euro-American settlement and to determine how natural forces such as fire, insect outbreaks, and windthrows periodically rearranged the pieces of the forest mosaic on the landscape. By determining the proportion of forest habitat types, their ages, and configurations across the landscape in response to various disturbances, the scientists have developed a range of natural variability that foresters can mimic in their management prescriptions.

Niemi and others argue that the health of forest bird populations should be used as a yardstick for measuring the success of these prescriptions. Birds make up 69 percent of the vertebrate species that breed in the SNF. Furthermore, their life stages involve nearly every forest niche. Hermit thrushes and winter wrens, for example, nest in brush piles or the overturned roots of downed trees. Brown creepers, on the other hand, are abundant in older forests, where they search out insects in the bark of big standing trees. Woodpeckers depend on the dead, standing trees for nesting cavities and insects. If the forest is diverse enough to meet the needs of the majority of the forest's vertebrate species, researchers point out, chances are that they will be habitable for many other species, such as salamanders and frogs, voles and squirrels, butterflies and beetles.

Saving Birds to Save the Forest

But there are other benefits to preserving the forest's full complement of birds. New research has shown that not only must we conserve the forest to save the birds but that also we must preserve the birds to save the forest. For centuries humans have speculated whether the predation of birds has any influence on their prey populations, most notably insects. In experiments conducted in the late nineteenth and early twentieth centuries by ornithologists from the Biological Survey of the U.S. Department of Agriculture, tens of thousands of birds were killed and their stomach contents analyzed in an attempt to answer this question. The scientists concluded that most damage done by birds to, say,

CONIFER-DEPENDENT BIRD SPECIES

Sharp-shinned hawk	Winter wren	Pine warbler
Northern goshawk	Golden-crowned kinglet	Palm warbler
Merlin	Ruby-crowned kinglet	Bay-breasted warbler
Spruce grouse	Swainson's thrush	Connecticut warbler
Great gray owl	Hermit thrush	Chipping sparrow
Boreal owl	Solitary vireo	Lincoln's sparrow
Three-toed woodpecker	Tennessee warbler	White-throated sparrow
Black-backed woodpecker	Nashville warbler	Dark-eyed junco
Olive-sided flycatcher	Northern parula	Rusty blackbird
Yellow-bellied flycatcher	Magnolia warbler	Purple finch
Gray jay	Cape May warbler	Red crossbill
Blue jay	Yellow-rumped warbler	Pine siskin
Common raven	Black-throated green warbler	Evening grosbeak
Boreal chickadee	Blackburnian warbler	
Red-breasted nuthatch		

Conifers in northern forests provide essential habitat for a diverse group of conifer-dependent birds.

crops, was outweighed by the good they do in eating harmful insects. Yet many scientists believed that birds were not a serious form of pest control. They were considered "frills" that feasted on nature's excess.

In 1994, however, Robert Marquis and Christopher Whelan published the results of a study demonstrating that birds have a profound effect on the health and vitality of forests. In the Missouri Ozarks, scientists draped a selection of oak trees in netting whose mesh size was big enough to allow the passage of insects but small enough to exclude birds. The scientists monitored the trees for two years. They found that the caged trees were plagued by twice as many insects as the trees visited by foraging birds. As a result, the caged trees lost twice as many leaves to insects than the uncaged trees. The insect defoliation stunted the productivity of these trees, which produced far fewer leaves in subsequent years.

Could birds exert similar effects on the northern forest? Pastor thinks so. Nearly one hundred species of birds in the SNF are insectivorous. They feed primarily on caterpillars, including those of commercially harmful tree pests such as spruce budworm. Severing this predatory link between birds and insects, Pastor says, can be catastrophic, as it has been in the forests of New Brunswick. When the pulp and paper industry there suddenly expanded in the 1930s, whole landscapes that contained older, diverse, and complex conifer-dominated forests were liquidated and converted to a landscape of younger, even-aged conifers. Foresters made these sweeping changes in a misguided attempt to produce stands that could be more easily managed for consistently high supplies of pulp.

To support warblers, conifer forests require structural diversity that offers multiple niches for essential activities such as feeding and breeding. This complexity only develops in forests that are eighty years and older. So as mature forests declined, so did

Palm warblers are among the many headwaters birds that perform critical ecosystem services, such as controlling forest pests. Photograph by Ryan J. Sanderson, M.D.

the populations of warblers. With them went a free and reliable means of pest control since warblers are one of the main predators of spruce budworm, a major defoliator of balsam fir and spruce.

Under normal conditions, spruce-fir forests are subject to outbreaks of spruce budworm every forty to sixty years. The infestations did not devastate the entire forest but killed stands in patches across the landscape. The outbreaks helped to maintain stands of various ages across the entire forest. Although the birds did not make significant dents in budworm populations once an outbreak began, researchers found that they suppressed insect numbers enough under normal conditions to lengthen the intervals between outbreaks. With far fewer birds, Pastor says, New Brunswick forests now are plagued by budworm outbreaks on an average of every twelve years.

Since the 1950s, the timber industry has attempted to tackle the problem primarily by spraying the forests with pesticides, a practice that carries with it environmental and economic costs. The chemical applications have not been enough, however, to break the budworm's grip on the forest. The outbreaks keep the forest in an unnaturally young condition—a state of perpetually arrested development. "Unfortunately, things are now in a state where it is difficult to get the forest back into the condition of good habitat for a healthy warbler population—the absence of warblers themselves prevents it," Pastor writes. As the New Brunswick predicament reveals, the fundamental necessity and challenge of truly sustainable forest management lie in harvesting forest products in a manner that conserves enough conifer habitat to support the warbler populations that keep them vigorous and healthy.

In 2013 Niemi and his colleagues at the NRRI completed some twenty years of bird monitoring in the Chippewa, Superior, and Chequamegon-Nicolet National Forests, demonstrating that maintaining forest diversity across the landscape is good news for bird populations. With more than a thousand point counts annually, the NRRI project is the longest-running and most detailed collection of bird data in the Midwest. Results show that while some birds species, such as Connecticut warblers, evening grosbeaks, Swainson's thrushes, and yellow-bellied flycatchers are declining precipitously, the populations of 90 percent of the species surveyed are holding steady or increasing. The researchers point to one big potential factor: increases in intermediate-aged and mature forests. Since the economic downturn that began around 2008, wood harvest totals in Minnesota have dropped to about two-thirds of the five-million-cord level identified as sustainable in the GEIS report. A 2013 report from the Minnesota Department of Natural Resources shows a corresponding 5 percent increase in forest cover in northeastern Minnesota. "We know that many bird species are associated with intermediate and older forests," observes Ed Zlonis, an NRRI ornithologist and lead author on the report. "When an area is logged it promotes a period of even-age and often single-species forests that overall support fewer bird species. It takes sixty to eighty years for more diverse layers and tree species composition to develop."

Managing forests to sustain birds in the long term also entails looking at forests differently. Foresters have traditionally labeled stands by the size and kinds of trees that grow in them. Green points out that these labels do not necessarily describe the characteristics of a forest that are important to birds, such as the density of the shrub layers, the amount of downed wood, and the number of layers of leaves in the canopy.

Forest ecologists such as Pastor admit that developing harvesting prescriptions that accommodate the needs of many different species of birds can be a complicated task. He maintains, however, that the problem is not silvicultural know-how. Technological aids, such as computer modeling programs, now can juggle a mind-boggling array of variables, enabling foresters to review the impacts of potential harvests on a site long before the first tree is felled.

So what is standing in the way of managing forests in a way that is more inclusive of all the creatures that depend on them? "It's politics, pure and simple, and partly economics," Pastor says. To have a healthy diversified forest, Pastor maintains that Minnesota must have a diversified forest-products industry. When the pulp market booms, he says, the industry tends to build more pulp and paper mills that "basically grind trees up and glue them back together again." The value-added forest-products industry—makers of furniture, cabinets, window frames, and veneer, for example—rely on forests of older, bigger trees and provide an economic constituency for a different kind of forest than fifty-year-old aspen trees. Unfortunately, Pastor charges, the sawmills that provide higher-quality woods and the manufacturers that utilize them are not as organized—and do not have as much political clout—as the pulp and paper businesses.

"When mills expand, they need a lot of timber, and they need it fast," Pastor says. "We can talk about the biology and the silvicultural techniques of maintaining bird species and structural forest diversity and why biologically and economically this is good, but when you have an elected official breathing down your neck saying, 'You supply the timber, damn it,' all of it gets flushed down the toilet. That's the problem. We can do this. We may not be able to have the timber industry that we want or that we now have. We may not be able to have the kind of timber industry that's politically palatable. But we can have a diverse timber economy and diverse habitat for birds."

The time to do this is now. "One thing you learn in forestry school," he says, "is that you can screw up the forest really fast and it takes a long time to fix it." ∞

SUGGESTIONS FOR FURTHER READING

Conkling, Philip W., ed. *From Cape Cod to the Bay of Fundy: An Environmental Atlas of the Gulf of Maine.* Cambridge, Mass.: MIT Press, 1995.

Ehrlich, Paul R., David S. Dobkin, and Darryl Wheye. *The Birder's Handbook: A Field Guide to the Natural History of North American Birds.* New York: Simon and Schuster, 1988.

Green, Janet C. *Birds and Forests: A Management and Conservation Guide.* St. Paul: Minnesota Department of Natural Resources, 1995.

———. "A Landscape Classification for Breeding Birds in Minnesota: An Approach to Describing Regional Biodiversity." *The Loon* 63 (1991): 80–91.

Green, Janet C., and Gerald J. Niemi. *Birds of the Superior National Forest.* U.S. Department of Agriculture, Forest Service, Superior National Forest, 1980.

Helle, Pekka, and Gerald J. Niemi. "Bird Community Dynamics in Boreal Forests." In *Conservation of Faunal Diversity in Forested Landscapes*, ed. Richard M. DeGraaf and Ronald I. Miller, 209–34. New York: Chapman and Hall, 1996.

Holmes, Richard T. "Ecological and Evolutionary Impacts of Bird Predation on Forest Insects: An Overview." *Studies in Avian Biology* 13 (1990): 6–13.

Jaakko Poyry Consulting, Inc. "Final Generic Environmental Impact Statement Study on Timber Harvesting and Forest Management in Minnesota." Technical Report prepared for the Minnesota Environmental Quality Board, St. Paul, 1994.

Marquis, Robert J., and Christopher J. Whelan. "Insectivorous Birds Increase Growth of White Oak through Consumption of Leaf-Chewing Insects." *Ecology* 75, no. 7 (1994): 2007–14.

Mladenoff, David J., and John Pastor. "Sustainable Forest Ecosystems in the Northern Hardwood and Conifer Forest Region: Concepts and Management." In *Defining Sustainable Forestry*, ed. Gregory H. Aplet, Nels. Johnson, Jeffrey T. Olson, and V. Alaric Sample, 145–80. Washington, D.C.: Island Press, 1993.

Pfannmuller, Lee A., and Janet C. Green. "Birds and Forests." *Minnesota Conservation Volunteer,* March/April 1999: 21–31.

Pollack, Andrew. "Lumbering in the Age of the Baby Tree," *New York Times,* 24 February 1991.

Stewart, Scot. "Scoping Out Lakeside Hot Spots." *Lake Superior Magazine,* August/September 2001: 21–23.

Waldbauer, Gilbert. *The Birder's Bug Book.* Cambridge, Mass.: Harvard University Press, 1998.

HARVESTING
THE FOREST'S BOUNTY

Wildcrafting

Ask a forester to name a few commercially valuable species that grow in the Northwoods, and chances are that trees, such as white pine or aspen, will appear near the top of the list. Ask a wildcrafter, and the catalog will include fiddlehead ferns, morel mushrooms, and old man's beard lichen.

In the past, forest management focused on trees to the exclusion of other valuable species. Wildcrafters point out that profits can be made from gathering a range of forest materials, from Christmas boughs and herbs used in alternative health therapies to edible plants sold in specialty-food markets to mosses and ferns for the floral industry.

The practice is catching on around the country, especially in regions hard hit by economic downturns in the logging industry, such as the Pacific Northwest and the northern Great Lakes states. And it is paying off. Minnesota, for example, leads the nation in the production of holiday wreaths, most of which are made of balsam boughs cut from live trees. According to the 2007 report *Balsam Boughs in Minnesota: A Resource and Market Study,* annual profits reaped by the wreath-making industry top $23 million. Even far smaller enterprises produce tidy returns. In 2010 sales of maple syrup

---

NONTIMBER SPECIALTY FOREST PRODUCTS HARVESTED IN NORTHEASTERN MINNESOTA

Product Category	Materials Legally Harvested
Wild Foods	Berries
	Mushrooms
	Ostrich and bracken ferns, such as "fiddleheads"
	Hazelnuts
	Maple and other tree saps for syrups
Decoratives and Floral Greens	Balsam, cedar, and pine boughs
	Black spruce, birch, and other hardwood tree tops
	Lycopodiums
	Mature ostrich and bracken ferns
	Conks
	Mosses
	Spruce roots
	Birch bark
	Sticks from a wide variety of tree and shrub species (e.g., diamond willow, aspen, birch, maple, and oak) for walking sticks and decorative purposes
	Twigs and vines from alder, sumac, dogwoods, mountain maple, and pussy willow
	Burls
	Cones
Medicinals	Highbush cranberry bark
Wild Native Seed	Cones
	Seeds of many native plants for sowing or seeding to grow new plants
Biofuels	Tree tops and branches left after timber harvest, small diameter trees, and shrubs

---

by the state's small group of commercial tappers and hobbyists exceeded $1 million, says Jerry Jacobson, vice president of the Minnesota Maple Syrup Producers' Association.

Of course, wildcrafting (referred to in bureaucratic circles as harvesting "special forest products" or "nontimber forest products") is not new. For Native Americans, the forest was a fully stocked pharmacy, hardware store, and grocery all rolled into one. When the early woodland people of North America needed diaper material, for example, they stripped mosses from trees, rocks, and downed limbs. Packing the plants snugly around their infants, they created a cushioned layer that could absorb ten times its weight in liquid.

Reliance on the forest for a wide variety of materials other than just wood is not limited to aboriginal cultures. The hill people of Italy, writes ethnobotanist Gary Paul Nabhan in his 1993 book *Songbirds, Truffles, and Wolves,* "still consume many times the amount of forest products that the average American does—in fact, more than most forest-dwelling Native Americans do today."

Proponents herald wildcrafting as an environmentally sound way of revitalizing rural economies without destroying forests. Nontimber forest products, they say, can be harvested without cutting a single tree. And the emphasis on collecting a wide variety of materials provides incentives for more holistic management that would protect the forest's biodiversity. So why are some environmentalists and forest managers cautious?

Take, for starters, the sheer quantity of some of the materials being extracted. As early as 1924, P. L. Ricker of the U.S. Department of Agriculture complained that the floral industry had picked mid-Atlantic state forests clean of plants such as laurel and trailing arbutus. Large remnant populations survived only in more remote northern outposts, such as Maine.

No less rapacious, today's horticultural trade has set its sights on another forest good: mosses. The ability of these plants to sponge up moisture makes them valuable as shipping material for drought-sensitive bulbs and plants. And with a little watering, mosses add long-lasting green accents to decorative floral wreaths, planters, and hanging baskets.

So widespread is the commercial use of wild-collected mosses that some U.S. forests are undergoing the botanical equivalent of strip mining. According to results from a study released in 2004, the U.S. floral industry alone consumes up to eighty-two million dry pounds of moss each year—both for domestic purposes and export. The study's author, Patricia Muir, a professor of botany and plant pathology at Oregon State University, Corvallis, calculates that this amount is enough to fill some twenty-four hundred semitrucks, most of which comes from the Pacific Northwest and Appalachian Mountains. Although she cautions that these figures are ballpark estimates—few accurate records are kept on permitted moss collecting, and illegal harvesting is rampant—Muir speculates that the commercial value of mosses can add up to $165 million annually.

And forests are showing the strain. Anecdotal reports by scientists and land managers in the Pacific Northwest describe whole swaths of forest that have been stripped of mosses, particularly those that lie within easy reach of roads. The effect of such wholesale removal is largely unknown since the life histories and ecology of most forest plants other than commercially valuable trees have been little studied. Consequently, few, if any, guidelines about sustainable harvesting levels have been established. Those that do exist often are based more on hunches than hard data.

Still, a growing chorus of botanists warn that the overcollecting of mosses may be doing serious harm to forest ecosystems. The small body of published research on mosses suggests that these plants may play a vital role in forest ecosystems by capturing nutrients and regulating humidity and water flow. Moss mats also provide nesting habitat for the threatened marbled murrelet and are home to some three hundred species of animals, primarily miniscule inhabitants such as mites, springtails, rotifers, and water bears.

Once removed, most mosses are slow to recolonize their former sites. Studies by Robin Wall Kimmerer at the State University of New York, Syracuse, show that some species recover at a painstaking rate of only 1 percent each year. Decades, if not more than a century, may pass before they regain their former plushness.

Regulating the collecting boom for these and other nontimber species, however, poses significant challenges for forest managers. Vast acres of public lands are impossible to effectively patrol, making it difficult for managers to monitor even those wildcrafters with legal permits. In some regions, plant poaching is rampant. In the mountains of Appalachia, populations of slippery elm trees are plummeting as wildcat collectors strip their bark for use in commercial herbal products that treat ailments such as coughs, the stomach flu, and skin rashes. In places such as northern Minnesota, the careless cutting of birch bark for use in crafts has disfigured forests and killed scores of individual trees.

Skeptics fear that without meaningful regulations based on sound science, wildcrafting could become another form of destructive consumption. But developing guidelines for these new forest practices is easier said than done. Take, for example, the issues raised by attempts to draft harvesting rules for *Lycopodium* species, or clubmosses, in the national forests of the upper Midwest.

Newcomers to the northern forest often mistake the low-growing, evergreen-like plants that grow in dense patches under the shady forest canopy for pine or cedar seedlings. In fact, some species of clubmosses are even referred to as ground pine. But *Lycopodium* (from the Greek *lycos,* meaning "wolf," and *pous,* meaning "foot") species are neither mosses nor conifers. They are known as fern allies, so called because their structural and reproductive characteristics resemble those of ferns.

Ferns and clubmosses also share an ancient lineage. Ancestors of *Lycopodium* species lived during the Carboniferous period about 350 million years ago, when they grew to the size of trees with 5-foot girths and heights of 120 feet. Although today clubmosses seldom top six inches, they can nonetheless reach grand old ages. Researchers have documented plants that lived to be twelve hundred years old.

Not surprisingly, clubmosses have long been used by humans. Native Americans steeped the entire plant in a tea to combat fever and weakness. The plants' reproductive spores, which are flammable, were collected to create flashes in early photography and the theater as well as in fireworks. Spores were also used as a powder to treat skin chafing.

Today *Lycopodium* species, in particular *Lycopodium dendroideum* and *L. hickeyi* (which grow in similar habitats, sometimes together, and are commonly referred to as princess pine), are coveted by the floral industry, where their long-lasting shoots are used in floral arrangements, including Christmas wreaths. Efforts to cultivate princess pine as an agricultural crop have failed, leaving wildcrafters dependent on gathering the plant from the wild. According to preliminary estimates, about one hundred tons of princess pine are harvested annually from the forests of Michigan's Upper Peninsula (thirty tons extracted from the Hiawatha National Forest alone) and another two tons from the Chequamegon-Nicolet National Forest in northern Wisconsin. How many more tons are harvested on private land or illegally on public land are anybody's guess. Very little is known about the long-term effect of collecting on princess pine populations since even permitted harvesting on public land is not monitored.

The stepped-up harvest prompted scientists, forest managers, and owners of wildcrafting businesses in the northern Great Lakes states to meet in 1995 to discuss picking regulations. Of concern to wildcrafters was the lack of research to justify potential harvest restrictions. They warned that curbing the supply of princess pine could prompt wholesalers to substitute princess pine with plant species from other countries. Local collectors would lose a critical source of supplemental income.

Lacking even the most basic information, such as the plant's abundance, forest managers too were frustrated. Despite this knowledge gap, clubmosses are nonetheless thought to play a vital role in the forest ecosystem. For example, compared with many other forest plants, clubmosses have a very complex—and extensive—underground root and rhizomal structure that enables them to colonize whole areas of the forest floor. Forming dense carpets on the surface, clubmosses slow the coursing of runoff, giving their root mats more time to sop up nutrients that would otherwise be lost to that part of the forest. The plants perform other valuable services as well, such as stemming soil erosion and providing thermal cover that helps to protect soil organisms and the root systems of other plants from extremes of temperature and moisture. This thermal and hydrological amelioration is particularly important in the winter and spring when the

This clubmoss, also known as princess pine, can form extensive patches on the forest floor and play a vital role in the forest ecosystem. Little is known about sustainable harvesting limits despite the plant's popularity as a decorative material in the floral industry. Courtesy of Superior National Forest.

canopy is leafless or in the aftermath of logging when the forest floor is opened to the elements.

Some researchers believe that these functions could easily be lost since the plants' structure, slow growth, and means of reproduction may leave them especially vulnerable to overharvesting. When it comes to *Lycopodium* species, for example, appearances may be deceiving. An area that appears carpeted with many individual clubmoss plants may, in reality, contain only one plant or a few individuals. That is because clubmoss shoots, known as ramets, sprout from a horizontal runner called a rhizome that threads along the forest floor or several inches underground. These ramets provide the plant with the photosynthetic energy it needs to survive. Slow-growing, a single plant produces only a few of these genetically identical shoots each year. In her research for a graduate thesis at the University of Wisconsin–Eau Claire, forest ecologist Colleen F. Matula found that it took four years for a plant to recover from having 50 percent of its aerial stems removed.

Scientists also are uncertain about effects of harvesting on plant reproduction. In the late summer and early fall, the ramets develop structures known as strobili. These miniature club-shaped antennae contain thousands of spores. When disturbed, the ripened strobili burst in a tiny cloud of yellow dust and scatter their spores, some of which travel on air currents to take up residence in unclaimed territory. Although a single clubmoss may release thousands of spores, only a precious 4 percent of them ultimately germinate. This wind-borne colonization is an important means by which new individuals—and vital genetic diversity—are introduced into an area that may be dominated by a few clonal plants.

How many ramets can be snipped off by eager collectors without harming the whole population remains an important question for further research. Of concern to scientists is the slow rate of clubmoss reproduction. For example, strobili begin to develop only on ramets that are older than four years of age. Bigger and bushier, these shoots are the ones most likely to be picked by collectors.

The research and monitoring of these forest activities cost money at a time when public-forest budgets already are stretched beyond their ability to meet existing needs. Lack of knowledge, however, has not—and likely will not—put a damper on the market demand for products such as clubmosses. The challenge to develop guidelines for sustainable harvests of nontimber resources falls to natural-resource agencies. Without science-based protocols for the management of forest products, neither human economies nor the long-term health of the forest on which they depend will prosper. ❧

SUGGESTIONS FOR FURTHER READING

Emery, Marla. "Rainforest Crunch from Michigan? Nontimber Forest Products and Household Livelihoods in a Northern Latitude." Paper presented at Annual Meeting of the Association of American Geographers, 1997.

Fischer, Adelheid. "Moss Conservation behind Bars." *Conservation in Practice* 6 (2005): 35–36.

Goldberg, Carey. "From Necessity, New Forest Industry Arises." *New York Times,* 24 March 1996.

Haapoja, Margaret A. "Enterprise in the Woods." *Minnesota Conservation Volunteer,* November/ December 2002: 21–31.

Jacobson, Keith, Mark H. Hansen, and Ronald McRoberts. *Balsam Boughs in Minnesota: A Resource and Market Study.* St. Paul: Minnesota Department of Natural Resources and U.S. Forest Service, 2007.

Jones, Eric T., Rebecca J. McLain, and James Weigand, eds. *Nontimber Forest Products in the United States.* Lawrence: University Press of Kansas, 2002.

Kimmerer, Robin Wall. *Gathering Moss: A Natural and Cultural History of Mosses.* Corvallis: Oregon State University Press, 2003.

Matula, Colleen F. "Growth Rate, Growth Pattern, and the Effects of Harvest of the *Lycopodium obscurum* Complex in the Lake Superior Region." Master's thesis, University of Wisconsin, Eau Claire, 1995.

Minnesota Department of Natural Resources. *Minnesota Special Forest Products: A Market Study.* St. Paul: Division of Forestry, 1994.

Muir, Patricia S. *An Assessment of Commercial "Moss" Harvesting from Forested Lands in the Pacific Northwestern and Appalachian Regions of the United States: How Much Moss Is Harvested and Sold Domestically and Internationally and Which Species Are Involved?* Final report to U.S. Fish and Wildlife Service and U.S. Geological Survey, Forest and Rangeland Ecosystem Science Center, 2004.

Thomas, Margaret G., and David R. Shumann. *Income Opportunities in Special Forest Products: Self-Help Suggestions for Rural Entrepreneurs.* U.S. Department of Agriculture, Forest Services, Agriculture Information Bulletin 666, 1993.

Tompkins, Joshua. "Moss Hunters Roll Away Nature's Carpet, and Some Ecologists Worry." *New York Times,* 30 November 2004.

Vance, Nan C., Melissa Borsting, David Pilz, and Jim Freed. 2001. *Special Forest Products: Species Information Guide for the Pacific Northwest.* General Technical Report PNW-GTR-513. Portland, Oregon: U.S. Department of Agriculture, Forest Service, Pacific Northwest Research Station.

INTERNET RESOURCES

Goods from the Woods, https://www.facebook.com/GoodsFromTheWoods

BE IT EVER SO HUMBLE

No Home Like That of a Pitcher Plant

ODIES OF WATER IN THE NORTHWOODS, LIKE OASES IN THE DESERT, ARE magnets for wildlife. By far the best studied are lakes, largely because of their recreational value to anglers. Only recently have scientists discovered the ecological importance of tiny forest ponds. Some of them, such as vernal pools, come and go with the seasons, filling with rain and snowmelt in the spring before gradually disappearing by midsummer. But before it evaporates in the July heat, a pool no bigger than a four-person hot tub will have cradled hundreds of species of animals. Many of them, such as insects and amphibians, mature in these temporary waters before emerging as adults and becoming dietary staples for small forest creatures, including songbirds, shrews, and snakes. (See also "The Secret Life of Salamanders.")

Some seasonal pools are even more inconspicuous. They can be so minuscule as to fill no more than a thimble with water, like the rain and dew collected in the upturned cap of a gilled mushroom. Others, such as holes in the exposed roots of trees, can hold an entire soup bowl of precipitation.

These small cavities are known as container habitats. Those formed by the structures of terrestrial plants—as opposed to empty snail shells, say, and roof gutters and tire swings—are called *phytotelmata* (from the Greek meaning "plant pond"). According to a review of the scientific literature by biologist Durland Fish, more than fifteen hundred species of plants worldwide possess water-trapping features of one kind or another. This number includes, however, only waters impounded by live plant structures. Too little is known, he says, about other kinds of plant-formed pools, such as those in fallen fruits and leaves, to even hazard a guess as to their numbers—or ecological importance.

Although phytotelmata may sustain fewer species of animals than larger seasonal pools, the concentration of life within their small confines is no less complex. These tiny reservoirs house representatives from all the major aquatic insect orders. The most common are the larvae of true flies, or Diptera (from the Greek, meaning "two winged"; dipterans have front wings but lack hind wings). By far the best studied of the dipterans are some four hundred species of mosquitoes, most of which are found only in container habitats. No home seems to be too humble. Biologist D. Dudley Williams points out that two species of mosquito live in the water captured in pineapple bracts, even though the volume of liquid in these tiny hollows may just barely fill two teaspoons.

But when it comes to packing a wallop of complexity into a very small space, few container habitats can match that of pitcher plants. The eastern North American pitcher plants belong to the genus *Sarracenia,* which includes eight species. Most are found in the southeastern United States, primarily in coastal-plain habitats. *Sarracenia purpurea,* the northernmost member of this family of plants, occurs from Alabama north along the Atlantic Seaboard into Labrador. From there the plant's range veers to the northwest into the northeastern reaches of British Columbia. *S. purpurea* can also be found throughout the Midwest, including Minnesota.

The northern pitcher plant, also called the purple pitcher plant, grows most commonly in waterlogged environments where the saturated—and often acidic—conditions prevent bacteria from fully breaking down dead plant matter and releasing essential nutrients such as nitrogen and phosphorus. These conditions starve out most other plants, but, amazingly, *S. purpurea* thrives. On peatlands in the northeastern United States and Canada, they may form densities of nearly 18,500 plants per acre.

How do they do it? *S. purpurea* have evolved a way to utilize a resource that seems to be available in endless supply around them—insects. To capture them, pitcher plants sprout a rosette of some half dozen modified leaves that are shaped like elegant champagne flutes. Among the *Sarracenia* species, says botanist Adrian Slack, "are found some of the most attractive of all carnivorous plants, for they combine graceful and often magnificently variegated foliage with great beauty of flower." Do not be fooled, however. The design of each exquisite detail—from the wine-red veins that branch throughout the foot-long leaf to its refined shape—has been carefully honed to make *S. purpurea* among the most effective death chambers on Earth. Field experiments conducted by Durland Fish and Donald Hall at the University of Massachusetts, Amherst, showed that one leaf in peak condition was able to snare forty houseflies in a single forty-eight-hour period. (The pitcher leaf is an extraordinary example of a phenomenon known as convergent evolution. So successful was its design in nature that plants in the tropical rain forest and in Australia evolved pitchers of their own even though they shared no common ancestor and were widely separated from one another and from other pitcher plants in North America.)

Sarracenia species' prowess in trapping insects attracted the attention of the early explorers of the American landscape. In the early eighteenth century Mark Catesby, the renowned artist of natural history subjects, theorized that insects used the pitcher leaf to escape from creatures that preyed on them, including frogs. In his 1793 book *Travels in North and South Carolina, Georgia and Florida,* the naturalist William Bartram also noted the occurrence of insects floating in the rainwater that collected in the bottom of the leaf pitcher. Although he suggested that the leaves might simply store water as a hedge against future drought, Bartram opined that the pitcher plant might also use the leaf pools to trap and kill insects. He expressed his suspicions about the predatory nature of the plant in a painting titled *The Sarasena.* In his view of the eat-and-be-eaten world of nature, an open pitcher leaf slides ominously along the ground, the curves of its upturned, almost animate, form blending into that of a raised snake with a half-swallowed frog in its mouth.

Not until 1829, however, would someone propose that pitcher leaves were organs of digestion much like the human stomach. In a likely reference to the presence of living midge and mosquito larvae in the pitcher fluid, British botanist Gilbert T. Burnett quipped that "even this simple digestive apparatus is not free from intestinal worms." Despite such lively interest, the science of carnivorous plants would not be advanced with any systematic rigor until Charles Darwin undertook a long series of investigations that culminated in his 1875 classic *Insectivorous Plants.*

Although it would take another century of scientific research to separate the fact from the fiction about northern pitcher plants, early observers nonetheless accurately described the plant's mode of insect entrapment. Pitcher plants secrete a sweet nectar that is irresistible to insects, especially ants, which in some populations comprise an estimated 20 percent of *S. purpurea*'s diet. The glands that produce this nectar are so numerous that they coat the exterior of the entire leaf. The amount of nectar increases near the lip of the leaf, luring crawling insects upward with promises of ever-greater bounty. Should the insect lose its way, it has only to follow the road map of maroon veins that develop in the leaf when the pitcher has reached its prime. All roads lead to the mouth of the leaf into an area known as the nectar roll, where the delectable liquor is so copiously produced that the surface glistens. To protect this ambrosia from being evaporated or diluted by rain, the leaf has evolved a delicately sculpted hood that shields the plant's precious stores like an awning. The investment in this protective device pays handsome dividends since the insects that step onto the slick platform of the nectar roll often slip and tumble into the water at the bottom of the pitcher.

Flying insects arrive on the plant using the hood as a landing pad. They follow a descending trail to the nectar roll, aided by stiff, downward-pointing hairs that thwart return trips. Like the crawling ants, those that step onto the slick platform of the nectar roll often slip and fall into the water at the bottom of the pitcher. Some insects take to the wing before contact with the bath, but their escape is complicated by the narrowness of the leaf's funnel, which prevents them from gaining altitude. Attempts to crawl out are soon foiled since the walls of the leaf just below the mouth are as smooth and polished as a freshly waxed ballroom floor.

The plant's nectar is so enticing that even the fortunate ones who manage to escape do not seem to learn from their experience. The insect that survives its brush with death, writes Slack, "seldom seems particularly dismayed by its experience, and often repeats the action several times, when its luck may run out." Death by drowning is a virtual certainty once the insect falls to the bottom of the pitcher and receives a thorough soaking. Even if the sodden insects manage to negotiate the funnel's slick walls, the down-pointing hairs make the climb back out as torturous as walking uphill through a brushy thicket while wearing a waterlogged woolen coat.

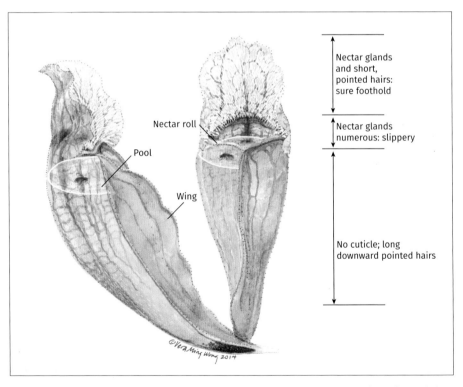

The elaborate leaves of the pitcher plant contain features that entice and trap prey, including a slippery nectar roll, downward-pointing surface hairs that thwart an easy exit, and a narrow funnel that prevents trapped insects from easily gaining flight. Illustration by Vera Ming Wong.

Scientists do not dispute that *S. purpurea* has evolved these and other mechanisms for attracting and "swallowing" prey. But whether the plants secrete enzymes that digest their catch is a fine point that has been disputed for decades. Unopened pitcher leaves are sterile, but subsequent chemical analysis of liquid that has collected in these same leaves once they have opened indicates elevated levels of digestive activity. Nonetheless, the current scientific consensus is that pitcher plants are unlike other insectivorous plants in the northland, such as the sundew or bladderwort. These plants not only trap their victims but also produce chemicals, most notably protease, to break down the protein of prey tissues completely on their own. True, researchers have identified protease in the pitcher's fluid, but it is not produced by the plant but rather by the bacteria in the broth.

Confusing matters further is the presence of acid phosphatase in the plant's capture pool. This enzyme strips phosphate molecules away from larger molecules that compose animal tissues, making them available for absorption by the plant. But acid phosphatase is a common component of cells and occurs on the exterior of the pitcher leaf as well as in the capture pool. As a result, University of Iowa ecologist Stephen Heard and other researchers have ruled out the possibility that *S. purpurea* secretes special quantities of the enzyme into its capture pool for the purpose of digestion.

Then there is the matter of the copious amounts of acid pumped by the plant into the capture pool such that the bath has a far lower pH than the already acidic waters in which most pitcher plants live. Researchers have found that this extreme acidity does not break down insect tissues directly but rather creates a beneficial environment that helps bacteria digest prey more easily. One curious footnote: These extreme conditions are lethal to all mosquito species except *Wyeomyia smithii,* also known as the pitcher plant mosquito. How *W. smithii*—or its cohorts, for that matter—thrive in the acid broth remains a mystery to science. In the case of the pitcher plant mosquito, one thing is clear: the species has had lots of time to adapt to its bitter natal waters. Scientists believe that *W. smithii* may have first turned to pitcher plants as nurseries during one of the Pleistocene glaciations many thousands of years ago.

The northern pitcher plant differs from sundews, bladderworts, and other insectivorous north-country plants by leaving the primary digestion of its victims up to a host of guest workers that have evolved the capacity to partake in the floating feast without becoming the main course. Making up the greatest biomass of living animals in the pitcher leaf is a trio of dipteran larvae: the flesh fly *Blaesoxipha fletcheri,* the pitcher plant mosquito, and a species of midge known as *Metriocnemus knabi.* When it comes to making nutrients available to pitcher plants, no organisms are more important. Together they not only cut up the meat on the plant's plate, so to speak, but also chew and digest the food for it too.

The question is, how do they keep from eating each other out of house and home, especially when home is no more than one-eighth of a cup of rainwater? According to the principle of competitive exclusion, if several species occupy the same niche and exploit the same resources, sooner or later only one of them will emerge victorious. In other words, two species make bad company; three are definitely a crowd. So how do the larvae form such a stable triad, so much so that these species are known as pitcher plant obligates; that is, they are found only in the tiny pool of water that collects in the base of the pitcher leaf?

The midge, mosquito, and flesh fly larvae have become specialists in what is known

Pitcher plants pump copious amounts of acid into their capture pools. The low pH creates a beneficial environment that helps bacteria digest drowned prey more easily. Photograph by Rod Planck/Dembinsky Photo Associates. Copyright 2014; all rights reserved.

in scientific circles as resource partitioning. Heard puts it more simply—they are essential links in a food-processing chain. It all begins, he says, with the flesh fly larva, a one-half-inch-long maggot that floats on the surface of the pool. (Adult flesh flies will lay only one live maggot per leaf. If by chance two maggots find themselves growing up in the same plant, the legless larvae will engage in a deadly full-body embrace until one is drowned.) The buoyant maggot tears into the carcasses of captured insects, leaving leftover chunks and crumbs to float down to the mosquito and midge larvae below.

Adult flesh flies lay their young as fully developed larvae rather than as eggs. Because of this reproductive head start, the maggot lives in the pitcher leaf for only a few weeks each year before emerging. Midges and mosquitoes, on the other hand, spend their entire first year of life in the pitcher, performing the bulk of the food processing.

This is where the plant's story gets particularly interesting. When the nursery for your offspring is a living—and, therefore, dying—structure, timing can be of utmost importance. So mosquitoes and midges synchronize their development cycles to coincide with the period in which pitchers are in their insect-trapping prime. Each growing season a pitcher plant produces an average of five to six new leaves. The leaves mature in sequence, with one leaf opening about every fifteen days. In late July and early August, adult mosquitoes and midges lay eggs in the newly opened leaves. Studies have shown that they prefer larger leaves over smaller ones, presumably for their greater food-capturing potential.

Mosquitoes deposit their eggs on the pitcher leaf shortly after it opens and before it even has the opportunity to collect water. By the time the eggs hatch, the capture pool will have formed, and food production will be in full swing, just in time to feed the hungry larvae. Furthermore, by laying eggs in new leaves rather than mature ones, the adults ensure that their offspring will not be outcompeted by older cousins that have had a chance to grow bigger and more food savvy.

Adult mosquitoes and midges prefer to lay their eggs in younger leaves for another reason. The larvae overwinter in the frozen pool at the bottom of the leaf, and new plant tissues are less likely to become brittle or fractured by freezing—and therefore to spill their precious contents. By late September shortened daylight hours, which signal the coming of winter, cause the larvae not only to stop feeding but also to discharge food particles from their guts that can form expanding ice crystals and rupture fragile tissues. Come the spring thaw, they resume feeding, leaving the leaf as adults in mid-July to begin the process of mating and egg laying all over again.

Although the development cycles of pitcher plant mosquitoes and midges run in parallel tracks, their feeding behaviors within the pitcher could not diverge more sharply. The differences are largely due to physiology. Midge larvae, on the one hand, have hemoglobin that allows them to draw oxygen directly from water. Because they do not need to visit the surface of the pool for a breath of fresh air, they lack body parts for floating or swimming and remain submerged throughout the larval stage. Not surprisingly, they can be found rooting through the pile of insect remains that has settled at the bottom of the pool. In fact, the midge larvae approach their meals with such gusto that when the carcass of a prey insect is "withdrawn from the bottom of the leaf," observes biologist William Bradshaw, it "is usually a veritable Medusa of writhing *M. knabi* larvae."

Mosquito larvae, on the other hand, can remain submerged for long periods of time but need to periodically visit the surface, like a snorkeler, to sip air through a breathing tube. As a result, mosquitoes are accomplished swimmers that spend most of their time in the water column. The structure of the pitcher also segregates the two larval populations in that midges occupy the base of the pitcher funnel away from the buoyant mosquitoes. In this case, invisible fences make good neighbors since laboratory experiments have shown that midge larvae aggressively attack other larvae that invade their turf.

Most critical to their mutual survival is their different feeding apparatuses. Think of the midge larva as a messy diner with a knife and a fork, and the mosquito as the waiter who tidies up the crumbs with a hand broom. Midges possess powerful mandibles that can shred large, solid particles of food. Equipped with filters composed of brushy hairs instead of sharp mouth parts, the mosquito larva must wait until the midges break the prey carcass down into decaying crumbs before it can sweep them into its mouth. Breaking tissues down into smaller particles also increases the surface area that can be colonized by bacteria, which provide mosquito larvae with another important source of nutrition. (Accomplished grazers as well as filter feeders, mosquito larvae will munch on bacteria wherever they find them, even on the surface of midge feces.) This feeding assembly line works so well that the presence of one larva does not negatively affect the other. In fact, quite the opposite is true: the more midges, the merrier the mosquitoes. Heard conducted experiments that showed that increasing the number of midges—and therefore the number of mouths working away to disassemble insect carcasses—resulted in fatter mosquitoes.

As it turns out, the fattening of mosquitoes in the pitcher leaf may have important reproductive consequences. Unlike *W. smithii* that live in pitcher plants in the southern United States, female adults of the pitcher plant mosquito in northern regions do not depend on blood meals in order to lay a healthy cluster of eggs. Instead, reproductive success depends on their conditioning during the larval phase. As in other animal populations, well-fed females are more likely to have built up the reserves needed to produce robust offspring than underweight ones.

Like the perfect host who waits until all the guests are fed before filling his plate, the pitcher plant is the last to partake in the floating feast. (Pitcher plants also inadvertently feed innumerable spiders that intercept their prey in webs strung across the mouth of leaves!) But gratuitous gestures are few and far between in nature. So therefore the question remains, what exactly does the pitcher plant get in exchange for this delayed gratification?

Researchers have found that the larval trio serves as a biological catalyst, accelerating the breakdown of prey and the release of important nutrients. Laboratory tests in glass jars have shown, for example, that the presence of mosquito or midge larvae, or a combination of both, dramatically increases levels of nitrogen. Pitcher plants readily take up these nutrients through special cells that line the bottom of the leaf cup. To expedite absorption, these cells lack a waxy coating known as a cuticle that is normally found on exterior plant cells. In fact, *S. purpurea* imbibes the digestive end products of its inhabitants so efficiently that the pitcher fluid usually lacks the rank odors normally associated with decay.

Scientists have focused on the nutritional services that the larvae provide for their pitcher plant hosts. Equally important, however, are photosynthetic boosts. Pitcher plants are low growers that often are embedded in, if not partially covered by, other plants such as thick mats of sphagnum moss. This crowding decreases the amount of carbon dioxide that is available to the plant for photosynthesis. At the same time, during the critical summer growing season, the hot, exposed conditions of pitcher plants cause temperatures within the pitcher leaf to rise, reaching anywhere from 77 to 104 degrees Fahrenheit. These high temperatures increase the rates of respiration and metabolism for the organisms that live inside the pitcher leaf—and, hence, raise the levels of carbon dioxide that are released as waste products. The plant absorbs the carbon dioxide and steps up photosynthesis. In turn, it creates its own waste product—oxygen—which is pumped into the pitcher leaf's water, where it fuels the food-processing assembly line.

To accelerate this gaseous exchange, pitcher plants have evolved another unusual adaptation. In most plants, the agents of photosynthesis, known as chloroplasts, are

Pitcher plants and their larval inhabitants have evolved sophisticated mechanisms for exchanging their own carbon dioxide and oxygen waste products, enabling the plants to thrive in crowded, nutrient-poor wetland habitats. Photograph by Hal Horwitz / Dembinsky Photo Associates. Copyright 2014; all rights reserved.

normally found in the middle layers of plant tissues. The chloroplasts of pitcher plants, however, are located both near the surface in the epidermal layer as well as sandwiched in the middle layers.

Together, pitcher plants and their larval inhabitants form one of the most elegant recycling systems found in nature. "The plant provides dinner (prey) for its inhabitants," Bradshaw observes, "and then cleans up the leftovers (carbon dioxide and nitrogenous end products of their metabolism) while enriching the environment with oxygen. In return, the plant obtains a ready supply of nitrogen and carbon dioxide as well as a sink into which to flush oxygen."

Yet even exquisitely synchronized systems like this one can go awry. Occasionally more insects fall into the pitcher than can be processed by its inhabitants. Decaying carcasses deplete the water of critical oxygen supplies, and the animals in the food-processing chain die of asphyxiation. Fortunately, when all else fails, pitcher plants can draw on the help of organisms that have evolved in the oxygen-poor conditions of peat-land environments, such as nitrogen-fixing anaerobic bacteria. These organisms, which live in the absence of oxygen, can remove nitrogen from the atmosphere and transform it into chemical forms that can be utilized by plants. With such a foolproof backup to an already beautifully honed system, it is small wonder that *S. purpurea* has become one of the northland's most common—and successful—peatland plants. ❧

SUGGESTIONS FOR FURTHER READING

Adams, Richard M., and George W. Smith. "An s.e.m. survey of the Five Carnivorous Pitcher Plant Genera." *American Journal of Botany* 64, no. 3 (1977): 265–72.

Bradshaw, William E. "Interaction between the Mosquito *Wyeomyia smithii,* the Midge *Metriocnemus knabi,* and Their Carnivorous Host *Sarracenia purpurea.*" In *Phytotelmata: Terrestrial Plants as Hosts for Aquatic Insect Communities,* ed. J. H. Frank and L. P. Lounibos, 161–89. Medford, N.J.: Plexus Publishing, 1983.

Fish, Durland. "Phytotelmata: Flora and Fauna." In *Phytotelmata: Terrestrial Plants as Hosts for Aquatic Insect Communities*, ed. J. H. Frank and L. P. Lounibos, 1–27. Medford, N.J.: Plexus Publishing, 1983.

Fish, Durland, and Donald W. Hall. "Succession and Stratification of Aquatic Insects Inhabiting the Leaves of the Insectivorous Pitcher Plant, *Sarracenia purpurea.*" *American Midland Naturalist* 99 (1978): 172–83.

Heard, Stephen B. "Pitcher-Plant Midges and Mosquitoes: A Processing Chain Commensalism." *Ecology* 75, no. 6 (1994): 1647–60.

Istock, Conrad A., Kyle Tanner, and Harold Zimmer. "Habitat Selection by the Pitcher-Plant Mosquito, *Wyeomyia smithii*: Behavioral and Genetic Aspects." In *Phytotelmata: Terrestrial Plants as Hosts for Aquatic Insect Communities*, ed. J. H. Frank and L. P. Lounibos, 191–204. Medford, N.J.: Plexus Publishing, 1983.

Juniper, Barrie E., Richard J. Robins, and Daniel M. Joel. *The Carnivorous Plants.* San Diego: Academic Press, 1989.

Slack, Adrian. *Carnivorous Plants.* Cambridge, Mass.: MIT Press, 1979.

Williams, D. Dudley. *The Ecology of Temporary Waters.* Portland, Ore.: Timber Press, 1987.

Zimmer, Carl. "The Processing Plant." *Discover,* September 1993: 93–95.

Healing a Watershed
for Coaster Brook Trout

Fish are an image of the once pure waterways of the continent.
They have followed the veins and arteries of the earth over an irretrievable past.
When fish disappear, as they are now doing at a terrible rate, and when
the age-old alliance between people and fish is all but lost to a world of
giant supermarkets, we lose our communion with a priceless heritage. . . .
Have we become accustomed to the idea of extinctions?

John Hay, *A Beginner's Faith in Things Unseen* (1995)

Coasters have an indefinable, perhaps sentimental, attraction all their own.
Their value is out of all proportion to their numbers and size.

George Becker, *Fishes of Wisconsin* (1983)

NOVEMBER 1997. IT IS LATE AFTERNOON WHEN WE PULL OVER ONTO THE shoulder of Highway 61 and park our vehicles to the left of a small stream on the Grand Portage Indian Reservation near the Canadian border. Biologists Lee Newman of the U.S. Fish and Wildlife Service (USFWS) and Gary Cholwek of the U.S. Geological Survey (USGS) have come from their offices in Ashland, Wisconsin, to conduct the year's final survey of coaster brook trout. They have been driving for nearly five hours and will log a few more before their workday is over. They are entitled to a little grumpiness, but Cholwek and Newman will have none of it. They revel in the autumn air, exchanging gibes as they grab cameras, buckets, nets, and waders from the back of their truck. They seem as lighthearted as convicts on furlough, having stretched the field season by another day, thereby postponing the inevitable routine of paperwork that chains field biologists to their desks each winter.

Loaded with gear, we join John Johnson, a biologist with the Grand Portage Band of Lake Superior Chippewa, and start briskly along a path that follows the stream. By the time we enter the woods, the light is already low. It reflects off the white trunks of the paper birch so that even on this overcast day, the forest glows. No one is taken in for long by the stillness of the forest and its silvery light. On a November afternoon this far north, they are nature's curtain call, a final warning that daylight is waning. We pick up the pace. There is another stream to visit before dark.

We drop the equipment on a rise overlooking a waterfall and a large pool. Cholwek assembles a makeshift lab on the stream bank. The setup is simple. Plastic buckets hold solutions of baking soda and acetic acid, a vinegary-smelling brew that will temporarily anesthetize the captured fish so that he can measure, photograph, and tag them as well as scrape a few scales for later DNA testing.

Newman and Johnson quickly don rubber-booted chest waders and head for the stream. Since 1991, Newman has collaborated with the tribe to bring coaster brook trout back from the brink of extinction in Lake Superior waters. In this stream and two others on the reservation, they have planted fertilized eggs and tiny fry in hopes that the fish will develop into juveniles and do what countless generations before them have done— head out into the lake and mature in the shallow reefs along the coast. A few will live out their lives exclusively in the lake, spawning on the nearshore reefs around their feeding grounds. Most of the others, however, eventually get a hankering for home and briefly return as adults to their natal streams to reproduce.

Coasters are all that is left of Superior's native species of anadromous fish (from the Greek, meaning "a running up"), and they too are in trouble. Coasters once migrated from the big lake to spawn in more than 106 tributaries of Lake Superior, including nine streams in Minnesota. Today, however, spawning runs are sparse and sporadic with reliable numbers occurring only in some of the biggest or most remote rivers on the lake, such as the Lake Nipigon watershed in Ontario, the Salmon Trout River in Michigan, and a few streams on Isle Royale.

Whether the Grand Portage waters will be added to this select list largely depends on the success of the scientists' efforts. Each year they measure their progress—the return rate of planted fish—by electroshocking some of the pools in the streams or around the mouths of the rivers in the lake and taking a careful census of each returnee. Although the spawning season for coaster brook trout is coming to a close, Newman and Johnson suspect that the meltwater from the previous day's snowfall may have carried the scent of the stream into the lake and lured a few latecomers into the narrow channel. Together they pitch a net across the outlet of the pool before Newman wades into deeper water and flips a switch. His rubber boots protect him from the 450-volt current that courses in a low hum from a control unit in his backpack through a metal pole into the water. Their bets pay off. No sooner does Newman slowly scribe his first underwater arc in the foamy, root beer–colored water than a twenty-inch fish darts to the outlet and plunges under the net. We watch it scuttle downstream, alternately diving into pools and skittering over shallow patches in the riverbed in the direction of safety—Lake Superior. Johnson adjusts the net to prevent other fish from taking a similar getaway route.

"We've seen this behavior before," Johnson says as he scans the surface of the pool while Newman makes another sweep. Stream-dwelling fish, he says, will dive down to a deep hole or under rocky overhangs and downed branches. "Coasters will head for the lake, perhaps because they're transients and not resident fish."

Eager to examine the legendary Lake Superior coaster up close, we gather near when Newman's second round of electrofishing brings three stunned brook trout to the surface. The scientists hoot with delight. Like the one that got away, these are large, mature fish. Small wonder that anglers once trekked from as far away as Europe for the chance to cast a fly on coaster waters. An adult stream-bound brook trout barely clears the scales at around a quarter of a pound. Mature coasters reflect life in their big lake environment. During the heyday of recreational angling for coasters in the nineteenth century, hefty 2- to 8-pound fish were the norm. On his reconnaissance of the North

In this North Shore stream, fisheries researchers use electroshocking to temporarily stun coaster brook trout as part of data gathering for a restoration project. Courtesy of U.S. Fish and Wildlife Service, Ashland, Wisconsin.

Shore in 1879, Minnesota state geologist Newton H. Winchell and his party fried up a whopper that measured more than twenty-four inches long and weighed in at 5.75 pounds. "As we had no jar in the museum large enough to hold a fish of this size," he reported, "we regretted the necessity that compelled us to make our breakfast that morning on a brook trout."

In 1916, a Nipigon River angler set the fishing world on fire with a record 14.8-pound coaster. (In September 2000, Grand Portage resident Charles Pederson made Minnesota fishing history when he pulled a coaster weighing nearly 6.5 pounds from the Pigeon River.)

Johnson skims off the immobilized fish with his net and submerges them in the anaesthetizing solution so that Cholwek can begin his measurements. The pair work quickly and carefully, aware that their handling, however necessary and well intentioned, is stressful to the animals.

Even though they have been through this routine hundreds of times, the biologists pause frequently to examine each fish and admire their beauty. "To get out here and handle the fish makes a world of difference in your attitude," Cholwek observes.

We could not help but share his enthusiasm. Affectionately dubbed the speckled trout, brook trout are beloved for their beauty above all. A maze of greenish-brown hues breaks into a shower of freckles along the fish's back and sides. There the mottling is punctuated by a line of red dots set like gems in blue halos. During spawning season, the brook trout takes on a tropical character, its pale underside glowing with colors ranging from saffron to sunrise yellow. The red color of the fish's fins and tail deepens and becomes more brilliant, forming a vivid contrast to their white edging, which looks as if it has been painted on the fish in a high-gloss enamel.

What you cannot appreciate about the fish unless you see it up close and in person,

Coaster brook trout, affectionately dubbed speckled trout, are beloved by anglers for their iridescent beauty. Photograph by Nick Laferriere.

Newman stresses, is the opaline sheen of its skin, as if the fish had passed through a shower of stardust. "Who could ever describe the iridescence," wrote the north-woods bard Sigurd Olson, "the mottling of gray and green and black, those flaming spots of crimson?"

Once their vital statistics are recorded, Newman carries the fish back down to the stream and gently settles them into a pocket of still water. Despite the temperature, which measures an icy 35 degrees Fahrenheit, he keeps his hands submerged, holding the animals upright as they slowly revive. Only a few years before, Newman plunged his hands into the same stream to plant fertilized eggs, one of which, he says, may have developed into the fish he now holds.

Serving as midwife to some powerful, ancient urge is only the first—and perhaps the easiest—step. Newman knows only too well that the coaster's future depends on abundant clean water, healthy forests, human goodwill. As he cradles the fish, he no doubt ponders the odds.

When the animals regain their vigor and alertness, he opens his hands and releases them back into the pool. And whether it is the day's fatigue finally catching up with him or just the enormity of his task, Newman lets the fish slip away in silence.

By Hook or Crooks: Historic Overfishing

In the last half of the nineteenth century, the growing sport of recreational fishing coincided with the proliferation of steamboats and railroads, which opened up ever-more remote areas of the country to the angler's rod. Among the waters that grew in popularity were those in the Nipigon drainage as well as rivers and streams on Lake Superior's southern shore. Anglers from around the world undertook the arduous journey to the Superior hinterlands in hopes of hooking a trophy-sized coaster. Fishing fever was fanned by such accounts as the 1865 book *Superior Fishing* by Robert Barnwell Roosevelt, uncle of Theodore Roosevelt, who rhapsodized that an angler could make the rounds of rivers that emptied into the lake and "visit pool after pool, try eddy after eddy, till he and his men and the boat are loaded, and satiety bids him rest."

Ironically, the very intoxication with abundance and ease of access to fish that inspired Roosevelt to write some of his finest lines contributed to the coaster's decline. By the mid-nineteenth century, Roosevelt lamented that the waters around the burgeoning settlement of Marquette, Michigan, already had been fished out. In the "Protection of Fish," a companion piece to his Superior travelogue, Roosevelt warned that without restrictions, the lakes and rivers of the frontier would be swept clean of game fish just as they already had been in the East. If the nation's natural bounty was to be ensured, he cautioned, provisions needed to be made "for the protection and preservation of the wild inhabitants of our woods and waters, a common heritage of beauty and sustenance, and the property of our citizens indiscriminately."

Despite its more remote and sparsely settled location, the Minnesota North Shore fared little better. In an 1880 report, C. W. Hall, a member of the state's Geological and Natural History Survey, observed:

> The brook-trout is an object of wanton destruction in northeastern Minnesota. This beautiful and universally admired species inhabits, in great numbers, the many

small rivers flowing into Lake Superior. These streams, in fact, have become one of the most famous fishing grounds on the continent. That they may continue so, they must be protected. Those within the State of Minnesota are visited annually by large numbers of amateur fishermen, who go in parties, and thus make most enjoyable vacation excursions. A boatman and a cook are engaged at Duluth or some other accessible point, who load into a sail-boat a store of provisions and other essentials to comfort and pleasure, and then take the excursionists to the best trout streams around the lake. One stream after another is visited. A camp is pitched beside each where it empties into the lake. Then, for several days, perhaps a week, the river banks are lined with the creeping, stealthy forms of the fishermen, throwing every temptation the ingenuity of man can devise before the eyes of the wary trout. By diligently and patiently continuing at their posts through every hour from daylight until evening, it is surprising if any fish are spared in the stream. . . . It is a very common thing for parties to fish out a stream and select only the very largest specimens for eating and salting, throwing all the rest, probably three-fourths of their whole number, back into the river. Such treatment of the fishing grounds causes much indignation among the people living in the northern part of the State, and who have a lively interest in the preservation of their fish and game.

Sportsmen, however, were not the only ones to blame for the coaster's demise. By the early nineteenth century, fur trading activity on Lake Superior was drawing to a close after the once-plentiful beavers had been trapped to the point of near extinction. Ramsay Crooks, president of the newly reorganized American Fur Company, spied other natural capital that could serve as a replacement venture in Lake Superior—fish. "We have great hopes of adding to the usual returns of our trade, a new and important item, in the Fisheries of Lake Superior," he wrote in a letter to a Washington supporter, General Charles Gratiot, in 1834.

Crooks set up two major fishing stations, at Grand Portage and La Pointe in the Apostle Islands. Augmenting them were numerous other outposts around the lake. For a time, his entrepreneurial instincts realized handsome returns. Lake Superior fish fed workers in the burgeoning lumber camps and iron and copper mines of northern Michigan as well as the growing industrial cities of the Midwest. Around Isle Royale alone, fishermen worked their nets from mid-June to mid-November, shipping an average of two thousand barrels of fish each season. It was not just the sheer volume of fish removed but also the indiscriminate targeting of age classes that harmed fish populations. The mesh sizes of most nets were small enough to trap not only mature adults but also young fish, which routinely were discarded in favor of bigger quarry.

The boom, however, would not last. In 1837 an economic panic in the United States sent the demand for fish plummeting. A company representative traveled as far as Indiana, Illinois, and Ohio to drum up business—but to no avail. The company folded in 1842. "It is interesting to speculate as to the probable outcome of the American Fur Company's venture, had the years from 1837 to 1842 not been the trough of a serious depression," writes historian Grace Lee Nute. "Perhaps in the interest of conservation of our natural resources it was fortunate that such extensive operations did not last longer." But the relaxation of fishing pressure, from both commercial outfits and recreational anglers, would be brief.

By the 1920s the once-abundant coaster brook trout had been nearly extirpated from Lake Superior, in part due to overfishing by recreational anglers, as seen in this 1878 photograph on the Devil Track River by William Henry Illingworth. Courtesy of the Minnesota Historical Society.

Despite the relatively long history of fishing on Lake Superior, developing something as basic as an accurate population count of coasters is difficult. In the absence of reliable data, fisheries biologists have had to depend on early fishing records and historic accounts by north-country adventurers to help them piece together a picture of the numbers and distribution of coasters in Lake Superior in the past. Newman warns, however, that their descriptions need to be taken with a grain of salt. Precise numbers are hard to come by since records of fish catches, for example, did not differentiate between lake trout and brook trout.

Furthermore, Newman points out that historic accounts like those by Roosevelt may present a skewed ecological snapshot of coaster productivity. Many anglers fished the mouths of streams during spawning season, when the coasters congregated in great numbers en route to spawning grounds. Scientists doubt that Lake Superior's waters—the least productive of the five Great Lakes—could have supported coaster populations as high as those suggested by early visitors.

Coaster numbers were kept in check not just by food scarcity but also by suitable habitat. According to Newman, waters that could support trophy brook trout were limited along Superior's shoreline. That is because coasters require not only stream spawning habitat but also a nearshore complex of reefs and shoals in the big lake. Newman estimates that where such combinations existed, large adults may have reached nearshore densities of twenty to thirty fish per mile.

What biologists do know for certain is that by the 1920s coasters were all but eliminated from Lake Superior. Only a few scattered populations survived. Those that survived fishing pressures were faced with a series of assaults—in the lake, in the streams, and from the land. For example, to transport cut timber more efficiently through river channels, commercial logging outfits cleared streambeds of the large rocks and woody debris that once dammed up deep pools as well as provided cover for fish. Many companies erected dams on rivers and streams that impeded the movement of spawning trout. During spring log drives, the dam gates were opened full throttle, unleashing torrents of water and logs, which scoured out streambeds, destroying prime spawning beds. Sediments from the denuded watershed poured into streams, choking many of the spawning beds that remained. Without woody debris and natural rocky barriers to slow the coursing of currents, riverbeds remained vulnerable to the obliterating deluge of floodwaters long after logging outfits moved on. Without the protective shade provided by streamside vegetation, the temperature of trout waters rose to lethal levels. Sawdust dumped into the lake smothered reefs in the coaster's nearshore haunts.

Many of the exotic fish species that were later introduced into the lake were better adapted to these degraded conditions and in some circumstances outcompeted coasters. The Lake Superior Fishery Management Plan drafted in 1995 by the Minnesota Department of Natural Resources (DNR) summed up the coaster's diminished standing: "Although a popular fishery from the mid-1800s through the 1920s, reports of brook trout catches in the lake have been rare since 1950."

A Hunch and a Hope: Early Attempts at Coaster Restoration

Before Newman began his collaboration with the Grand Portage tribe in the early 1990s, "You never heard about coasters," Cholwek observes. Things changed when Newman caught "coaster fever" following a presentation by Rob Swainson, a charismatic young biologist and coaster devotee from Ontario's Ministry of Natural Resources, who led his agency's efforts to revive the famed coaster fishery in the Nipigon River. Swainson's message was simple: without human intervention, coasters would disappear from Lake Superior.

Newman decided to take up the coaster's cause in U.S. waters. In many ways, he was an unlikely standard-bearer. Tall, graying, and soft-spoken, he has the demeanor of an introverted, slightly absentminded professor rather than a firebrand of fisheries conservation. After a day's work in the field, Newman, an avid angler, would more likely grab his fly rod and beat the bush to fish the Brule River near his home in northern Wisconsin than attend a meeting.

But Newman's modest, laid-back style belies his intensity. When asked about anything related to coasters, his speech would become urgent, as if it had taken on the tumbling, irrepressible cadence of the northern streams that became his passion. When Newman was not in the field conducting fish censuses, he was out pleading the coaster's cause to fly-fishing groups in Denver and suburban Chicago or meeting with fisheries biologists around the lake as chair of the Brook Trout Subcommittee for the Great Lakes Fishery Commission, an organization that was founded in 1955 by the United States and Canada to oversee fisheries research and management in binational waters.

Perhaps it was his love for coasters—for the world in which they live and the environmental values they represent—that made Newman such a persuasive advocate for the coaster's cause. Indeed, largely due to his campaigning, coaster rehabilitation began to receive financial and political support in the United States from numerous organizations, including the influential citizens group Trout Unlimited, whose goal is to conserve, protect, and restore North America's coldwater fish and the watersheds in which they live.

As a template for how the enormous task of coaster restoration might be accomplished, Newman looked to Swainson's innovative work—in the field as well as in the boardroom. With the blessing of his agency and Canadian sports anglers, in 1990 Swainson negotiated a groundbreaking contract with Ontario Hydro. The power company, which operates several hydroelectric dams on the Nipigon River, agreed to maintain water levels during critical spawning periods in the fall and in the winter when the fish hatch and swim up from the nests dug into gravel streambeds. Swainson's rehabilitation plan also called for clearing sediment-choked springs, protecting spawning grounds, and setting number and size limits on catches so that adult coasters could reproduce at least once before being bagged.

His efforts paid off. Coaster numbers in the Nipigon watershed began increasing, especially those of spawning adults. Nonetheless, Swainson knew that a healthy population of coasters in Canada alone would not stave off the fish's extinction in Lake Superior. He urged that similar restoration be undertaken on other tributaries around the lake. Rivers and streams that were deemed hospitable to coasters could then be stocked with fertilized eggs and young fish from the Nipigon strain.

The possibility that the coaster—which had played such an important role in the history and ecology of Lake Superior—could be brought back from the brink of extinction in U.S. waters captivated Newman's imagination. The task logically fell to Newman's agency, the USFWS, whose mission is to work with others "to conserve, protect, and enhance fish, wildlife, and plants and their habitats for the continuing benefit of the American people."

In 1991 he began a collaboration with the Grand Portage Band of Lake Superior Chippewa to launch several pilot projects on tribal streams. Their task would not be easy. Although records indicate that these waters once hosted coasters, they provided only marginal trout habitat. Typical of North Shore streams, they are characterized by steep gradients and predominantly rocky streambeds. To make matters worse, their waters are subject to considerable variability in temperature and flow.

But in other respects, the reservation streams were ideal for coaster rehabilitation. For example, the USFWS and the tribe share a stated goal to restore native species in Lake Superior, including coasters. "Many of the older tribal members caught coasters when they were kids and have seen these fish disappear from tribal waters," says Johnson, who is a member of the Grand Portage band. "Everything that tribal tradition teaches is to keep the land and waters whole. That's all we're trying to do. We're trying to make the lake whole again."

Because the projects were carried out on tribal waters, Newman was also able to avoid the political wrangling of powerful interest groups and the time-consuming research protocols, budgetary procedures, regulations, and paperwork involved in carrying out projects on waterways controlled by state natural resource agencies. Perhaps most important, Newman was able to sidestep their skepticism. Citing a string of failed rehabilitation efforts between the 1970s and early 1990s, most agencies had given up on stocking brook trout long ago, deciding instead to devote resource dollars to cultivating fish that produced a higher rate of return for their constituents' investment in fishing licenses. Minnesota was no exception. Beginning in 1970, the DNR stocked a variety of cultured strains of brook trout in streams and protected bays in the lake. According to follow-up surveys, most of the fish wound up in anglers' creels within a few months of their release. Furthermore, biologists found no evidence that the fish had established themselves in the lake and were reproducing on their own.

When hatchery-raised fish from the Nipigon strain became available in the 1980s, the DNR launched a renewed three-year stocking program in the French River. The goal—to establish a self-sustaining population of fish—fell far short of projections. DNR surveys conducted on the river from 1987 to 1989 came up with only seven returning fish. In response, the agency scuttled further plans for coaster rehabilitation.

Critics of coaster rehabilitation charged that a woeful lack of scientific knowledge about the species seriously hampered the odds of success in reintroduction projects like these. Since the fish were all but exterminated by the time rigorous scientific studies began to be carried out on the lake, little hard data were available about even fundamental coaster biology, such as morphology (the fish's form and structure), life history, the composition of its population, habitat use, life span, and, perhaps most important, genetics. Some fisheries biologists further argued that coaster numbers might have dwindled to the point where genetic variability was too low to rebuild sustainable populations, the equivalent of losing too many letters from the alphabet to be able to construct meaningful sentences, much less coherent paragraphs. Historic accounts of daily catches from the Salmon Trout River around 1900, for example, bragged of creels bulging with hundreds of fish. By contrast, a census of coasters migrating upstream during the July–October spawning season in 2000 turned up 161 individuals. In a study of population data from five major sites around Lake Superior published in the 2008 *North American Journal of Fisheries Management,* fisheries biologist Casey J. Huckins and colleagues point out that coasters number some 300 individuals in U.S. waters. (Population estimates, they say, were not available for the Nipigon River and Nipigon Bay, where coasters have made the biggest gains.) At the very least, scientists charged, finding enough fertilized eggs or young fish to plant around the lake would be a significant challenge. (Only three strains of coaster brood stock currently are available for transplantation: the Tobin Harbor and Siskiwit Bay strains from Isle Royale, and the Lake Nipigon strain from Ontario, the most commonly used strain in coaster-restoration efforts in Lake Superior.)

Newman was the first to admit that such misgivings were amply justified. But he also knew that time was running out for the coaster. If there was any hope of pulling it back from the brink of extinction, fisheries biologists would have to make educated guesses about coasters based on the results of research conducted in the Nipigon watershed and the more exhaustive studies of other fish such as stream-bound brook trout and lake trout.

And they would have to do what scientists are expressly trained to avoid, that is, act quickly based on promising, but preliminary, results that have not yet run the gauntlet of rigorous—but time-consuming—scientific testing and peer review. Which is exactly what Newman did when he implemented Swainson's experimental stocking techniques in the Grand Portage rivers.

After reviewing the innovative protocols devised by Swainson and his colleagues, Newman suspected that fundamental flaws in management philosophy, as well as fish-stocking practices, were responsible for the failure to rehabilitate coasters in the past. By and large, fisheries managers had been using techniques developed over decades in Lake Superior for what is known as a put-and-take fishery. In the 1930s, for example, Minnesota historian Nute observed that commercial fishermen around Isle Royale replenished lake-trout fish stocks by stripping eggs from netted female fish and fertilizing them on board with sperm from males. These eggs were reared in a fish hatchery for one or two seasons. Then the fingerlings were released into the lake for the sole purpose of dangling from an angler's hook or being caught in a fishing net when they grew up.

Since then advances in fisheries science have led to a greater sophistication in fish rearing. But the put-and-take approach persisted, including its dependence on the reproductive chamber of the hatchery—the surrogate "womb" and nursery for fish such as coasters.

In recent years, however, this practice has been challenged by scientists and members of the public who maintain that fisheries management efforts should be focused on helping native species reestablish and sustain themselves, that is, nurturing fish that are genetically adapted to their environment and are able to reproduce on their own, thereby reducing our costly dependence on fish hatcheries. "The emphasis used to be on the numbers and pounds of fish that were raised and released," Newman observes. "Fish that grew fast and would eat pellets were selected. In the mid-1990s some biologists began to ask about what happens after you put the fish into the water and to judge success by how many fish survive to maturity and spawn instead of just how many fish are caught as adults."

However, biologists are just now learning how anadromous fish complete this circle of life. A 2013 issue of *Current Biology,* for example, published research showing that wild Pacific salmon hatched in the Fraser River in British Columbia memorized the local magnetic fields around their natal river. Later as spawning adults, the fish used their ability to sense subtle changes in the intensity of these fields to successfully guide them on their journey home. Biologists suspect that Lake Superior fish, such as lake trout and coaster brook trout, may employ similar navigational aids. Brook trout, which have a keen sense of smell, may also rely on olfactory cues to locate the tributaries in which they were born.

In most traditional fish-stocking practices, fish have been released into the lake and its tributaries as fingerlings and yearlings under the assumption that animals stocked at older life stages would have a better chance of survival than, say, fertilized eggs, which would have to complete their early development in the wild without the benefit of the hatchery's protection. Once they reach sexual maturity, however, these fish lack the imprinting that would guide them back to suitable spawning grounds. Ironically, a few home in on cues in the waters in which they were released. Observers have witnessed hatchery-raised lake trout make futile attempts to spawn in the shallow waters around boat ramps or in rivers around hatcheries, unlike their wild cousins, which head for the cobble substrates of offshore reefs where they emerged from eggs. (In the case of lake trout, Lake Superior biologists developed a simple, but elegant, technological solution. They constructed a "sandwich" of artificial turf. In its layers they embedded fertilized lake trout eggs. This incubator could be fixed onto rocky offshore reefs, allowing the emerging fish to become imprinted on spawning areas that could, in turn, serve as hospitable quarters for their own offspring.)

The new fisheries philosophy—measuring success by how many adult fish live to spawn rather than how many survive simply to fill an angler's creel—became Newman's mantra for the rehabilitation of coaster brook trout as well. He started with a close examination of the coaster's life history. In the Nipigon River, for example, researchers found that coasters spawn only over groundwater upwellings. "Springs and cold, undammed water flows are the lifeblood of trout streams," Newman observes. "Take those away and brook trout are gone." This intimate relationship is even encoded in the animal's scientific nomenclature, *Salvelinus fontinalis,* meaning "little salmon of the springs."

Indeed, so important are springs to the offspring of brook trout that adult fish will put themselves at considerable risk in order to spawn around them. The reason is simple. As eggs develop, they consume increasing amounts of oxygen. Depositing them over upwellings ensures a steady supply of this vital ingredient.

So instead of following the traditional release procedure—unloading the fish from hatchery trucks at locations convenient to the driver—Newman and tribal biologists walked the banks of the reservation's rivers in winter looking for spots of open, free-flowing water in the river ice that marked the location of springs.

Newman was further convinced that previous stocking efforts failed to produce self-sustaining coaster populations because the hatchery-raised fish lacked the necessary physical fitness or suitable genetic makeup for survival in the big lake. And many simply were too old to have developed appropriate homing instincts. To avoid these pitfalls, the team planted native Nipigon-strain coasters at their earliest life stages—as fertilized eggs and fry—in the Grand Portage streams.

But depositing eggs on the surface of the streambed leaves eggs vulnerable to predators or being washed away. The eggs must be buried—but not in a way that would cut off the flow of oxygen. Brook trout have solved this dilemma by digging nests, known as redds, into gravel riffles, fanning away silt and other fine debris with their tails. The excavated gravel is then used to cover the nests. From there, the stream takes over. Tucked into the crevices between the tiny rocks, the eggs are protected from predators. At the same time, they are bathed by flowing water, which delivers oxygen and carries away the fine sediments that can accumulate and smother them.

Placing nests around springs carries other life-saving benefits as the fish mature. Streams with upwellings have a consistent year-round volume of water, whereas streams that are fed by rain and snowmelt can surge in spring but shrink to a trickle by late summer.

Spring-fed tributaries also have greater thermal stability. Regardless of air temperatures, which can fluctuate from extreme highs to icy lows, North Shore groundwater temperatures hover around 40 degrees Fahrenheit. In winter, flows from underground springs moderate surface water temperatures, preventing redds from turning to ice when the thermometer drops below freezing. And the comparative warmth promotes a greater abundance of stream invertebrates such as the larvae of aquatic and terrestrial insects, crustaceans, and worms, a boon to tiny fry that emerge from their redds in March with big appetites. In summer, underground seeps help to keep the water well below 77 degrees Fahrenheit, the upper lethal threshold for young trout.

Taking their cues from nature, Newman and fellow biologists mapped the location of upwellings in the Grand Portage streams and returned to them in spring to dig artificial redds, measuring some three feet in diameter, into the gravel. Fertilized eggs or

To encourage natural reproduction, new fish-stocking techniques mimic behaviors used by coaster brook trout in the wild. Fisheries biologists dig nests in clean gravel riffles and then insert fertilized eggs. The emerging fish become imprinted on their natal streams and, as adults, leave Lake Superior to re-enter their birth waters to spawn. Courtesy of U.S. Fish and Wildlife Service, Ashland, Wisconsin.

young fish were poured into the redds through ordinary PVC pipes and buried at depths ranging from one to six inches. The process is extremely economical, Newman says. Even when field labor is factored in, the price tag for stocking eggs and fry is a fraction of the cost of raising and releasing older fish from hatcheries.

But not all stocking procedures would be so straightforward. To rebuild a self-sustaining population of coaster brook trout, Newman knew that biologists would need to do more than find an adequate supply of fish eggs or develop appropriate planting techniques. In many cases, they would need to repair some of the damage done to the land.

Waters whole enough to support anadromous fish require that the frayed bonds between rivers and the land be replaited. In her book *In Service of the Wild,* Stephanie Mills writes about efforts to restore salmon to one California river. She observes that anadromous fish "bind together the land and the sea in the course of their existence." If you want to restore their populations, she adds, "you have to heal a whole watershed: vegetation, erosion, social fragmentation—the works."

To make one Grand Portage stream hospitable to coasters, the tribe and the USFWS undertook a $30,000 reconstruction of its headwaters. Before the fierce postlogging fires of the late nineteenth and early twentieth centuries, the water from springs had collected in a low-lying bog at the source of the river. The upwellings were protected by a mantle

Female coasters clear nests, known as redds, in clean gravel, where eggs *(top)* and hatchlings *(below)* can find protection from predators while receiving the abundant oxygen they need from free-flowing waters. Photographs by Robert Michelson / Photography by Michelson, Inc.

of peat, which kept the flow of water into the stream channel consistently cold. It also filtered the water, leaving it crystal clear.

The natural features that sustained these crystalline waters, however, were damaged beyond repair. Clear-cut logging operations that moved through the region at the turn of the century left great slash piles of trimmed branches and dead and unusable trees. Although regular wildfires were integral to the land's vitality, the intensity, frequency, and scale of conflagrations fueled by these heaps of dry vegetation were unmatched in nature. In 1936 a wildfire swept through the Grand Portage area, burning so hot that it consumed the bog's peat. Without this protective sponge, which absorbed and released water in a consistent year-round flow into the stream channel, water levels began to fluctuate wildly. The stream dwindled to a trickle during summer droughts. Flash floods scoured the channel during heavy rains and snowmelts. Water temperatures during summer and fall rose to lethal levels. In the winter the streambed regularly froze, killing eggs and young fish in their redds. Without the protective filter of the bog, more sediments flowed into the stream, clogging spawning beds.

It would take centuries, if not a millennium, for the protective peat layer to accumulate to the depths that had existed prior to the slash fires. If coasters were ever going to successfully reproduce in the Grand Portage stream again, Newman and his colleagues would have to find a way to mimic the life-enhancing functions of this headwaters habitat. The biologists drew up a plan that called for damming the outflow of the headwaters lake that fed the stream. At the deepest point of the lake they installed a pipe that siphoned water from its cold twenty-foot depths and directed it into a reconstructed channel below the dam. The plan worked. A sensor that records daily stream temperatures on a computer chip has shown that the diversion keeps temperatures within a range that is more hospitable to coasters. The lake has provided an added benefit: by serving as a holding tank for excess stormwater, it has helped to even out the flow of water into the stream.

To measure the rate of hatching success in the artificial redds, the scientists also planted experimental incubators that were stocked with eggs from the same strain of Nipigon coaster. The results were impressive: between 84 and 100 percent of eggs in properly placed incubators produced young fish.

In 2002 the team completed their restoration of the stream with the construction of a new fish passage that replaced an old culvert under Highway 61. The new passage opens up three miles of precious spawning habitat to migrating fish.

Electrofishing for five consecutive years in the early 1990s turned up significant runs of spawning adults. To test whether they were reproducing successfully, the scientists skipped a year of stocking in all Grand Portage streams in spring 1997. That fall electrofishing surveys turned up young fish, which the scientists say are likely the offspring of adults that had found their way back from Lake Superior to reproduce in their natal streams.

Coaster Rehabilitation: Nurturing Nature

Isolated interventions like those on the Nipigon River and Grand Portage streams would not be enough, however, to reestablish self-sustaining coaster populations in Lake Superior. Success over the long haul required a coordinated effort among organizations, citizen groups, and government agencies around the lake. This was a daunting task. Lake Superior covers more than twenty-seven hundred miles of shoreline. The governance and management of this vast territory are overseen by two countries, one Canadian province, three U.S. states, and several tribes, all of which have goals and legal obligations that sometimes clash with one another, complicating brook trout restoration efforts. A study published in 2008 in the *North American Journal of Fisheries Management* outlines just how complicated the human dimensions of fishery management can get. Laura Hewitt of Trout Unlimited and her colleagues noted that since fishery programs of state and provincial natural resource agencies "are often funded in large part by license or stamp fees, they are by design responsive to angler constituencies and legislators." As a result, the authors observed, programs tended to put "stronger emphasis on utilitarian goals related to angling and economic benefits." On the other hand, the study pointed out that federal agencies in Canada and the United States, such as the USFWS, "were more supportive of an ecological approach." The USFWS lacks the authority, however, to carry out its own management projects and must rely on its powers of persuasion to develop collaborations with state, tribal, and citizen partners. In general, the authors found that fisheries managers in Canada included viewpoints of a greater array of stakeholders in decision making and tended to "emphasize safeguarding native species," whereas U.S. fisheries managers favored "approaches that often involved the stocking or management of nonnative species."

But despite thorny differences, coaster fever was catching on. In 2003, shortly after the completion of the Grand Portage stream project, several organizations joined forces with the Great Lakes Fishery Commission to sponsor a gathering of parties with an interest in coasters. Some forty experts from the United States and Canada convened at the University of Minnesota's Cloquet Forestry Center to present and discuss the latest research in coaster biology, management, and restoration. In 2004 this research found a national platform in a symposium at the annual meeting of the American Fisheries Society. Symposium papers that focused on coaster biology and ecology were published in 2008 in the *Transactions of the American Fisheries Society*, while a second series, which examined management issues, followed in an issue of the *North American Journal of Fisheries Management*.

These efforts began to break the logjams to coaster restoration by answering some of the most fundamental—and pressing—questions about the fish's biology and lifeways in Superior's streams and the open waters of the lake. Among the foremost issues to be resolved was this: Were the burly coasters a subspecies of brook trout—that is, genetically distinct from their diminutive, stream-bound cousins—or were they simply more adventurous, growing larger because they ventured farther afield into greener pastures?

Turns out that the brook trout within each stream were closely related to one another. Furthermore, each tributary hosted its own clan of genetically allied fish. Within each clan, however, the research uncovered no pronounced genetic distinctions that could account for the great variation in size and behavior among its members. Nonetheless, studies showed that the life histories of coasters and their stream-loving relatives were so different that they could just as well be separate species. For reasons that remain unclear, at around two years of age, some brook trout leave their natal streams for life in the big lake, where they take on a silvery color more befitting their new home environment. The authors of a study published in the 2008 *Transactions of the American Fisheries Society* offered an intriguing look at the behavioral patterns of these migrants once they set up housekeeping in their new digs. Canadian scientists Jamie Mucha and Robert Mackereth outfitted a group of coasters with radio transmitters, and from June 1999 to October 2000, they tracked their movements within Nipigon Bay and four of its tributary streams. In general, the data showed that coasters largely stuck within five hundred feet of the shores of Nipigon Bay in waters less than twenty-three feet deep. Here, the researchers surmised, large boulders, shoal edges, and aquatic plants provided more ample cover from predators.

But the fish's behavior was marked by pronounced seasonal differences. After ice-out in early spring, the fish haunted the nearshore shallows where waters were the first to warm up. As midsummer progressed and lake waters approached the fish's tempera-ture tolerances, coasters sought refuge around groundwater upwellings along the lake bottom or in the deeper, cooler waters of the lake. But "rather than move further from shore to deeper areas of preferable temperature," the study's authors write, "brook trout moved laterally, selecting adjacent nearshore areas with steeper slopes that provided deep, cool waters closer to shore." Minimizing the distance between its thermal refuges and preferred feeding grounds is important for a fish that is not built for efficient long-distance travel. As Mucha and Mackereth point out, the brook trout's morphology—such as a square tail that generates a larger turbulence wake as it swims than that of a fish with a strongly forked or crescent-shaped tail—is "not conducive to high-speed swimming and pelagic cruising."

Beginning in late July, the fish exhibited what the scientists called a "probing pattern," in which they swam in and out of the tributary streams, staying only a few days at a time. The researchers conjectured that these restless excursions could be triggered by early spawning urges or by a need to find a respite from Superior's warming waters. By mid-October, the fish were ready to reproduce, faithfully returning to the area in the stream that they had used as spawning grounds the year before. By early November, the tagged fish had exited the streams and taken up residence on their wintering grounds—the same shallow waters they occupied after ice-out in the spring.

Swimming into a Sea of Challenges

Generating this kind of knowledge about the coaster's life history is critical to the success of efforts to reintroduce the species in Lake Superior and its tributaries. But it is just the first of many challenges. Native brook trout now share their stream nurseries and compete for forage in the lake with brown trout imported from Europe as well as fish from the Pacific Ocean, including coho and chinook salmon and rainbow trout, not to mention human-engineered hybrids such as splake (a cross between native brook trout and lake trout). Already by 1935, rainbow trout, which had been stocked in Superior at the turn of the century, had became so abundant that George Shiras, an accomplished amateur naturalist and U.S. congressman whose family had vacationed on the Michigan shore for generations, expressed the following concern for the future of native coasters:

> After frequent plantings, the rainbow trout have become permanently established in Lake Superior. During the spawning, in May, they are to be found in considerable numbers in most of the streams entering the lake.
>
> As an admirer of the speckled trout, I have wondered just what effect these alien fish would have upon its abundance and perpetuation, and fear for the result. The maintenance of all fish naturally depends upon the food supply. This seems imperiled by the presence of the newcomers, which, being larger, will necessarily cause a greater depletion of the water life.
>
> It is not unusual in spring to find streams heretofore occupied exclusively by brook trout crowded with rainbow trout, many of which exceed 10 pounds in weight. Even the smaller creeks harboring the immature speckled trout have been sought by these large fish.

Were the streams invaded only during the spawning season, the situation would not be so bad, but the young of the introduced species remain a long time in the streams before seeking Lake Superior and thus become year-long competitors for the food of the brook trout. Not only this but they raid the brook trout spawning beds.

In the years since Shiras made these observations, a multimillion-dollar sport-fishing industry—from charter boat fleets and tourist lodges to influential angler associations—has been built around such exotic fish. And since fees generated by the sale of fishing licenses are the chief source of funding for Lake Superior's fisheries, managers are likely to continue stocking exotics even if they hamper native coaster reintroduction. Splake, for example, may interbreed with coasters, threatening their genetic integrity.

Even if reintroduction efforts are successful, it may be decades before coaster numbers rebound enough to support a robust fishery. Currently, anglers in both U.S. and Canadian waters are limited to a daily catch of one fish measuring twenty inches or more. "It may take several generations (15–20 years)," write Huckins and colleagues, "before any noticeable increase in coaster numbers is observed."

According to some biologists, the biggest obstacles to coaster rehabilitation, however, may be the fish themselves. What will not change, they say, are the behaviors that made coasters vulnerable to overexploitation in the first place. They have evolved to survive within the nearshore reaches of the lake (hence the name "coasters"), which leaves them within easy reach of anglers out on the big lake.

They also are vulnerable in rivers and streams. During spawning runs, coasters can be found in clear, relatively shallow water. And when the fish are in the act of spawning, Newman says, coasters seem oblivious to all but the task at hand. As a result, coasters are easy targets for a variety of predators—including poachers.

Finally, there is the matter of the coasters' prodigious appetite. Newman points out that the fish thrive in oligotrophic waters using skills they developed over thousands of years while swimming in the low-productivity waters following glacial advances and retreats. "Their whole environment dictates their lifestyle," Newman observes. "It's eat everything you can get your mouth on or starve to death. They've evolved to be aggressive feeders, so they're very easy to catch." Those that get away, biologists say, are slow to learn from their experience and readily pursue an angler's lure again and again. Unfortunately, these adaptive behaviors, which allowed them to persist so successfully through the comings and goings of glacial epochs, have helped to bring coasters closer to the edge of extinction.

Live and Let Live: Coaster Rehabilitation and Development

In his 1964 book *Familiar Freshwater Fishes of America,* Howard T. Walden wrote that "one is inclined to use the past tense in writing of the wild brook trout. Constitutionally incompatible with the advance of civilization, this exquisite fish is dying. Where man has dried up his springs by deforestation, polluted his waterways, straightened streams into ditches and denuded them of their natural cover, the wild brook trout has vanished. And with it has gone an essence of that early America which somehow it symbolized: rural peace, unmachined enterprise, and nature left to herself."

Coasters can survive, skeptics say, only in places like the Grand Portage reservation where settlement is sparse and where people have reached a consensus about their value. What about people who work, live, or vacation around brook trout habitat? How many, they ask, would be willing to make changes necessary for the fish's continued existence? Each time the land is opened up for the construction of new roads or housing developments, more sediments pour into streams and into the lake. Pollution from septic systems that inadequately treat the waste of burgeoning shore populations affects the nearshore habitats occupied by coasters. Ironically, even the increasing use of water, albeit in a place surrounded by water, presents a problem for successful coaster reintroduction. "Everyone's favorite sport is trying to drill a well down to find artesian water so that they can squirt it up on their lawns," Newman points out. "Every one of those wells is stealing from springs that, in some cases, provide brook trout spawning habitat."

Increasingly, write Hewitt and her colleagues in their article for the *North American Journal of Fisheries Management,* fisheries managers may need to bring even more people to the table, including "those who shape and influence land development. Along with traditional fishery stakeholders, individuals involved in forestry and watershed land use planning and management must be engaged in the process of developing coaster rehabilitation strategies." Some threats to coaster restoration may lie in the far inland reaches of the watersheds and adversely affect their stream habitats from many miles away.

Take, for example, the boom in exploration for nonferrous metallic minerals such as copper and nickel and precious metals such as gold, palladium, and platinum in both the U.S. and Canadian territories of the Lake Superior watershed. This includes the vast Duluth Complex, a massive bedrock formation arcing from Duluth to Grand Portage in northeastern Minnesota. These kinds of mining operations require large-scale conversions of land (in some cases, more than a thousand acres) for both the ore mining and waste disposal. When exposed to air and water, the enormous volumes of waste rock generate toxic effluent laced with heavy metals that must be contained and managed for many generations. As demonstrated in countless examples from around the world, such mines have, without exception, contaminated ground and surface waters with pollutants that are lethal to aquatic and terrestrial life and pose a serious health risk to humans.

Given this industry's potential to impact water quantity and quality, international mining companies will need to take their place at a table already crowded with recreational fisherman, real estate developers, Indian tribes, local tourism economies dependent on a high-quality natural environment, and managers from natural resource agencies charged with managing healthy watersheds and fisheries while their colleagues across the table are tasked with generating money from public lands through mining. It remains to be seen how the needs of coasters will fare midst these competing and conflicting values.

Another wild card for the future of coaster rehabilitation is global warming. Under some climate-change scenarios, the levels of Lake Superior are expected to drop sharply, causing sandbars to form at the mouths of rivers. These barriers cut off fish from critical spawning grounds. Changes in stream temperatures are anticipated as well, say Bill Herb of the University of Minnesota's St. Anthony Falls Laboratory and his colleagues at the University of Minnesota's Natural Resources Research Institute. The group developed a climate model showing a warming of 3 degrees Fahrenheit in North Shore streams. "That's pretty significant," he observes. Hardest hit will be the larger rivers that coasters favor since they are less shaded and also include more reaches of sluggish waters that can better absorb heat.

Brook trout provide us with a yardstick for measuring the health of our waters and the wholeness of our land. As such, they present us with a dilemma, as well as a choice. The observations that writer Steve Grooms made about efforts to restore native lake trout to the waters of the lower Great Lakes could just as easily apply to the efforts to rehabilitate coaster brook trout in the Lake Superior watershed. "The issue," he says in an article for *Trout* magazine, "is whether managers should continue trying to restore an ecosystem destroyed by centuries of abuse, or whether they should accept the loss of some ecological integrity as a *fait accompli* and do the best job of managing what's left for human benefit."

In other words, how accustomed have we become to the idea of extinctions? ❧

SUGGESTIONS FOR FURTHER READING

D'Amelio, Silvia, Jamie Mucha, Rob Mackereth, and Chris C. Wilson. "Tracking Coaster Brook Trout to Their Sources: Combining Telemetry and Genetic Profiles to Determine Source Populations." *North American Journal of Fisheries Management* 28 (2008): 1343–49.

D'Amelio, Silvia, and Chris C. Wilson. "Genetic Population Structure among Source Populations for Coaster Brook Trout in Nipigon Bay, Lake Superior." *Transactions of the American Fisheries Society* 137 (2008): 1213–28.

Fraser, Dylan J., and Louis Bernatchez. "Ecology, Evolution, and Conservation of Lake-Migratory Brook Trout: A Perspective from Pristine Populations." *Transactions of the American Fisheries Society* 137 (2008): 1192–1202.

Gorman, Owen T., Seth A. Moore, Andrew J. Carlson, and Henry R. Quinlan. "Nearshore Habitat and Fish Community Associations of Coaster Brook Trout in Isle Royale, Lake Superior." *Transactions of the American Fisheries Society* 137 (2008): 1252–67.

Grooms, Steve. "The Enigma of the Lake Trout." *Trout,* Spring 1992: 21–49.

Heath, Daniel D., Corwyn M. Bettles, Sara Jamieson, Iga Stasiak, and Margaret F. Docker. "Genetic Differentiation among Sympatric Migratory and Resident Life History Forms of Rainbow Trout in British Columbia." *Transactions of the American Fisheries Society* 137 (2008): 1268–77.

Hewitt, Laura E., Karen G. Mumford, Donald R. Schreiner, and Gregory J. Fischer. "Coaster Brook Trout Rehabilitation in Lake Superior: A Human Dimensions Perspective." *North American Journal of Fisheries Management* 28 (2008): 1365–72.

Huckins, Casey J., and Edward A. Baker. "Migrations and Biological Characteristics of Adfluvial Coaster Brook Trout in a South Shore Lake Superior Tributary." *Transactions of the American Fisheries Society* 137 (2008): 1229–43.

Huckins, Casey J., Edward A. Baker, Kurt D. Fausch, and Jill B. K. Leonard. "Ecology and Life History of Coaster Brook Trout and Potential Bottlenecks in Their Rehabilitation." *North American Journal of Fisheries Management* 28 (2008): 1321–42.

Lawrie, A. H., and J. F. Rahrer. "Lake Superior: Effects of Exploitation and Introductions on the Salmonid Community." *Journal of the Fisheries Research Board of Canada* 29 (1972): 765–76.

Mucha, Jamie M., and Robert W. Mackereth. "Habitat Use and Movement Patterns of Brook Trout in Nipigon Bay, Lake Superior." *Transactions of the American Fisheries Society* 137 (2008): 1203–12.

Newman, Lee E. "The Grand Portage Project, a Successful Model for the Reintroduction of Lake Superior Coaster Brook Trout Populations." In *Wild Trout* VII, *Management in the New Millennium: Are We Ready?,* ed. Dan Schill, Steve Moore, Pat Byorth, and Bob Hamre, 149–54. Bozeman: Montana State University, 2001.

———. "Movement and Range of Coaster Brook Trout of Tobin Harbor, Isle Royale." Ashland, Wis.: Ashland Fishery Resources Office, U.S. Fish and Wildlife Service, 2000.

Newman, Lee E., Robert B. Dubois, and Theodore N. Halpern, eds. *A Brook Trout Rehabilitation Plan for Lake Superior.* Ann Arbor, Mich.: Great Lakes Fishery Commission, 1999.

———. *A Brook Trout Rehabilitation Plan for Lake Superior.* Miscellaneous Publication. Ann Arbor, Mich.: Great Lakes Fishery Commission, 2003.

Newman, Lee E., and John Johnson. "Development of a Reintroduced, Anadromous Brook Trout Population at Grand Portage, Minnesota, 1991–1996." Ashland, Wis.: Ashland Fishery Resources Office, U.S. Fish and Wildlife Service, 1996.

Newman, Lee E., Ray G. Johnson, John T. Johnson, and Richard J. Novitsky. "Defining Habitat Use and Movement Patterns of a Reintroduced Coaster Brook Trout Population in Lake Superior." Ashland, Wis.: Ashland Fishery Resources Office, U.S. Fish and Wildlife Service, 1999.

Putman, Nathan F., Kenneth J. Lohmann, Emily M. Putman, Thomas P. Quinn, A. Peter Klimley, and David L. G. Noakes. "Evidence for Geomagnetic Imprinting as a Homing Mechanism in Pacific Salmon." *Current Biology* 23 (2013): 312–16.

Ridgway, Mark S. "A Roadmap for Coasters: Landscapes, Life Histories, and the Conservation of Brook Trout." *Transactions of the American Fisheries Society* 137 (2008): 1179–91.

Roosevelt, Robert Barnwell. *Superior Fishing: The Striped Bass, Trout, and Black Bass of the Northern States.* Reprint, St. Paul: Minnesota Historical Society Press, 1985. Originally published New York: Carleton, 1865.

Schreiner, Donald R. "Coaster Brook Trout Rehabilitation in Lake Superior: An Introduction." *North American Journal of Fisheries Management* 28 (2008): 1305–6.

Schreiner, Donald R., Ken I. Cullis, Michael C. Donofrio, Gregory J. Fischer, Laura Hewitt, Karen G. Mumford, Dennis M. Pratt, Henry R. Quinlan, and Steven J. Scott. "Management Perspectives on Coaster Brook Trout Rehabilitation in the Lake Superior Basin." *North American Journal of Fisheries Management* 28 (2008): 1350–64.

Shiras, George, III. "The Huron Mountain District. Fishes of Lake Superior." Chap. 20 in vol. 1, *Lake Superior Region,* of *Hunting Wildlife with Camera and Flashlight: A Record of Sixty-Five Years' Visits to the Woods and Waters of North America.* Washington, D.C.: National Geographic Society, 1935.

———. "The Wild Life of Lake Superior, Past and Present." *National Geographic Magazine* 40, no. 2 (1921): 113–204.

Sloss, Brian L., Martin J. Jennings, Ryan Franckowiak, and Dennis M. Pratt. "Genetic Identity of Brook Trout in Lake Superior South Shore Streams: Potential for Genetic Monitoring of Stocking and Rehabilitation Efforts." *Transactions of the American Fisheries Society* 137 (2008): 1244–51.

Walden, Howard T. *Familiar Freshwater Fishes of America.* New York: Harper and Row, 1964.

Ward, Matt. "Status of Coaster Brook Trout in the Minnesota Waters of Lake Superior." Minnesota Department of Natural Resources Fisheries Division, Lake Superior Area. Minnesota F-29-R (P)-28. 2008.

Wilson, Chris C., Wendylee Stott, Loren Miller, Silvia D'Amelio, Martin J. Jennings, and Anne M. Cooper. "Conservation Genetics of Lake Superior Brook Trout: Issues, Questions, and Directions." *North American Journal of Fisheries Management* 28 (2008): 1307–20.

INTERNET RESOURCES

Trout Unlimited, www.tu.org

U.S. Fish & Wildlife Service, www.fws.gov/midwest/fisheries/coaster-brook-trout.html

HIGHLANDS

Thex are known in the tourist trade as leaf peepers, people who indulge in long meandering drives for the sole purpose of gazing at forests in the full blaze of autumn color. Come September, the North Shore is hit with a tsunami of peepers. According to 2010 figures from the Cook County Lodging Tax Report, lodging taxes generated by the stepped-up tourist traffic in September and October alone accounted for 22 percent—nearly $116,000—of the annual total in lodging levies collected by Cook County in the Tofte–Lutsen area, and 27 percent—nearly $53,000—in the Grand Marais area. Purchases of gasoline, food, and other goods by leaf peepers add millions more to the regional economy.

What fuels this economic engine—and the enthusiasm of leaf peepers—is a swath of forest covering the tops and lake-facing slopes of a ridge that parallels the Superior shore from Duluth to the Canadian border. Measuring some six to ten miles wide, this diverse region of bedrock outcrops, sweeping hills, and soggy lowlands is known as the North Shore Highlands. Here in the space of an hour, a hiker can descend into a valley grove

The highlands landscape represents the midsection of northeastern Minnesota's Lake Superior watershed. Large patches of sugar maple or birch and aspen forest cover the massive hills of glacial end moraine and bedrock ridges near Lake Superior. These prominent landforms are separated by deeply incised streams flanked by lowland white cedar forests and swamps. This portion of the watershed includes considerable state, federal, and county land. Nonetheless, private land is much more common in the highlands than in the headwaters, particularly on inland lakes, along rivers, and on the ridges above Lake Superior, bringing a different suite of land uses and potential impacts to the watershed.

In the highlands, white cedar forests can be found both on steep slopes and in riparian lowlands and swamps in the valleys between the hills. Copyright Minnesota Department of Natural Resources–Lynden B. Gerdes.

Sugar maples, which form a parti-colored autumn cloak on the highlands' bedrock ridges and morainal hills, thrive in the moist, loamy soils and milder climate near Lake Superior. Their prodigious rain of leaves every autumn creates a deep, spongy organic layer that supports a distinctive community of wildflowers. Photograph by Chel Anderson.

Unlike the poor black spruce swamps typical of the headwaters, rich black spruce swamps in the highlands receive significant amounts of minerals and nutrients from groundwater that has percolated through the deep forested soils of surrounding slopes. These inputs support greater plant diversity. Copyright Minnesota Department of Natural Resources–Ethan Perry.

Forests of paper birch and aspen that regenerated after wildfires in the late nineteenth and early twentieth century can still be seen in parts of the highlands. In time, they were joined by shade-tolerant species, such as balsam fir, white spruce, white cedar, white pine, and red maple. Photograph by Chel Anderson.

White cedar swamps are common along the river valleys of the highlands. Their deep, cool moss carpets provide a home for a wide variety of plants and animals and promote the slow release of cold water to North Shore trout streams. Photograph by Chel Anderson.

of northern white cedar and climb up to a ridgetop forest of sugar maple before dropping down into the dim and ragged interior of a black spruce and tamarack swamp.

According to a 1993 ecological classification system developed by the Minnesota Department of Natural Resources (DNR), University of Minnesota, and U.S. Forest Service (USFS), this hybrid forest with its commingling of broadleaf and needle-leaf trees is part of a broad climatic zone referred to as the Laurentian Mixed Forest (LMF) Province. Sandwiched between 41 and 48 degrees latitude, the LMF's main range cuts a path from the tallgrass prairie to the Atlantic Ocean, covering northern Minnesota and southern Canada before veering into northern New England. In Minnesota, it covers the northeastern third of the state and also includes the headwaters forest just to the north of the highlands.

This broad classification obscures, however, fine distinctions in the character and ecology of the regions nested within its boundaries. The headwaters forest is composed primarily of black spruce, white spruce, balsam fir, tamarack, and a suite of pine species that includes red, white, and jack pines. These conifers are joined by a deciduous component of boreal hardwoods such as trembling aspen, paper birch, and balsam poplar. The highlands forest contains all of these very same trees. What distinguishes it from its headwaters cousin is a handful of additional species, such as sugar maple, basswood, northern red oak, and yellow birch. In the Great Lakes region, the highlands is part of the northernmost outpost of these species, whose ranges extend at least as far south as Tennessee.

Such "in-betweenness" makes the North Shore Highlands an ecologically interesting place. On dry, shallow soils and bedrock, pine trees such as red pine, white pine, and jack pine take hold. On the moist, richer loams and clay soils, broad-leafed trees such as yellow birch and northern red oak find hospitable accommodations within a flourishing forest of sugar maple. In wet pockets or on the moister lower slopes, you are likely to find a thriving cohort of northern white cedar trees that may have taken root when Europeans first set eyes on Lake Superior some 350 years ago. Because of this diversity, declares Canadian scientist E. C. Pielou, the mixed conifer-hardwood forests of the Great Lakes are "by far the most diversified of any now growing in a once glaciated area."

But as any Minnesota leaf peeper will tell you, the real draw of the North Shore Highlands is its spectacular beauty. This is due in no small measure to the sugar maples that dominate large areas within its reach. Stand on any overlook in the crisp, clean air of autumn, and you can pick out their

Native people are said to have learned to make maple syrup from red squirrels, which puncture sugar maple twigs and wait until the water has evaporated from the sap before eating the concentrated sugars or crystallized candy. Photograph by Bernd Heinrich.

blocks of gleaming gold and brilliant red in the parti-colored canopy of forest that unrolls at your feet.

Forests such as these were treasured long before the advent of botanists and leaf peepers. Take one of the maple forest's most enterprising residents—the red squirrel. During late winter and early spring, Maine biologist Bernd Heinrich has photographed red squirrels puncturing sugar maple twigs with their teeth. But the animals do not immediately lap up the sap that oozes from these tiny wounds. Instead, they wait until the water has evaporated, a process that concentrates the sap into maple syrup or crystallized candy.

Humans may have learned the fundamentals of sap processing from red squirrels, but Ojibwe cosmology credits the trickster Naniboujou with giving mortals the gift of maple sugar. Naniboujou, it is said, once stood under a maple tree when it began to rain sap. He decided to pass the knowledge on, but not before requiring humans to offer prayers and expend energy to procure the much-coveted sweetener.

So important did maple sugar become in the lives of the Ojibwe that nearly every early historical account mentions its production and use. The Minnesota ethnographer Frances Densmore concluded that maple sugar and wild rice were the two most important vegetable foodstuffs among the bands of Ojibwe that she studied in northern Minnesota, Wisconsin, and Ontario in the early twentieth century.

The Minnesota North Shore was no exception. On his 1823 travels to the north country, which he chronicled in the 1850 book *The Shoe and Canoe,* Dr. John Bigsby noted two locations along the shores of Lake Superior that were especially rich in maples, including a stretch from the Pigeon River to present-day Duluth. These "extensive groves of sugar-maple," he wrote, "are highly prized by the Indians." On an 1879 botanical investigation of the shore between Duluth and the Devil Track River, scientist Thomas Roberts noted the same phenomenon. "Sugar bushes [i.e., maple sugar processing sites] are common, and a large amount of maple sugar is

Each spring native Ojibwe family groups established seasonal camps to tap the sap of sugar maple trees. So important was maple sugar in the lives of the traditional Ojibwe that nearly every early historical account mentions its production and use. Neg. 315506, American Museum of Natural History, Library.

made by the Indians," he wrote in the eighth annual report of the *Geological and Natural History Survey of Minnesota*.

The process of maple sugaring is described in detail by Densmore in her 1928 report *How Indians Use Wild Plants for Food, Medicine, and Crafts.* Beginning in mid-March, she relates, groups ranging in size from one to three families typically decamped to familial grounds in the maple sugar forest. The women opened the season by unpacking and repairing utensils stored in lodges that had been erected on the site in previous years. Once preparations were complete, both men and women took up their axes and headed into the woods. They began by making cuts near the base of the tree into which they inserted spikes to funnel sap into birch-bark containers. An average-size group set about nine hundred plugs, with mature trees bearing multiple taps. Larger groups might install up to two thousand. (In the absence of catastrophic disturbance, many of these trees could live to ripe old ages. On a 1960–61 reconnaissance of old-growth mixed-forest stands in the highlands, botanists Edward Flaccus and Lewis Ohmann discovered diagonal scars near the bases of sugar maples on Maple Hill near Grand Marais. They were scribed by Indian people probably more than a century before.)

The thin, watery sap was transported back to camp and then boiled overnight. In some cases, the concentrate was poured into decorative wooden molds, where it hardened in the shape of human figures, flowers, animals, the moon, or the stars. These candies were stored and given away primarily as gifts throughout the year.

To the delight of the children, some of the syrup was emptied into cones made of birch bark or into the hollows of mandibles taken from ducks. The syrup could also be flicked directly onto the snow, where it formed coagulated lumps that could be chewed much like gum.

But the bulk of the syrup was poured into basswood troughs and worked with wooden paddles into a granulated form for use in day-to-day cooking. (The Ojibwe, like the food industry before the advent of plastic containers, valued basswood because it did not contaminate the taste of food with resinous or woody odors.) This granulated sugar served "as the universal and almost only condiment in Indian cookery," observed the nineteenth-century German ethnographer Georg Johann Kohl. Finding salt distasteful, the Ojibwe seasoned fruit, rice, vegetables, and meat dishes with maple sugar, even sweetening fish soups with it. One traveler to northern Wisconsin in 1852 noted that to make the warm waters of the Namekagon River more palatable for drinking, area Indians would add a pinch of maple sugar. Densmore wrote that bitter medicines often were dissolved in maple sugar before they were administered to children.

The Ojibwe palate was attuned to far more complex properties of maple sugar than just its texture and sweetness. Densmore wrote that the "best sugar was made when the early part of the winter had been open, allowing the ground to freeze deeper than usual, this being followed by deep snow. The first run of sap was considered the best. A storm usually followed the first warm weather, and afterwards the sap began to flow again. This sap, however, grained less easily than the first and had a slightly different flavor. Rain produced a change in the taste and a thunderstorm is said to have destroyed the characteristic flavor of the sugar." Indeed, when European trade made cane sugar from the West Indies available, area tribes eschewed it, despite the labor involved in making their own sweetener. When Kohl asked the reason why, they told him that maple sugar "tastes more fragrant—more of the forest."

Sugar maple trees also provided Native people with other prized amenities. Kohl observed how Indians in the Chequamegon Bay region cut burls from maple trees and then hollowed out their rounded forms for use as soup bowls. And when low on tobacco or other smoking materials, Indians utilized the bark of maple trees in their pipes.

Small wonder that the Ojibwe vetoed a government-led attempt in the early nineteenth century to relocate the tribe to the treeless prairie. "The closeness of the relationship between northern lakes Indians and the maple tree," writes anthropologist Charles E. Cleland, "is illustrated in a report by James Schoolcraft of an Ottawa scouting party sent to explore land in Kansas as a possible site for the removal of Indians, under the treaty between the Ottawa and Ojibwa and the United States in 1836. Although the Kansas lands were generally found to be acceptable for agricultural purposes, the scouting party reported that they were 'disappointed at not seeing the sugar tree.' The Ojibwa and Ottawa vigorously resisted removal and never did part company with the 'sugar tree.'"

Back to the Beginning: The Aftermath of the Wisconsin Ice Sheet

Ecologically speaking, it is difficult to pinpoint the time when the relationship between humans and maples first began. If you had stood on an overlook in the highlands as the last glacier was making its final retreat some ten thousand years ago, you would have surveyed a landscape that was nearly as blasted and barren of plant and animal life as the surface of the moon. It would have seemed that nothing could have made the land fecund enough to produce today's bounty of maple sugar. Yet within several thousand years—the blink of an eye by geological reckoning—a diverse forest dominated by maples would grow up from the glacial rubble.

Just how forests were able to reclaim deglaciated land after being routed time and again by glacial advances has been a subject of debate among paleoecologists since the 1930s. (Paleoecologists reconstruct a vegetational history by studying concentrations of microfossils. These include microscopic pollen grains and macrofossils such as seeds, cones, fruits, leaves, needles, twigs, and bark. Knowledge of present-day ecology and distribution of plants provides the context for interpreting these fossil findings. Since the 1950s with the development of certain scientific techniques and the refinement of others, including pollen analysis and radiocarbon dating, scientists have been able to map ancient vegetation with even greater accuracy.)

Despite the huge knowledge gaps that remain, most scientists seem to agree on one point: the formation of the forests that we enjoy today is the result of processes far more haphazard, chaotic, and chancy than previously believed. Given this natural serendipity, the existence of maple forests seems nothing short of miraculous.

To understand just how remarkable, you need to go back to the region's glacial beginnings. In the past two million years, known as the Quaternary period, Earth has undergone about twenty glaciations. During periods of glacial advances in North America, huge walls of ice pressed southward, covering Canada and the northern third of the United States. These long, cool periods lasted between sixty thousand and ninety thousand years. Together, they accounted for 90 percent of the duration of the Quaternary.

This "cold, glaciated epoch," writes the renowned University of Minnesota paleoecologist Margaret Bryan Davis, was "interrupted periodically by catastrophic warm events."

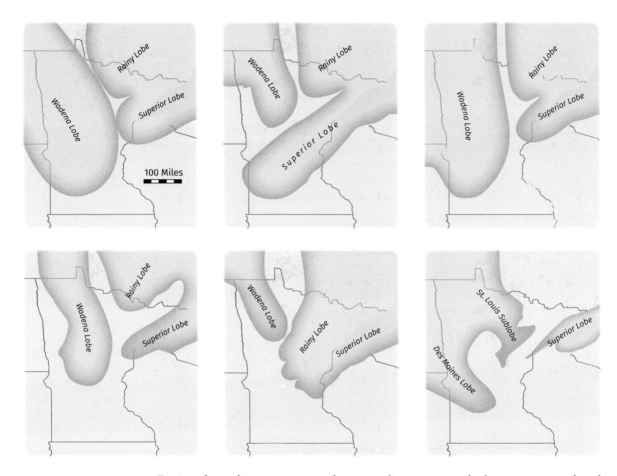

During the early advance of the Wisconsin glacier about 100,000 years ago (upper left), Minnesota was invaded by several distinct lobes that protruded off the main ice sheet. Over time these fingers of ice advanced and retreated multiple times (upper left to lower right), carving complex patterns that remain visible on the region's land and waters to this day.

During these short, comparatively warm climatic intervals, known as interglacial periods, the glaciers largely retreated and exposed a sterilized land. Lasting between ten thousand and twenty thousand years, interglacials allowed plants and animals from refuges south of the ice sheet to migrate northward to reclaim the ice-free terrain. Little is known about the vegetational history of these interglacials since each ice advance largely wiped out evidence of prior colonization.

Much of the evidence that scientists have gathered dates from the most recent advance of the Laurentide ice sheet known as the Wisconsin glaciation. About 30,000 BP the ice sheet began to expand, reaching its greatest extent, known as the Last Glacial Maximum (LGM), about 22,000 BP. (BP stands for "before present.") At its farthest extension, the Wisconsin glaciation ice sheet, whose thickness grew to two miles in some places, covered half of North America from the Canadian Rockies east to the Atlantic Ocean and from the Arctic south to central Illinois. During this time, nearly all of Minnesota (with the exception of its southeastern and southwestern corners) was overrun by ice.

Around 22,000 BP, the ice loosened its grip and began to recede along most of its southern margin. The melting accelerated after 15,000 BP. Between 11,000 and 15,000 BP (a period known as the late glacial), average temperatures soared by 9 to 14 degrees Fahrenheit, with a spike in the rate of climatic warming occurring about 11,000 BP. By

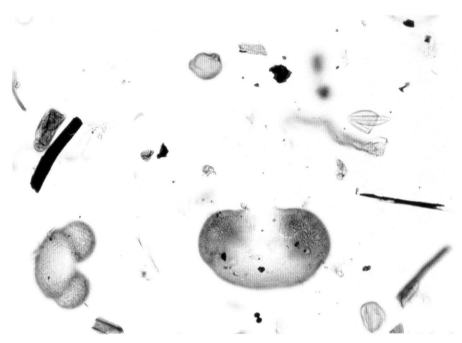

Paleoecologists reconstruct a vegetational history of the past by studying concentrations of microfossils. Here, a microscopic view of the pollen grains of birch, cedar, pine, spruce, and oak. The long narrow black particles *(left and right center)* are fragments of charred wood. Photograph by Edward J. Cushing.

about 11,500 BP, the great ice margin that once penetrated as far south as central Illinois had receded from the North Shore Highlands, retreating into the northeast corner of what is now the Lake Superior basin. The northern tier of the United States had finally emerged from under the main body of the Wisconsin ice sheet. The land was once again ready to be colonized by living organisms.

How this process unfolded sparked a now-classic debate among biogeographers. What intrigued scientists was this: where did the trees that are now common in deglaciated regions find sanctuary during periods of glacial advances? How and when did they make their way northward?

In the 1950s, E. Lucy Braun, a plant ecologist at the University of Cincinnati, Ohio, offered a series of hypotheses about postglacial plant migrations. Put simply, she argued that the glaciers exerted negligible effects on the forests south of the ice sheet. According to Braun, not only did these forests survive the Wisconsin glacier roughly in their present-day locations, but they also persisted with their species composition intact. Indeed, Braun contended that some forests, most notably the deciduous woodlands of the southern Appalachians, had not changed appreciably since the mid-Tertiary period some thirty million years ago. As proof of their antiquity, she pointed to their extraordinary evolutionary flowering: today the region hosts the most species-rich forests on the entire North American continent. The lowlands of the southeastern United States, for example, are populated by more than forty species of oak alone. Indeed, Braun and others have suggested that mountain chains trending north to south, such as the

Appalachians, served as a route for plants to escape the denuding ice sheets and, once they had melted, as a highway for recolonization.

According to Braun's hypotheses, these distant forests were not the only ones to have eluded the influence of the Wisconsin ice sheet. Undisturbed too were forests that lay relatively close to the ice margin, she argued. An analysis of a late-glacial site in Minneapolis, for example, revealed that a boreal forest of conifers and paper birch grew in a wet lowland within 150 miles of the active ice front. Waiting in the wings, so to speak, in nearby dry uplands grew the plant species of the deciduous forest that would soon quickly succeed it. Noting that the temperate forest appeared to follow closely on the heels of the boreal forest, Braun postulated that "forest displacement even at latitude 45 degrees N. [the latitude of Minneapolis] was not extreme."

Based on such evidence, Braun proposed that as the forests adjusted their ranges slightly northward, tree species did not migrate individually but as a group. Then, as now, she argued, the major forest biomes of North America fell into a geographical pattern—a kind of parfait of vegetational zones—from the boreal forest in the far north to the mixed-forest belt followed by mesophytic deciduous forests to the south. These floristic provinces and their associated complement of species, she suggested, moved in lockstep to the south when the glaciers advanced and largely maintained rank in their northward migrations after the ice sheets melted.

But armed with new techniques for detecting and dating ancient plant remains, plant geographers began to uncover evidence that raised questions about Braun's construction of postglacial events. In 1949 the plant geographer Edward S. Deevey offered what Braun called "an unusual viewpoint, one very different from that held by the majority of plant geographers." Deevey argued that the coming and going of the glaciers resulted in radical—and sometimes abrupt—displacements of the forests south of the ice sheet. Furthermore, he countered that trees recolonized deglaciated terrain in a far more individualistic and unruly manner than the one described by Braun. In scenarios put forth by Deevey and others, seeds took root whenever they happened to land in suitable microhabitats. Plants moved around on the landscape less like a disciplined army and more like players in a competitive game of musical chairs fed by a constant supply of new contestants. Some species lost their place while others scrambled to fill the void, only to lose dominance in time to yet another newcomer. The result was a continually changing ensemble of players that formed associations that would have appeared strange to modern eyes. In this vegetative free-for-all, success was as much a matter of luck as of fitness.

Among the evidence used to support such hypotheses were cores taken from bogs as far south as Louisiana and Texas that showed that spruce and pine trees once grew alongside warm-weather species during the height of the Wisconsin glaciation. Braun conceded that some northern species did selectively infiltrate southern forests. But she postulated that in the mild climate of the time, such species were a minor component of the overall forest. She proposed that boreal species were restricted to microhabitats such as frost pockets or along the shorelines of rivers and lakes where cold waters produced localized fogs that lowered summer temperatures.

So which argument eventually settled the great paleobotanical debate? Since the 1960s, scientists have expanded the database of fossil samples taken from the sediments of lakes and wetlands and fine-tuned techniques for their analysis. In the 1980s, for example, a large interdisciplinary consortium of scientists in what is known as the Cooperative Holocene Mapping Project launched an ambitious effort to assemble a more definitive chronology of vegetational change during the height of the Wisconsin glaciation and in its aftermath. As a result, scientists now have a far better understanding of how and when plants moved across the landscape.

Paleoecologist Davis, a member of the mapping team, reviewed evidence from an extensive sampling of radiocarbon-dated sites, including layers of ancient pollen from lake bottoms in the eastern United States. After weighing the findings, she ruled in favor of Deevey's point of view. According to Davis, the mountains of southern Appalachia during the Wisconsin glaciation were dominated not by an ancient assemblage of broad-leafed trees but by a far younger forest of conifers, most notably jack pine and red pine. Less abundant, but still present, was spruce. The nearest modern location in which pine and spruce coexist is Maine, observed Davis's fellow paleoecologist William A. Watts. From this cohabitation, Watts inferred that the climate of the southern Appalachians during the Wisconsin maximum was characterized by cold winters and a short growing season. Although climatic conditions were not as severe as those found in the boreal reaches of eastern North America today, they were nonetheless far too frigid to have supported the lush forests that Braun had envisioned. Perhaps more important, conditions were also too dry. According to estimates by some paleoclimatologists, precipitation rates plummeted by as much as 50 percent in some regions during this time.

New evidence showed that the extraordinarily diverse complement of tree flora in the southern Appalachians developed only within the past fifteen thousand years. Paleoecologists suggest that the region's proximity to glacial refugia that harbored many different species may account for the rich concentration of plants that have found a home there in modern times.

While pine remained the most abundant tree in the forests of the unglaciated southeastern United States during the LGM and may have extended as far south as the Gulf Coast of Florida, spruce ruled the forests of the continental interior where temperate deciduous trees now dominate. Indeed, spruce formed the major component of forests of the Mississippi valley, from southern Minnesota to the coastal boot of Louisiana, and grew as far south as Texas.

So where did other conifers, not to mention the deciduous species that cover much of the eastern United States today, find refuge during the height of the Wisconsin glaciation? According to Davis, the paleoecological data show "virtual elimination of deciduous forest" during this time. Far from being widespread and common as they are today, she observes, deciduous trees "endured much of the Pleistocene epoch [the most recent ice age] as rare and endangered species, in small areas of favorable climate. Most of these areas have not been identified with certainty." Paleoecologists suspect that boreal and cool-temperate conifers and hardwoods, including such common mixed-forest species as balsam fir, hemlock, white pine, paper birch, sugar maple, and basswood, persisted in isolated, scattered groves on south-facing slopes or in the rich soils of protected habitats among the more extensive forests of jack pine and red pine in the southern Appalachians and central United States.

The dramatic climate change that caused the rapid wasting of the glacier also rearranged the position of vegetation across the landscape. Contrary to Braun's assertion that such changes amounted to little more than a geographical tweaking, the vegetation of eastern North America was literally turned on its head. About 15,000 BP, spruce—and the species associated with the forests that it dominated—began to move northward. As the climate warmed, boreal forests that once flourished in Kansas ultimately shifted their main range into central and northern Canada. In the southern United States, formerly vast forests of spruce and pine were replaced by deciduous trees. Now-common species that once had narrowly escaped extinction, such as oak and maple, broke free from their scattered outposts to dominate huge tracts of forest.

Much of their success depended upon an ability to rapidly disperse. To reach the northern edge of its current range (a line from New Brunswick to southern Manitoba), sugar maple, for example, traversed some 1,180 miles. Paleoecologists estimate that maples accomplished this migratory feat by expanding their range at a rate of 656 feet per year.

Such sparse taxa gained ground while far more abundant ones underwent extinction. Take, for example, the species of spruce *Picea critchfieldii*. It occurred in the lower Mississippi valley around the LGM and disappeared from its former range around 15,000 BP. How many other taxa suffered the same fate is anyone's guess since plant remains, particularly pollen, often can be identified only to genus and sometimes to an even less precise family level. The trees that we see in the deglaciated portions of North America today, Davis says, evolved during times of biotic instability and are adapted to change. Indeed, species survived to the present day by becoming "successful invaders." As for those that "have been unsuccessful at moving or in becoming abundant," she adds, "presumably most such species have become extinct."

In the wake of the retreating glacier, each species traveled not only north to south but also east to west at its own speed and departure times. Each took different migratory pathways, making their way across the landscape not as a group but as individuals that were able to take hold and grow depending on their dispersal capabilities and on the suitability of the physical and biotic characteristics of the site in which seeds landed. As a result, paleoecologists have searched in vain for ancient analogues to the modern forest. The spruce forest, for example, that covered the continental interior of the United States in late-glacial times contained modest amounts of oak pollen and other temperate deciduous trees. On the other hand, only minute quantities of pine pollen have been detected, suggesting that pine had little or no presence in the forest despite the fact that jack pine is abundant in today's boreal forest.

From the painstaking task of counting grains of pollen under a microscope and sifting through sediments for bits of ancient plant fragments, paleoecologists have constructed a picture of the past that is astounding in its implications: By the very nature of their haphazard formation, the forests that we see today are unique in space and time, the product of serendipitous natural forces. The vegetational deck has been reshuffled not just once during the most recent glaciation but multiple times throughout the Quaternary. According to Davis, "The flora of the eastern United States is a fortuitous grab bag of species, the survivors of 16 to 18 glacial-interglacial cycles." Each time the ranges of vegetation contracted and expanded, Davis points out, "it can be supposed that many genetic lineages have been lost through extinction at times of low population size or through failure to expand quickly enough to exploit favorable habitats as they became available."

The next glaciation is likely to reset the clock and initiate yet another new game of floristic musical chairs. "During the Quaternary Period," Davis observes, "the forests with which we are familiar seldom maintained a constant species composition for more than 2,000 or 3,000 years at a time." What we see today, she says, "are chance combinations of species without an evolutionary history."

After the Ice: The Forest Returns to Minnesota

Thanks to the work of Davis and the equally prodigious investigations of her University of Minnesota colleagues, most notably Herbert E. Wright, the return of life to deglaciated terrain in Minnesota has received more careful study than almost any other region in North America.

Researchers have found that the earliest botanical community to invade Minnesota's newly ice-free terrain was not a forest but a tundra-like assemblage of low-growing plants. At the height of the Wisconsin glaciation in the eastern United States, a relatively narrow and irregularly shaped belt of tundra snaked its way between the ice and the northern front of the forest in Minnesota and Wisconsin. In central New York State and Pennsylvania, the tundra widened to more than sixty-two miles. It also persisted at higher elevations in the Appalachian Mountains, a range that extended hundreds of miles south of the forest's northern boundary.

That tundra plants should be the first to encroach on the deglaciated land is not surprising: they grew on or near the ice and therefore provided a ready seed source. Once established, their persistence was remarkable. The tundra held out—in some places for thousands of years—against the spruce-dominated forest, which was one of the most aggressive and widespread colonizers in postglacial times. For example, scientists studying the sedimentary record of Wolf Creek in central Minnesota, which may reach back as far as 20,000 BP, found that a tundra flora of dwarf shrubs and herbs grew on the site for several thousand years before trees were able to invade their territory.

Poised along the southern flank of the tundra's wide-open spaces, what prevented forests from quickly overtaking this botanical frontier? Unstable conditions created by the ice sheet, for starters. Even as it receded, the glacier did not mount a clean, orderly retreat but continued to make periodic incursions into the surrounding terrain. During four different periods between 20,500 and 12,000 BP, a great tongue of ice known as the Superior Lobe oozed into the lowland that today holds the waters of Lake Superior. As the ice advanced, it rose higher and higher until it overtopped the basin's rim and flowed out onto the land that lay behind the ridge to the north and west. When the climate warmed, the ice began the stop-and-start process of melting, pausing for varying lengths of time on its retreat back into the basin. "If a time-lapse moving picture could have been taken from a stationary satellite, the great ice sheets would have looked like active amoebas," Pielou writes, "with undulating outlines, wobbling unsteadily as they contracted to nothing. The advance of the biosphere, as more and more ice-free land became available, must have been correspondingly irregular. It was not a continuous advance; there were innumerable delays, retreats, and readvances. Conditions for life close to the ice margins were sometimes harsh and sometimes benign but rarely constant for long."

Often these glacial incursions were abrupt and violent. As Pielou points out, the buildup of a dome of ice at various points in the glacier could create enough pressure to cause the leading edges of the glacier, located thousands of miles away, to suddenly surge. Like a blunt frontal plow, such outward flows crushed and buried whole forests.

Even forests that lay beyond the direct reach of the ice were not immune to its decimating effects. In 1976 while constructing a tailings basin about ten miles southwest of Marquette, Michigan, workers for the Cleveland Cliffs Iron Company exposed portions of a white spruce forest that had been buried for some ten thousand years. In a reconstruction of events, the scientists who excavated the site theorize that during a brief readvance, the Wisconsin ice sheet halted in its tracks a few miles north of the site. A period of climatic warming caused the ice to rapidly melt, sending water and debris pouring into the forest in a series of floods. Drowned and then entombed in silt and fine sand were hundreds of trees, some of them up to 150 years of age. Many of them died standing with their bark and twigs still attached. And on the forest floor were the perfectly preserved remains of needles, cones, and mosses. The intact conditions suggested to scientists that the catastrophic burial occurred quickly. Now known as the Gribben Forest, the ancient site represents one of only a handful of fossil forests that have been uncovered intact. Most were simply crushed and scraped away by the ice sheet.

Even the mere presence of the glacier would profoundly affect the newly vacated landscape. At the very least, the proximity of the ice would dramatically lower temperatures. The fact that paleoecologists have found evidence of plants that now grow in alpine, arctic, or subarctic regions suggests that temperatures in late-glacial times were colder than at present. In studies of vegetative zones in present-day northern Canada, for example, scientists discovered that a mean annual temperature of 24 degrees Fahrenheit or below separates continuous tundra from the coniferous forest. They hypothesize that winds streaming off the ice sheet resulted in similarly cold conditions during postglacial times, which either killed trees outright, kept the ground frozen so that they could not take root, or inhibited reproduction by killing tender blossoms or preventing the ripening of seeds.

Within this periglacial zone, the earth was in a constant state of upheaval. Permafrost (ground that was frozen year-round) lay adjacent to the ice margin. The upper layers of the permafrost were subject to dramatic changes throughout each seasonal cycle. During summer the surface of the earth would thaw, forming great pools of standing water over the underlying layer of impermeable frozen soil. On slopes the melting caused water-saturated sediments to slowly slump downhill, a process known as solifluction. With the onset of winter, frost heaving, like tiny invisible plows, churned up the earth. These natural processes, along with the massive erosion from glacial meltwaters, continually prevented and disrupted the development of plant communities, such as forests, that required more stable conditions.

Even when the glacier finally disappeared from a given area, it would continue to exert a destabilizing influence for thousands of years to come. Littering the land, for example, were great masses of stranded ice, known as dead ice, that once were part of the main body of the glacier. As the top layer of dead ice melted away, it left behind a mantle of sediments and rocky debris. So effective was this layer as insulation that, in some cases, millennia would pass before the buried portion of these blocks disintegrated. One of the best studied examples of this phenomenon lies in central North Dakota in the Missouri Couteau. The mass of dead ice that produced this distinctive prairie ridge covered an area of more than thirty-seven miles and measured some 328 feet thick at the time it was formed around 14,000 BP. It was so well protected from the warming rays of the sun that the buried ice persisted for three thousand years before melting away.

The thick mantle of sediments covering such masses of dead ice was readily colonized by plants and, in time, supported fledgling forests. But their tenure would be relatively brief. Rainwater seeped into cracks and began to melt portions of the underground ice until they were riddled with caverns. The ground's surface began to subside. "Patches of forest sank into hollows, which soon filled with water, drowning the terrestrial plants," Pielou explains. "Much of the forest was 'drunken forest,' with trees leaning in every direction owing to the instability of the ground. On a small scale, the topography was continually changing. Sometimes the slumping exposed a cliff of ice, which would begin to melt as soon as it was exposed to sunlight and air. Innumerable little superglacial puddles and pools were formed because of the subsidences. They were icy cold and probably, like glacial lakes today, milky with rock flour, as environments for life, they were unpromising."

On first appearance, the sterile new habitats seemed wholly unsuited to supporting vascular plants and animals. But while trees may have had a difficult time taking root and perpetuating themselves, there were other plants, many of them found today on the tundra, that not only tolerated but thrived in extreme and volatile environments. The pollen record shows a periglacial flora rich in sedges, wormwood, ragweed, and grasses. Macrofossils reveal the presence of tundra plants, including mountain avens, arctic mosses, alpine bilberry, and dwarf willow, which gained a footing in the frost-churned permafrost, gravel river bars, and slopes of loose rock. The small seeds of these arctic plants were perfectly engineered to be whisked along the glazed surface of the ice. They may have been carried to present-day Minnesota on the prevailing westerly winds. Or they may have hitched a ride on more distinct "wind highways," such as those that were discovered in 2004 to carry seeds and bits of plant parts from mosses, liverworts, and lichens for thousands of miles in the islands around Antarctica. Without the presence of trees and other tall vegetation to act as seed traps, these species were widely disseminated. Once they took root, they possessed numerous coping strategies for life in a periglacial environment. By growing low to the ground, they took advantage of protected pockets and the heat given off by sun-warmed earth. And since wind speeds normally are slower down near the ground, such plants grew in a less turbulent environment, avoiding the strong, incessant winds that dried the plant tissues of taller plants, abraded them with blowing ice and sand, and in some cases toppled them.

These plants adapted to extreme conditions by evolving successful strategies for reproduction or procuring nutrients in the nitrogen-deficient conditions that, according to arctic plant researchers, are a defining characteristic of nearly all arctic soils. Take mountain avens, for example, which once was common in the tundra-like conditions of postglacial times in Minnesota. Like plants such as legumes and alders, mountain avens possesses bacterial nodules on its roots that are capable of capturing abundant nitrogen from the air and converting it into an organic form that plants can use. This tiny arctic dweller, in other words, developed its own on-site fertilizer factory.

But like other tundra plants, it created the conditions that led to its own demise. Mountain avens is a pioneer plant that favors newly exposed soils. With its long taproot, it began to stabilize and enrich the substrates it colonized. As these and other pioneer plants grew in density and numbers, they helped to build organic soils by capturing blowing dust and sediments and then leaving behind their decayed remains, thus creating more hospitable sites for the establishment of taller vegetation. Trees and shrubs began to take root in scattered clumps, forming shelterbelts from the wind. In time, a forest was born.

The Land and the Lake: Superior's Influence on the Highlands' Forest

Researchers have studied just such a succession of species in terrain newly exposed by retreating glaciers in Glacier Bay, Alaska. Scientists found that in modern times, as in the postglacial period, mountain avens was the first to invade deglaciated surfaces. Within three decades, however, the shade-intolerant tundra herb had lost its competitive edge to willows and alders. Eventually, these too succumbed, and a forest dominated by spruce trees emerged.

Paleoecologists have constructed a similar sequence in Minnesota following the glacier's retreat. A rapidly warming climate melted the permafrost and allowed a forest to invade the tundra. A few species of remnant arctic plants can still be found on cliff and talus habitats within the forest as well as along the shoreline of Lake Superior and its islands. But the main ranges for these species now lie hundreds of miles to the north in arctic and subarctic North America. (Mountain avens, which has disappeared from Minnesota, can still be found growing along the Canadian North Shore.)

Displacing the tundra in Minnesota were forests of spruce. Once these forests gained a secure toehold, their takeover was swift. For example, the pollen record shows that by 15,000 BP spruce trees had arrived at the Wolf Creek site in central Minnesota. By 13,000 BP, the pioneer community of herbs and shrubs in central and southern Minnesota already had given way to a boreal forest dominated by spruce.

No species colonized as wide an area during the LGM and late-glacial times as spruce. Growing from the Great Plains to New England, spruce populations reached their greatest density—and purity—in the forests of the Midwest south of the ice sheet. In the continental interior, spruce formed the basis for what could be called, in human terms, the Roman Empire of late-glacial forests, covering a latitudinal distance of approximately twelve degrees, a geographical spread that is roughly equivalent to the extent of its range today. Aggressive colonizers, spruce trees expanded their northward range onto deglaciated land at a rate of 820 feet per year. So successful was the boreal forest that it formed a vegetative swath that stretched from the Rocky Mountains to Nova Scotia. (Today, South Dakota's Black Hills are host to outlying stands of white spruce that likely are remnants of a much wider distribution during the Pleistocene.)

Spruce trees enjoyed several advantages over their competitors. They grew adjacent to and even on top of the receding ice sheet and therefore supplied an immediate stock of lightweight seeds that could be blown long distances by the wind. Black spruce, in particular, was tolerant of many different kinds of habitats, including sun-drenched, nutrient-poor conditions. And it has extremely wide climatic tolerances, growing today from the edge of the Manitoba prairie to the northern tree line, a distance of some seven hundred miles.

With the exception of the proglacial Great Lakes, spruce had few geographical obstacles to surmount on its northward migration, unlike trees such as jack pine.

According to Wright, the long period of cooling (70,000–20,000 BP) that caused the Laurentide ice sheet to expand into much of the northern United States wiped out pines and deciduous trees in the Great Lakes region. During the LGM, jack pine was largely eliminated from the continental interior south of the ice sheet and restricted to a sanctuary in the southern Appalachian Mountains. Despite its ability to expand its range faster than any other tree species in postglacial times (the species moved northward at an average rate of one-quarter mile annually), jack pine did not arrive in Minnesota until about 11,000 BP, some three thousand years after it began moving from its Appalachian sanctuary. Paleoecologists hypothesize that its northward progress was hindered by a series of mountainous barriers, and its westward migration into the Great Lakes region was blocked by glacial ice and the large proglacial lakes that formed as the ice retreated.

Scientists are quick to point out that the spruce-dominated forest in late-glacial times bore little resemblance to the boreal forest of today. High pollen counts of nonarboreal pollen (NAP; the pollen of plants other than trees and shrubs) suggest that spruce trees initially grew in tandem with tundra plants. Scientists speculate that the trees grew in open parklands with trees clumped in hospitable microhabitats. As the climate ameliorated, the density of spruce and other species slowly increased, creating inhospitable conditions for shade-intolerant tundra plants.

In the continental reshuffling of vegetation following the retreat of the ice sheet, the spruce forest formed temporary associations with other tree species unlike any found in the boreal forest today. The pioneer forest not only contained a large percentage of tamarack but also included small numbers of temperate trees. Indeed, 10 to 25 percent of the total pollen count in many parts of the late-glacial boreal forest was deposited by hardwoods such as oak, black ash, and ironwood. Other trees may also have been present, including elm, box elder, balsam poplar, and juniper.

Wright points to the great diversity of microhabitats in the postglacial landscape—with their varying degrees of exposure, soil moisture, and stable substrates—as the reasons for this strange coexistence of boreal and temperate species in the late-glacial forests of southern Minnesota. Boreal species such as spruce and tamarack, he hypothesizes, grew on stagnant ice as well as beyond the ice margin. Temperate species, on the other hand, were probably found growing on well-drained uplands with southern exposure. The high representation of wormwood pollen in the palynological record suggests, however, that the landscape was still in a period of great flux. Wormwood species colonize well-drained, bare soils. From their presence, scientists infer that even as the pioneer forest was taking root, much of the land was still open, as is the case today at the forest-tundra border.

The spread of the boreal forest in Minnesota was time transgressive; that is, the same species arrived in different places at different times. Indeed, spruce made its appearance in southern Minnesota some three thousand years before it reached northeastern Minnesota. In several sites around Duluth, for example, tundra communities lingered until about 12,000 BP, even though the northern front of the boreal forest lay only fifty miles to the south at the time. But here too the standoff between trees and tundra did not last long. Sediment samples taken from Weber Lake in Lake County show that by 11,000 BP the boreal forest had begun to displace the tundra.

The surging global temperatures that helped bring about the boreal forest's rise to continental prominence would, however, also contribute to its downfall. The forest that had displaced tundra around Duluth reigned for only a few centuries before disappearing in what paleoecologist Watts says was nothing less than a "population collapse." Wright points to the palynological record in which the percentage of spruce pollen drops from 50 to 5 percent in a mere twelve inches of sediment. To scientists, this indicates the virtual disappearance of spruce in only a few centuries. In some places, the spruce forest may have come and gone in a matter of only fifty years, a change that would have taken place in the course of a lifetime for some of the Paleo-Indians who were there to witness it. In fact, so abrupt is the drop in spruce pollen after 11,000 BP in the palynological record from Manitoba to New England that it is used as one of the indicators to mark the official end of the Pleistocene and the boundary between the late-glacial and postglacial period.

In northern Minnesota spruce populations continued to decline so that by 6,000 BP, paleoecologists say, spruce was likely eliminated from the region. Scientists hypothesize that the spruce forest may have been routed to more northerly latitudes by a longer growing season and a climate that was too warm and dry for spruce regeneration. These conditions likely favored instead the existing components of the forest that could tolerate rising temperatures, such as balsam fir, speckled alder, and white birch, as well as new colonizers such as jack pine, which had arrived in eastern Minnesota about 11,000 BP from the eastern United States and quickly dominated the forest by filling the gaps left by spruce. By this time sugar maple too had made an appearance in northeastern Minnesota. Also present, to a lesser extent, were other trees now common in today's northern Great Lakes forest, such as trembling aspen and black ash. They formed the template of species from which the modern Great Lakes forest ultimately evolved.

The Holocene opened with temperatures roughly similar to today. Then a warming period began that peaked between 8,000 and 5,000 BP. As the climate continued to warm and grow drier, new species were added to this basic mix, not the least of which was white pine *(Pinus strobus),* which expanded its range to the west from the eastern highlands of the Appalachians and adjoining coastal plain at a rate of 984 to 1,641 feet per year. The retreat of the glacier into Canada opened a pathway north of the Great Lakes for *P. strobus* to enter Minnesota from the east about 7,000 BP.

At 8,000 BP the distributions of yellow birch, beech, hemlock, and sugar maple first coincided. After 6,000 BP the composition of the mixed forest began to solidify, and many species reached their northernmost limits during this time. Some species began to increase in abundance, including sugar maple and yellow birch.

During the past four thousand years (the very recent spike in global warming excepted), the climate swung into a cooler, wetter phase. Scientists base this supposition on several important changes: the westward incursion of forests into former prairie lands, higher lake levels, an expansion of bogs, and increases in the populations and ranges of trees that favor mesic environments, such as yellow birch, sugar maple, and hemlock. By 3,000 BP spruce had moved back into northeastern Minnesota. "Only temporarily at the time of the thermal maximum were conditions suitable for an invasion by the hardwoods and some of the prairie species," write botanists Fred K. Butters and Ernst C. Abbe. "These enjoyed but a brief period of expansion and have been pinched off into very limited areas by the southward surge of the coniferous forest during the current period of mild refrigeration." And so, in the chaos of flux and chance, the modern mosaic of the Laurentian mixed forest was born.

Here to Stay

The swing to a cooler, wetter climate expanded the ranges of some plant communities while "pinching off" others, including the maple-dominated mixed forest. But in the highlands, remnants held their ground. Along the Minnesota North Shore, the maple-dominated mixed forest in Euro-American presettlement times (in this region, the period prior to the 1850s) grew as patches of up to thousands of hectares within two different matrices, much like pieces within a larger puzzle. Southwestward of Illgen City in Lake County, the forest was embedded primarily in a matrix of Great Lakes pine forest dominated by white pine, red pine, aspen, and birch. Northeastward of Illgen City a mix of aspen, birch, white spruce, white cedar, balsam fir, and white pine grew around maple-dominated patches.

But how could species such as sugar maple, whose main range lies in the Big Woods of southern Minnesota, persist in these cooler northern latitudes? After all, writes forester Eric A. Bourdo, "It is seldom realized that the Arrowhead Country of Minnesota and the northern shore of the Upper Peninsula of Michigan are closer to the true arctic (not the Arctic Circle) than any other part of the United States except Alaska."

Maples owe their existence in the highlands as much to their species' adaptive life strategies as to the laws of physics and accidents of geography, not the least of which is proximity to Lake Superior. Among the most powerful influences on the maple-dominated forest is the pronounced temperature differential between Lake Superior and the surrounding land. Forever out of sync, land and water temperatures are engaged in a perpetual game of catch-up—for several reasons. For one thing, water is capable of storing more heat energy than land. At the same time, its rate of heat gain and release is far slower. Like a deep inhalation, the lake takes in heat, warming from an average low of 32.7 degrees Fahrenheit at the beginning of April to an average maximum temperature of 35.2 degrees Fahrenheit at the beginning of October. During the cold season, the lake exhales, slowly shedding its thermal burden.

Because of its large size, Superior governs the climatic regime of the adjoining shore. So pronounced is the lake's effect that coastal residents and their inland neighbors to the north often seem to be living in separate universes. During the daytime in the summer, the predominant surface winds blow from the lake to the land, cooling the coast such that temperatures routinely register 10 degrees Fahrenheit lower than in the interior reaches of the watershed. In July, it is not unusual for temperatures in downtown Grand Marais to hover around 70 degrees Fahrenheit while those on Gunflint Lake soar to 80 degrees Fahrenheit or higher. In Duluth, meteorologists have recorded shoreline temperatures as much as 30 degrees Fahrenheit cooler than those at inland stations around Duluth International Airport.

On wind-still days and nights, the warmer air of the land is wafted to the cold waters of the lake, creating nearshore fogs. Shore dwellers frequently wake up to a bone-chilling fog bank that causes them to reach for their fleece, while hikers in the headwaters' forests are stripping down to T-shirts and shorts in hot, dry, and sunny conditions.

In winter, Superior's effects on coastal conditions are just the opposite. Once the lake is heated up, it takes a long time to cool down. This results in mean winter water temperatures that are 30 degrees Fahrenheit higher than the surrounding land. Like a sun-warmed brick wall that continues to radiate heat long into the night, the lake emits heat energy as it cools down, keeping the adjacent land warmer than inland locations.

As a result, winter temperatures along the lakeshore typically register at least 10 degrees Fahrenheit higher than those farther inland.

Superior's tempering of hot and cold extremes is good news for many species since the moderated temperatures translate into a longer growing season. Gardeners in Grand Marais, for example, enjoy an average of 138 days between the first killing frosts of autumn and the last killing frosts of spring. Inland, this frost-free period drops to only 100 days at best. Such climatic mediation is especially important for sugar maples, which grow at the northern edge of their range in the North Shore Highlands. Maples are especially vulnerable in the spring when the trees undergo an early flowering before their leaves emerge. A late frost can blast the tree's tender flowers and eliminate seed production for the year.

Ancient Highs and Lows: The Highlands' Geology

Species such as sugar maples benefit also from the highlands' rugged topography. The dramatic landscape is defined by a more or less continuous ridge of ancient Precambrian bedrock that roughly parallels the Superior shoreline from Duluth to the Canadian border. The ridge first took shape during the Midcontinent Rift period, about 1.1 billion years ago, when great waves of molten rock poured out of the earth. The movement of the glaciers plowed away surface deposits and the softer rock, exposing a series of

The elevational changes found in the North Shore Highlands, shown here in the hills of the Sawtooth Mountains, allow warmer and snowier microclimates to form on the uplands, where species such as sugar maples can find protection from temperature extremes. Photograph by Beau Liddell, Images by Beaulin.

high bedrock peaks and knobs. Indeed, the greatest elevation change in Minnesota can be found near Finland, just west of the boundary between Lake and Cook Counties. Towering above Lake Superior, which lies 602 feet above sea level, are the tallest of the highlands' hills, with summits of 1,950 feet.

Some of the magma sandwiched its way into and between earlier eruptions, forming what geologists call intrusive rocks. Along the North Shore, these incursions of molten rock solidified as gabbro or diabase. At times, the intrusive flows rafted up great blocks that had hardened deep underground, including an especially erosion-resistant variety known as anorthosite. As the glaciers pulverized the softer rock surrounding it, they exposed vast outcrops of this light, coarse-grained mineral. Today, some of the ridge's most prominent features are made up of anorthosite, including Carlton Peak.

Many of the promontories of the bedrock ridge are not only steep and rugged but also visually distinctive. Among the most well known is a lineup of jagged peaks to the southwest of Grand Marais. These prominent summits are commonly referred to as the Sawtooth Mountains. (In the state's ecological land-classification system, the Sawtooth Mountains officially include the portion of the ridge that contains the highest concentration of outcropping bedrock. This area roughly stretches from Beaver Bay to the Temperance River.)

To the north and west of the Sawtooths lie areas that share the same bedrock foun-

dation but contain fewer exposed outcrops. Here, the glaciers molded till (unsorted glacial deposits that include everything from fine sediments and cobbles to boulders) into long, parallel ridges known as the Highland Flutes. These features form a linear ridge-and-valley system that trends in a northwest to southeast orientation marking the direction of ice flow out of the lake basin.

The glaciers would sculpt other topographic extremes in the highlands as well. During its retreat from the basin, the Superior Lobe, for example, shed some of its burden of sediments and rocky debris, creating landforms known as moraines. When the ice retreated uniformly and without pauses, it deposited broad, gently undulating blankets of till known as ground moraines. Varying in thickness of up to hundreds of miles, these moraines filled in depressions and evened out the landscape.

At times, however, the retreating glacier would come to a standstill, heaping loads of till into long ridges that lay parallel to the ice front. The size of these features, known as end moraines, depended on both the magnitude of the glacier and the duration of its pause.

In the North Shore Highlands, ground moraines cover bedrock hills and ridges to varying depths. By far the most prominent moraine—and the largest example of an end moraine—is the Highland Moraine, which lies just behind the bedrock ridge some six to fourteen miles from the lake edge. It parallels the Superior shoreline from the east side of Duluth northeastward for more than sixty miles and measures between four and five miles wide. Nestled in its large, complex mosaic of rolling hills and ridges are moderately sized lakes and ponds that are drained by numerous small rivers.

The Highland Moraine was created during one of the readvances of the Superior Lobe, a period known as the Automba phase (whose maximum occurred approximately 17,000 BP). During this time, the ice reached an extraordinary thickness as it inched out of the basin and climbed up the surrounding bedrock hills, eventually spilling over the top. Measuring from the crest of the highlands north of Silver Bay to the bedrock bottom of the trough that lies just offshore, Minnesota paleoecologist Wright calculates that during the Automba phase the Superior Lobe in this area was probably more than one-half mile thick.

The varied terrain provides particularly hospitable circumstances for the maple-dominated mixed forest, including soil conditions that are conducive to its establishment. In the aftermath of glacial retreat, hillsides served as windbreaks that trapped blowing dust and other particles. Over time soils accumulated to greater depths on moraines than on flatter land. These windblown clay and silt particles, which became incorporated into the moraine's coarser materials, also increased the moisture-retaining capacity of the soil, unlike in areas dominated by outwash deposits of sand and gravel.

In time, pioneer species colonized the moraines, the decay of plant matter further improving the quality of their soils. As the climate grew warmer and wetter, species that thrived on deep, moist soils, such as maples, yellow birch, and basswood, moved into the morainal terrain.

The elevational changes found in both the bedrock ridge and Highland Moraine also influence climate and precipitation patterns in ways that are favorable to the perpetuation of the maple-dominated mixed forest. Because cold air is denser—and therefore heavier—it drains down their long slopes, allowing warmer microclimates to form on the uplands. In the North Shore Highlands, maples occupy these more clement sites and are seldom found growing in low-lying valleys, where frost pockets are common.

The highlands' sugar maple forests (visible in this satellite image as large golden patches on the Sawtooth Mountains near Tofte and Lutsen) favor the morainal hills and the south-facing slopes of bedrock ridges. The deeper snows and milder temperatures near the lake allow the maple forests to persist at a more northerly latitude than elsewhere in the region. Courtesy of Minnesota Department of Natural Resources, Forestry Resource Assessment.

When the cold season does arrive, the rugged topography helps to ensure that the maple-dominated mixed forest is protected from temperature extremes. In winter, trees growing within the climatic embrace of Lake Superior enjoy milder temperatures. In addition, onshore winds drive warmer, moisture-laden air from the lake onto the cold land, where it hits the highlands' slopes and rises, cooling as it enters higher altitudes. Because cold air holds less moisture than warm air, much of the water vapor falls to earth as snow. For this reason, not only does snow arrive earlier in the highlands than along the shore, but accumulations are deeper. Since the roots of sugar maples are especially vulnerable to freezing, an early and reliable snowfall that helps to insulate them against the cold is undoubtedly an important reason why maples have been able to persist in these northern extremes of their range.

Strategies for Persistence: The Natural History of Sugar Maples

The credit for sugar maples' persistence in the North Shore Highlands cannot simply be attributed, however, to a geological luck of the draw. Sugar maple forests have evolved a number of aggressive feedback loops that benefit their own survival and that of their offspring while discouraging incursions by other plant species. In other words, they create—and actively perpetuate—habitats that assure their own future.

To accomplish these goals, they have resorted to some shady means—literally. One way in which maples maximize the chances of their own survival is by creating a dense canopy of leaves that allows very little light to penetrate to the forest floor. Indeed, shadiness is the hallmark of this forest. On a visit to an old-growth mixed forest in northern Wisconsin in 1946, ecologist Forest Stearns observed, "As one enters the forest in the

In the deep shade of mature trees, maple seedlings blanket the forest floor in numbers of up to one million per acre. Surviving on the sunflecks that dapple the forest floor, these seedlings undergo a growth spurt when wind or old age topples the canopy trees, flooding them with sunlight. Photograph by Virginia Danfelt.

summer the most noticeable change is the tremendous reduction in light intensity, so that, until the eyes become accommodated, the stand seems almost dark. Light seems to filter down from a great height reaching the forest floor in small patches." Indeed, research has shown that the amount of light that falls on the forest floor is a mere 1 to 5 percent of that which falls on an open meadow. Among temperate and tropical forests, the light levels in the understory of a sugar maple forest register among the lowest.

To maximize the absorption of energy that stokes their photosynthetic machinery, maple seedlings leaf out in spring before the taller canopy trees and are slower to shed these leaves in fall. These strategies allow them to take full advantage of the light that floods the forest floor during these seasons. Indeed, the time of maximum growth for sugar maples in the understory is spring, when light is abundant.

Researchers hypothesize that during the growing season, maples supplement this feast-or-famine photic diet by harvesting energy from small patches of light known as sunflecks. So scarce is light in the maple forest understory that even these tiny dapples may be critical to seedling survival.

Such extreme light deprivation discourages the establishment of most other species of trees. If you looked around a sugar maple–dominated forest in mid-July, it would have what Stearns called a "roofed appearance." (The canopy of a mature maple-dominated mixed forest can reach heights equivalent to a nine-story building.) The woodland floor would feature only a smattering of saplings among a sea of maple seedlings. Although littered with logs of varying sizes and in various stages of decay, the forest would be easy to walk through and would feel soft and spongy underfoot. In the Wisconsin maple forests he visited, Stearns found "relatively little air movement, resulting in an atmosphere of remarkable quiet and heightened humidity."

Many species find such conditions inhospitable. But maple seedlings thrive. In the northern Wisconsin forest, Stearns found that maple seedlings were one hundred times more abundant than basswood and more than thirty times more plentiful than their nearest competitor, yellow birch. In his seminal book *Vegetation of Wisconsin,* University of

Wisconsin botanist John T. Curtis points out that during peak seed-production years, a maple forest can shed between four million and five million seeds per acre. Under the cover of deep shade, maple seedlings ranging in height from four to twelve inches may blanket the forest floor in numbers up to one million per acre.

But as is often the case in nature, organisms evolve a strategy of reproductive abundance to offset high mortality. In studies of maple-dominated stands in Michigan's Upper Peninsula Experimental Forest, for example, researchers found that only 20 percent of maple seedlings escaped the perils of frost, drought, and severe competition to survive their second year. Of these, only 3 percent lived long enough to reach heights of more than five feet. A sapling might very well be forty years old but measure less than one-half inch in diameter. That is because under such an extreme degree of light suppression, individual trees may grow only a tiny fraction of an inch annually. Indeed, according to forest ecologists Craig Lorimer and Lee Frelich, many of the saplings in the understory may be the same age as the large trees in the forest's canopy.

Sugar maples can bide their time in the shade better than any other tree in the maple-dominated mixed forest. Frelich points out that maple seedlings and saplings may spend more than one hundred years in suppression. Their growth, even for the great patriarchs of the forest, is fitful in the extreme. Dendrologists, those who study the growth rings of trees, have discovered that maples undergo an average of three separate growth spurts, each followed by long periods of suppression, before finally reaching the forest canopy.

This ability to persevere has great reproductive advantages. The toppling of individual or small groups of overstory trees allows light to penetrate to the forest floor, where the near-dormant seedlings and saplings are quick to shift gears, resuming normal rates of active growth. Having this critical jump start, maples can outcompete a neighboring species that is slower to capitalize on such sudden photic opportunities because it must first germinate from seed.

And for the seeds of most other species, the odds of landing on mineral soil are nearly impossible in an undisturbed maple-dominated forest. That is because maple leaves create an obstacle course between seedling roots and mineral soil. The forest floor under a sugar maple–dominated woodland comprises three distinct layers of organic material, which includes a topdressing of largely intact fallen leaves and a bottom layer of heavily decomposed organic fragments. In a mature forest, the rate of litter that accumulates, such as fallen leaves, dead plants, woody bark, twigs, branches, and tree trunks, equals the rate of decomposition, known as a steady-state process. Thus, the physical structure and levels of chemical nutrients of the forest floor remain relatively constant from year to year, creating an effective barrier to the ready establishment by other species.

Seeds that germinate in the coarse upper layers of this forest duff, which measures between two and five inches thick, often dry out or choke in leafy debris before they are able to sink their roots into mineral soils. For species such as yellow birch, whose seedlings are very sensitive to extremes of temperature and moisture, germination on the surface of the duff layer can be a death trap.

But maples have evolved to turn this handicap into an asset. Maple seeds ripen and fall just before the trees let loose their autumn blizzard of leaves. The timing ensures them an insulating winter blanket. Their relatively large seeds germinate early in the

Carpenter Ant

Pillbugs, Sow Bugs, or Isopods

Soil Bacteria

Actinomycetes

Molds

Fungal Mycelium

Blue-spotted Salamander

©Vera Ming Wong 2014

The deep, organic layer of the forest floor teems with the early-stage recyclers of the nutrients in leaves and wood, including bacteria, fungi, arthropods, and insects. Blue-spotted salamanders and other amphibians find food and shelter in the moist leaf litter and spongy wood. Illustration by Vera Ming Wong.

spring with swiftness and vigor. Maples rapidly establish themselves, aided by seed radicles (the roots that sprout from the germinating seed) that speed their way to the humus-rich topsoil that underlies the coarse leaf litter.

The nutrient cycling of maple forests ensures that these tender young plants are well provisioned with the abundant supplies of nitrogen on which they depend. This is accomplished via a process that Curtis calls "nutrient pumping." The deep roots of trees tap minerals such as calcium, magnesium, and potassium from the subsoil, where they lie beyond the reach of other forest plants. Instead of storing them in their trunks, like many other hardwood species, including oaks, maples sequester these nutrients in their leaves. In the 1980s, scientists discovered evidence to suggest that oaks retain their leaves until spring in order to siphon as many nutrients as possible before abscission (the hormone-induced process that causes the leaf stem to break off from the twig). By contrast, maple leaves blanket the forest floor in autumn. Their high nutrient and mineral content supports robust populations of microbial decomposers such as bacteria and fungi as well as the micro- and macrofauna that feast on them. These activities speed up the release of nitrogen and minerals that are bound up in plant tissues. Indeed, maple-dominated mixed forests make nitrogen available to plants at a rate that is two to four times faster than in conifer forests. As the organic content of the forest floor builds, its water-holding capacity increases, slowing the infiltration (leaching) of these vital elements downward through the soil beyond the reach of most plants, thereby making them more available to plant roots that occupy the upper layers of the soil.

Scientists have long assumed that maples outnumbered yellow birch and other tree species because together they control many environmental variables, such as light, moisture, nutrients, and the availability of suitable seedbeds for germination and growth. But it turns out that there are additional adaptations that give sugar maples an edge over other species. In experiments that investigated competition between sugar maples and yellow birch, botanist Carl Herman Tubbs found that not only do maple seeds germinate much earlier in the season than yellow birch (caused by the species' differential response to temperature), but they also quickly develop a root system well before seedlings unfurl their first leaves. Their peak period of root growth occurs before that of yellow birch, which waits until its leaves are opened before expanding its root network. The disparity in timing gives maples a leg up when growing periods for the two species overlap.

Natural Disturbances in Sugar Maple Forests

In uncut forests, maples may grow to be three feet in diameter and reach grand old ages in excess of 350 years. Combine their ecological aggressiveness with their formidable longevity, and you can understand why the maple-dominated mixed forest has been so persistent as a cover type.

But there is one final ingredient to their success. These forests are remarkably resistant to large-scale disturbances that in boreal forests, for example, frequently reset the successional clock. After conducting tree-ring studies, forest ecologist Frelich and his colleagues identified the temporal and spatial patterns of disturbance in the maple-dominated mixed forest. Saplings that have been suppressed by shady conditions, for example, will register light deprivation by producing narrow growth rings. When a gap opens overhead, the width of these rings suddenly increases as the tree grows in response to the influx of sunlight. Examining these rings, the scientists were able to date the canopy disturbance. By counting the number of trees that registered a similar increase in growth within the same year, they also were able to gauge the size of the gap as well as offer a hypothesis regarding the kind of natural event that caused the gap to form.

The researchers found a disturbance pattern in the maple-dominated mixed forest that differed sharply from the boreal forest to the north. The boreal forest contained comparatively little old-growth forest. Indeed, less than 10 percent of the stands survived more than 120 years. In much of the Boundary Waters Canoe Area Wilderness, for example, either stand-killing or maintenance fires occurred every fifty to seventy-five years. Stand-killing conflagrations also tended to be sweeping in their reach. On average, they burned 12,000 acres, and large fires mushroomed to 181,000 acres. The result: vast tracts of the landscape were largely occupied by even-aged stands of trees. Like their predecessors', their life spans typically were short. Trees and other plants no sooner seeded a patch of forest that had been destroyed by a catastrophic fire than they too were susceptible to being consumed by another fire.

According to forest ecologists, there were several reasons for such frequent and large conflagrations in presettlement boreal forests. For one thing, their shallow soils dried out easily, a condition that was exacerbated by frequent droughts. Add to this a land dominated by conifers with resinous foliage that was easily combustible, and you have the recipe for regular and severe fires.

The opposite was true for the presettlement maple-dominated mixed forest. Then, as in the second-growth maple forests of today, fire was able to make few inroads. For this reason it is known among foresters as the "asbestos forest."

When it comes to fire, why are these two forest types so different? One reason is that the effects of drought are less pronounced than in the boreal forest since the maple-dominated mixed forest is protected from dry conditions on multiple fronts. The decomposition of mineral-laden leaves creates a humus-rich topsoil that is characterized not just by its fertility but also by its moisture content. The forest is mesic, meaning that it falls in the middle of the moisture gradient. Water percolates quickly through the loose, fluffy upper layers of the soil. Protecting them from desiccation is a thick overlying layer of duff that keeps the soil moist even during dry periods. Research has shown that soils in a woodland of mature jack pines, which colonize well-drained, sandy terrain, retain moisture equivalent to an average of 120 percent of dry weight. In the maple-dominated mixed forest, on the other hand, this moisture content soars to 250 percent.

In many maple-dominated mixed forests, such as those throughout the highlands, the deep, loamy soils are afforded drought protection from below as well as from above. Underlying these soils is a layer of clay or other material that slows down the drainage of water within the lower portion of the root zone, keeping it moist but not waterlogged. This serves as a kind of reservoir for plants during the growing season. The soil's moisture retention and high mineral content help to ensure that plant tissues are well hydrated even during droughts, thus enabling the forest to hold severe surface fires at bay.

Like the forest's living plants, the downed material of the maple-dominated mixed forest also is relatively fireproof. During summer, the litter and woody debris are kept moist under the shade and high humidity of the dense forest canopy. The autumn and spring accumulation of litter is readily dampened by precipitation as well as by the moisture in the underlying organic and soil layers. And unlike the pitch-filled conifers of the

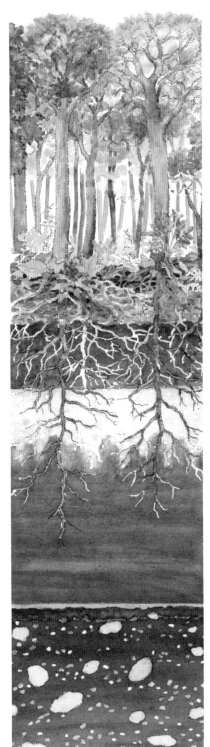

Leaf litter
& humus

Mineral
soil

Zone of
leaching

Clay, iron,
aluminum,
and humic
components
leached
down from
above

Fine-
grained
water-
slowing
layer

Weathered
Glacial Till

boreal forest, which are resistant to decay, the fallen trunks and branches of maples are readily degraded by the especially diverse and abundant populations of fungi and bacteria that thrive in this forest type. These decomposers rapidly transform even the largest maple trunks into pulpy, moisture-retaining sponges that are very resistant to burning.

In his study of a virgin maple-dominated mixed forest in northern Wisconsin, ecologist Stearns observed that in the course of twenty years, several hot fires had burned their way to the forest's edge, barely crossing its borders before dying out. When fires do gain a foothold, they are more likely to slowly meander around the forest floor and smolder in the duff rather than explode into crown fires. Even during the driest periods, such as the year 1976, during which a prolonged drought gripped the upper Midwest, no severe fire activity was reported for maple-dominated mixed forests in the Lake Superior basin. Indeed, when lightning sparked a blaze in the maple-dominated forests of Isle Royale in August of that year, it was allowed to burn. After smoldering for two and a half months, it grew to only five acres in size.

But despite its elaborate defenses, the maple-dominated forest was not immune to cataclysmic fires. These fires usually occurred after major blowdowns flattened whole forests and produced a buildup of dead and drying fuels. When severe droughts succeeded such events, high winds could sweep decimating fires from tracts of downed timber into healthy, intact hardwood stands. During the cutover period of the nineteenth century, large piles of logging slash fueled similarly intense blazes. Like fires caused by windfall debris, these conflagrations could turn back the successional clock and usher in a new vegetational regime.

Maple-dominated communities were extremely vulnerable to such conflagrations. Maple and basswood trees, which are capable of resprouting after moderate fires, were consumed in the bigger blazes, as were the understory plants, the seedbed, and layers of thick duff and humus. Opportunistic species, such as aspen and birch, quickly filled the vacuum. Such disturbance also allowed white pine to invade.

Maple-dominated forests have a soil profile distinctly different from that of other forests in the Minnesota Lake Superior watershed. Deep loamy mineral soil is sandwiched between a thick, organic-rich humus of decomposing, nutrient-rich leaves. They are often underlain by clay, which slows the drainage of water, keeping the lower portion of the root zone reliably fertile and moist even during dry periods. Illustration by Vera Ming Wong.

Canopy

Subcanopy

Sapling/shrub

Seedling/
herbaceous

The structural diversity of a mature sugar maple forest is maintained by the small-scale windthrow of individual or small groups of mature trees, resulting in a multilayered canopy of different ages. Saplings and shrubs occupy the forest floor in low densities, while maple seedlings and herbaceous plants form dense carpets.

Because of its moderate tolerance for shade, white pine could survive in the understory of aspen or birch stands for twenty to forty years. In time, this long-lived species could become the dominant tree in the canopy. In other cases, white pine became a prominent member of a mixed canopy containing early-successional hardwoods or sugar maple whose seeds spun into scorched tracts from adjacent unburned forests.

These trees undoubtedly served as a source of seed for colonizing the small gaps that were burned down to mineral soils by occasional lightning strikes. Such hot spot fires cleared the ground for species such as white pine that required these bare soils for germination.

White pine also was quick to exploit other small opportunities in the maple-dominated mixed forest. Scientists have located white pines growing on low, crib-sized mounds within the forest. Many of these trees, they hypothesize, got their start in the mineral soils that were exposed in the upturned roots of large, windthrown trees. Once established, pines gained an edge over their sugar maple competitors by being fast growers.

Before 1900, individual trees and small clusters of white pine towered over the leafy domes of the maple-dominated mixed forest. Although they were not as numerous as in the stands of pure white pine that covered other parts of Minnesota, Wisconsin, and Michigan, they were much coveted by loggers. The straightest and largest pines grew in hardwood forests. In colonial New England, for example, these isolated white pines were so valuable that the British Navy laid exclusive claim to them for use as ship masts. They retained their value over time. According to Curtis, who in the 1950s drafted ecological maps of Wisconsin based on the work of government surveyors nearly a hundred years earlier, even stands "with only 2–3 pine trees per acre were highly profitable since these few trees were likely to be forest giants from 3–6 feet or more in diameter." But these pines are rare today since many were high-graded (the targeted removal of valuable timber species) during the Great Lakes cutover period.

The Freshening Blow: Wind as an Agent of Renewal in the Highlands' Forests

Nearly as rare—but equally as far-reaching in their effects as fire—were catastrophic blowdowns caused by intense windstorms. Whereas fire visited the maple-dominated mixed forest in intervals of two thousand years or longer, windstorms flattened whole forests on the order of every one thousand years or longer.

Although tornadoes are known to occur in the maple-dominated mixed forest, researchers now suspect that thunderstorm downbursts cause most of the catastrophic blowdowns that occur in the Great Lakes forests. Only identified in the Great Lakes region after a killer storm hit northern Wisconsin on July 4, 1977, downbursts are caused by frigid drafts of air that shoot out the base of anvil-shaped thunderheads at speeds well in excess of 100 miles per hour. (During the 1977 storm, experts estimate the temperature of the air at the top of the anvil to have been an icy −101 degrees Fahrenheit.) When these parcels of air slam into the ground, in what some have described as "avalanches of air," they spread out horizontally. These storms leave an easily identifiable footprint of devastation in the forest: great oval-shaped gaps in which toppled trees lie facing in the same direction.

Downbursts have the potential to create significant damage over a wider area than tornadoes since they hopscotch across the land, flattening forests like the steps of a great lumbering beast. In a single four-hour period, the 1977 storm, one of most severe ever recorded, produced twenty-five separate downbursts whose wind speeds ranged between 100 and 157 miles per hour. Because the downbursts occurred at relatively short intervals, they caused a contiguous area of damage measuring some 166 miles long and 17 miles wide. Trees were broken and uprooted in an estimated 850,000 acres of forest land.

These windstorms were the curse of the region's earliest explorers. In his geological survey of northern Wisconsin, published in 1880, R. D. Irving described cutting a path though a tract of forest that had been hit by a major windstorm eight years earlier.

"The traveling is rendered yet more difficult by the frequent areas of fallen timber, known as 'windfalls.' These are well enough known in all of the Wisconsin forests, but never become so well known as to be welcome to the 'packer' through the woods. They are like great swaths having been prostrated by hurricanes, and are often sharply defined from the standing timber around. The trees at times lie all in one direction, but in other cases are wholly without such arrangement, having been prostrated by a tornado. The great windfall of September, 1872, so far as my knowledge goes, is as much as forty miles in length, though it is reported to have a greater length than this. . . . This windfall had been partly burned over, but was, for the most part, made more impenetrable than ever on account of the new and dense growth of bushes and small trees. In crossing it on the Penokee Range it was found necessary to cut the way with an axe for nearly a mile and a half."

Although downbursts leveled vast tracts of forest, the devastation zones still only amounted to a small fraction of the larger presettlement landscape at any given time. Far more frequent and wider in their effects are milder storms known as cyclones whose winds can range between 62 and 75 miles per hour. According to Frelich and Lorimer, the Lake Superior region lies in "one of the most active weather zones in the northern hemisphere. More cyclones pass over the area than any other part of the continental U.S."

What makes the region so vulnerable is its position between two major atmospheric highways. Severe weather occurs when cold, dry air from the polar jet stream collides with another, more southerly jet stream that brings warm, moist air from the Pacific Ocean and Gulf of Mexico. "Major cyclone tracks cross the region every month of the year," say Frelich and Lorimer. In summer, however, the cold waters of Superior generally snuff out or divert these storm systems. Not so in autumn, when the incidence of cyclonic episodes, commonly referred to as November gales, reaches its peak. The fury of these gales is fed by the differences in temperature and air pressure between the two continental air masses. But like hurricanes that build a head of steam from passing over warm, tropical waters, the storms also gather energy as frigid, northerly airflows blast across Lake Superior's warmer surface.

These winds may topple individual trees or a small group, resulting in common, widely scattered alterations known as gap-phase disturbance. Frelich and his colleagues discovered that relatively catastrophic windstorms that removed 10 to 50 percent of any given cohort (group of similarly aged trees) occurred every one hundred to two hundred years. The area affected was fairly small, averaging 240 acres with an approximate maximum size of 9,600 acres.

Winds of lesser severity are a far more common agent of change in the maple-dominated mixed forest. Collectively, minor tree-fall openings have the biggest effect on the forest's composition. More than half of the trees found in the canopy have been able to reach such prominent positions through gap disturbances in which less than 20 percent of the canopy trees have been removed. Ecologist Stearns referred to these continual small-scale reconfigurations as an "island type of replacement." Joining the understory maple seedlings that stand poised for a growth spurt are the seeds of various other tree species, he wrote, that drift "into wind-produced clearings . . . and form an island of young growth surrounded by the mass of the mature forest."

Older trees that are hit by heart rot, fungi, or insect infestations, damaged by ice glaze, or stressed by drought are more likely to be toppled by such winds than healthy trees. The result is a constant, but localized, turnover in the canopy. The average time that an individual tree spends in the canopy of Great Lakes mixed forest is 145 to 175 years. Even in the old-growth maple-dominated stands in Michigan's Upper Peninsula, researchers found no trees older than 540 years. Unlike great tracts of even-aged trees in the fire-dependent forests to the north, the highly variable patterns of disturbance in the maple-dominated mixed forest—from catastrophic windfall to the toppling of individual trees—result in a tree population of many different ages. Indeed, Frelich and Lorimer point out that the forest is likely to contain at least ten different age classes.

And unlike the Great Lakes' forest communities that are regularly strafed by fire, wind disturbance in the maple-dominated mixed forest does not radically change its species composition. In the shade-tolerant forest, seedlings are already present in the understory. Neither these seedlings nor the seed bank in the soil are as altered by wind as they are by fire.

Thriving in Shady Circumstances: Life under the Canopy of the Sugar Maple Forest

To survive in the deep shade of the maple forest, understory plants have evolved a variety of adaptations. By comparing the leaves of shaded maple seedlings with those plucked from the sun-bathed canopies of mature trees, scientists have uncovered tantalizing differences, including a whole suite of distinctive leaf characteristics. On the one hand, leaves directly exposed to the sun absorb red and blue wavelengths of light. The shaded leaves of seedlings, on the other hand, possess the ability to carry out photosynthesis using the remaining array of light in the spectrum. In addition, their stomates (minute breathing pores through which water and carbon dioxide are exchanged) are able to open and close more rapidly in response to changes in the low-intensity light of the forest floor. Shaded leaves also tend to be relatively thin, suggesting that plants do not bother to invest precious energy in the development of structural cells or surface coatings that protect leaves from drying winds. Equipped with such adaptations, shaded leaves usually can reach their peak photosynthetic capacity at a mere 25 percent of full sunlight. When sunlight does manage to penetrate to the forest floor in dappled sunflecks, these leaves can quickly take full advantage of the small windows of light.

For many species, the key to survival in the maple-dominated mesic forest is to exploit the windfall gaps that open up the forest floor to life-giving sunlight or to maximize growth and development during fall and spring when canopy branches are bare. Some have been more successful than others. Take shrubs, for example. Scientists attribute the paucity of shrubs, which make up just 2 to 10 percent of the plants found in the understory of old-growth forests, to the dim light conditions.

For those that have found a home under the maple-dominated canopy, mobility often is key. Many of the shrubs that do survive in the maple-dominated mixed forest, as well as a whole suite of ground-layer plants, including common ferns, clubmosses, sedges, and flowering herbs (also known as forbs), persist by mounting their own active "search-and-grow" campaigns. Dewberries, for example, "walk" through the forest by sending out flexible stolons—horizontal plant stems or runners. The plant roots at intervals along these runners and then sends up shoots. In this way, dewberries are able to take advantage of sunlit gaps they encounter and generate additional leafy stems that can capture more energy for flowering and fruiting. The common shrub pagoda dogwood accomplishes its journey through the forest by growing prostrate in the duff when light conditions are low. But when it encounters a light gap, it straightens into a tall shrub that can flower and set fruit.

Other plants are able to expand their territories by reproducing vegetatively. Plants that propagate via rhizomes, for example, are common in the maple-dominated mixed forest. These rootlike underground stems grow laterally, where they can slowly roam the forest floor for years at a time. When they sense a chemical or temperature change caused by the opening of a sunlit gap, they send up aerial shoots.

Rhizomes also function as a means of energy storage that can boost plant reproduction. These shoots, which sometimes form large clonal colonies, transfer excess photosynthetic energy back to the mother stem, where it can be stored to fuel growth when opportunities arise. With limited opportunities to attract pollinators or to photosynthesize enough energy for growth and reproduction in the dim light, rhizomatous plants draw on these communal reserves to beat the seasonal clock and bloom before maples

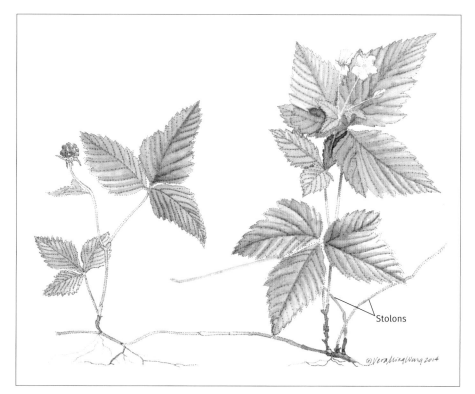

Plants have developed a variety of strategies for growing in the dense shade of a sugar maple forest. Dewberries, for example, "walk" through the forest by sending out horizontal plant stems or runners. The plant roots at intervals along these runners and then sends up shoots whenever it locates favorable patches of light. Illustration by Vera Ming Wong.

Stolons

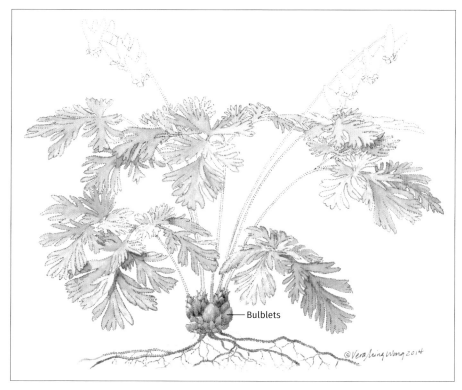

Bulblets

Dutchman's breeches are one of several shade-adapted wildflowers, including spring beauties, bloodroot, trillium, and wild ginger, which seques-ter energy reserves in fleshy tubers and bulblets. Illustration by Vera Ming Wong.

have completed their spring leafout. Not surprisingly, the three most widespread species of spring bloomers (all members of the lily family) are rhizomatous plants: Canada mayflower, Solomon's seal, and rose twistedstalk.

Other plants sequester energy reserves in structures such as fleshy tubers and bulblets, including such woodland favorites as spring beauties and dutchman's breeches. These plants, along with others such as trillium, bloodroot, and wild ginger, are known as "spring ephemerals," so named for their brief but sweet springtime appearance.

Helping to spur the growth of these plants are hospitable conditions on the forest floor. In spring, a thick leaf litter that has been compressed by the winter's snowpack covers the duff, keeping it warm and moist. High temperatures within the humus prior to leafout initiate bacterial activity that releases nutrients and breaks the dormancy of buds and seeds. And the more clement temperatures on the forest floor in winter protect reproductive structures such as roots, rhizome networks, and the vegetative buds that form in the fall.

So successful are such strategies that of the common species of woodland herbs, Curtis found that 70 percent bloomed before June 15; less than 5 percent flowered after August 15. By comparison, in the boreal forest 47.2 percent of prevalent species bloomed between June 15 and August 15.

Vegetative reproduction helps trees, as well as woodland herbs, to flourish in the maple forest's dim light. Basswoods, for example, rarely reproduce by seed. Instead, the trees send up shoots from the base of their trunks. When the main trunk dies and is toppled, these clones form a perfect circle known as a fairy ring. Each, in turn, is capable of producing its own clone circlet, thus creating patches of basswood trees in the maple-dominated forest. In this way, they gain a reproductive edge over maple seedlings, which must invest time in the process of germination and establishment.

Other trees, such as yellow birch, use a combination of strategies to compete with maple dominance. Germinating birch seeds that fall in undisturbed leaf litter typically die because they cannot reach the mineral soils they need to become established. Even those that sink tentative roots into the earth often cannot match the speed and vigor of maple. In time, maples usually outcompete birch for nutrients and light.

If they can grow to the seedling stage, however, yellow birch is the faster growing of the two species. In his experiments, botanist Tubbs measured the total dry weights of both species and found that those of yellow birch were double those of sugar maple, even under the lowest light level (13 percent of full sun). In the most favorable light conditions (45 percent of full sun), the dry weight of yellow birch was three times that of maple.

The chances of yellow birch establishment increase substantially if seeds sprout on rotting stumps, tip-up mounds (the soil cavity and corresponding root mass of toppled trees), the mossy surface of well-decayed downed logs, or at the wet margins of small ephemeral wetlands within the forest. Because such places lack an insulating layer of leaf litter, they are inhospitable to maples. But how do birch seeds find these propitious microsites when they lie scattered throughout the forest?

Birch trees increase their odds of success by timing their seed fall. Although birch produce catkins in the spring, the seeds do not ripen until September and October. The majority of the seeds do not fall until winter, when powerful winds skid them long distances over the glazed surface of the snow, effectively scattering them through the forest. Along the way they encounter a variety of seed traps, including the rotted cavities

The seeds of species such as yellow birch struggle to germinate in the maple forest's deep leaf litter. These seedlings find refuge in forest microsites, such as tip-up mounds and the decaying trunks of trees known as "nurse logs." Photograph by Ray Asselin.

and mossy skin of exposed logs and stumps or even dimples in the surface of the snow that are formed by falling balls of snow from tree branches.

Seeds trapped in the mossy outer sheath of decaying logs in spring, in particular the logs of sugar maple or other yellow birch, stand the greatest chance of survival. Scientists hypothesize that the moss and decaying wood offer the seedlings a consistent supply of moisture. Mature yellow birch that have begun life under these conditions are easily identified in the forest. As birch seedlings grow, their roots straddle their nurse logs. In time the decaying logs disintegrate, leaving birch trees with long, leggy roots, making the trunk look as if it were propped up on stilts. In 1877, F. L. King, a surveyor in northern Wisconsin, was so impressed by such growth that he recorded it in his journal: "On the North Fork [of the Flambeau River], there was seen a yellow birch sixteen inches in diameter, growing with its roots astride a white pine log more than two feet in diameter and 50 feet long. . . . Of course the seed of the birch must have fallen upon the pine log after it had become moss-clad and had decayed sufficiently to furnish nourishment and support for the young birch until its roots could penetrate the earth." Occasionally, a single log may support several birch seedlings, which grow up to form a straight line, a uniformity that is jarring in the random scatter of understory plants in the maple-dominated mixed forest. Once firmly rooted, yellow birch may reach an age of three hundred years and a height of one hundred feet.

Some plants, however, are not dependent on disturbance, because they have dispensed with the need for photosynthesis altogether. Frequently encountered on the forest floor in the North Shore Highlands are a class of plants known as saprophytes, which include three species of coralroot orchids. These orchids possess no chlorophyll, hence no green plant parts, since they do not rely on the energy of sunlight to fuel the production of food. Instead, coralroots obtain nourishment from decaying organic matter via mycorrhizal fungi in the soil. Their branching underground network of rhizomes intertwines with the filaments of mycorrhizae. The fungi serve as the go-between that transfers nutrients from decaying matter to the orchids.

Other plants are epiparasites as well as saprophytes, including Indian pipe, whose white and waxy appearance has earned it such nicknames as corpse plant and ghost flower. Like the coralroots, Indian pipe too relies on mycorrhizal fungi for sustenance, but the nutrients come to it in a one-way flow via an associated mycorrhizal fungus it "shares" with a live green plant.

Some plants adapt to the deep shade of the maple forest by dispensing with photosynthesis. Saprophytes and epiparasites, such as Indian pipe (pictured here), rely instead on associations with mycorrhizal root fungi, which supply nutrients from decaying organic matter and green plants. Photograph copyright Jay Cross (CC BY 2.0).

Crosscuts to Ploughshares: The Highlands' Early Logging and Farming History

A stroll through the hushed interior of an old-growth mixed forest, like the one ecologist Stearns visited in northern Wisconsin in 1946, is an uncommon experience for most people nowadays. According to Frelich, remnants of unlogged mixed forest "are among the rarest vegetation types in the Lakes States." Only 0.6 percent of the presettlement mixed forest remains, much of it located in forest reserves in Michigan's Upper Peninsula. Most forests today, he observes, are composed of even-aged stands ranging from sixty to ninety years of age. Compared to unlogged forests, such second-growth forests appear tame, wrote Henry David Thoreau. Writing about one New England woodland, he described the difference: "It has lost its wild, damp, and shaggy look. The countless fallen and decaying trees are gone, and consequently that thick coat of moss which lived on them is gone too. The earth is comparatively bare and smooth and dry."

In Minnesota's Arrowhead country, the highlands contain a higher proportion of old-growth forest than any other subsection within the Northern Superior Uplands (the part of the ecological classification scheme that covers all of the state's northeastern tip). Here, corkscrewing out of the wet hollows are cohorts of cedar that sunk their first tentative roots into the highlands' soil just as William Shakespeare was polishing the final passages of *King Lear*. And in a protected pocket of a ridge that was too steep for the logger's skidder grows a solitary yellow birch, the girth of its trunk as wide as a telephone booth, the bark of its upper branches polished and bronzed. It was a sapling when French traders delivered the first bundles of Lake Superior furs to their superiors in Quebec.

Deep in these forests, protected from harsh light and drying winds, thick mounds of mosses and liverworts grow on the rocks and the trunks of live trees. Here too in the dim, humid interior can be found some of the state's rarest lichens. The lucky—or extremely observant—may spot such uncommon wildflowers as the yellow-green globes of flowering moschatel gleaming in the forest litter or the fronds of Braun's holly fern, which retain their glossy deep-green color in winter even as the snow gathers around them.

But it is not just these old-growth remnants that make the highlands so rich and ecologically diverse. In contrast to the tracts of second-growth mixed forest in the lower Great Lakes states, many of which have been cut up into smaller fragments by cities, farms, and tree plantations, the highlands' tree canopy is largely unbroken by human development. Consequently, the structural and compositional diversity of the forest remains mostly intact, supporting a wide variety of animal species. For instance, together with the rest of Minnesota's northern forests, the highlands host more species of breeding birds, including the largest complement of songbirds, than any other major landscape in the state, and more species than most other parts of the continent north of Mexico. (For an in-depth discussion of forests and birds, see "As Good As It Gets: Bird Diversity on the North Shore.")

The highlands are rich not only in terrestrial endowments but also aquatic features. Twenty-eight streams cut through the volcanic rock, creating important riparian habitats for numerous plants and animals. The region also contains lakes and ponds that supply critical feeding and breeding locales for waterfowl. Added to these are more subtle features, such as ephemeral pools, whose pockets of standing water on the forest floor provide temporary nurseries for a host of insects and amphibians. So important is its natural heritage that the Minnesota DNR's Biological Survey has designated 147 sites in the highlands as having "statewide biodiversity significance." In addition, private organizations, including The Nature Conservancy, have made the highlands a priority for conservation efforts.

Despite such ecological accolades, the highlands' conditions are much changed from those that existed prior to Euro-American settlement. When the Lake Superior Ojibwe ceded Minnesota's Arrowhead region and parts of Wisconsin in 1854, prospectors poured into the Minnesota North Shore. To expose veins of valuable ore, they resorted to a method practiced by mineral seekers throughout the Lake Superior region—they torched the forest. Minnesota biologists Lloyd Smith and John Moyle observe that widespread wildfires on the North Shore were recorded in 1850 and 1878, both of which were likely related to prospecting activities.

How many hectares of the highlands' forest went up in these intentional flames is anyone's guess. The miner's pick was followed by the logger's axe, which created the conditions for the further spread of wildfires. In the commercial logging activities that began in earnest on the North Shore in the 1890s, loggers began by high-grading the forest; namely, they extracted the most valuable timber—white pine—which was much in demand by the burgeoning cities on the midwestern prairie, where building materials were scarce. First to be cut were the forests that grew near rivers, since log drives to Lake Superior were the only available means of transportation. Later, railroads and their complex network of spur routes allowed timber companies to penetrate the interior and exploit a second tier of timber. Decay-resistant species such as northern white cedar and tamarack were harvested for railroad ties and fence posts; balsam and spruce for the growing pulp trade; birch, oak, and maple for fuel woods and interior finishing; basswood for packaging; and ash, whose strength and pliability made it especially valuable for handles.

Forests were decimated not simply by rampant cutting but also by the fires in its aftermath. Many blazes could be traced to locomotives whose smokestacks launched sparks into the woods as far as three hundred feet from the tracks or whose ash pans dropped live coals and cinders onto tinder-dry brush along tracks. Huge quantities of slash—cut branches as well as the trunks of deformed trees or nonmerchantable species that were cleared to make way for skid trails, haul routes, and railroad spurs—were left behind in the woods. During dry periods, the slash fueled destructive wildfires. Before European settlement, stand-replacing conflagrations on the lakeshore and coastal ridge were rare, occurring only once every 150 to 1,000 years. But after the cutting of the forest began, Smith and Moyle point out, fires of varying intensity raged off and on until 1936, when the last major fire burned a large portion of the Pigeon River watershed.

The imprints of early fires persist in the highlands to this day. On a walk through a mature maple forest, you can suddenly stumble upon a younger forest of large aspen trees or a woodland swath containing a mixture of even-aged sugar maple and paper birch interspersed with the charred snags of white pine and white cedar. In these fire-swept forests, birch trees often grow in clumps from a single stump rather than as separate trees, a resprouting response commonly seen in birches that have been visited by fire. Instead of harboring species that are dependent on the deep drift of litter found in unscathed deciduous forests, such as spring beauties, twistedstalk, hairy Solomon's

Patches of bracken fern and young white pines within a mature maple forest are often legacies of historic wildfires in the nineteenth and early twentieth centuries. Photograph by Bob Firth/ Firth PhotoBank.

seal, dutchman's breeches, and wild ginger, the forest floor in these fire-scoured patches contains such fire-adapted herbs and shrubs as large-leaved aster, raspberry, wild sarsa-parilla, pale vetchling, and bracken fern.

Among the habitats most heavily impacted by the fires in the highlands were riparian areas. Smith and Moyle point out that the blazes destroyed vegetation that shaded streams and kept water temperatures cool. The fires also altered hydrological regimes by denuding the trees on forested slopes that helped to slow runoff. And they burned out centuries' worth of peat accumulation in the swamps and muskegs that supplied rivers with a steady source of cool water. Without these protective vegetative sponges that evened out flows between high and low water, streams became "flashier," or more prone to violent flooding, which further eroded stream banks and washed heavy loads of silt into waterways. During droughts, many streams dwindled to a trickle.

The long-term damage caused by such logging activities is difficult to assess since no baseline measurements of North Shore rivers were made prior to Euro-American settlement. But scientists have cited waterways such as the Temperance River and its Heartbreak Creek tributary as bearing the scars of wounds that were most likely inflicted during this frontier era of exploitation. The damage includes widened stream channels, loss of coarse woody debris within the streambed, and heavy sediment loads that were likely scoured from river channels and their flanking banks during high water flows and then later redeposited in more sluggish reaches.

Timber companies cut the forest without consideration for the future. In its 1934 report *Land Utilization in Minnesota* (an analysis of problems ranging from logging and public tax policy to recreational development), the Committee on Land Utilization, a group of experts from the University of Minnesota, pointed out that three-fourths of the standing timber in the state had been cleared between 1880 and 1915. In the span of a mere thirty-five years, relentless cutting had so depleted cheap and plentiful supplies of local timber that many sawmills were forced to close their doors. Those businesses that survived paid a premium for trees imported from pine forests in the South and West.

Greed and public policy colluded to create what government officials said was nothing less than a "timber famine." According to the authors of the report, "The American land policy of the nineteenth century may be summed up in two phrases, *private ownership* and *immediate exploitation*. Natural resources seemed to be unlimited and inexhaustible, and the prevailing political and economic thought favored individualistic development of resources as the quickest, surest, most democratic way of getting the country developed, increasing its wealth, and promoting its welfare."

Little thought was given to sustainable harvesting practices or reforestation, in part because forest clearing was considered the necessary first step in the agricultural settlement of the country. The movement had philosophical roots in the Jeffersonian ideal of creating a nation of self-sufficient farmers. But there were less noble and more self-serving motivations in the urgency to put settlers on the land. Many railroads, land companies, timber corporations, and speculators stood to profit twice, first by stripping the timber from the land (much of which was obtained by fraud, snapped up from Civil War veterans who were given land as a reward for their military service or purchased for a nominal fee from the U.S. government) and then selling the denuded acres to hopeful farmers. To attract settlers, they placed advertisements in newspapers and published an array of promotional literature that praised the virtues of Minnesota to potential land buyers in the Northeast as well as immigrants abroad. Translated into several languages, these materials were placed on board passenger boats and in the taverns and hotels that newly arrived immigrants were likely to frequent.

Finding itself in stiff competition for settlers with neighboring states, Minnesota mounted a public relations campaign by establishing a state office, known as the Commissioner of Immigration, in 1864. One of the appropriations made by the first state legislature was money for an advertising pamphlet geared to enticing prospective farmers.

But private companies as well as the newly minted states of the upper Midwest would leave nothing to the chance of the printed page. They hired representatives to intercept new arrivals just as they took their first steps on American soil. "When a ship docked," writes historian Theodore C. Blegen, "a hundred or more agents, runners, and peddlers were at hand to make prey of the immigrants." Indeed, faced with such extreme portside competition, some agents actively traveled abroad to recruit new settlers. The high population of Scandinavian immigrants in Minnesota in the 1870s, for example, is largely attributed to the work of Swedish American Hans Mattson, who served as secretary of the state's Board of Immigration while employed as a land agent for a railroad selling parcels in southern Minnesota. So persuasive was he that in 1869, on one of his many trips to Europe, he induced eight hundred Swedish immigrants to accompany him to the United States in a single voyage.

Many of the newcomers who founded homesteads on the state's rich prairie soils attained the prosperity promised by land agents and their brochures. Unfortunately, the cutover lands of northern Minnesota were not evaluated according to their suitability for agriculture, and settlers were lured by fantasy-charged prose and rosy photographs to places where many of them would find nothing but heartbreak and ruin. The pamphlet *Northeastern Minnesota: Land of Certainties,* published by the Minnesota Board of Immigration in 1919, is typical of the deceptive advertising used to sell the state's cutover acres. "This was the Indian Paradise!" the authors crooned. "Pioneered by the lumberman, invaded by the miner, now witnessing the beginning of an era of agricultural development destined to produce more wealth from the surface than ever existed above or below." For the immigrant audience who could not read the text, the authors purposely plied its pages with tinted photographs of produce heaped into bushel baskets, lakeside pastures dotted with fattened livestock, and tidy farmsteads framed by woodlots and rows of lush crops.

Even the University of Minnesota's Agricultural Experiment Station at Grand Rapids joined in, lending an air of scientific credibility to what would become the state's most disastrous land policy experiment. "'More farmers and better farming for northeastern Minnesota' has come to be the watch word," stated a 1909 station bulletin. "Pass it along till this section of our fair state is contributing its rich share to agriculture. Till there is a farmer on every eighty acres of land. Till every swamp is drained and every needed road is built."

More fancy than fact, duplicitous salesmanship than sober appraisal, these publications exaggerated the agricultural potential of cutover lands in northeastern Minnesota and in some cases promulgated outright lies. This was particularly true in their discussions of the state's winters. Even before Minnesota's establishment as a territory in 1849, the region had developed a reputation as the American Siberia, writes historian William E. Lass. Consequently, more print was devoted to discussing—and defending—the region's climate than to any other aspect of its geography. In a subsection titled "Really Delightful Winters," for example, the authors conceded that the season was indeed cold, but, they argued, the climate was dry and windless. As a result, "when the thermometer registers low below the zero mark, the days and nights in Northeastern Minnesota are infinitely more comfortable and bearable than where the temperature is twenty to forty degrees higher and the atmosphere is marrow-penetrating with its dampness, or where high winds bite through the heaviest garments and the snow whirls ceaselessly when blizzards rage."

The promoters were especially adept at characterizing such liabilities as assets. True, they conceded, the heavy snowfall shortened the growing season. But once the snow melted,

The pamphlet *Northeastern Minnesota—Land of Certainties,* published by the Minnesota Board of Immigration in 1919, was typical of the deceptive advertising that oversold the agricultural promise of the northland's cutover acres to unsuspecting immigrants in search of productive farmland. Collection of Adelheid Fischer.

the "latent heat in the soil" induced "grass and other vegetation [to] spring into life as if by magic," they exclaimed.

The frigid winters were said to not only invigorate plants but livestock. As for people, the Minnesota climate made for sounder, more robust constitutions. To underscore this point, many treatises on climate were written by physicians or by recovered invalids such as Girart Hewitt, who purportedly experienced the curative powers of Minnesota firsthand and championed them with missionary zeal. In the 1867 pamphlet *Minnesota: Its Advantages to Settlers,* Hewitt attributed health benefits to a dry wind that blew not in "strong, continuous currents" but in "lively agitations" that swept away "noxious vapors and effluvia" and imparted a "clearness and purity . . . which give tone to the system and invigorate nutritive functions." (In the land of ten thousand lakes, however, wet locations could not always be avoided. But as one doctor advised, "Those who dress in flannel or its equivalent are less liable to be attacked by [miasmic] diseases.")

Climate was said to impart superior moral fiber as well as physical hardiness. Promoters even argued that the climate of temperate zones, such as the one in which Minnesota was located, supplied the necessary conditions for great civilizations. "There seems to be a certain zone of climate within which humanity reaches the highest degree of physical and mental power," wrote J. W. Bond in the 1853 pamphlet *Minnesota and Its Resources.* "That zone included Rome in her great day when the eagles sat upon her seven hills and the Tiber was frozen to its bottom. . . . It is the good fortune of this [Minnesota] Territory to lie not only within this zone, but within its very apex."

Along the North Shore, most of the land cleared for cultivation lay in the highlands on hectares formerly occupied by mixed hardwood-conifer and maple-dominated forests. In his report to the eighth annual *Geological and Natural History Survey of Minnesota,* published in 1880, C. W. Hall opined, "The wooded and broken character of the country is often mentioned as highly favorable to the development of a wheat-producing region, as the danger of loss from windstorms is thereby materially lessened. The proximity of the lake would have a tendency to keep the temperature low during the season when the wheat berry is forming and ripening; so it seems hardly possible that such a discouraging blight as swept over the southern part of the State in 1878 could ever afflict St. Louis and Lake counties. Oats and barley should by no means be omitted if one were to make out a list of those cereals whose successful cultivation here has been placed beyond a doubt."

Moreover, Hall claimed that in some parts of the country Lake Superior produce was considered a delicacy. "It is claimed in Duluth," he wrote, "that in the Chicago market the deliciousness of the Lake Superior potatoes is appreciated so highly as to make them preferred above those from any other locality in the West or Northwest."

But promoters were quick to point out that Lake Superior growers would not be forced to rely on faraway markets. Growing potatoes for the laborers in the mines and lumber camps alone, never mind the growing population of Duluth, was said to be enough to ensure a stable, adequate income. The authors of *Minnesota—The Land of Opportunity for Agriculture, Horticulture, Live Stock, Manufactures, Mining, Education and Everything That Attracts the Immigrant,* published by the Minnesota State Bureau of Immigration around 1912, assured the new settlers of Lake County that "there are undoubtedly opportunities in this county for hundreds of poor men who are living a hand to mouth 'treadmill' life in a congested city. Here is an opportunity to get a piece of land for almost nothing on which he can make himself a comfortable home and become an independent citizen. True there will be a few years of 'roughing it' but after a small tract is cleared he will have steady income."

The authors were no less enthusiastic about the agricultural potential of Cook County. For a small down payment, settlers could buy land. Proceeds from the sale of the timber growing on it, they instructed readers, were enough to cover the outstanding portion of their mortgage. Beneath the trees lay a "vast area of excellent agricultural land," whose products could be speeded to market via the construction of an inland railroad and the extension of a highway, now known as Highway 61, along the lakeshore. Joining the chorus of shameless boosterism was C. H. Carhart, the county's notary public. "I am sincere," he wrote, "when I say that there are splendid opportunities in Cook county for a poor man to make and own a home, for any county that will produce vegetables and grasses is bound to bring profit to the farmer."

High wartime prices for farm goods, as well as the demand for wood used in the manufacture of fence posts and railroad ties, masked the signs of a brewing catastrophe until about 1920, say the authors of *Land Utilization in Minnesota*. By 1931, however, the State of Minnesota stood to take possession of nearly seven million acres of tax-forfeited or abandoned land in the cutover counties of northern Minnesota, an area encompassing more than 20 percent of the total area of the state and measuring larger than Massachusetts, Rhode Island, and Connecticut combined. On January 1, 1932, Cook County tallied $581,400 in delinquent taxes ($8.6 million in 2010 dollars). Lake County registered $652,976 ($9.7 million in 2010 dollars), about two and a half times the total of the most tax-delinquent county in southern Minnesota.

To survive, cutover counties relied on government subsidies. In 1932, for example, the residents of three northern townships in St. Louis County paid a combined total of $658 in county and state taxes (about $10,000 in 2010 dollars). Together, these same counties received more than $25,000 in state and county aid (nearly $370,000 in 2010 dollars). Such expenditures did little, however, to reverse the fortunes of most cutover communities. Many of those who remained on the land were destitute and demoralized. According to the land-utilization report, the living conditions on cutover farms were "far from good. The population is sparse, and homes are frequently remote from any trading center. Many are attempting to carry on subsistence farming while working part time at trapping or guiding, or on road or forest jobs. Incomes are low, in some cases almost nothing. Housing conditions are often very poor. Schools are far apart, as are the trading centers. To attend church services and social events, to seek the aid of a doctor or the services of a hospital for themselves and their families, men must travel far over little-used roads. To pay for such services has been very difficult, often impossible. When the depression began in 1929, it is these people in the cut-over woods areas whose little reserves were first exhausted and to whom government aid had first to be extended when it became available."

In 1935, University of Minnesota professor Darrell H. Davis published a study of the tiny hamlet of Finland in Lake County, offering a rare portrait of a prototypical cutover community in the highlands. The settlement got its start in 1895 with a small cluster of Finnish families who arrived at boat landings between Beaver Bay and Little Marais and then packed their belongings nine miles overland on narrow forest trails into the highlands. At the time of his study, Davis tallied a total of forty farms, of which an average of twenty acres per farm were cleared. These clearings were used primarily to grow hay, with an assortment of field crops ranging from oats and potatoes to barley.

But as Davis points out, profits on such agricultural goods were meager. Many farmers supplemented their incomes through subsistence hunting, logging trees on their own land, or trapping fur-bearing animals. They also took seasonal jobs such as building highways, fighting fires, and working at area resorts. "Even with such additional sources of income," Davis points out, "some of the families are on relief." As a reminder of the community's poverty, he pointed to a local cemetery whose "unkempt condition suggests that life is much too hard to give much thought or time to the dead."

The fortunes of the land mirrored those of its human inhabitants. The forests of tall timbers that once helped to build America's great midwestern cities were reduced to brush and blackened stumps. Of the trees that remained standing, few were large enough to fuel a profitable saw timber trade. Instead, most of the forest was composed of pioneer species such as aspen and paper birch, which, if they could be sold at all, fetched only low stumpage fees in the pulp trade. Streams were choked with silt, and analysts warned of the near extinction of once-abundant animals such as moose and woodland caribou and the dwindling waterfowl populations, whose wetland habitats had been drained.

The Committee on Land Utilization was unsparing in its blame. The deplorable human and natural conditions in northern Minnesota were not simply the result of the Great Depression, they observed. Tax delinquency and land abandonment had begun long before the economic collapse of 1929. "One could make dark and disagreeable charges," wrote the committee's chair, "charges that would not point to any single

Farms in the cutover areas of the highlands often supplemented their meager agricultural profits with subsistence hunting, logging, trapping, seasonal jobs, and, in many cases, government relief. Courtesy of the Cook County Historical Society, PH-B232.

administration. To a greater or less extent they run through the entire history of the state. They show that land has been given away or practically given away, that timber has been ruthlessly destroyed, and that agriculture has been stimulated in unsuitable areas. Inefficiency and incompetency, a palpable neglect of public welfare, a careless and unsystematic conduct of the business of the state, unpardonable ignorance, and sometimes downright dishonesty are revealed in the manner that land has been disposed of or exploited in the past."

The committee singled out as especially misguided the practice of moving settlers onto lands where conditions were unsuited to agriculture or where the costs of draining swamps or removing farming obstacles such as tree stumps and boulders far outweighed the profits that could be reaped from marginal soils. The report cited, for example, that in 1929 the average statewide value of a farm's produce was more than $2,300. In the cutover region, a typical farm netted less than $1,200. Such a policy of land settlement was especially cruel to the poorest settlers, who, attracted by the cheap price of land in northern Minnesota, often managed only to scrape together a small down payment but then lacked the cash reserves to improve their farms beyond the grubstake stage. "The simple fact is that, for thousands of people in the area, any work to which they can now put their hands offers so little reward and so little security that discouragement prevails everywhere," the authors lamented. "Many have settled upon land which has never yielded them a good living, even in better times. In short, they simply cannot earn incomes needed to give them a standard of living comparable to that of friends and relatives in other sections of the state." The report noted that even in the few places with productive soils, such as portions of the highlands, settlements often were scattered and located so far from roads that the cost of transporting produce to market cut deeply into profits.

To finance expensive new services such as schools, roads, fire protection, mail delivery, and law enforcement, northern Minnesota counties not only levied high property taxes but also borrowed heavily. Northern landowners were taxed at two and three times the rate of landowners in southern counties and shouldered the largest per capita local debt burden in the state.

Settlers were living on borrowed time. When the supply of trees dwindled and crop prices fell, their time had run out. In 1932, the tax delinquency rate for the Finland community, for example, was 62 percent. But compounding the financial woes of this small rural community—and other communities just like it—was the high rate of tax delinquency countywide, especially among large landowners. That same year Lake County had a tax delinquency rate of 28 percent. Most of the landowners were absentee lumber companies that had creamed off the valuable timber on their holdings and then abandoned the near-worthless hectares when no bidders came forward to purchase the cutover land. According to 1932 records, Lake County had collected no taxes on 80 percent of landholdings of more than one thousand acres. Even though officials offered bargain settlements for back taxes, few delinquent owners responded. In some cases, lands were sold for a fraction of their back taxes in a deferred-payment plan, which allowed corrupt owners to quickly liquidate the second-growth forest on their holdings for a minuscule sum before they let the land slip into tax delinquency once again. Those who remained on the land were left holding the bag, contributing further to the spiraling abandonment of land. "Every tax-delinquent acre narrows the tax base and adds to the burden of those who pay," Davis observed.

Preserving Ecological Wholeness in the Highlands

Faced with ownership of millions of hectares of tax-delinquent and abandoned land, the State of Minnesota recognized the need to ward off another economic, social, and environmental catastrophe by initiating centralized planning in the northern counties. Authors of the land-utilization report urged the state to adopt a land-classification scheme similar to the one proposed for the cutover regions of Michigan and Wisconsin. Attempting to categorize the land according to "its highest and best use," the state devised a system that placed land into five divisions: agricultural, forestry, mining, recreational, and water storage.

With the exception of a small corner of southwestern Lake County, which was identified as suitable for agricultural purposes, most of Lake and Cook Counties was designated for forestry and recreational purposes. The report's authors explained that because the North Shore was prone to surface runoff and soil erosion, the watershed would be best protected by keeping the land tree covered. Not only would trees help to protect water quality and replenish the nation's timber supply, but they would also provide the scenic backdrop for a growing new industry—tourism. Davis was part of a growing chorus of voices to suggest that "attractive campgrounds and adequately furnished cabins should be provided for tourists. The numerous streams and lakes, the rocky landscape, the wild life, and the gorgeous coloring of the forest in the early fall make this an exceptionally attractive resort area." (For a more in-depth discussion of the growth of North Shore tourism, see "Nearshore.")

The debt-ridden status of northeastern Minnesota, as well as the growing awareness that rebuilding the forest was fundamental to national security and local economies, spurred expansion of the Superior National Forest and a patchwork of state and county forests along the North Shore. But recognizing that the land was better suited to forestry and recreation than growing lettuce and potatoes has not necessarily restored the full complement of species to this altered ecoregion. True, sugar maples form the basis for one of the most intact forest types in the watershed. Unlike maples in more southerly forests, which were harvested for saw timber and veneer, the highlands' maples escaped wholesale cutting by the logger's ax largely due to their poor quality. "In general it is an unquestionable and well recognized fact that the hardwood timber becomes smaller and scrubbier toward the north," stated an 1898 report from the U.S. Department of Agriculture. Growing at the northern end of their range, maples not only were stunted but also damaged by the cold. Cracks and thickened ridges riddled their bark, indicating frost damage. Such injuries often marked the entry wounds through which insects and fungal diseases penetrated the tissues of the trees. Not surprisingly, among the northern hardwoods, the report noted, mature maples were the most defective, holding "but third rank and less as a timber tree."

As the availability of saw timber declined, the wood-products industry began to focus on species that could be utilized as pulp for manufacturing high-quality paper. Initially, sugar maple and other hardwoods were largely ignored because, pound per pound, their comparatively short fibers yielded less pulp than other species. Instead, from the 1930s through 1960s in northern Minnesota, the industry targeted long-fibered woods like that of black spruce. By the 1970s, new technologies that utilized pulp made from aspen allowed the midwestern wood-products industry to produce other kinds of wood products, such as oriented strand board. Aspen was far more abundant than other species in the late-twentieth-century forest, and supplies were located near pulp mills. And compared to other tree species, aspen was also faster growing and easier to regenerate.

Nonetheless, the highlands' maples would not escape the logger's saw. In the 1970s and 1980s, the highlands' rich soils attracted the attention of federal and state forest managers. Many maple-dominated woodlands were cleared to make way for plantations of commercially valuable species such as spruce and red pine. Heavy-handed practices, such as intensive mechanical scarification and herbicide treatment, were used to suppress the regrowth of the maple forest in favor of conifer monocultures.

As a result, many of the fundamental processes in the forest ecosystem were altered, functional relationships severed, and diversity diminished. Since then, such conversions have been uncommon largely due to changes in forestry markets and products, shrinking budgets for expensive plantation-style reforestation, and a greater appreciation of the native forests for their recreational and ecological values. Nonetheless, decades after their ecological alteration, these lands remain one of the major causes of habitat fragmentation in the highlands.

By the 1990s, another threat to the ecological integrity of the highlands was on the horizon—clear-cut logging. The Generic Environmental Impact Study (GEIS), a study of statewide timber harvesting completed in 1993, identified the highlands as one of two regions from which future increases in logged wood would come to fuel the needs of expanding pulp mills. This signaled a pronounced shift in the kind of raw materials targeted by the pulp industry. As supplies of aspen, the longtime staple of Minnesota's pulp industry, entered a long-predicted decline, stumpage prices rose. Having developed methods for pulping hardwood species more efficiently, mills could quickly turn toward more readily available—and therefore cheaper—wood such as maple. In 2002, Saapi Ltd. (a South African company that purchased the Potlatch paper mills) set payments for aspen that were lower than those at which loggers could profitably deliver. To make up the difference in its pulpwood demand, the company bought as much available maple as possible. In the future, such market forces, along with expansions of Minnesota's pulp-mill capacity, could increase pressure on the highlands' maple forests.

And if forest-products researchers at the Natural Resources Research Institute (NRRI) have their way, maple's physical defects will no longer disqualify it from use in the construction of houses and light-commercial buildings. NRRI researchers developed techniques that transform red and sugar maples into structural components such as I-joists and trusses. NRRI wood engineer Brian Bradshaw estimates that such applications could increase the value of this low-grade lumber by 100 percent. In a different economic climate or during shortages of other wood resources, maple trees growing in the North Shore Highlands may attract far more attention from wood product manufacturers than ever before.

The result could be catastrophic. Unlike the boreal-hardwood and conifer forest, which is adapted to great stand-replacing disturbances such as fires and therefore has been more forgiving of drastic changes such as clear-cuts, the maple forest most commonly is regenerated through small gap openings in the forest caused by the fall of individual or small groups of trees. Northern hardwood forests, Minnesota ornithologist Janet Green says, have been managed successfully elsewhere by a technique known as uneven-aged harvest in which individual trees are selected for cutting, but "we don't have a tradition for that in northern Minnesota. The question is, Can people in this state figure out a silvicultural technique for harvesting northern hardwoods besides clear-cutting?"

Preserving the integrity of the North Shore Highlands is even more critical when you examine the larger ecological and economic context. In many parts of northern Minnesota, the northern hardwood forest already is declining. Eager to increase populations of deer and grouse for hunting on their lands, private landowners are converting their forests to aspen. The maple ridge is "one of the last areas in northern Minnesota for large, contiguous blocks of mature deciduous forest," Green says. The forest is vital to North Shore tourism, drawing thousands of visitors each year for the fall leaf tour. How important is it to birds? "It's largely responsible for the health of the bird populations that use northern hardwoods," Green says. (See also "As Good As It Gets: Bird Diversity on the North Shore.")

Given their prolific reproduction, sugar maples have been able to recover in many cases from human-caused disturbances. Not all species have been so lucky. Largely logged out of the mixed forest, white pine is having a difficult time making a comeback. The lack of mature trees that provide a seed source and browsing by deer are major barriers. Seedlings that escape the mouths of hungry deer frequently succumb to white-pine blister rust, a fungal disease that was introduced into the United States around 1908.

Yellow birch regeneration also appears to be failing in many parts of the highlands. Like white pine, yellow birch was high-graded from the forest (during World War I yellow birch was much in demand for use in airplane propellers). In Minnesota yellow birch grows only in the highlands and in a few sheltered inland sites. Its many desirable qualities render yellow birch the most commercially valuable species in the maple-dominated mixed forest. Solitary giants, the trees can grow up to three feet in diameter, a size that is well suited to veneer production. And its beauty—a fine-grained, honey-colored wood—makes it highly coveted in the crafting of cabinets, furniture, and paneling.

But the cutting of yellow birch has stymied regeneration. Gone are many of the big, isolated trees that served as vital sources of seed. Just as important is the absence of toppled giants decaying on the forest floor. Most yellow birch seedlings gain their first root hold in life on downed nurse logs. The importance of preserving and fostering these specialized microhabitats to serve the needs of tender seedlings has been largely ignored by foresters in harvest plans. At the same time, they have been unable to promote yellow birch regeneration through other means.

Like logging, recreational development has not been an ecologically benign use in the highlands. As lakeshore lots grow increasingly scarce and expensive, people are building vacation and retirement homes on the highlands' ridges, fragmenting the forest's canopy as well as introducing a landscape aesthetic that often is at odds with the ecological vitality of the forest. The floor of the maple-dominated forest, for example, is characterized by coarse woody debris and logs in various stages of decay. Many homeowners "tidy up" the forest, clearing out the jumble of organic matter that provides homes for small—but ecologically important—animals such as voles and salamanders and brooding sites for ground-nesting birds. Many of the small forest dwellers that escape the leaf blower and rake fall prey to domesticated animals, especially cats. In 2013, a trio of scientists, including Peter Marra of the Smithsonian Conservation Biology Institute in Washington, D.C., published a review of the scientific data from multiple studies on the predation of birds and mammals by cats. The scientists estimated that between 1.4 billion and 3.7 billion birds are killed each year by both feral and household cats as well as from 6.9 billion to 20.7 billion mammals. "Our findings suggest that free-ranging cats cause substantially greater wildlife mortality than previously thought," the study authors write, "and are likely the single greatest source of anthropogenic mortality for U.S. birds and mammals."

Subdividing the woods also carries with it other threats to healthy ecological functioning. As Michigan writer Stephanie Mills observes, "The effect of this land-use pattern is the usual: fragmentation of the landscape, reducing its biotic integrity, stability and beauty. Commonplace suburban conditions—lawns, prowling house cats, septic systems, gaps in the forest canopy, vehicles unpredictably crossing what used to be the nocturnal animals' paths to the water, rank vegetation flourishing in the sunny paths of soil disturbance, noise during the birds' nesting seasons, and additional tykes making a pastime of catching frogs and fish, in competition with the herons—are hard on all but the weediest plants and animals; *sic transit* paradise."

And with the burgeoning interest in sports and physical fitness has come an increased demand for recreational trails. Users including hikers, mountain bikers, cross-country skiers, snowmobilers, and ATV riders all are demanding their own trails through the woods. Wide trails, in particular those that are needed to safely accommodate high-speed travel by snowmobiles and ATVs, multilane ski traffic, and trail-grooming machinery, can not only create serious erosion problems but also provide migratory corridors for nonnative species, which may proliferate in otherwise hard-to-reach interior forest habitats and threaten native species.

A tidy landscape aesthetic can threaten the healthy functioning of forest ecosystems by fragmenting the canopy with open lawns and clearing woodlands of dead snags and woody debris. Photograph by Paul Stafford. Copyright Explore Minnesota Tourism.

With the greater influx of people comes a need to find local and cost-effective means for dealing with waste, not simply landfill space for garbage but for sewage sludge from water-treatment facilities and soils contaminated by petroleum and other chemicals. The extraordinarily rich microbial life of soils in maple-dominated forests, particularly microscopic fungi, have made the highlands an increasingly attractive place for a practice known as land treatment. In this process, land is cleared and tilled. Soils contaminated with petroleum products ranging from gasoline and kerosene to heating and diesel fuels are spread over the surface of the exposed earth at a depth no greater than four inches and then worked into the top layers of the soil. The petroleum contaminants attach to soil particles, stabilizing them so that microbes, insects, and worms can degrade them into benign substances, including carbon dioxide, water, and fatty acids. Chemical breakdown also is speeded up by regular tilling, which exposes the petroleum components to sunlight and oxygen. Most sites are given a clean bill of health— that is, they contain contaminant levels within a certain allowable limit—within one to two years. Land in Cook County has been used to treat contaminated soils for government agencies such as the USFS and the Army Corps of Engineers as well as for privately owned businesses in Grand Marais.

Such practices, however, can contaminate groundwater and surface waters and cause other environmental problems in the future. While most sites are permanently converted to industrial uses, some parcels are abandoned. Since no monies are available for restoration, these tracts often introduce conditions that make the forest vulnerable to biological pollution. While native trees and shrubs may reclaim these tracts, the recovery of native herbaceous plants is unlikely since the prescribed tilling regime normally destroys the seed bank stored in the soil. Those that do germinate are often outcompeted by exotic species that are quick to invade open, disturbed areas and prevent the establishment of native plants.

Eating the Forest Out of House and Home: The Invasion of Earthworms

But some of the greatest threats to the highlands may come about as the indirect result of human activities. In the mid-1980s, employees of the Chippewa National Forest were working in a forest near Leech Lake when they noticed bare patches of black soil on the forest floor where the duff had disappeared. Scientists largely dismissed these "bald spots" as localized occurrences until the early 1990s when a project to map the ecological zones of the forest revealed additional swaths of exposed earth. Finding so much bare soil in places that should otherwise have had a thick layer of litter and cushiony duff prompted scientists to take a closer look.

What they discovered was both shocking and dismaying—an army of earthworms was eating the forest right out from under its roots—literally. The black soil was all that remained of the forest duff after it had passed through the guts of these small but powerful eating machines. In worm-infested study plots, researchers tallied an average of twenty-eight to thirty-seven worms per square foot; in others that number soared to more than sixty-five individuals per square foot. (For a more in-depth discussion of forests and earthworms, see "The Case of the Missing Duff: Earthworm Invasions.")

Contrary to popular perception, earthworms are not native to the parts of North

America that once were covered by the Wisconsin ice sheet, which includes northern Minnesota. Earthworms were introduced to the region from Europe and Asia in a wide variety of imported materials, ranging from ship ballast to the root balls of plants. These exotic invaders now are finding their way into North Shore forests via anglers who discard fishing bait on the shores of inland lakes, in the tire treads of ATVs, mountain bikes, and logging equipment, and in the lugs of a hiker's boot.

In some forests, earthworm damage seems to be compounded by the presence of another species that invaded in the wake of European settlement—white-tailed deer. "Most people don't know that white-tailed deer are not native to most of Minnesota," says Cindy Hale, one of the principal investigators of worm impacts. Changes in vegetation that followed Euro-American settlement favored the expansion of deer into the more northerly territories that were once dominated by elk, moose, and woodland caribou. In fact, many researchers hypothesize that white-tailed deer may not have even ventured into northeastern Minnesota prior to the early 1900s. By 1938, researchers recorded deer densities of ten to forty-one individuals per square mile. Since then average densities have dropped only slightly. In the past decades, wildlife biologists have registered mean densities of eight to twenty-six animals per square mile.

For many forest species, worms and deer can be a deadly combination. The researchers hypothesize that invading worms reduce plant populations but do not completely eliminate them. When freed from the pressure of hungry deer, many plants are able to recolonize former territories. But once worms have reduced plant populations below a critical threshold, deer browsing is like the final nail in the coffin—many species are driven to local extinction.

One of the potential strategies for helping the forest recover in some areas is to intensively control or exclude deer for a period of time until plants can rebuild their populations. Unfortunately, any plan to remediate the forest is complicated by the fact that each species has its own population threshold, most of which have not yet been determined since few plants have been thoroughly studied.

But even without the aid of earthworms, deer have suppressed—and in some cases, virtually eliminated—the regeneration of many plant species in the Lake Superior basin. Forest ecologist Meredith Cornett spent five years studying the effect of deer browsing on northern white cedar trees along the Minnesota North Shore. "Virtually no regeneration is happening, at least within one-third mile of the North Shore," she observes.

The problem is twofold. Cedar seedlings rely on safe sites, such as large downed logs on the forest floor, for germination and establishment. Logging has changed the species composition as well as the age structure of the forest such that these seedbeds are few and far between. But even when seedlings do get a start in life, chances are good that they will be cropped by deer, which favor cedar as browse. This is particularly true along the North Shore. During winter, milder temperatures and lower snow depths draw many deer down to the shore. In 1998 biologists cataloged as many as 324 animals per square mile in deer yards within a few miles of Superior. Many deer seek out the protective microclimates of cedar forests, where the dense growth cuts the sharpness of the wind and the closed canopy keeps temperatures higher than the surrounding terrain by slowing the radiation of heat out into the winter sky.

Such concentrations of deer only increase the vulnerability of seedlings in the understory. Indeed, nineteenth-century surveys show that northern white cedar occupied 8 percent of the canopy cover in the presettlement forest. Today that number has fallen to 4 percent. If browsing continues at current levels, forest ecologists, like Cornett, predict that the northern white cedar "will not persist as an important forest type in the Great Lakes Landscape."

Cedar is not the only species to suffer declines due to pressure from deer in the Lake Superior region. Populations of mountain maple, yellow birch, mountain ash, and Canada yew all have plummeted. Dwindling too are herbaceous species such as wild sarsaparilla, Canada mayflower, trillium, and bluebead lily. Some species, such as blue-bead lily and Canada yew, are especially sensitive to browsing and have difficulty making a comeback even when protected by deer exclosures. Deer, for example, have decimated populations of Canada yew in the Apostle Islands (just as excessive consumption by moose has resulted in the near elimination of the species on Isle Royale). Researchers suspect that the absence of bluebead lily on Madeline Island, the largest in the Apostle Island chain, is attributable to the presence of deer. Recolonization by many species, even in the face of relaxed browsing pressure, is complicated by habitat fragmentation and by the influx of exotic species that have invaded areas formerly occupied by native species.

Solutions to the problems of intensive browsing by deer include building exclosures that would allow sensitive species to recover. But such fencing is expensive to erect and maintain. Exclosures are particularly impractical for protecting species that require extensive forest habitats for their long-term survival.

Controlling deer numbers is key to alleviating their negative effects on sensitive plant populations. Fundamental to any deer-control strategy, however, is the question of forest management. The pattern and frequency of cutting have subdivided a significant portion of the forest into patches of young aspen and birch, a browse favored by deer. Patterns of timber cutting, along with those of human settlement, have created openings and edges in the forest that make for prime deer habitat. At the same time, the warmer wintering grounds near Lake Superior still retain enough conifer cover to provide refuge for deer. These conditions work together to sustain high numbers of deer and make it easier for deer numbers to rebound after periods of high mortality during severe winters.

Deer numbers are also controlled by a powerful and vocal constituency that demands that the population be artificially inflated for the purposes of hunting. "Deer herds are currently maintained at high levels largely as a consequence of predator elimination and legislation that restricts the taking of deer by hunting," write biologists Roger C. Anderson and Alan J. Katz. "At issue is a conflict of purposes between game mangers, who maximize herd size for hunting, and those desiring to maintain biotic diversity by restricting deer numbers."

The Highlands' Future

The maple-dominated mixed forest is a remarkably resilient, long-lived forest type. The architecture of the forest resists large-scale blowdowns. Its high moisture content holds fires at bay. And because its dominant species—sugar maple—is represented in a variety of age classes, from saplings to granddaddy trees, the forest is free from widespread decimation by insects and disease.

If we abide by its ecological rules, such woodlands may long outlive us, providing a critical foil to the neighboring near-boreal forest to the north, which thrives on more frequent catastrophic changes for its health and longevity.

Indeed, the forests of the North Shore Highlands offer us the possibility of a fresh start. Despite the unprecedented changes that have swept the highlands during the past 150 years, much of this great linear belt of forest remains intact—so too its ecological function.

Whether we preserve this ecological function depends on how well we have learned our lessons from the past. Forest ecologists now know, for example, that the maple-dominated mixed forest is renewed by gap-phase disturbances, in which individual or small groups of trees are toppled, creating gap-sized openings within the larger forest matrix. By mimicking this natural disturbance pattern, timber-harvest techniques, such as thinning or uneven-aged management, preserve the pulse of change to which the forest's plants and animals have become adapted, thereby achieving ecological goals as well as commodity production.

Such cutting techniques also allow a greater diversity within the forest-products industry. Large, individual hardwood trees can be utilized in quality veneers, saw timber, and value-added products such as cabinetry and furniture. Unfortunately, whenever the balance of the forest-products industry has tipped in favor of the state's pulp mills, commitments to sustaining ecological and economic diversity have wavered. The pulp industry and developing industries in the energy sector feed on vast quantities of wood that are procured primarily using practices that run counter to the ecological processes and self-sustaining strategies of the maple-dominated hardwood forest.

Place-based development models must govern other land-use activities as well. The beauty of the highlands undoubtedly will continue to draw people seeking to build vacation and retirement homes. Unfortunately, developers and homeowners have few guidelines for good management practices that preserve the ecological underpinnings of the forest, including limits on forest clearing, retention of the tree canopy over openings such as driveways, homes, and yards, and strategies for reducing the introduction of exotic species.

The future of the highlands' forest depends on our willingness to play by its ecological rules. The choice is ours. ❧

SUGGESTIONS FOR FURTHER READING

Amundson, Donna C., and Herbert E. Wright Jr. "Forest Changes in Minnesota at the End of the Pleistocene." *Ecological Monographs* 49, no. 1 (1979): 1–16.

Anderson, Roger C., and Alan J. Katz. "Recovery of Browse-Sensitive Tree Species Following Release from White-Tailed Deer *Odocoileus virginianus* Zimmerman Browsing Pressure." *Biological Conservation* 63 (1993): 203–8.

Baker, Richard G. "Late-Glacial Pollen and Plant Macrofossils from Spider Creek, Southern St. Louis County, Minnesota." *Geological Society of America Bulletin* 76 (1965): 601–10.

Balgooyen, Christine P. "The Use of *Clintonia borealis* and Other Indicators to Gauge Impacts of White-Tailed Deer on Plant Communities in Northern Wisconsin, USA." *Natural Areas Journal* 15, no. 4 (1995): 308–18.

Balogh, Anne L., Thomas B. Ryder, and Peter P. Marra. "Population Demography of Gray Catbirds in the Suburban Matrix: Sources, Sinks and Domestic Cats." *Journal of Ornithology* 152 (2011): 717–26.

Bennett, Edward B. "Characteristics of the Thermal Regime of Lake Superior." *Journal of Great Lakes Research* 4, nos. 3–4 (1978): 310–19.

Blegen, Theodore C. "The Competition of the Northwestern States for Immigrants." *Wisconsin Magazine of History* 3 (1919): 3–29.

Bourdo, Eric A., Jr. "The Forest the Settlers Saw." In *The Great Lakes Forest: An Environmental and Social History*, ed. Susan L. Flader, 3–16. Minneapolis: University of Minnesota Press, 1983.

Braun, E. Lucy. "The Phytogeography of Unglaciated Eastern United States and Its Interpretation." *Botanical Review* 21 (1955): 297–375.

Brown, Ralph H. "Fact and Fancy in Early Accounts of Minnesota's Climate." *Minnesota History* 17, no. (1936): 243–61.

Butters, Fred K., and Ernst C. Abbe. "A Floristic Study of Cook County, Northeastern Minnesota." *Rhodora* 55 (1953): 21–85.

Cleland, Charles E. "Indians in a Changing Environment." In *The Great Lakes Forest: An Environmental and Social History*, ed. Susan L. Flader, 83–95. Minneapolis: University of Minnesota Press, 1983.

Committee on Land Utilization. *Land Utilization in Minnesota: A State Program for the Cut-Over Lands*. Minneapolis: University of Minnesota Press, 1934.

Conover, Adele. "Foreign Worm Alert." *Smithsonian,* August 2000.

Cornett, Meredith W., Lee E. Frelich, Klaus J. Puettmann, and Peter B. Reich. "Conservation Implications of Browsing by *Odocoileus virginianus* in Remnant Upland *Thuja occidentalis* Forests. *Biological Conservation* 93, no. 3 (2000): 359–69.

Curtis, John T. *The Vegetation of Wisconsin.* Madison: University of Wisconsin Press, 1959.

Davis, Darrell H. "The Finland Community, Minnesota." *Geographical Review* 25 (1935): 382–94.

Davis, Margaret Bryan. "Holocene Vegetational History of the Eastern United States." In *Late Quaternary Environments of the United States.* Vol. 2, *The Holocene,* ed. Herbert E. Wright Jr., 166–81. Minneapolis: University of Minnesota Press, 1983.

———. "Pleistocene Biogeography of Temperate Deciduous Forests." *Geoscience and Man* 13 (1976): 13–26.

———. "Quaternary History and the Stability of Forest Communities." In *Forest Succession: Concepts and Applications,* ed. Darrell C. West, Herman H. Shugart, and Daniel B. Botkin, 132–53. New York: Springer-Verlag, 1981.

Densmore, Frances. "Uses of Plants by the Chippewa Indians." In *Forty-Fourth Annual Report of the Bureau of American Ethnology to the Secretary of the Smithsonian Institution, 1926–1927,* 275–397. Washington, D.C.: U.S. Government Printing Office, 1928. Reprint. *How Indians Use Wild Plants for Food, Medicine, and Crafts.* New York: Dover Publications, 1974.

Eichenlaub, Val. *Weather and Climate of the Great Lakes Region.* Notre Dame, Ind.: University of Notre Dame Press, 1979.

Ellsworth, David S., and Peter B. Reich. "Leaf Mass Per Area, Nitrogen Content, and Photosynthetic Carbon Gain in *Acer saccharum* Seedlings in Contrasting Forest Light Environments." *Functional Ecology* 6, no. 4 (1992): 423–35.

Flaccus, Edward, and Lewis F. Ohmann. "Old-Growth Northern Hardwood Forests in Northeastern Minnesota." *Ecology* 45, no. 3 (1964): 448–59.

Frelich, Lee E. "Old Forest in the Lakes States Today and before European Settlement." *Natural Areas Journal* 15, no. 2 (1995): 157–67.

Frelich, Lee E., and Peter B. Reich. "Old Growth in the Great Lakes Region." In *Eastern Old-Growth Forests: Prospects for Rediscovery and Recovery,* ed. Mary Byrd Davis. Washington, D.C.: Island Press, 1996.

Gates, Gordon E. "Miscellanea Megadrilogica." *Megadrilogica* 1 (1970): 1–14.

Gundale, Michael J. "Influence of Exotic Earthworms on the Soil Organic Horizon and the Rare Fern *Botrychium mormo.*" *Conservation Biology* 16 (2002): 1555–61.

Hale, Cindy. "Earthworms May Be Threatening Biodiversity of Hardwood Forests." *Minnesota Plant Press* (Minnesota Native Plant Society Newsletter) 20, no. 2 (Winter 2001): 1, 3.

Hale, Cindy M., Lee E. Frelich, and Peter B. Reich. "Changes in Cold-Temperate Hardwood Forest Understory Plant Communities in Response to Invasion by European Earthworms." *Ecology* 87, no. 7 (2006): 1637–49.

———. "Effects of European Earthworm Invasions on Soil Characteristics in Northern Hardwood Forests of Minnesota, U.S.A." *Ecosystems* 8 (2005): 911–27.

———. "Exotic European Earthworm Community Composition in Northern Hardwood Forests of Minnesota, U.S.A." *Ecological Applications* 15, no. 3 (2005): 848–60.

Hall, Christopher W. "Report of Professor C. W. Hall." In *The Geological and Natural History Survey of Minnesota: Eighth Annual Report for the Year 1879*. St. Paul: Pioneer Press Company, 1880.

Hendrix, Paul F., ed. *Earthworm Ecology and Biogeography in North America*. Boca Raton, La.: Lewis Publishers, 1995.

Hendrix, Paul F., and Patrick J. Bohlen. "Exotic Earthworm Invasions in North America: Ecological and Policy Implications." *BioScience* 52 (2002): 801–11.

Hewitt, Girart. *Minnesota—Its Advantages to Settlers*. St. Paul: Press Printing Company, 1867.

Jackson, Stephen T., and Chengyu Weng. "Late Quaternary Extinction of a Tree Species in Eastern North America." *Proceedings of the National Academy of Sciences of the United States of America* 96, no. 24 (1999): 13847–52.

Kohl, Johann Georg. *Kitchi-Gami: Life among the Lake Superior Ojibway*. London: Chapman and Hall, 1860. Reprint, St. Paul: Minnesota Historical Society Press, 1985.

Lass, William E. "Minnesota—An American Siberia?" *Minnesota History* 49 (1984): 149–55.

Loss, Scott R., Tom Will, and Peter P. Marra. 2013. "The Impact of Free-Ranging Domestic Cats on Wildlife of the United States." *Nature Communications* 29 January 2013. http://www.nature.com/ncomms/journal/v4/n1/abs/ncomms2380.html.

Marks, Peter L. "The Role of Pin Cherry (*Prunus pensylvanica* L.) in the Maintenance of Stability in Northern Hardwood Ecosystems." *Ecological Monographs* 44, no. 1 (1974): 73–88.

McKee, Russell. "Tombstones of a Lost Forest." *Audubon,* March 1988: 62–73.

McLean, Mary Ann, and Dennis Parkinson. "Soil Impacts of the Epigeic Earthworm *Dendrobaena octaedra* on Organic Matter and Microbial Activity in Lodgepole Pine Forest." *Canadian Journal of Forestry Research* 27 (1997): 1907–13.

Milius, Susan. "Wind Highways: Mosses, Lichens Travel along Aerial Paths." *Science News* 165, no. 21 (22 May 2004): 324.

Minnesota Pollution Control Agency. "Land Treatment Fact Sheet." St. Paul: Minnesota Pollution Control Agency, n.d.

Minnesota State Bureau of Immigration. *Minnesota, The Land of Opportunity for Agriculture, Horticulture, Live Stock, Manufactures, Mining, Education and Everything That Attracts the Immigrant*. St. Paul: State Bureau of Immigration, circa 1912.

———. *Northern Minnesota—Land of Certainties*. Minneapolis: Great West Printing Co., 1919.

Mortensen, Steve, and Carol Mortensen. "A New Angle on Earthworms." *Minnesota Conservation Volunteer,* July/August 1998: 20–29.

Morton, Ron, and Carl Gawboy. *Talking Rocks: Geology and 10,000 Years of Native American Tradition in the Lake Superior Region*. Duluth, Minn.: Pfeifer-Hamilton Publishers, 2000.

Muñoz, Jesús, Ángel M. Felicísimo, Francisco Cabezas, Ana R. Burgaz, and Isabel Martínez. "Wind as a Long-Distance Dispersal Vehicle in the Southern Hemisphere." *Science* 304, no. 5674 (21 May 2004): 1144–47.

Phillips, David W. "Environmental Climatology of Lake Superior." *Journal of Great Lakes Research* 4, nos. 3–4 (1978): 288–309.

Pielou, E. C. *After the Ice Age: The Return of Life to Glaciated North America*. Chicago: University of Chicago Press, 1991.

Potzger, J. E. "Phytosociology of the Primeval Forest in Central-Northern Wisconsin and Upper Michigan, and a Brief Post-Glacial History of the Lake Forest Formation." *Ecological Monographs* 16 (1946): 211–50.

Pregitzer, Kurt S., David D. Reed, Theodore J. Bornhorst, David R. Foster, Glenn D. Mroz, Jason S. Mclachlan, Peter E. Laks, Douglas D. Stokke, Patrick E. Martin, and Shannon E. Brown. "A Buried Spruce Forest Provides Evidence at the Stand and Landscape Scale for the Effects of Environment on Vegetation at the Pleistocene/Holocene Boundary." *Journal of Ecology* 88 (2000): 45–53.

Proulx, Nick. Ecological Risk Assessment of Non-indigenous Earthworm Species. Prepared for U.S. Fish and Wildlife Service, 2003.

Reynolds, John Warren, Dennis R. Linden, and Cindy M. Hale. "The Earthworms of Minnesota (Oligochaeta: Acanthodrilidae, Lumbricidae and Megascolecidae)." *Megadrilogica* 8 (2002): 86–100.

Roberts, Thomas S. "Plants of the North Shore of Lake Superior, Minnesota." In *The Geological and Natural History Survey of Minnesota: The Eighth Annual Report for the Year 1879*. St. Paul: Pioneer Press Co., 1880.

Smith, Lloyd L., Jr., and John B. Moyle. "A Biological Survey and Fishery Management Plan for the Streams of the Lake Superior North Shore Watershed." Technical Bulletin No. 1. St. Paul: Minnesota Department of Conservation, Division of Fish and Game, 1944.

Stearns, Forest. "The Composition of the Sugar-Maple-Hemlock-Yellow Birch Association in Northern Wisconsin." *Ecology* 32, no. 2 (1951): 245–65.

———. "Forest History and Management in the Northern Midwest." In *Management of Dynamic Ecosystems*, ed. James M. Sweeney. West Lafayette, Ind.: The Wildlife Society, 1990.

Stewart, Amy. *The Earth Moved: On the Remarkable Achievements of Earthworms*. Chapel Hill, N.C.: Algonquin Books of Chapel Hill, 2004.

Tubbs, Carl Herman. "The Competitive Ability of Yellow Birch (*Betula alleghaniensis* Britton) Seedlings in the Presence of Sugar Maple (*Acer saccharum* Marsh) with Specific References to the Role of Allelopathic Substances." Ph.D. diss. University of Michigan, 1970.

Watts, William A. "Late-Glacial Plant Macrofossils from Minnesota." In *Quaternary Paleoecology*, ed. Edward J. Cushing and Herbert E. Wright Jr., 89–97. New Haven, Conn.: Yale University Press, 1967.

———. "Vegetational History of the Eastern United States 25,000 to 10,000 Years Ago." In *Late Quaternary Environments of the United States*. Vol. 1, *The Late Pleistocene,* ed. Stephen C. Porter, 294–310. Minneapolis: University of Minnesota Press, 1983.

Wehrwein, George S. "A Social and Economic Program for the Sub-marginal Areas of the Lake States." *Journal of Forestry* 29 (1931): 915–24.

Wright, Herbert E., Jr. "Glacial Fluctuations and the Forest Succession in the Lake Superior Area." *Proceedings of the 12th Conference on Great Lakes Research,* 397–405. Ann Arbor, Mich.: International Association of Great Lakes Research, 1969.

———. "The Roles of Pine and Spruce in the Forest History of Minnesota and Adjacent Areas." *Ecology* 49 (1968): 937–55.

———. "Tunnel Valleys, Glacial Surges, and Subglacial Hydrology of the Superior Lobe, Minnesota." *Geological Society of America Memoir* 136 (1973): 251–76.

Wright, Herbert E., Jr., and William A. Watts, "Glacial and Vegetational History of Northeastern Minnesota." Minnesota Geological Survey, SP-11, 1969.

INTERNET RESOURCES

Great Lakes Worm Watch, http://www.nrri.umn.edu/worms/research/publications

Minnesota Worm Watch, www.nrri.umn.edu/worms

Earthworm Invasions

WHEN BIOLOGIST CINDY HALE TAKES PEOPLE ON A TOUR OF HER STUDY PLOTS in Minnesota's Chippewa National Forest, she asks them to get a feel for the sites—literally, with their feet. As they leave the forest's edge along the roadside and plunge deeper into the dim, humid interior of a maple-dominated mixed woodland, visitors invariably comment on the change underfoot. Beneath the shade cast by the dense tree canopy, the forest floor feels soft and spongy, as if they had slipped out of worn-out sneakers into a pair of cushy new running shoes.

Hale kicks back the dry leaves on the surface, exposing a dense mat of blackened leaves. She peels a layer from what looks like a lasagna of slick leaf bits. A close examination of the earthy-smelling wad reveals a furring of whitish fuzz. Under a microscope, the cottony mass would resemble a mesh of fine threads. These pale filaments, known as hyphae, make up the mycelium, or vegetative part, of a fungus. Scientists estimate that up to 465 miles of these threads can wind their way through one square foot of leaves measuring 1.5 inches thick. Numbering in the hundreds of species, these fungi team up with equally diverse swarms of microscopic bacteria to break down the detritus on the forest floor. Together they drive the metabolic engine that consumes everything from leaves and bird carcasses to bits of tree bark, shriveled herbs, and the carapaces of dead beetles.

Their digestive powers are prodigious but slow, especially given the forest's abundant litterfall. In autumn, maples let loose a blizzard of leaves that are as wide as the span of a grown man's open hand. More leaves fall each year, Hale explains, than can be broken down by even legions of bacteria and fungi, and so the forest digests this bounty in stages. A fallen leaf in this deciduous forest, for example, can take between three and five years to decay. The layering of leaves in various phases of decomposition is what lends a springy feel to the forest floor.

But whole swaths of this plush-pile carpet, particularly near roadways and riverbanks and lakeshores frequented by anglers, have become threadbare. Instead of a thick covering, the forest floor is only thinly mantled with fresh-fallen leaves or is riddled with exposed patches of black earth that resembles potting soil. The cause? A newcomer introduced from Europe whose ravenous appetite is matched only by its innocuousness: the lowly earthworm.

When Hale delivers the double ecological whammy—that not only are earthworms considered exotic species in the formerly glaciated regions of North America but that they have also become what many scientists call the faunal equivalent of invasive weeds in woodland environments—she is met with exclamations of surprise and more than a few eyebrows raised in disbelief. Surely the scientists have got it wrong, some visitors wonder out loud. After all, for those who spent their childhoods in the upper Midwest, sifting the forest floor for fishing bait up at the cabin was as routine as watching robins tugging at nightcrawlers in the rain-soaked lawns of spring. Earthworms were as common as, well, dirt. So how is it that the lowly earthworm, the prized helpmeet of farmer and victory gardener alike, has suddenly been recast as a greedy woodland villain?

It turns out that researchers—albeit in the tiny and extremely specialized field of oligochaetology (the study of worms)—have long known about the alien status of earthworms in northern forests, largely due to the work of Gordon E. Gates (1897–1987), the father of modern oligochaetology. While working for the U.S. Department of

In worm-free sugar maple forests, layers of leaves in various stages of decay create a soft and spongy forest floor, supporting a dense and diverse carpet of herbaceous plants adapted to living in the deep leaf litter. Photograph by David L. Hansen, University of Minnesota.

In forest floors invaded by earthworms, the leaf litter is completely consumed or severely depleted, resulting in the loss of the life-support system for many organisms and the healthy function of the maple forest ecosystem. Photograph by David L. Hansen, University of Minnesota.

Agriculture's Bureau of Plant Quarantine, Gates intercepted thousands of specimens in imported plant-related materials from around the world. From 1950 to 1982 he assembled an exhaustive catalog of both native and exotic earthworm species and mapped their distributions around the United States.

Gates noted that earthworm species of European origin dominated formerly glaciated regions, while native earthworms controlled the unglaciated areas of the United States. In 1966 he published his conclusion that native earthworms naturally occur only south of the glacial margin, a line that roughly runs from Washington State to New Jersey with a southerly jog around the Great Lakes. To date, researchers have tallied nearly one hundred species, approximately two-thirds of which reside south of the glacial margin in the eastern United States, and the remainder occupying terrain throughout the Pacific region.

Whether earthworms once inhabited the formerly glaciated reaches of North America remains a matter of conjecture. Earthworms have left behind no fossil record of their existence largely because their soft tissues decompose before the process of mineralization can begin. But experts agree that even if earthworms managed to make their way north during an interglacial period, they were extirpated during the last glaciation by the Wisconsin ice sheet, portions of which pushed as far south as southern Indiana and Ohio during its maximum extent some twenty-two thousand years ago.

Because earthworms invade new territory at a relatively slow rate—averaging about thirty-three feet per year—scientists estimate that native species have only covered about 155 miles on their northward migration since the retreat of the Wisconsin glacier. As a result, native earthworms have been ruled out as members of the hungry hordes that have assailed forests as far north as Alberta.

Instead, researchers believe that the trouble-making worms arrived in the formerly glaciated regions of North America as early as AD 1500 by a far speedier means—in the hoofs and bedding of animals, in the root balls of plants that were part of the horticultural trade, and in the ballast of ships from Europe. In oligochaetology circles, this explanation is known as the Post Quaternary Introduction Theory. Although many adult worms may not have survived the temperature and moisture extremes of an ocean voyage, the more durable cocoons that contain their offspring likely arrived intact. A single cocoon, which is produced by the clitellum (the reproductive structure that looks like an overinflated tire encircling a worm's body near the head), can harbor up to several thousand hatchlings. In one season alone, an adult can produce at least three or four cocoons, and sometimes more. Their new habitat conditions—from soil types to ground temperatures—closely resembled those of their overseas homelands, enabling many transplanted species to thrive. Researchers estimate that at least forty-five species of exotic earthworms have been introduced into the United States and Canada, mostly from Europe and Asia. Today, forests in Canada, New England, some mid-Atlantic states, and much of the upper Midwest have been colonized by nonnative species of earthworms, predominantly by the European lumbricids (the nightcrawlers).

It was not until the 1950s and 1960s that scientists began to document changes wrought by worms. In 1964 researchers reported findings of a study of forests in the University of Wisconsin Arboretum, Madison, showing that the common nightcrawler species, *Lumbricus terrestris,* had consumed an entire year's worth of litterfall, stripping the forest floor of an estimated seven million leaves per acre. So efficient were they that the depth of the organic layer of forests infested with earthworms measured little more than one-half inch and weighed an estimated 393 pounds per acre. By contrast, the organic layer in worm-free forests was nearly three times as deep and weighed an estimated 3,530 pounds. Around the same time, similar changes were reported from forests in New Brunswick where worms had been deliberately introduced for fishing bait or had hitchhiked in the treads of truck tires and hooves of horses used in logging operations.

So why didn't such findings set off alarm bells throughout the scientific community? In part, the bad news was eclipsed by the good press given to worms in agricultural environments. Among the first to sing the praises of the humble earthworm was Charles Darwin, whose 1881 book *The Formation of Vegetable Mould, through the Action of Worms, with Observations on Their Habitats* sold more than eight thousand copies in three years, becoming the hottest-selling book of any that he published during his lifetime. Darwin countered the perception that earthworms were agricultural pests that competed with humans for precious food crops. In his numerous experiments with worms (he even played the piano to some of them to ascertain whether they could hear), Darwin discovered that worms served as tiny tillers of the soil. From calculations taken from one of his study fields, Darwin discovered that worms could deposit two inches of new topsoil every ten years. Championing the powers of worms to one skeptic, Darwin declared, "It will be difficult to deny the probability that every particle of earth forming the bed from which . . . old pasture land springs has passed through the intestines of worms."

Since then, researchers have discovered other benefits that result from the activity of earthworms in agricultural soils. If left up to microbial action alone, plant residues on the surface of farm fields would decompose very slowly, delaying the release of nutrients that could fuel the growth of new plants. Earthworms, on the other hand, feed on the bacteria and fungi that grow on decaying organic matter. In the process, they consume vast quantities of everything from hay stalks to animal droppings to satisfy their nutritional needs. As a result, they speed up the pace of decomposition and the release of chemical nutrients.

Worms simultaneously ingest huge amounts of soil, which, in the absence of teeth, serve to grind their food. The soil that passes through their digestive tracts is rich in microscopic plants and animals, including mycorrhizal fungi. These fungal filaments attach to plant roots, extending their reach and their ability to scavenge water and nutrients from the soil. Worms further enrich the contents of their guts with their own microbes. As they excrete their nutrient-rich castings, they also disperse these beneficial organisms throughout the soil.

Worms also excavate tunnels that loosen compacted soils, giving plant roots space to expand while increasing the amount of available oxygen and water. By allowing a greater infiltration of water, these vertical burrows also help to stem soil erosion. Last but not least, worms control populations of agricultural pests while serving themselves as a high-protein staple for animals ranging from birds and moles to salamanders.

"Everything you've ever heard about worms being good for the soil is still true in agricultural and garden soils," says biologist Hale, who began monitoring plant diversity and the abundance of tree seedlings in four hardwood forests in the Chippewa National Forest in 1998. "But the reason they can be good in one place and bad in another is how they fit into the ecosystem. Deciduous forests, including the maple-dominated mixed forest, have no defensive mechanisms to cope with earthworms.

These sugar-maple-dominated hardwood forest ecosystems and all the strategies that plants and animals use to perpetuate themselves have evolved in the ten thousand years since glacial retreat in the absence of earthworms. So all the dynamics and rules of the ecosystem were established given the ground rules that there were no earthworms."

Hale's research focuses on the sugar maple–dominated mesic forest, which along the North Shore occurs in large patches only in the highlands. This woodland type is part of a larger swath of forest known as the Laurentian Mixed Forest, which covers 147,300 square miles in the northern Great Lakes region and parts of New England and Maine. According to the U.S. Fish and Wildlife Service, this mixed forest of coniferous and deciduous trees is the habitat type presently at highest risk from earthworm invasion.

Earthworms damage the maple-dominated hardwood forest where it is most vulnerable—the forest floor. According to Hale, "The forest floor, especially in sugar-maple-dominated hardwood forests, is *the* central feature of the ecosystem. It's the centerpiece around which all the plants and animals revolve. It's where all the plant germination and growth occur, where all the seeds are, where all the fungi are. It's what provides cover and protection for all the critters too."

In worm-free deciduous forests, the structure of the forest floor resembles a parfait of decomposing organic matter. It ranges from a topdressing of largely intact fresh-fallen leaves on the surface, called the litter layer, to underlying layers in various stages of decay known as the duff, or humus. In very rich sites dominated by sugar maple and basswood trees, this organic debris can be up to five inches thick.

A combination nursery, mess hall, and safe house, the duff serves as life support for a wide range of organisms including dozens of understory plant species. Like an insulating blanket, the thick duff provides protection from freezing temperatures in winter and desiccation in summer. Such a buffer from the elements is especially important to certain plants that germinate on the forest floor. Hale points out that many plants native to the deciduous forest have lengthy seed-germination strategies that require seeds to undergo two, and sometimes three, freeze-thaw cycles before they can fully germinate. In the early stages of growth, most plant species, including tree seedlings, are rooted in the duff layer, rarely penetrating it to reach mineral soil. During this two- to three-year period of establishment, tender seedlings require protection not only from the elements but also from the mouths of hungry voles, birds, and deer.

Consummate detritivores, worms interrupt the regeneration of the forest not by eating live plants but by eating their seeds or, more important, grazing the medium in which young plants are rooted, leaving them vulnerable to disturbances that can be as simple as a gust of wind. According to scientists, earthworms serve as "ecosystem engineers," that is, species that cause changes in the physical environment, directly or indirectly affecting the habitat of fellow organisms. The floor of a woodland infested with worms may have only a thin ground cover of the previous year's fallen leaves mantling a layer of worm castings. The radical simplification of the tiered litter and duff layers of the forest floor is similar to the collapse of a multilevel parking structure where the stories are compressed and jumbled in a heap of rubble. And if you cut a cross section from the underlying soil, you would find that it too has undergone a similar reduction and mixing of its normally stratified layers.

The list of changes to the forest following the arrival of worms is staggering. In a study of one plot on the leading edge of earthworm invasion, Hale and her colleagues

The upper soil profile in an earthworm-free sugar maple forest *(top)* features a layer of leaves from the previous fall on top of an older layer, known as the duff or humus, which contains leaf litter in various stages of decay. In forests dominated by sugar maple and basswood trees, this organic debris can measure up to five inches thick above the dark mineral soil. Exotic earthworms consume the duff *(bottom)*, leaving behind a far more simplified soil horizon. Photographs by Chel Anderson.

noted a fourfold increase in the worm population, tallying an average of twenty-eight to thirty-seven worms per square foot, and finding the populations in other plots had soared to more than sixty-five individuals per square foot, numbers that rivaled or exceeded those found in agricultural settings. As worms advanced at a rate of sixteen to thirty-three feet per year into a previously healthy forest, the plush, spongy duff disappeared. Within one to two years, it was replaced by a dense layer of black soil. Not surprisingly, the percentage of plant cover on the forest floor plummeted from 100 percent to less than 25 percent. The number of native plant species, including springtime favorites such as large-flowered and nodding trilliums, bloodroot, wild ginger, downy yellow violet, large-flowered bellwort, and Canada mayflower, dwindled from more than a dozen to only one or two species. Of the spring wildflowers, only jack-in-the-pulpit appears to be able to withstand earthworm grazing.

Even though mature sugar maple trees did not slow their prodigious output of seeds, virtually no sugar maple seedlings germinated on the forest floor. Such a paucity of new growth is especially striking since in healthy forests scientists have documented maple seedlings covering the forest floor in numbers of up to one million per acre. The surviving plant community was far more simplified and included weedy or disturbance-tolerant species. In forests that have been invaded for longer periods of time, worms may have paved the way for invasive species such as garlic mustard and buckthorn. The fact that these species have coevolved with earthworms in their native European habitats suggests that they are better acclimated to earthworm-worked soils and could eventually become dominant in worm-infested forests.

Researchers also saw dramatic changes in populations of small mammals that inhabited worm-infested areas. A healthy forest floor, Hale points out, is dominated by red-backed voles and one or two species of shrew, species that rely on abundant plant cover and fungi in their diets. The shrews had disappeared and were replaced by mice in the genus *Peromyscus.* Voles in the genus *Microtus* supplanted the red-backed voles.

Amphibians, from wood frogs and spring peepers to blue-spotted salamanders, too are at risk. In studies of heavily worm-infested forests in Wisconsin, Hale says, populations of species such as salamanders that require thick duff for food, moisture, and cover have plummeted. Ornithologists are concerned about already imperiled songbirds, such as ovenbirds and thrushes, that feed or nest on the forest floor. For the decline or loss of every species that has been studied, Hale observes, there are "hundreds and perhaps thousands of fungi species that have never even been documented, much less monitored for population changes. The same goes for invertebrates such as insects, spiders, and other animals that are affected by the loss of the forest floor."

Indeed, scientists continue to provide startling new insights into even some of the most routine and widely known ecological phenomena of the maple-dominated forest. For example, researchers have long known that ants play a critical role in dispersing the seeds of some of the maple forest's most beloved springtime wildflowers, including trilliums, violets, spring beauties, bloodroot, and hepatica. These species have evolved a small, fatty appendage, known as an elaiosome, on the outer shell of their seeds. Attracted to the elaiosome, ants tote the seeds back to their nests, where they feed the nutritious nub to their hungry larvae. The undamaged remainder of the seed is left to germinate in the colony's refuse pile, which is not only rich in nutrients but also relatively safe from the jaws of hungry rodents.

The deep leaf litter and thick duff of worm-free forests provide food, moisture, and cover required by many amphibians, such as this wood frog, and by birds, countless invertebrates, and fungi. Photograph by Steve Mortensen.

Researchers have cited the ant–spring ephemeral relationship as a classic case of mutualism; that is, species engaging in liaisons that confer mutual reproductive advantages. But until recently, scientists did not fully understand the benefits to ants. In 1998 scientists at Kenyon College in Ohio showed that although the populations of ant colonies fed on bloodroot seeds did not grow substantially larger than colonies deprived of seeds, they produced 3.5 times as many reproductive females.

Earthworm invasions not only sever such finely honed connections but also institute a new set of ecological rules that work against the recovery of these relationships in the future. By increasing the density of the forest's soils, for example, earthworms appear to be altering its hydrological regimes. In heavily compacted agricultural soils, the action of worms helps to loosen the soil and improve water filtration. Indeed, one of the measures of the health and fertility of a farm field is the abundance of worms that till its soils. But in maple-dominated forests, worms do just the opposite—they increase the compaction of the soil. That is because forest soils are already very loose and porous. As a result, the rate of water percolation into the soil is very high. By comparison, measurements of the bulk density of worm-worked forest soils, which are composed largely of worm castings, show that they are twice as dense as natural soils in the hardwood forest. Not only does the action of worms decrease water filtration into the forest floor and increase surface runoff, but by consuming the protective duff and leaf litter, the worms also leave soils vulnerable to erosion.

Such changes also subvert long-established nutrient cycles. Maple-dominated forests supply tender young plants with abundant nitrogen via an ingenious process known as "nutrient pumping." The deep roots of maple trees tap minerals such as calcium, magnesium, and potassium from the subsoil, where they lie beyond the reach of most other forest plants. These nutrients are stored in the leaves of the tree. The autumn litterfall of maple leaves supports robust populations of microbial decomposers such as bacteria and fungi that slowly release the abundant stores of nitrogen and minerals that are bound up in plant tissues. As the organic content of the forest floor builds, so does the forest's water-holding capacity. The thick layers of litter slow the infiltration of vital elements downward through the soil beyond the reach of most plants, thereby making them more available to plant roots that occupy the upper layers of the soil. So, as soon as the nutrients are produced, they are absorbed again by plant roots. In this closed loop, little to no nutrients are lost to the system.

Researchers studying earthworm invasions in the maple-dominated forests of New York have found that they speed up the breakdown of plant materials and flood forests with nitrogen, disrupting the measured pace of nutrient release and plant uptake to which the forest has evolved. Without the duff and litter layers to slow the coursing of water runoff, much of this precious nutrient is washed out of the system. As the nitrogen flows into wetlands, rivers, and lakes, it contributes to water-quality problems. Another essential nutrient—phosphorus—remains in the soil, but it is leached down below the rooting zone so that it becomes inaccessible to most plants.

Perhaps most worrisome is the loss of the bacteria and fungi that form the base of the food web in the mixed forest, especially the mycorrhizal fungi whose great networks of filaments serve as highways of nutrient exchange upon which the future productivity of the forest depends.

The disruption of these fungal arterials appears to already be driving one plant species closer to the brink of extinction. In a study published in the December 2002 issue of *Conservation Biology,* biologist Michael J. Gundale surveyed twenty-eight separate populations of the rare goblin fern in the Chippewa National Forest. The fern, which is restricted to forests with a strong sugar maple component in the upper Great Lakes region, is considered endangered in Wisconsin and threatened in Michigan and Minnesota.

Little is known about this elusive plant. Indeed, the goblin fern was only officially described in 1981. Such oversight is not surprising since mature plants, which can measure less than one-half inch tall, often grow under the leaf litter. Even when the goblin fern does emerge from the forest duff, it appears not to rely much on sunlight since its tiny stalk possesses little plant surface with which to photosynthesize, hence its pale-green color. The fact that its roots are shallow with few hairs suggests that the goblin fern derives much of its carbon and nutrients from the filaments of mycorrhizal fungi attached to its roots. This connection to the larger food-distribution system in the forest, researchers say, may be crucial to the survival of the fern at every stage in its life cycle. Diminish the duff layer, Gundale conjectured, and the goblin ferns would disappear.

Gundale's grim hypothesis was borne out in surveys on the Chippewa National Forest. The plant had disappeared in nine out of twenty-eight sites that had once hosted the tiny fern. In all nine sites, the leafy cushion on which goblin ferns depend was noticeably thin. Gundale suspected that the 1- to 1.5-inch-long earthworm *L. rubellus,* which infested eight out of the nine plots, played a role in the thinning of the forest floor.

Wherever it was present, the forest litter was half the thickness of that in worm-free areas. Indeed, when Gundale conducted laboratory experiments in which *L. rubellus* was reared in a simulated forest environment, he found that the worms consumed the surface litter, reducing its thick, spongy layers into castings that were plowed into the underlying mineral soils.

Although the goblin fern does not occur in the highlands, the invasion of worms into its maple-dominated forests could imperil the future of similarly rare species. Especially at risk are narrow triangle moonwort, Carolina spring beauty, and moschatel.

An Ounce of Prevention, a Pound of Cure: Halting the Spread of Earthworms

When it comes to worms, an ounce of prevention is worth pounds of cures, Hale says. Once earthworms have advanced into an area, they cannot be eradicated. But they can be controlled, Hale says, by raising awareness about the human activities that spread worms. In Alberta, Canada, earthworm invasions into lodgepole pine and aspen forests have been blamed on logging trucks, which may seed new areas with worms via the dirt that is lodged in tire treads. Logging equipment in Minnesota appears to serve as a similar vector of contamination, since Gundale's research shows that worm invasions in the Chippewa National Forest were most pronounced near logging access roads. Also to blame, Hale says, are all-terrain vehicles, whose tires can readily transport earthworms from infested areas of the state into worm-free zones.

Anglers too have become the target of public-education campaigns. While all of southern Minnesota's mesic hardwood forests are infected with worms, largely because of their proximity to the most intensive land uses, such as urbanization and agricultural fields, most experts believe that up until the late 1970s, they were uncommon in the state's undisturbed northern forests. Worm researchers have traced epicenters of cropped duff to lakeshores, boat ramps, and fishing resorts where anglers routinely dump unused bait on land. They speculate that this influx of earthworms into northern forests may have been exacerbated by the growing imports of nightcrawlers from Canada. By 1980, Canada was exporting one-half billion nightcrawlers to the United States, where their size—they grow up to ten inches in length—makes them prized as fishing bait. Today, the sale of worms is a $20 million industry.

Some have suggested the hiring of worm pickers in U.S. forests since scientists have no safe chemical or biological means to control worm populations. Research on how to eliminate worms without inflicting further damage on the larger forest ecosystem has yielded some promising results. Progress is slow, however, since safe, effective, and practical approaches in one forest community are not necessarily transferable to others. Further studies in a wider variety of forest ecosystems are needed.

In the meantime, scientists like Hale emphasize the need to raise awareness among recreationalists, the forest industry, and public agencies because human activities are the prime vectors for the spread of worms. To sustain worm-free areas and slow the spread of earthworms, road and trail building should be kept to an absolute minimum through careful planning and coordination among landowners. Reexamining how people use their tools and toys is equally important since many of them can carry and disperse earthworm eggs along with other exotic species. For example, instituting new regulations that would require operators to scour recreational and timber-harvesting equipment prior to entering forests would help prevent earthworm encroachment and further damage to the complex beauty and finely honed processes of northern forests.

"There are still worm-free areas, particularly in northern Minnesota and northern Wisconsin," Hale points out. "If we can prevent the transportation and the inoculation of worms into these currently worm-free areas, we can buy ourselves hundreds of years to do the research that is needed." ∞

SUGGESTIONS FOR FURTHER READING

Conover, Adele. "Foreign Worm Alert." *Smithsonian,* August 2000.

Gates, Gordon E. "Miscellanea Megadrilogica." *Megadrilogica* 1 (1970): 1–14.

Gundale, Michael J. "Influence of Exotic Earthworms on the Soil Organic Horizon and the Rare Fern *Botrychium mormo*." *Conservation Biology* 16 (2002): 1555–61.

Hale, Cindy. "Earthworms May Be Threatening Biodiversity of Hardwood Forests." *Minnesota Plant Press* (Minnesota Native Plant Society Newsletter) 20, no. 2 (Winter 2001): 1, 3.

Hale, Cindy M., Lee E. Frelich, and Peter B. Reich. "Changes in Cold-Temperate Hardwood Forest Understory Plant Communities in Response to Invasion by European Earthworms." *Ecology* 87, no. 7 (2006): 1637–49.

———. "Effects of European Earthworm Invasions on Soil Characteristics in Northern Hardwood Forests of Minnesota, U.S.A." *Ecosystems* 8 (2005): 911–27.

———. "Exotic European Earthworm Community Composition in Northern Hardwood Forests of Minnesota, U.S.A." *Ecological Applications* 15, no. 3 (2005): 848–60.

Hendrix, Paul F., ed. *Earthworm Ecology and Biogeography in North America.* Boca Raton, Fla.: Lewis Publishers, 1995.

Hendrix, Paul F., and Patrick J. Bohlen. "Exotic Earthworm Invasions in North America: Ecological and Policy Implications." *BioScience* 52 (2002): 801–11.

McLean, Mary Ann, and Dennis Parkinson. "Soil Impacts of the Epigeic Earthworm *Dendrobaena octaedra* on Organic Matter and Microbial Activity in Lodgepole Pine Forest." *Canadian Journal of Forestry Research* 27 (1997): 1907–13.

Mortensen, Steve, and Carol Mortensen. "A New Angle on Earthworms." *Minnesota Conservation Volunteer,* July/August 1998: 20–29.

Proulx, Nick. *Ecological Risk Assessment of Non-indigenous Earthworm Species.* U.S. Fish and Wildlife Service, 2003.

Reid, Walter V. C., and Kenton R. Miller. *Keeping Options Alive: The Scientific Basis for Conserving Biodiversity.* Washington D.C.: World Resources Institute, 1989.

Reynolds, John Warren, Dennis R. Linden, and Cindy M. Hale. "The Earthworms of Minnesota (Oligochaeta: Acanthodrilidae, Lumbricidae and Megascolecidae)." *Megadrilogica* 8 (2002): 86–100.

Stewart, Amy. *The Earth Moved: On the Remarkable Achievements of Earthworms.* Chapel Hill, N.C.: Algonquin Books of Chapel Hill, 2004.

Trombulak, Stephen C., and Christopher A. Frissell. "Review of the Ecological Effects of Roads on Terrestrial and Aquatic Communities." *Conservation Biology* 14 (2000): 18–30.

Westbrooks, Randy G. *Invasive Plants: Changing the Landscape of America.* Washington D.C.: Federal Interagency Committee for Management of Noxious and Exotic Weeds (FICMNEW), 1998.

INTERNET RESOURCES

Great Lakes Worm Watch, http://www.nrri.umn.edu/worms/research/publications

Minnesota Worm Watch, www.nrri.umn.edu/worms

The Secret Life
of Salamanders

We have it from many authorities that a snake may be born
from the spinal marrow of a human being. For a number of animals
spring from some hidden and secret source, even in the quadruped class,
for instance salamanders, a creature shaped like a lizard, covered with spots,
never appearing except in great rains and disappearing in fine weather.

Pliny the Elder, *Natural History* (first century)

April may indeed be Minnesota's cruelest month. Winter, like an ousted dictator, does not relinquish its grip on the north country without mounting a scorched-earth retreat. Slant rains and wind lash the land. Ground fogs lodge their stubborn chill deep into the bone. Most Minnesotans succumb to the siege and simply wait out the changing of the guard indoors. They throw another log on the fire and hunker more deeply into comforters and seed catalogs.

But cruelty, particularly with regard to the weather, is a matter of perspective in the North. Come the first soaking rains of the season, when temperatures rise above freezing but hover around 40 degrees Fahrenheit, waves of amphibians emerge from underground burrows and the depths of the forest litter. Among spring's earliest risers are blue-spotted salamanders. Together with their cousins, the tiger salamanders, they are the northernmost members of the eastern North American family of *Ambystomatidae,* or mole salamanders.

Blue-spots take spring's earliest wake-up calls seriously. So eager are these animals to get a reproductive jump on their fellow amphibians—as well as to avoid major predators such as snakes while they are still deep in their winter torpor—that males will cross great expanses of snow and ice in their single-minded treks to breeding pools. Undeterred by the temperature of nuptial waters that is several degrees colder than that of the surrounding air, they begin nosing around for mates.

Getting an early start is not risk free, however, in a season known for its sudden cold snaps. But blue-spots come prepared—particularly the males. Because they set out for breeding ponds earlier than females and are therefore exposed for a greater period of time, male salamanders are especially hardened to severe weather. They are so adapted to the sudden freeze-thaw cycles of spring that when scientists warmed ambystomatids that had been frozen in buckets of water, the animals revived. With the icy nuggets of their bodies thawed tender and pliable as hard-boiled eggs, the blue-spots then walked away as if nothing had happened.

Mole salamanders even look hearty. Unlike many of their longer, leaner, more elegant relatives, mole salamanders are characterized by stout bodies, meaty limbs, and squat, rounded heads. In form, ambystomatids are the Jesse Venturas of the salamander world.

Yet blue-spotted salamanders are the very paradox of delicacy and durability. They more than make up for their family's burly lineaments with coats of almost tropical hues. Their moist skin is a glistening black field. Spattered on their sides, tail, legs, and belly is a random pattern of turquoise-blue flecks. The effect resembles the surface of fine spongeware porcelain.

So how does such an extraordinary creature elude even the most devoted hiker, who can spend a lifetime on the Superior Hiking Trail without sighting a single animal? Well, it is not a question of rarity. How many inhabit the Lake Superior watershed is anybody's guess, but of the seven species of terrestrial salamanders that occupy northeastern Minnesota, blue-spots are the most common and the most widely distributed. They can be found under rocks and rotting logs and deep in the forest's litter, from the interior boreal forests to the deciduous woodlands of the North Shore Highlands.

Why then don't they make a wildlife watcher's life list as often as red squirrels, say, or ovenbirds? For one thing, blue-spotted salamanders are small. Unlike their cousins the tiger salamanders, which can grow to about thirteen inches long, blue-spots attain a maximum length of about five inches.

Furthermore, they overwinter in underground burrows known as hibernacula. Breeding adults engage in a great spurt of activity only in the early spring when they make nocturnal round-trip excursions from woodland territories to breeding pools. (Another movement takes place in the fall when newly metamorphosed salamanders leave their aquatic nurseries and head for drier ground. Scientists suspect that these youngsters spend the two years they need to mature out of sight in underground burrows.)

After the spring breeding season, adults return to their forest habitats, where they largely stay put under leaves, logs, and rocks. Darkness and moisture are salamanders' active ingredients—which explains why they are so rarely sighted. Few people are willing to dig out rain gear and flashlights to pace the forest floor on a rainy night, even during the height of summer. Lacking an impermeable skin layer such as scales to protect them against water loss, salamanders are dependent on ambient humidity to remain moist. As a result, they are most apt to forage within view—on top of the leaf litter, on the stalks of tall plants, or even on the bark of trees—during and shortly after rainfall and at night, when humid conditions keep them from dehydrating in the open air. Research suggests that these conditions may also flush potential prey out of the forest's nook and crannies. It is no surprise then that the greatest salamander diversity in the United States can be found in two regions, the Southeast and the Pacific Northwest, both heavily forested places with abundant rainfall.

Because salamanders spend so much of their time in the damp recesses of the forest, they have been rightly described as "inconspicuous and fossorial." This very modesty has led, however, to neglect if not outright misunderstanding. Even the most elemental fact of a salamander's existence—water—is curiously at odds with the roots of its name. The word *salamander* comes from the Arab-Persian word for "lives in fire." Ancient writers, including Aristotle, perpetuated this misconception by reporting that not only were salamanders incombustible, but they also extinguished fires by running through them. The yoking of salamanders and fire persisted in early European legends. To the casual observer, the link appeared to be corroborated by the behavior of the fire salamander *(Salamandra salamandra),* which fled its woody refugia when logs were placed into flames.

Salamanders do, however, possess powers far more miraculous—and nearly as implausible. "Salamanders are the superstars of regeneration," writes Andrew Pollock in a 2002 *New York Times* article. When limbs, tails, retinas, eye lenses,

Blue-spotted salamanders are the most common and widely distributed salamanders in northeastern Minnesota. These delicate yet durable denizens of the forest prowl the moist forest floor for invertebrate prey and, in turn, provide a stable source of nutritious food for larger animals. Photograph by D. Lawson Gerdes.

Decomposing woody debris is an important microhabitat of the forest floor that provides a moist refuge for cryptic salamanders and other amphibians. Photograph by Chel Anderson.

and parts of their hearts are lost or damaged, salamanders can regrow them, sparking hopes in those studying regenerative medicine that the same capacities can be engineered in humans someday.

Nevertheless, outside of the laboratory the championing of salamanders has been far less enthusiastic. The public's focus on edible game species, such as deer and fish, and its fear of "creepy-crawly" critters, especially those perceived to be slimy (which salamanders are not), undoubtedly have contributed to a general underappreciation of salamanders among natural resource agencies. Not surprisingly, salamanders often have earned little more than a footnote in most forest-management plans.

Ecologists have done little to remedy this scientific oversight. "Until the mid-twentieth century," writes herpetologist James H. Harding, "amphibians and reptiles were frequently assumed to be comparatively insignificant members of most North American wildlife communities, even by some biologists."

While much has been discovered about salamander reproduction and development (like frogs, they are easily studied in the laboratory), only the most rudimentary information is available about how they actually live in the wild, especially once adults and their offspring leave breeding pools for the forest. Researchers and land managers lack even some of the most basic information about salamanders' habitat needs, population dynamics, and their roles as both predator and prey in the larger forest ecosystem.

One of the first studies to demonstrate that salamanders are not simply luxury items in the forest ecosystem but an integral part of its daily functioning was a 1975 study of salamander populations in New Hampshire's Hubbard Brook Experimental Forest. Inventories revealed an average of about 1,200 per acre, nearly 94 percent of which were red-backed salamanders *(Plethodon cinereus)*. So numerous were they that their numbers outstripped that of birds and mammals combined. Salamander biomass (as calculated by the weight of live animals) was found to be 2.6 times greater than birds (even when measured during the peak nesting season!) and about equal to that of the resident small-mammal population. "This finding is somewhat surprising," say lead researchers Thomas Burton and Gene Likens, "as most ecologists have ignored amphibians in ecosystem energy flow and nutrient cycling studies while considering birds and mammals in detail."

The Hubbard Brook research prompted subsequent researchers to take a closer look at salamander ecology. Despite this greater focus, knowledge about salamanders remains sketchy. A 1980s review of ecology journals, for example, revealed that less than 5 percent of major research articles were devoted to the study of amphibians (a group that also includes scores of frog species), making them the least-studied vertebrates. Most of what we now know about amphibians and forests, say wildlife ecologists Phillip deMaynadier and Malcolm Hunter, has been gleaned from studies published since the mid-1980s.

For many species of amphibians, the attention may have come too late. Several species of frogs, from the Cascade Mountains to the highland rain forests of Panama, already have gone extinct. According to a study published in the journal *Science* in 2004, 1,856 species of amphibians—32.5 percent of those known to science—are globally threatened. Declines have been blamed on a variety of factors from pesticides, habitat destruction, and exposure to excessive levels of UV-B radiation to deadly viral and fungal infections that may be linked to fishing bait and hatchery-raised fish.

The little information that we have about salamanders only underscores the need for more extensive research, particularly at a time when each new study seems to throw our knowledge about salamanders into yet another state of flux. Even classic amphibian studies, such as those conducted in the Hubbard Brook Experimental Forest in the 1970s, are not immune to startling revisions. Based on research carried out in streamside habitats in an old-growth forest in southern Appalachia, University of North Carolina biologist James Petranka concluded that both the density and biomass of red-backed salamanders were grossly underestimated in the Hubbard Brook experiment. That is because scientists had based their calculations on a single reconnaissance of their study site. And they had not factored in a key piece of the life history of the animals they were studying: that at any given point in time, more salamanders are likely to be sequestered underground than actively traversing the surface. In an article published in a 2001 issue of the *Journal of Herpetology,* Petranka and fellow biologist Susan S. Murray provide estimates of salamander density and biomass based on repeat visits to their study plot over the course of many consecutive nights. According to their calculations, the salamander population in the Appalachian study site was seven times greater than that in the Hubbard Brook plot. In addition, their estimates of live biomass were fourteen times that of salamander biomass in the Hubbard Brook study.

If Petranka and Murray are correct, salamanders may play a far greater part—indeed may even have a starring role—in some of the forest's most fundamental ecosystem processes. The need for more information about these elusive creatures will only intensify as those charged with overseeing natural resources, such as timber and wildlife, continue to pursue more inclusive management prescriptions. These new strategies attempt to account for the welfare of not only game species, such as grouse and deer, but also the forest's tiniest inhabitants, including decomposers and nutrient cyclers such as soil fungi, and pest controllers such as songbirds.

The Function of Salamanders in the Forest

Based on the preliminary evidence, to ignore the existence of salamanders would be to imperil the health—and quite possibly the productivity—of the forest ecosystem. Much of this may have to do with the links that salamanders forge in a variety of food chains—between aquatic and terrestrial systems and microscopic and macroscopic animals.

Pool-breeding species such as blue-spots begin their ecosystem services at birth. According to biologist Petranka, they "tend to be explosive breeders." Depending on conditions and the location, he says, they may engage in up to three major breeding episodes. This behavior is largely determined by the nature of their reproductive habitat. The seasonal pools that are used as nurseries by blue-spots—depressions around the roots of overturned trees, for example, or shallow woodland basins that are filled by rain and snowmelt only in the spring—are fecund but chancy environments. In dry years they may disappear before salamander larvae have a chance to complete their metamorphosis. As a hedge against an uncertain future, seasonal pool breeders produce large numbers of eggs. Each female lays an average of 225 eggs. Only 10 percent of eggs and larvae (the equivalent of the gilled tadpole phase in frogs), however, successfully run the gauntlet of survival and go on to live as adults on land. The remaining 90 percent become food for other animals.

Despite this high initial mortality, enough salamander young survive to exert a profound influence on these small but life-teeming pools. The larvae of blue-spotted

Female blue-spotted salamanders lay an average of 225 eggs each breeding season. Only 10 percent of these eggs and larvae (shown here), however, survive to live as adults on land. Photograph by Allen Blake Sheldon.

salamanders are considered the top predators of the fishless pools they occupy during the two to three months in which they make their transition from water-bound youngsters to their terrestrial phase. One of the advantages of early breeding is that the larvae are able to enjoy hefty helpings from a smorgasbord of tiny aquatic crustaceans, including water fleas, scuds, and the diaphanous fairy shrimp, which is active in all stages of its development only when the pool's waters are very cold.

Studies of ponds in the Colorado Rockies that are inhabited by tiger salamanders, the blue-spot's cousin, showed that the hearty appetites of ambystomatid larvae may play an important role in structuring the invertebrate food webs of aquatic communities. Research suggests that salamander larvae may derive a growth advantage by selectively preying on the biggest and most nutritious morsels in the zooplankton pool, leaving smaller crustaceans to their fellow pool inhabitants. Ponds inhabited by tiger salamander larvae, for example, contained only two species of aquatic crustaceans. Those without salamander larvae supported twice as many species. The presence of the larvae also influences the size of individuals in the zooplankton community. In pools with larvae, the largest crustaceans measured only 0.06 inches in size. In the absence of salamanders, the specimens grew up to 0.5 inches in size.

As both predator and prey, salamanders may serve a similarly important role on land. Among their many virtues, salamanders provide the larger animals that eat them, such as birds, small mammals, snakes, and other amphibians, with a stable source of nutritious food. Burton and Likens discovered, for example, that salamanders convert 50 to 80 percent of the food they eat directly into living tissue, unlike birds, who put a mere 0.5 to 3 percent of ingested energy into biomass. Birds, like humans, are endotherms (a word meaning "inside heat")—they regulate their own body temperatures. Stoking an internal furnace comes at great metabolic cost since much of the energy from food goes into operating the body's thermostat. Salamanders, on the other hand, are ectotherms ("outside heat"). They derive body heat and cooling from free sources such as sunlight

Diaphanous fairy shrimp and other aquatic invertebrates provide food for blue-spotted salamanders as they make their transition from water-bound youngsters to terrestrial adults. Photograph by Chris Buelow.

and shade and therefore can devote a greater share of their energy budget from food into building tissue. And they are easy to digest—no feathers, scales, or hair to pick through—and high in phosphorus. Salamanders are also a better source of protein than birds or small mammals.

In addition, they are a food item with a long shelf life. Slow-maturing and long-lived (they may reach grand old ages of thirty years), salamanders provide a steady food supply in the forest's boom-and-bust cycles. In this respect, say biologists Robert Stebbins and Nathan Cohen, salamanders are dependable forest staples and "like plants, act as an energy reserve within ecosystems."

The discovery of salamander abundance by Burton and Likens has led many researchers to conjecture about their influence as predators on the larger forest ecosystem. The most tantalizing theory—yet unproven—is that salamanders may play a vital, albeit indirect, role in the productivity of the forest by structuring the food web of the forest floor, known as the detrital food web. Because of their pliable bodies and relatively small size—an average of 0.1 ounces for adults versus 0.7 ounces for small birds and mammals—salamanders can insinuate themselves into the tightest corners of the forest, such as the cracks in rotting logs or mere slits of space under rocks. They are able to consume organisms that are too small or too inaccessible for many other forest creatures. They have been known to dine voraciously on everything from beetles, spiders, and snails to earthworms and sow bugs. As a result, they serve as a population check on the organisms that prey on the bacteria and fungi responsible for decomposition and the release of important nutrients such as carbon, nitrogen, and other essential elements into the forest ecosystem.

If, as theorized, salamanders are the top predators in detrital food webs, their elimination could impair the healthy functioning of whole ecosystems. Burton and Likens point, for example, to studies of marine rock pool communities that depend on predatory starfish for maintaining their biological diversity. Researchers found that when the starfish were removed from these systems, populations of barnacles exploded. In time, they literally crowded out most other rock-pool species. Salamanders may similarly serve as arbiters of biodiversity, known as keystone species. "The importance of salamanders in nutrient cycling," Burton and Likens observe, "may be in their role as top predators in the detritivore food web and in the maintenance of diversity of prey populations." That, they say, might lead to "more efficient breakdown of litter by these prey. An increase in the breakdown of litter would result in greater availability of the nutrient pool to vegetation and perhaps could have some regulating effect on the function of the [forest] ecosystem," such as the growth rate of trees.

Ancient Pasts, Uncertain Futures for Salamanders

About 360 million years ago a group of fish developed lungs and limbs and crawled from the sea onto land. This small evolutionary turn of the key opened the door to one of the most long-lived lineages of animals—amphibians, Earth's first terrestrial vertebrates.

Mole salamanders share this distinguished heritage. Paleontologists have identified fossils of ambystomatids in North America that date to about thirty million years ago. Remarkably resilient, several of the species found in the fossil record have survived to the present.

Although these mole salamanders endured conditions of the late Pleistocene that drove nineteen genera of birds and forty-six genera of mammals into extinction in eastern North America, their adaptability in the face of threats from cornfields and clear-cuts, global warming, and acid rain is far less certain. Even the numbers of blue-spotted salamanders, which Petranka says, "is a relatively common species in many areas of its range," have declined "as native forests have been replaced by urban and agricultural regions and thousands of natural breeding sites have been destroyed."

Unlike some of their relatives that live either exclusively on land or in water, mole salamanders maintain a dual citizenship in both areas. Consequently, they are vulnerable on two fronts: the forested territories they feed and rest in and the seasonal pools they rely on for the propagation of young.

One of the most destructive impacts on salamanders has been the practice of clear-cutting, which removes the forest's shady canopy, exposing the forest floor to sun and winds that elevate soil temperatures and zap its moisture. Precious leaf litter dries out and crumbles. Some kinds of clear-cutting and whole-tree harvesting may also strip the forest floor of coarse woody debris—the rich crosshatch of downed tree trunks, rotting snags, and branches in various stages of decay that harbors the insects that salamanders eat as well as provides the moist places they use for cover.

Dry conditions take their toll on salamanders, even in forests with a well-structured canopy and shrub layer and an intact detritus. During periods of drought, salamanders seek out moist depressions in the forest floor or in the deep layers of leaf litter. They wait out extended rainless cycles tucked into the cool, damp undersides of rocks and decaying logs with sizable cracks. When dry conditions are extreme, salamanders withdraw to underground burrows.

Petranka charges that clear-cutting exacerbates these extreme conditions and causes the localized extinctions of whole populations of salamanders. In a 1993 study he surveyed populations of plethodontid salamanders in both forested and recently clear-cut sites in North Carolina. Petranka and his colleagues found that salamanders were largely absent from logged tracts. Extrapolating from this research, he estimated that more than fourteen million salamanders died each year in western North Carolina due to the destruction of their habitats by clear-cut logging. Furthermore, he argued that it would take fifty to seventy years for salamander populations to regain their predisturbance levels.

Petranka's claims touched off a heated debate among scientists. Research on clear-cuts and salamanders in other parts of the country supported the group's findings that clear-cuts were not habitable abodes for salamanders. In fact, the number of amphibians in general has been shown to be 3.5 times higher on control plots of intact forest than on clear-cut sites. Field studies of New England forests also corroborated Petranka's estimated salamander recovery rates. DeMaynadier and Hunter point to research showing that the detrital matter on the floor of northern hardwood forests decreased in depth for one to two decades after cutting, largely because of the increased decomposition rates of existing organic material as well as the absence of trees that replenish the forest floor with leaves and woody debris. In these northern hardwood forests, which are similar to the ones found in the Lake Superior Highlands, it can take up to eighty years for the buildup of the detrital layer to reach predisturbance levels.

But these seemed to be the only points on which most scientists could agree. After reviewing the North Carolina study, biologists Andrew N. Ash and Richard C. Bruce concluded, "We know [salamanders] disappear from clearcuts, but that is all we know."

Skeptics such as Ash and Bruce charged that the salamanders could have dispersed from the logged sites and set up new territories in adjacent forests or found safety in underground burrows. Petranka countered these possible explanations with several facts about the life history of plethodontids. Studies have shown, for example, that when these salamanders are faced with the prospect of dehydration, they seek safety underground instead of fleeing into surrounding areas. The dry conditions of sun-drenched clear-cuts, Petranka maintains, would have sent stressed salamanders belowground, where they have been shown to occupy burrows up to nearly thirty-six inches in depth for considerable periods of time.

Yet even in these climate-buffered havens, salamanders find only limited shelter. Subterranean chambers are not foolproof safe houses since often they are patrolled by their creators—the salamander's arch enemy, the short-tailed shrew, and other hungry mammals. Furthermore, the most abundant food for salamanders lies in the leaf litter. Underground retreats can become starvation chambers when salamanders are consigned to them for too long and are forced to hunt from burrow entrances on whatever scant prey chances along. And the parched conditions of clear-cut tracts may not only dry up small seasonal pools but also discourage breeding-age salamanders from even attempting to leave their burrows in the first place.

As for the dispersal theory, many salamanders are quintessential homebodies. Stebbins points out that "some small terrestrial salamanders probably spend their entire lives in and around a single proven protected site such as a rotting log, rock pile, root tangle, or other shelter." This home range may measure less than ten feet in diameter.

Establishing a new domain can be difficult, if not impossible, for many animals since they are philopatric, or extraordinarily faithful, to their narrowly circumscribed home grounds. So vital are these ranges that salamanders have developed complex internal compasses, which researchers believe may be largely based on the sense of smell, to guide them back whenever they leave home. In experiments conducted with red-backed salamanders in Michigan's Upper Peninsula, researchers found that fourteen out of fifteen salamanders were able to locate their home ranges after having been displaced at distances of up to 295 feet. Studies of spotted salamanders showed that animals could return to breeding pools following displacements of 1,640 feet. Researchers found that many of these animals were able to make their way back by locating migratory pathways on the forest floor that measured thirty-three to ninety-eight feet wide.

Stebbins observes that there are several evolutionary advantages to this philopatrism. Repeat visits to tried-and-true breeding pools, for example, maximize the rate of reproductive success by preventing salamanders from wasting precious time and energy in wandering the forest floor, where their exposure to predators would also be increased. It helps to eliminate unnecessary competition by distributing individuals in a stable matrix around available resources such as breeding pools. And when environmental conditions become unfavorable, individuals have the security of a home base from which to explore new territories that offer greater food supplies and cover or better protection from predators.

Yet, even if salamanders from the clear-cut sites had managed to disperse into neighboring forests, Petranka argues, it would have been difficult for them to have gained a toehold. He points out that many species, such as the salamanders that he observed occupying the forest in areas adjacent to his clear-cut study sites, are very aggressive in defending their territories from newcomers.

Highways of Doom: The Dangers of Road Crossings for Amphibians

Salamanders suffer from alterations of habitat that include more than just the removal of trees. For many small animals, road building can present insurmountable obstacles, if not outright death traps. On some stretches of rural and suburban roadways across the United States, amphibians setting out on their spring migrations are killed in such numbers that their carcasses build up on the surface of the pavement until it is as slippery as ice. Researchers point out that on some roads the mortality of salamanders is far higher than frogs, largely because salamanders move much more slowly. In a study of one paved rural road in New York, scientists tabulated average mortality rates for road-crossing salamanders at 50.3 to 100 percent.

The open road could prove just as harmful to populations that decline these crossings as for those that undertake them. According to research by biologist H. J. Mader, roads built through forested areas mark a distinct break in microclimate conditions. Roads have higher temperatures and levels of evaporation as well as less humidity than the surrounding habitat. "Roads which cross a forest," he points out, "have a resemblance to a savanna or rocky habitat rather than to the microclimate of a moist wood." Moisture-dependent animals such as salamanders may be reluctant to cross such inhospitable terrain and therefore lose access to breeding pools.

Over time, these barriers can impact whole populations of animals in a given locale. As Mader observes, they "cut off the gene flow by dividing animal populations into fractions on either side of the road." For populations of animals with poor dispersal capabilities, such as salamanders, this isolation can lead to inbreeding and localized declines in genetic fitness, making animals less resilient as a group to cope with disease or changes in environmental conditions.

For animals that do attempt a crossing, traffic may not be their only danger. On April 22, 1989, Lawson Gerdes, a wildlife biologist with the U.S. Forest Service (USFS), was taking a hike on her day off along a forest service road near her home in Isabella, Minnesota, when she came across hundreds of dead blue-spotted salamanders in the roadbed. The road, which ran deep through the heart of the Superior National Forest, was isolated and lightly traveled. Not surprisingly, the salamanders showed no signs of injuries that could have been caused by vehicles. Lawson observed one peculiar detail that seemed to provide a clue about the cause of this mass mortality: there was a shallow circular depression around the bodies of the salamanders, as if the animals had writhed in the loose gravel before dying. The following spring salamanders again were found dead in large numbers on the road.

Gerdes brought the incidents to the attention of Richard Buech, a fellow wildlife biologist with the USFS. They suspected that a chemical agent—calcium chloride (CaCl)—was responsible for the salamanders' demise. Calcium chloride, a salt compound that is commonly applied to gravel roadways throughout the United States, is hygroscopic; that is, it draws moisture from the air, dampening the road surface to

The hostile microclimate of forest roads may pose a barrier to salamanders traveling between their home territories to breeding grounds, such as this pool along a forest road in the Superior National Forest. Photograph by D. Lawson Gerdes.

control dust clouds. Buech and Gerdes theorized that on their seasonal migrations to several nearby wetlands, the salamanders crossed the treated roadway. Once in contact with the animals' semipermeable skin, the chemical would have leached the moisture out of the salamanders' bodies through osmosis, leaving them to die of desiccation. The pair consulted roadway maintenance records and found that the stretch of road that was littered with the dead salamanders had indeed been sprayed with a 38 percent solution of calcium chloride. In the process they uncovered one especially disturbing detail: the chemical had been administered in July 1987, nearly two years before Gerdes came upon the first episode of dead salamanders. The discovery led researchers to fear that deadly effects of the chemical could persist for years following its application.

Gerdes and Buech followed up their hunches with a pilot study in 1991 that examined salamander activity along two treated and two untreated sections of roadway, all four of which were located within a larger complex of the animals' habitat. In their analysis of the data, the researchers uncovered an unexpected finding. Ninety blue-spotted salamanders were retrieved from traps set up along untreated segments of the road, whereas only three individuals were in traps along treated sections. The data led Buech to propose "that perhaps CaCl presented a barrier to salamander migration."

The study confirmed strong suspicions that treating roadways with calcium chloride in salamander territory could stop the animals—literally—in their tracks. Fortunately, when it comes to ensuring the welfare of blue-spotted salamanders, erring on the side of caution is a feasible management strategy. Trucks that dispense calcium chloride are equipped with on-off mechanisms that precisely control the application of the chemical. In many cases, operators release spot applications of the compound only around intersections of streets and driveways or on blind curves where decreased visibility due to dust clouds could result in traffic accidents.

Operators could easily avoid spraying the roadway sections of seasonal routes that are used by blue-spotted salamanders to travel to and from breeding pools. The problem

Pit traps along roadsides treated with calcium chloride have been used to investigate the effects of this chemical on blue-spotted salamanders that cross roads en route to their breeding pools. Photograph by D. Lawson Gerdes.

is that most of these pools—known under the technical term of small embedded wetlands—have not been mapped. Gerdes says that because they often are less than 0.1 acre, they can be difficult to locate using existing forest- and wetland-inventory techniques. Under the U.S. Fish and Wildlife Service's National Wetlands Inventory, for example, ponds under 3 acres are considered "undetectable."

In addition, these pools may come and go depending on the season or year-to-year climatic variations. Some are classified as vernal pools; that is, they are filled with water primarily in the spring. Others are known as autumnal pools, meaning that they exist in spring and fall. Moreover, in the winter, when most of the road building and cutting of the northern forest takes place, many of them simply disappear under the snow, making it difficult for loggers and construction workers to avoid damaging pools and adjacent forest habitats even when management guidelines are in place.

What Is Good for the Salamander Is Good for the Forest

When the last glacier receded from the north country about twelve thousand years ago, it left behind a landscape pockmarked with pools of standing water. Scientists credit salamanders' ready access to water as fundamental to their survival and widespread dispersal in the glacier's wake. "Local populations of wetland species often are small and isolated and thus vulnerable to extinction," observes researcher James P. Gibbs. Outbreaks of disease, inbreeding, or climatic variations that cause fluctuations in water levels can wipe out whole populations in a given area. Tiny pools—and lots of them dotting the landscape—function as stepping-stones that survivors can use to travel to new territories.

Today, human activities such as road building and forest clearing are drying up the ancient pathways that salamanders have used to hopscotch across the landscape. At stake is not just the survival of salamanders. If the needs of salamanders are accommodated, ecologists argue, chances are that the forest will be a healthier place as well. According to Buech, tiny forest basins not only serve as nurseries for salamanders and other wildlife but also "likely contribute substantially to the hydrology and biodiversity of the forests in which they occur." They help slow the coursing of runoff so that it can percolate into the ground and recharge groundwater reserves.

Gibbs points out that these small aquatic islands often support flora and fauna that are rarely found in other parts of the forest. Many of these unique aquatic plants are able to colonize new parts of the forest only by hitchhiking on the bodies of mobile animals such as salamanders, turtles, and birds. And the intense concentration of life around seasonal pools, particularly insects, indicates that these pools function as oases for forest animals, including many songbirds, which have been shown to seek out poolside habitats as nesting sites.

But before the future of salamanders—and some say the forest—can be secured, much work remains to be done on some of the most fundamental aspects of salamander ecology, such as taking a census of their numbers and mapping their habitats. Some of this work has already begun. To make the mapping of small embedded wetlands less expensive and time-consuming, Buech is investigating ways to streamline methods for their detection, including the use of infrared aerial photography before spring leafout. Such surveys are the first step in implementing regulatory protections for seasonal pools, such as those pioneered by Massachusetts and Maine.

Protecting the complement of forest habitat around these ponds is just as important. Research in the forests of Massachusetts, for example, has shown that a salamander's choice of a particular breeding pool was determined by the percentage of forest that lay within a five-hundred-foot radius. Judging what constitutes a good forest habitat from a salamander's perspective also requires a change in the way foresters traditionally describe and evaluate a patch of woods. Instead of simply measuring the age of a stand of living trees, forest managers are learning to value the number of downed trunks and decaying stumps.

The tasks for scientists now is to move from a reactive stance—assessing the damage of management actions on salamander populations—to developing proactive blueprints that ensure their survival. "The challenge for future researchers of amphibian-forestry relationships" say deMaynadier and Hunter, "has grown from simply documenting harvest impacts, to identifying realistic harvest prescriptions that best maintain those components of the forest's biological legacy that are essential for healthy amphibian populations and ecosystem integrity as a whole." ∽

SUGGESTIONS FOR FURTHER READING

Ash, Andrew N., and Richard C. Bruce. "Impacts of Timber Harvesting on Salamanders." *Conservation Biology* 8, no. 1 (1994): 300–301.

Beebee, T. J. C. *Ecology and Conservation of Amphibians*. New York: Chapman and Hall, 1996.

Blaustein, Andrew R., David B. Wake, and Wayne P. Sousa. "Amphibian Declines: Judging Stability, Persistence, and Susceptibility of Populations to Local and Global Extinctions." *Conservation Biology* 8, no. 1 (1994): 60–71.

Burton, Thomas M., and Gene E. Likens. "Energy Flow and Nutrient Cycling in Salamander Populations in the Hubbard Brook Experimental Forest, New Hampshire." *Ecology* 56 (1975): 1068–80.

———. "Salamander Populations and Biomass in the Hubbard Brook Experimental Forest, New Hampshire." *Copeia* 1975, no. 3 (1975): 541–46.

Cohen, Jeffrey P. "Salamanders Slip-Sliding Away or Too Surreptitious to Count?" *Bioscience* 44, no. 4 (1994): 219–23.

DeMaynadier, Phillip G., and Malcolm L. Hunter Jr. "The Relationship between Forest Management and Amphibian Ecology: A Review of the North American Literature." *Environmental Review* 3 (1995): 230–61.

Douglas, Michael Edward. "A Comparative Study of Topographical Orientation in *Ambystoma*." *Copeia* 1981, no. 2 (1981): 460–63.

Fahrig, Lenore, John H. Pedlar, Shealagh E. Pope, Philip D. Taylor, and John F. Wegner. "Effect of Road Traffic on Amphibian Density." *Biological Conservation* 73 (1995): 177–82.

Fraser, Douglas F. "Empirical Evaluation of the Hypothesis of Food Competition in Salamanders of the Genus *Plethodon*." *Ecology* 57, no. 3 (1976): 459–71.

Gibbs, James P. "Importance of Small Wetlands for the Persistence of Local Populations of Wetland-Associated Animals." *Wetlands* 13, no. 1 (1993): 25–31.

Harding, James H. *Amphibians and Reptiles of the Great Lakes Region*. Ann Arbor: University of Michigan Press, 1997.

Heatwole, Harold. "Environmental Factors Influencing Local Distribution and Activity of the Salamander, *Plethodon Cinereus*." *Ecology* 43, no. 3 (1962): 460–72.

Kleeberger, Steven R., and J. Kirwin Werner. "Home Range and Homing Behavior of *Plethodon cinereus* in Northern Michigan." *Copeia* 1982, no. 2 (1982): 409–15.

Mader, H. J. "Animal Habitat Isolation by Roads and Agricultural Fields." *Biological Conservation* 29 (1984): 81–96.

Moriarty, John J. "Six Slippery Salamanders." *Minnesota Conservation Volunteer* May/June 2001: 28–37.

Moriarty, John J., and Carol D. Hall. *Amphibians and Reptiles in Minnesota.* Minneapolis: University of Minnesota Press, 2014.

Paine, Robert T. "Food Web Complexity and Species Diversity." *American Naturalist* 100, no. 910 (1966): 65–75.

Petranka, James W. "Response to Impact of Timber Harvesting on Salamanders." *Conservation Biology* 8, no. 1 (1994): 302–4.

———. *Salamanders of the United States and Canada.* Washington, D.C.: Smithsonian Institution Press, 1998.

Petranka, James W., and Susan S. Murray. "Effectiveness of Removal Sampling for Determining Salamander Density and Biomass: A Case Study in an Appalachian Streamside Community." *Journal of Herpetology* 35, no. 1 (2001): 36–44.

Pollack, Andrew. "Missing Limb? Salamanders May Have Answer." *New York Times,* 24 September 2002.

Stebbins, Robert C., and Nathan W. Cohen. *A Natural History of Amphibians.* Princeton, N.J.: Princeton University Press, 1995.

Stuart, Simon N., Janice C. Chanson, Neil A. Cox, Bruce E. Young, Ana S. L. Rodrigues, Debra L. Fischman, and Robert W. Waller. "Status and Trends of Amphibian Declines Worldwide." *Science* 306, no. 5702 (2004): 1783–86.

INTERNET RESOURCES

Amphibia Web, http://elib.cs.berkeley.edu/aw/

North American Amphibian Monitoring Program, https://www.pwrc.usgs.gov/naamp

Partners in Amphibian and Reptile Conservation, http://www.parcplace.org

Savannah River Ecology Laboratory Herpetology Program, http://www.uga.edu/~srelherp/

Plant Galls and Their Makers

The Entomologist does not indeed pretend to understand the language of Insects,
for, as they all breathe through spiracles or branchiae, their mouths are everlastingly dumb.
But from signs and tokens well known to him he can interpret their actions, and recognize
at a glance what object they are pursuing, whether sport, or love, or war, or food
for themselves, or food for their future progeny, or the construction of habitations
either for themselves or for that future progeny which they are doomed never to behold.
Under every stone, under every clod, and even under the most despised substances, there is
a little world in miniature opened to his eyes. And there scarcely grows a plant but what
contains, in Nature's own hieroglyphs, a whole volume of Natural History written
by the finger of the Great Author of our being.

Benjamin D. Walsh, "Insects Inhabiting Willow Galls," 1864

I T IS THE STUFF OF HOLLYWOOD SCIENCE-FICTION THRILLERS. IMAGINE THIS scenario: insects—as small, light, and inconspicuous as a mosquito—that reproduce by laying eggs on or just under your skin. Within a week, the developing larvae begin to feed. Like some insurgent SWAT team, they take control of the command center that regulates the growth mechanisms of skin cells. Their first order of business is to grow their own nurseries and then stock their new quarters with food purloined from your own bloodstream. As the nurseries develop, growths begin to appear that are unlike any your body would produce under normal circumstances. Depending on the insect species, the balls of your feet or your fingertips may sprout little more than a few bristled polyps or an elaborate rosette here and there. In some cases, whole swaths of skin could be covered by a series of raised structures that resemble cones, lozenges, or tiny clubs. Housed in each are larvae, often no bigger than a grain of rice, that silently gnaw away at their nursery walls until they are ready to emerge as adults—and start the process all over again.

Too ghoulish even for fiction? Well, in reality, it happens all the time—to plants, that is. The strange growths are known as galls. These atypical structures are formed in response to the feeding or oviposition (egg laying) of a wide variety of invertebrates, including nematodes and mites. By far the most important group of gall inducers, however, is insects—some thirteen thousand species. (Although galls occur worldwide, the greatest diversity is found in the world's temperate zone, which includes the Lake Superior region.)

The purpose of galls is to provide the offspring of gall-inducing organisms with good food and safe lodgings. To fulfill their reproductive mandate, these predatory houseguests leave few plants untouched, not even algae and fungi. In fact, the gall insect's motto might read something like, "You grow it, we gall it." Galls can surface on just about any part of a plant, from flower heads and bark to root tips and seeds. Their host material of choice, however, is leaves. In fact, 80 percent of galls are found on leaves, where, as entomologist May Berenbaum points out, they occur on just about every leaf part—stipules, petioles, veins, leaf blades. "On leaf blades alone," she writes, "there are Filzgalls, fleshy galls, kammergalls, fold galls, roll galls, roll and fold galls, pit galls, pocket galls, oak-apple galls, and lenticular galls, to name but a few."

This astonishing variety is due to the extreme specialization of gall inducers. They tend to be highly host specific, singling out oaks, say, to the exclusion of other trees, ignoring red-osier dogwood in favor of arctic willow.

But this only begins to describe the extent of their choosiness. Gall inducers even differentiate among plant species within the same genus. For example, two races, or subgroups, of *Eurosta solidaginis* inhabit the Mid-Atlantic states. Each race lays its eggs on the flower buds of goldenrod (the larvae eventually make their way to the plant stems, where they instigate gall formation). But not just any goldenrod will do. The region hosts five related species of goldenrod, all of which are similar in appearance. Yet one race of *E. solidaginis* prefers the buds of late goldenrod. The other favors tall goldenrod. How can the adult flies tell the difference? Experiments conducted by goldenrod gall expert Warren G. Abrahamson and his colleagues show that *E. solidaginis* crawls over the flower buds of the various species, relying on chemical sensors in her feet to alert her to the presence of the exact goldenrod that she is seeking.

So important is it that *Eurosta* locate the correct goldenrod species that the fly possesses backup strategies to safeguard against cases of mistaken identity. When researchers wrapped the bud of another goldenrod species with the leaves of tall or late goldenrod, the associated *Eurosta* fly initially was fooled into piercing the plant material with her ovipositor, a syringe-like organ used in egg laying. But additional sensors located in the fly's ovipositor detected the underlying goldenrod species, and *Eurosta* quickly withdrew before laying any eggs. Such divining skills are essential to the survival of *E. solidaginis*. Even when scientists succeeded in tricking the fly into laying eggs in plants other than their preferred tall or late goldenrod host, the larvae died.

Gall inducers not only favor a particular species of plant but also target the plant organ on which they will lay their eggs. Each produces its own telltale gall much like a famous architect who can be identified by the signature style of her buildings. These relationships are so precise that you often can pinpoint the identity of the gall inducer just by looking at the gall. Among the largest—and most familiar to north-woods berry pickers—are galls on the stems of thimbleberry, nurseries for the gall wasp *Diastrophus kincaidii*. The larvae of this quarter-inch wasp (family Cynipidae) can colonize whole thickets, producing gnarly swellings on stems, sometimes twisting them in what look like crippling knots. The spent galls often are pitted with tiny holes marking the exit tunnels made by emerging adult wasps in the spring or pockmarked with cavities caused by the pecking of birds, especially overwintering chickadees and downy woodpeckers, who raid their stores of juicy larvae.

The gall midge *Rhabdophaga strobiloides,* on the other hand, lays its eggs on leaf buds at the ends of the branches of pussy willows. The hatched larvae induce elegantly tapered structures composed of numerous overlapping scales—in some cases up to seventy-five of them. Resembling rosebuds or closed pine cones, they have caused more

Using chemical sensors in her feet to guide her to the right species of goldenrod, an adult female fly of *Eurosta solidaginis* lays her eggs in the plant's flower buds. The larvae eventually make their way into the stem tissue, where they instigate the formation of a telltale marble-shaped gall *(below)*. Photograph by W. G. Abrahamson.

U.S. Fish and Wildlife Service Digital Library.

Confined and defenseless, gall residents are targeted by foragers, such as this downy woodpecker, as well as by chickadees and nuthatches. Photograph by Warren Uxley.

than one north-woods hiker to do a double take since they look as out of place on this deciduous shrub as horns on a house cat.

Equally startling are the tiny clusters of reddish spines that are induced by the cynipid wasp *Diplolepis polita* on the leaves of prickly wild rose. Resembling miniature pincushions, these galls refute the conventional wisdom that says only the stems of roses are thorny.

Sometimes the variations among galls can be extraordinarily subtle. In the north woods, two different kinds of rounded galls occur on the stems of goldenrod. Seasoned winter anglers can readily tell them apart, however. The maggot of *E. solidaginis* induces a marble-shaped gall about one inch in diameter. The larva of a moth (*Gnorimoschema gallaeosolidaginis*), on the other hand, causes a spindle-shaped gall to form. Anglers who search out the dried stalks of goldenrods along the banks of frozen lakes bypass these elongated galls in favor of the more perfectly rounded ones. Why? The moth larvae pupate in August and emerge as adult insects in fall, leaving behind an empty chamber. The larvae of the goldenrod gall fly, however, overwinter in their tiny globes, providing anglers with a ready source of plump white grubs perfect for dangling through the ice before a hungry trout.

Plants and Gall Makers: An Ancient Coexistence

Galls and their makers have had ample opportunity to evolve these finely calibrated relationships—in fact, hundreds of millions of years. In 1996 paleontologists discovered evidence of a gall in the fossilized remains of a tree fern that lived some 302 million years ago. Taken from a coal mine in Illinois, the fossil contained fecal pellets that scientists say may have been excreted by the insect larva that caused the gall to form in the first place.

No less ingenious are the ways in which humans have learned to exploit these ubiquitous growths. In China, India, and Europe, galls have been used as food, medicine, and industrial materials for thousands of years. Perhaps the most famous—and multipurpose—example is the Aleppo gall, which occurs on several species of oaks in eastern Europe and western Asia. These "oak apples," so called for their apple-like

Willow cone galls induced by the midge *Rhabdophaga strobiloides* are distinctive, elegantly tapered structures composed of numerous overlapping scales. Photograph by Warren Uxley.

forms on the leaves of oak trees, are high in tannic acid. This substance was used extensively in the dyeing of textiles and leather. Just as important, when mixed with iron salts, it forms a durable, high-quality ink, which has been used by humans since 2000 BC, including in the illumination of medieval manuscripts and later in official documents for the U.S. and several European governments.

Yet despite this long human and natural history, surprisingly little is known about the precise dynamics that cause galls to form. It was not until 1687 that the Italian physiologist Marcello Malpighi linked the formation of galls to the presence of certain insects. Despite such early insights, today's gall scientists, known as cecidologists, have yet to completely crack the biochemical code that animals use to communicate with their host plants. That is because the majority of research dollars are devoted to studying pests in economically important crops. And with a few famous exceptions, gall insects do not damage plants with great commercial value. Furthermore, "to understand galls," say gall researchers Joseph D. Shorthouse and Odette Rohfritsch, "one must be aware of the life strategies of both the host plant and insect," a double expertise that few researchers have mastered. "Our knowledge of galls is so poor," writes British gall scientist Michele A. J. Williams, that even amateur naturalists have many opportunities "to make valuable scientific contributions."

New discoveries are reversing long-held theories and offering tantalizing conjectures about some of the most fundamental processes in gall formation. For example, scientists long believed that galls were formed primarily in response to the secretions of insects. It turns out that enzymes in the saliva of feeding insects or substances in the fluids of egg-laying adults are only part of the story. The biomechanical stimulation that plants receive from gall insects may be just as important as these chemical signals. According to some theories, the complexity of the gall structure depends on the mouth parts and feeding mode of the larvae. The young of sawflies and beetles, for example, possess mandibles that enable them to tear off and chew chunks of food. This feeding style is thought to result in galls that are structurally simple and the least differentiated from the normal organs of the host plant. Some researchers suggest that more elaborate galls, such as the willow-cone gall with its large teardrop-shaped mass of overlapping scales, are instigated by larvae with mouth parts that allow them to puncture the walls of plant cells and drain their juices. The most structurally complex galls are formed in response to the feeding activity of the larvae of cynipid wasps. Using their mandibles to slash the cells lining their chambers, the larvae then slurp the cellular fluids from their faces. Researchers theorize that the damaged cells send chemical signals to nearby intact cells. Within these communiqués are instructions that direct the formation of cynipid galls.

Even less understood is the potential genetic component of gall formation. Some researchers have even suggested that the insects may introduce viruses that deactivate the plant's own growth-regulating genes, allowing the larvae to take control of the gall's formation, everything from its size, color, and metabolism to its structure and shape. Their ability to induce forms that are foreign to their host plants has earned gall insects a reputation as the horticulturists or genetic engineers of the insect world.

What scientists *do* know about galls, however, demonstrates just how remarkable some of the most common processes in nature can be. The adults of most gall insects, for example, forego laying eggs on mature plant tissues in favor of young, actively growing tissues. Consequently, the seasonal emergence of adult gall insects in search of

The larva of the fly *Eurosta solidaginis* within its distinctive marble-shaped gall. Photograph by Warren Uxley.

egg-laying sites most often coincides with spring's flush of new growth or occurs later in the summer when fruits form or a second round of leaves sprout.

Once the eggs are deposited, galls normally undergo four stages of development. During the first phase, known as initiation, the act of egg laying by adults or feeding by the larvae isolate cells from the host plant. These cells change in response to the physical and chemical stimulation by the larvae and, like a batch of "starter" cells, serve as both the template and the stimuli for the cells that form the gall.

At first the young gallers feed only intermittently, waiting until the surrounding plant cells become enlarged before they begin to feast. They do not abstain for long. During the next phase—gall growth and development—the plant produces abnormal cells that are not only swollen (hypertrophy) but also rapidly divide (hyperplasia).

In the third stage, known as maturation, the larvae settle into their nurseries and focus on consuming the gall cells that surround them, called nutritive cells. They provide the sedentary larvae with a replenishable and easily accessible food supply that is highly concentrated in starches, sugars, fats, and proteins.

The gallers direct their host plants to provide not only food but also protection for themselves and their larders. During the maturation stage, the cells surrounding the nutritive stores become differentiated as sclerenchyma cells. Plants commonly produce these thick-walled cells as a way to protect their vulnerable tissues and to provide the support they need to stay upright. In the case of galls, sclerenchyma cells lend toughness and rigidity to the gall structure—and sometimes an added measure of protection. In many galls, for example, sclerenchyma cells are composed of hard-to-penetrate substances such as lignin and bitter-tasting tannin, which some scientists believe discourages invasion by bacteria and fungi. Together with an outer, third layer of cells (composed of cells normally found on the exterior of plants), the gall provides a kind of localized microclimate that encapsulates the larva and helps to protect it from extremes of temperature and humidity as well as from many predators.

In time, the gall ceases to grow as the mature larvae stop feeding and prepare to leave their nurseries. The emerging adults of cynipid wasps are equipped with mouth parts that enable them to chew their own escape tunnels. Many other gallers, however, must wait for the gall to undergo a final phase—gall dehiscence—in which the mature insects are released after they cause the walls of their former nurseries to split open.

Live and Let Live: Gallers and the Health of Their Host Plants

How vulnerable plants are to being weakened or killed by the growth of galls depends on the species. Even the presence of one spherical gall on a goldenrod stem, Berenbaum says, can tap enough of the plant's energy to cut seed production by half. For a few commercial crops, gall infestations have been devastating. For example, in the mid-nineteenth century, an aphid-like insect known as *Phylloxera* was accidentally introduced into France. Native to North America, this tiny insect induces galls on the leaves and roots of grapevines, which can shut down the flow of water and nutrients to the host plant. So devastating was *Phylloxera* that by 1885 nearly one-third of the country's vineyards were lost. Vintners were able to save France's wine industry by importing *Phylloxera*-resistant rootstocks from the United States and grafting them on to their own grapevines. A gall midge known as the Hessian fly *(Mayetiola destructor),* so called because scientists believe it was introduced to the United States in the bedstraw of

Hessian soldiers who fought on behalf of the British during the Revolutionary War, stunts the growth of wheat, reducing yields in infested crops.

But many plants host legions of gall larvae with seemingly no ill effect. The question remains, why after more than a billion years of evolution haven't plants—from algae and trees to fungi—developed a resistance to being strong-armed by gall insects in the first place, especially since they do not seem to derive any apparent benefits? Well, some of them have. Research by Abrahamson and his colleagues revealed that individual tall-goldenrod plants have developed the capacity to starve gall inducers by shutting down the growth of their galls. Curiously, female *Eurosta* flies often are able to distinguish vulnerable plants from resistant ones.

Still, plants are losers in the numbers game with gall insects. A single oak tree, for example, can host five hundred thousand galls induced by one species of cynipid wasp alone. According to one theory, plants have given up on winning the war against gall insects and instead have called for a truce. By concentrating abundant food around particular nodes, plants quarantine gall makers, thus protecting other vital tissues. To further minimize their impact, plants have forced gall insects to become feeding specialists on particular plant organs. It might also be that gall insects have tricked their host plants into becoming more efficient in their photosynthetic and nutrient-accumulating abilities. As a result, they can satisfy the needs of the gall larvae without causing harm to the overall health of the plant. Regardless of the strategy, over millions of years of evolution, both plants and gall insects appear to have accommodated one another through a complex and finely tuned relationship that allows each to thrive.

But in nature even free room and board come at a cost—and gallers are no exception. Most gall-inducing larvae share their cramped quarters with a host of freeloaders. One thimbleberry gall the size of a peanut, for example, can support fourteen different species of inhabitants. Willow-cone galls host an even greater variety of organisms. In 1905 Roy L. Heindel published a study that revealed the astoundingly diverse community of insects that lives in these common galls. After capturing the insects that emerged from a collection of willow-cone galls, he cataloged ten inquilines (insects that feed on the cells of galls produced by other insects), sixteen parasites or hyperparasites (parasites that feed on other parasites), and five vagrant insects that likely used the gall only as a temporary shelter. Add the resident larvae of a gall-inducing midge to this list of occupants and Heindel tallied thirty-two different kinds of insects that utilize the galls on one plant species alone.

Although the sturdy walls of most galls help to protect gall inducers from many of the ills that beset other insects that feed externally on plants—such as disease, predators, and desiccation in times of drought—they have their own vulnerabilities to contend with. The larvae of cynipid wasps belonging to the genus *Diplolepis,* for example, are not immune from attack, not even those of *D. polita,* which lie tucked inside the prickly galls they induce on the leaves of wild roses. Females of inquiline wasps manage to penetrate the gall's thick walls and fatally stab the *Diplolepis* larvae before laying their own eggs within the gall chambers. Making themselves right at home, the interloper larvae will issue new chemical signals to redirect the gall growth, thereby "remodeling" the quarters designed by the original occupant.

Because the larvae of gall inducers are largely sedentary, confined, and defenseless, they also make easy pickings for parasitoids, especially parasitic wasps. In fact, many types of parasitic wasps have evolved a dependency on gall inducers as a dietary staple.

After studying large collections of rose galls in Canada that were induced by *D. polita,* Shorthouse commonly found that up to 90 percent of the gall inducers were killed by the combined pressure of inquilines and parasitoids.

In temperate regions, many cynipid larvae that survive the pressures of inquilines and parasitoids must surmount additional hurdles, such as winter. Unlike the larvae that inhabit leaf galls, which drop to the ground in fall and often are protected by blankets of snow, the inhabitants of stem galls are far more exposed and vulnerable to the cold, especially during winters with frigid temperatures and scant snow. The combination can be deadly. Shorthouse studied overwintering larvae of *D. triforma* following a severe winter that hit central Ontario in 1977–78. About 75 percent of the larvae sampled had succumbed to the cold within the spiny galls they induce on the stems of prickly wild rose.

When galls appear each spring, whole communities of other organisms are housed and fed. As to the ecological importance of these many dependencies, however, today's gall experts can hazard a guess with little more certainty than their colleagues more than 150 years ago. "If this one little gall and the insect which produces it were swept out of existence, how the whole world of insects would be convulsed as by an earthquake!" exclaimed the nineteenth-century naturalist Benjamin D. Walsh. "How many species would be compelled to resort for food to other sources, thereby grievously disarranging the due balance of insect life! How many would probably perish from off the face of the earth, or be greatly reduced in numbers! Yet to the eye of the common observer this gall is nothing but an unending mass of leaves, of the origin and history of which he knows nothing and cares nothing!" ∾

SUGGESTIONS FOR FURTHER READING

Abrahamson, Warren G., and Arthur E. Weis. *Evolutionary Ecology across Three Trophic Levels: Goldenrods, Gallmakers, and Natural Enemies.* Princeton, N.J.: Princeton University Press, 1997.

Ananthakrishnan, T. N., ed. *Biology of Gall Insects.* London: Edward Arnold, 1984.

Berenbaum, May R. *Bugs in the System: Insects and Their Impact on Human Affairs.* New York: Addison-Wesley Publishing Company, 1995.

Fagan, Margaret M. "The Uses of Insect Galls." *American Naturalist* 52, no. 614 (1918): 155–76.

Frost, S. W. *Insect Life and Insect Natural History.* New York: Dover Publications, 1959.

Heindel, Roy L. "Ecology of the Willow Cone Gall." *American Naturalist* 39, no. 468 (1905): 859–73.

Shorthouse, Joseph D. "Adaptations of Gall Wasps of the Genus *Diplolepis* (Hymenoptera: Cynipidae) and the Role of Gall Anatomyin Cynipid Systematics. *Memoirs of the Entomological Society of Canada* 165 (1993): 139–63.

Shorthouse, Joseph D., and Odette Rohfritsch, eds. *Biology of Insect-Induced Galls.* New York: Oxford University Press, 1992.

Weis, Arthur E., and Warren G. Abrahamson. "Just Lookin' for a Home." *Natural History* 107, no. 9 (1998): 60–64.

Williams, Michele A. J., ed. *Plant Galls: Organisms, Interactions, Populations.* Oxford: Clarendon Press, 1994.

INTERNET RESOURCES

The *Solidago Eurosta* Gall Homepage, http://www.facstaff.bucknell.edu/abrahmsn/solidago/main.html

Black Bears and the Tettegouche Oaks

Here is a creature that can weigh a quarter ton or more, hang by its teeth,
and haul down a fully grown Rocky Mountain bull elk in deep snow, but which
prefers to climb trees to escape danger, and teases hazelnuts, one by one, from the
forest litter. A predator by formal definition, a carnivore by taxonomic rank,
it has evolved further away from carnivorous traditions than man himself.
Due to this choice of foods—a choice made, in the evolutionary sense, twenty-five
million years ago and which first made the bear line distinct—the black bear's
entire ecology, including breeding, productivity, survival, population density,
even social organization, is a function of the annual distribution and abundance of
tiny fruits and nuts. The production of these foods is controlled and delimited in part
by the same cold season which shaped bear physiology.

Jeff Fair, *The Great American Bear* (1994)

Wildlife biologist Lynn Rogers heads up a steep trail into Tettegouche State Park, climbing the rise with a strong, even gait. It is a practiced step honed over more than four decades of field research in which Rogers trailed the footsteps of wild black bears from the Boundary Waters Canoe Area Wilderness to the shores of Lake Superior.

The morning is unusually brisk, and even though June is prime nesting season for many birds in the north, their calls are distant and half-hearted. We are here to search for evidence of bears, but Rogers cannot pass up the opportunity to first check out the birds. He pauses to try on a new set of earphones—thick pads with tiny protruding microphones—and fiddles with a knob on a volume-control box as he scans the tops of nearby maples. After years of conducting aerial reconnaissance of his study bears, in which he strained to tease out the faint bleat of radio signals from a background of static and the roar of airplane propellers, Rogers has lost the ability to hear high-pitched sounds, including the upper register of many bird songs. He hopes that the amplified earphones will help him catch the buzzy staccato of the black-throated blue, a warbler that experienced birders know is heard in this part of the shore's northern hardwood forest.

But the results are mixed. The bird's call can barely be made out above the heightened hiss and crackle of the surrounding forest. Rogers stuffs the earphones into his backpack and seems keenly disappointed by the failure of the device. It is clear that he has far more than just a professional interest in the outdoors. He says as a child growing up in Michigan, he was frequently laid low by bronchitis. The birds in the mulberry tree outside his window kept him connected to the world beyond the confines of his small bedroom. Today, it is hard to know what pleases him more—birds, bears, or the great remnant white pines whose future he has come to vigorously champion in the later years of his career.

Although he may no longer be able to hear all of the forest's sounds, Rogers has lost none of his expert woodsmanship. We leave the trail and bushwhack across a maple forest to a rocky overlook, then follow the ridgeline to the east until we step into a grove of northern red oak trees. No sooner do we enter its deep shade than he retrieves a dark lump from the forest floor. Bear scat. Dry and corky, smelling of earth. It is a sign that Rogers reads like a calling card indicating that bears fed on the acorns of these trees the preceding fall. He points to "nests" of branches up in the crotches of the trees that were broken by the bears' powerful grip as they lounged in the trees and reached out and bent the acorn-laden limbs to their lips.

Many visitors to the forest would miss these cues, not because they lack care or vigilance, but because in the north country the most profound natural events often do not call great attention to themselves. What appears to be an ordinary woods is in fact a rare island of oak trees whose ancestors took hold perhaps thousands of years ago and held their ground as a sea of sugar maples rose around them.

Generations of *Ursus americanus* have walked great distances to feed in this oak forest, their toddling cubs growing up to carry on the tradition by bringing their own families. The meaty nuggets that rain down from the trees have enabled countless bears to build up the critical stores of fat that give sows the energy boost for bearing a new litter and help yearlings in lean berry years to survive the winter. The oak forest is an immovable feast in an environment not overly provident to bears.

Biologists have long recognized the value of oak trees to wildlife. Acorns are a staple in the diet of an estimated ninety-six species of birds and mammals in North America. Black bears so prize the high-energy and easily digestible nuts that they undertake what biologists call a "fall shuffle"—treks to nodes of abundant food. Annual bear migrations to acorn-producing areas have been documented in Michigan, Wisconsin, Tennessee, and Ontario.

Along the Minnesota North Shore, however, the phenomenon was largely unknown until the late 1980s when Rogers began to notice a pattern in the movements of radio-tracked bears. Many of the bears in Rogers's study—even sows with cubs that vigorously defended their territories throughout the spring and summer—abruptly broke rank in the fall once they had exhausted the backcountry's supplies of early-ripening fruits such as blueberries, juneberries, sarsaparilla, pin cherries, and chokecherries. The bears headed south to the North Shore Highlands to feed on later-ripening crops of hazelnuts, acorns, highbush cranberries, dogwood berries, and mountain ash berries. After having eaten their fill of the nutritious mast, the bears returned to their inland territories to den up for the winter.

One spot seemed particularly attractive—the Tettegouche oak stand. Rogers had been tipped off to the potential importance of the site in the 1970s when he radio-tracked a mother bear (known as Female 320) and her cubs. The bear family had traveled twenty miles outside its territory to feed on the acorns produced by the Tettegouche oaks. Rogers discovered that Female 320 was joined by other bears, which congregated from many points inland to partake in the bounty.

In late fall 1987 Rogers flew over parts of northeastern Minnesota in an attempt to identify the locations of area oak stands. By then the leaves of most other deciduous trees had fallen, exposing the orangey-brown foliage of the oaks, which would not be shed until late winter. From the air the outlines of the oak forests were perfectly legible.

As he suspected, oak stands were rare in the region. Rogers counted twenty-two patches of northern red oak ranging from 5 to 154 acres in size. Along the North Shore, red oak forests generally grow no farther east than Lake County and lie within one to seven miles of Lake Superior. Although patches of northern red oak crop up farther inland near Ely and north of the Iron Range, their acorn crops are not nearly as prodigious as the stands near the shore.

By far the largest was the Tettegouche stand. The tract, comprising 154 acres, ran for about a mile in shallow soil along a dry, south-facing ridge. It took Rogers one hour to walk its entire length. Along the way he found evidence of abundant wildlife. On the ground were scuffles in the dirt where deer had pawed through the leaf litter in search of acorns. As he walked through the forest, he flushed fifteen ruffed grouse, a number that Rogers calls a "hunter's dream." Most exciting of all, he counted thirteen trees that had recently been climbed by black bears. It was a diner's paradise for critters with an appetite for acorns. Noting its abundance of mast-producing trees over an exceptionally large area, Rogers concluded that it was the best xeric northern red-oak community he had ever visited in northeastern Minnesota.

Adding to the ecological value of the stand was the vigor of the trees. Although short and stocky compared to the oaks that grew in more hospitable conditions among maples in a nearby forest, many trees were up to twenty-four inches in diameter and estimated to be more than one hundred years old. Studies of northern red oaks in warmer climates show that oaks are long-lived but slow to reach maturity. The trees do not begin to produce abundant acorn crops until they are about fifty years old. Mature trees can yield crops for two hundred years or more.

Knowing the importance of this rich food source in a land not often given to surplus bounty, Rogers drafted a letter to the director of the Minnesota chapter of The Nature Conservancy outlining its ecological merits. The Conservancy, which already had considered securing adjacent lands on the basis of the rare species and other ecologically valuable communities they contained, used Rogers's endorsement to include the oak parcel in its conservation objectives. In 1992, through the efforts of private organizations, state agencies, and county government, 2,873 acres of land were added to Tettegouche State Park, preserving the oak forest and nearly doubling the size of the existing park.

Bears in the Land of Fruits and Nuts

The presence of the oak stands and their relationship to area black bears remain among the many biological mysteries of the North Shore region. The three species of oak that occur in northeastern Minnesota, all of them relatively rare and found in widely scattered patches, commonly grow in more southerly hardwood forests. Rogers theorizes that about five thousand years ago when the climate was warmer and drier, the oaks were able to expand the northernmost limits of their territories into the North Shore Highlands, where they colonized well-drained ridges. When the climate swung into a cooler, wetter phase, the oaks may have been largely outmaneuvered by a fellow hardwood species—sugar maple—and relegated to south-facing ridges with shallow soils, where they could hold sway over competing species such as maple, birch, and aspen.

Their rarity makes them all the more important to wildlife, especially bears. According to Rogers, one of the biggest limitations bears face in northeastern Minnesota is the lack of fall food. Although most gain some weight in the spring, they really only put on the pounds they need to survive the winter after carbohydrate-rich berries ripen in midsummer. Mother bears are particularly dependent on these midsummer crops since their food stores must carry them through not only the winter but also the following spring. During the spring, nursing bears lose weight and supplement their energy needs with the fat accumulated from the previous year as they forage on whatever foods are in season within their territories.

During the spring green-up, bears may graze on tender young grasses, skunk cabbage, aspen, willow catkins (cattail-like flowers), newly emerged aspen leaves, flowers, and clover. Bears obtain maximum benefits from consuming new vegetation. Not only is the protein content higher in young plants, but also bears are able to better absorb their nutrients. As many plants mature, their cell walls stiffen into cellulose, a plant component that bears cannot digest. (Although taxonomists classify bears as carnivores, they have developed several physiological features that reflect their evolution toward vegetarianism. For example, compared to other carnivores, black bears and grizzlies have a longer gastrointestinal tract relative to body size—8:1 for bears versus 4:1 for wolves, says biologist Fair. This allows them to digest a wide range of plant materials. But unlike other north-country animals, such as snowshoe hares, bears lack a special intestinal pouch known as a cecum that harbors bacteria capable of breaking down the cellulose found in mature woody plant materials.)

Bears obtain another major and dependable source of protein from insects and their larvae. During the height of tent caterpillar outbreaks, bears supply valuable pest-control services, Rogers says, by consuming an average of twenty-five thousand caterpillars daily! As opportunistic omnivores, they also eat the occasional fawn and moose calf, beginning with the curdled, high-fat milk in its stomach.

As biologist Fair observes, "The black bear is a full-time professional food consumer. He lives to eat, as the saying goes, and eats to live. Except for his winter sleep and those few complicated weeks around late June [when attentions are focused on mating or defending territories], he spends his waking hours occupied with nutrient input. The voyageurs had a name for black bears—*Cochons de bois,* 'pigs of the woods.' George Laycock once described the bear as an animal of 'perpetual hunger.' And why not? Any creature of this bulk that needs to lay down enough fat in two seasons, by pushing vegetable matter through its short gut, to sleep away half the year or more had better be serious about his feed."

Early studies by biologist Lynn Rogers revealed that in the fall many black bears leave the backcountry and head south to the North Shore Highlands in search of later-ripening crops of hazelnuts, acorns, highbush cranberries, dogwood berries, and mountain ash berries. Photograph copyright Stan Tekiela / NatureSmartImages.com.

Bears become especially intent on feeding as the summer progresses. By mid-August, they step up their consumption of nuts and berries and begin a period of almost round-the-clock eating known as hyperphagia. On September 5, 1988, Rogers sat in a hazelnut patch with one of his study bears and watched her put away 4,081 nuts. The next day she upped her consumption to 4,225 nuts. Wild hazelnuts are especially prized by bears since their contents—60 percent fat, 25 percent protein, and 15 percent carbohydrates—provide them with the kind of food they desperately need. In addition, each nugget packs a caloric wallop. On day two, Rogers's study bear consumed between 12,500 and 19,000 calories in hazelnuts, the equivalent of two to four gallons of Häagen Dazs ice cream—the high-fat variety!

Putting on weight—and lots of it in the form of fat—is critical to a bear's survival. Bears get the biggest bang from their energy buck by burning fat instead of protein or carbohydrates during their winter fast. The picture gets even more complicated for female bears of reproductive age. Without ample fat stores, a bear will not conceive. Bears mate in midsummer, but the fertilized ova, known as a blastocyst, remains suspended in the female bear's uterus for about five months until she dens up for the winter. If the sow has put away a sufficient fat supply, the blastocyst will become implanted in the uterine wall, and the pregnancy begins. In underweight bears, the blastocyst is aborted. If the pregnancy is more advanced, the growing fetus is absorbed by the undernourished animal's body. This survival mechanism, closely keyed to the boom-and-bust cycles of the black bear's northern habitat, saves bears from being subjected to a costly, if not fatal, pregnancy.

Having enough body fat also determines the physical condition of her offspring. Rogers found that female black bears that enter their winter dens weighing less than 148 pounds rarely give birth. Those that weigh between 148 and 175 pounds produce cubs with high mortality. The bears that stand the best chance of producing healthy offspring tip the scales at 176 pounds or more in the fall. Just how important are these fat reserves to mother bears? Rogers's studies show that by the time they emerge from their dens in the spring, some sows that have borne and nursed cubs during their midwinter slumber may have lost up to 40 percent of their weight, all of it trimmed from their reserves of fat, a metabolic path that leaves muscles and vital organs untouched.

The scarcity of oak in much of northeastern Minnesota, Rogers says, has important developmental and reproductive consequences for the region's resident bears. Compared to black bears in the forests of the eastern United States, which have access to a veritable smorgasbord of fall mast from oak, beech, and hickory trees, bears in some parts of the north country face a shortage of food after berry crops disappear in late summer. As a result, northern bears tend to grow more slowly and mature later, producing their first offspring at much later ages than their counterparts in the eastern and southern United States. Rogers's research showed that bears in northeastern Minnesota produce their first cubs at 6.3 years of age versus cub production at 3 to 5 years in more mast-rich areas of the country. The scarcity of fall foods also prompted these bears to conserve energy by dramatically slowing their pace in September and entering their dens in late September or October. Their well-fed cousins in nut-rich regions such as Pennsylvania, on the other hand, are just beginning a feeding frenzy and a period of rapid weight gain that can last well into November.

Rogers's studies of bears that visited the highlands' oak stands, however, revealed

It is said that bears live to eat and eat to live. Putting on weight is critical not only to a mother bear's survival but also to the physical health of her offspring. Photograph copyright Stan Tekiela / NatureSmartImages.com.

surprising similarities to bears in mast-rich areas. Like them, the study bears postponed hibernation to early November to take advantage of the acorn feast. The additional feeding time also had important physiological consequences. In 1970 Rogers compared the weight of Female 320 with seven other females that did not visit oak stands. During September and early October, Female 320 fed on acorns, nearly doubling her weight. The other female bears lost weight during the same period.

The benefits of the additional nutrition extended to Female 320's offspring, which remembered the location of the oak stand after visiting it for only one season with their mother. One of the cubs born to Female 320, for example, gave birth to her own cubs at only four years of age. She was the only female in Rogers's study area to match the reproductive rate commonly found in bears that live in more mast-rich areas of the country. Rogers attributes her early motherhood to the better conditioning that resulted from her carrying on the fall shuffle tradition to the Tettegouche oaks.

The Seasonal Food Traditions of Black Bears

The question remains, How do bears know when the acorns are ripe for the picking? Furthermore, how do they find their way through miles and miles of unmarked terrain?

In fall 1976 Rogers tracked an eleven-year-old male that had traveled 119 miles across northeastern Minnesota to an oak stand near Palisade. After feeding at the site for about five weeks, the bear returned home to den up for the winter.

Scientists theorize that bears may use multiple navigational systems to negotiate unfamiliar or little-visited terrain. Some suspect that bears may be able to orient themselves by sensing changes in the intensity and direction of Earth's magnetic field. This receptivity to geophysical cues might help to explain how "nuisance" bears (those that have become too acclimated to the presence of humans) are able to find their way back to their familiar haunts after having been trapped and released into new, faraway terrain.

Some animals that navigate using magnetic forces have been shown to possess internal compasses. For example, in 1975, scientists discovered a chain of magnetite particles (the same material used in compass needles) in the midsection of a marine bacterium *Aquaspirillum magnetotacticum.* The magnetite allowed the organism to align itself in a position parallel to the local magnetic field. If they became dislodged by ocean currents, the mud-dwelling bacteria could rely on these particles as a kind of compass for locating the ocean bottom again. The discovery prompted scientists to examine other organisms with reported sensitivities to magnetic forces, known as magnetoreception. Their investigations revealed magnetite particles in animals ranging from homing pigeons and spiny lobsters to Pacific salmon and blind mole rats.

Bears also may be able to detect and utilize localized magnetic anomalies, caused by deposits of magnetic iron oxide. In the Lake Superior watershed, magnetic fields of dramatically varying intensity are common. As was discovered by many early mariners along the North Shore and nineteenth-century surveyors working farther inland, these powerful anomalies so skew the readings of magnetically based compasses as to render them useless in navigating northern waters and backcountry forests. The hypothesis that bears are able to sense geomagnetic forces in northern Minnesota, or anywhere else for that matter, remains unproven since scientists as yet have not located biogenic compasses, such as particles of magnetite, in the tissue of bears.

On the other hand, more extensive research has been conducted into the ways in which bears rely on memory and smell to get around. Rogers points out that bears have a phenomenal capacity to store information. This is in no small part due to the size and structure of their brains. Among carnivores they have the heaviest brains in relation to body length. Their brains also bear some striking similarities to those of humans and primates, including neurological structures that resemble the temporal lobe of the human brain.

Also, the animal's sense of smell may be the most refined of any North American mammal. Rogers points out that the area of mucous membrane in a bear's nose is one hundred times that of a human's. Their sensitivity to odors may also be heightened by a vomeronasal organ, called the Jacobson's organ, located in the roof of the mouth. This sensory apparatus is found in a variety of animals, including rattlesnakes, common house cats, and, in a surprising recent discovery, humans.

In the process of using their Jacobson's organ, an act known as *flehmen,* animals touch an object of interest with their nose or tongue. With their heads raised and mouths agape, they suck in air but refrain from breathing. The effect is to bathe the Jacobson's organ with the airborne scent particles. Curiously, the nose and vomeronasal organ function independently of one another. The sensory information collected by the

Jacobson's organ is relayed to a completely different part of the brain than the place where the olfactory nerves are wired, giving animals a kind of sixth sense with which to apprehend a greater array of environmental cues.

Rogers has seen the bear's olfactory prowess firsthand. Bears commonly amble from tree to tree, sniffing out middens of hazelnuts cached by squirrels under their roots. One fall Rogers watched a research bear stop dead in her tracks and suddenly dig up the gleaming white root of a coralroot orchid even though there was no vegetation above the ground to tip the bear off to its presence. Another research bear once took off at a trot and headed forty miles south to the North Shore Highlands after a strong southeasterly wind shift. Rogers had tracked the bear from infancy and knew that she had never ventured that far outside of her home turf before. Rogers suspects that the wind may have swept over the highlands' ripe hazelnut crop and lured the bear south with the promise of its bounty.

To understand just how discriminating the bear's gustatory powers are, you would have to examine the forest floor under the Tettegouche oak trees. Many nuts lie unbroken, apparently overlooked by the bears. Not so, says Rogers. Tiny holes bored in the husk of some nuts indicate that insects consumed their contents. Somehow the bears know—perhaps by smell—which nuts have been eaten by insects, and do not waste time and energy cracking open any of the empty shells.

For an animal its size, bears also eat with extraordinary dexterity and precision. Even the most nimble blueberry picker has marveled over a bear's gentle yet efficient technique: lips swiftly plucking clumps of ripe blue orbs without breaking the plant's slender branches or stripping its foliage.

The bear's light touch is legendary. Fair observed a bear family nip off the blossoms of a bearberry bush while sparing the similar diminutive pink bells of blueberry flowers that were intermingled in the shrubbery, leaving them to develop into a delectable midsummer fruit. Later, while watching another bear consume a midden of hazelnuts, Fair saw that the bear picked up nuts "one by one between tongue and extended upper lip, transferred each to the rear of her mouth, cracked it, and chewed it carefully, discarding hull fragments out the front and sides of her mouth. Sometimes while chewing she allowed chunks of meat to fall onto the back of a forepaw for later retrieval. . . . The combination of excellent sense of smell, good close-range vision, and dexterity of lips, tongue, and claws renders the bear surprisingly clean and delicate in its feeding."

The bear's sense of smell may be the most refined of any North American mammal. To heighten their sensitivity to odors, bears hold their mouths agape to bathe their Jacobson's organ in the roof of the mouth with airborne scent particles. Courtesy of the Northwoods Research Center, Wildlife Research Institute, Ely, Minnesota.

This unexpected juxtaposition of power and refinement in black bears is exemplified in the following anecdote from John Henricksson's *A Wild Neighborhood*. The Gunflint Trail writer tells of his neighbors' failed attempt to discourage a group of bears from visiting their yard after one animal took a fancy to lounging in a porch chair. The home owners planted balloons in their garden that were filled with a harmless but unpleasant ammonia gas. They smeared the outside of the balloons with honey in hopes that the bears would discharge the lot with their sharp teeth and claws and receive a dose of negative reinforcement. The bears licked each balloon clean alright—without puncturing a single one!

Natural Foods and the Future of Bears

According to Rogers, bears will turn to garbage or raiding gardens and bird feeders when natural foods are scarce. Preserving the bears' natural food supply, including critical feeding grounds such as oak forests, is key to their long-term survival. When human-bear encounters become too frequent or overly familiar, bears are often shot and killed.

Unfortunately, the ecological and cultural issues surrounding this preservation are extremely complicated. Many oak stands, for example, are less than five acres in size. Taken together, northern red oak, pin oak, and bur oak make up less than 0.05 percent of the total acreage of Superior National Forest. Like other small, but ecologically vital, features of the forest, such as embedded wetlands, these oak stands are tricky to detect and to map by existing conventional means. Forest maps may fail to differentiate the oak stands from the larger surrounding forest type.

In addition, the stands grow at the northern limits of their range and often are too stunted and deformed by exposure to the cold to be commercially valuable as saw timber. Unlike white pine or aspen, they do not show up on the management or research agendas of most foresters. As a result, many stands are losing out to competition from other species or are purposely being converted to trees such as spruce and pine that deliver higher economic returns. Other stands have deliberately been cut to create openings for such wildlife as moose and deer.

Because the oaks are so widely scattered, the loss of a single stand can leave animals in a large area without a critical food source. And once eliminated, northern oak stands may prove extremely difficult to regenerate. Despite scores of academic studies about the tree's ecology, U.S. Forest Service foresters were unable to reestablish a stand that was mistakenly cut in a firewood sale near Two Island River in the early 1980s.

Rogers adds that identifying and saving bear pathways are just as important as preserving bear destinations. Keeping intact and allowing free access to the earth's "spatial and temporal mosaic," says biologist Talbot H. Waterman, are essential to the survival of many animals. Waterman observes that

> most animals (including humans) spend much of their time in transit—whether swimming, flying, walking, creeping, burrowing, jogging, running, galloping or jetting, from one place to another. All this movement depends mostly on the interaction of two factors. First, the earth is a mosaic of localized sharply different places: mountaintops, meadows, hot springs, glaciers, beaches, forests, deserts, deep seas and so on. Second, although many animals may prefer one of these habitats, they must

move through the larger environmental patchwork to meet their complex needs. . . . Furthermore, the suitability or availability of these special regions often changes regularly with the seasons, the time of day, the tides, and even the organism's age. Accordingly, a mobile animal's habitat typically consists of a number of subhabitats. Each of these has certain features that satisfy particular biological needs at a specific time or during a certain phase of the life cycle.

Having room to roam is especially critical in the North, Rogers says, where generally poor habitat and cycles of extreme food shortages force bears to forage greater distances than bears in more dependable mast-rich forests. From one year to the next in northeastern Minnesota, the abundance of berry crops, for example, can vary by an astounding 10,000 percent. It is the human equivalent of having ten thousand bushels of tomatoes show up at the grocery store one summer versus the delivery of only one bushel. Extended periods of famine can send bear populations plummeting. During a period of nut and berry shortages from 1974 to 1976, for example, Rogers's research revealed that the bear population in northeastern Minnesota dropped by about 35 percent.

Preliminary research suggests that minimizing human-made obstructions, such as highways and residences, along the bears' seasonal food rounds is critical to their long-term survival. Studies have shown that bears shy away from crossing interstates. Even less-trafficked roads are intimidating to bears. A North Carolina study of the relationship between bears and roads by biologist Allan Brody revealed that bears were scarce in areas with more than one-half mile of improved road per one square mile of forest.

In Minnesota the influx of greater numbers of people to the forest poses a long-term threat to bears. In lean food years human developments are little more than baiting stations—and inadvertent death traps—for hungry bears. Rogers calculated that for every home in his study area, one bear was shot at an average interval of nine years. According to biologist Fair,

> Forest homes and "nuisance" bear kills are each increasing over much of the black bear range as the human population grows older and more people build forest retirement homes and vacation homes. These homes and other human developments become particular problems in years when natural food shortages force bears to forage farther than usual. Long movements are difficult when travel corridors are cut down, paved over, fenced off, built up, and otherwise developed. Bears are shot for being attracted to garbage, gardens, dog food, bird feeders, bee hives, apple trees, livestock, farm crops, or simply passing through. In its extreme, this phenomenon is called "habitat fragmentation." Bears are confined to the fragments. When the fragments are subdivided and consumed to the point at which bears and other species can no longer exist within, the wildlife disappears.

Farms and cities have eliminated black bear habitat in many parts of Minnesota, leaving the large blocks of contiguous public lands in the northeastern part of the state as the most whole and healthy remaining home for bears. Yet even here bear habitat is dangerously girdled by human development. In 1985, for example, major failures in nut and berry crops forced hundreds of bears from all over the Canadian Shield to migrate

From the attentiveness they shower on their offspring to their comical postures, bears have long reminded humans of themselves. Courtesy of the Northwoods Research Center, Wildlife Research Institute, Ely, Minnesota.

south in search of food. Following the shoreline of Lake Superior, many bears never made it past the urban choke points of Thunder Bay on the east and Duluth on the west. Reluctant to cross major thoroughfares, the desperate bears raided whatever food reserves they could find, including garbage and apple trees. On the outskirts of Thunder Bay seventy bears were shot. In Duluth ninety bears were killed. Included in the body count were three of Rogers's study bears that had never before left the sparsely populated Isabella area since the study began in 1969.

Rogers also cautions that the tie between the bears and the oaks could be severed by overhunting. Wild bears that manage to avoid the hunter's gun or highway automobile can live to thirty years of age or more. In Minnesota, however, hunting accounts for 90 percent of all bear fatalities. The average male bear is shot at three years of age, and the average female is shot at five years of age. Rogers worries that in time the bears that know about the oaks could be killed before they have the chance to reproduce, much less to pass the fall ritual on to their offspring. (When it comes to the amount of time invested in teaching their young the ways of the world, bears are second only to humans and the great apes.)

Bears "R" Us

Many visitors to the North Shore have had their first—and perhaps only—glimpse of black bears at town garbage dumps, the last of which was closed to bears in the mid-1980s. The popularity of driving to the local landfill to watch the animals making their dinner rounds reflects the fascination people have had with this magnificent creature from the earliest encounters between bruins and humans.

Unlike many of today's observers, however, aboriginal people viewed bears as more than gullible gobblers of leftover Big Macs and stale marshmallows. In the mythologies of many northern American Indians, bears were considered the most powerful and most intelligent animal in nature's pantheon and were emulated for their cunning, resourcefulness, and ability to survive.

This reverence was not surprising. Like bears', the fortunes of Native people rose and fell with the wildly fluctuating bounty of the forest. Humans, like bears, spent a part of the year fasting, and like them faced the terrifying prospects of starvation. The central role of bears in many myths and rituals attests to the deep respect that aboriginal peoples had for the way this extraordinary animal retreats into a kind of sleeplike state and, without eating for months, not only triumphs over adversity but also emerges in the spring with new life in tow.

Like these early peoples, those who choose to live in the forest today must learn the ways of the bear. That means mapping the highways that bears use to travel from seasonally shifting feeding grounds and ensuring that they remain open and unobstructed. It means identifying nodes of plenty and preserving their ability to sustain bears and other forms of life. By doing so, we restore bears to themselves. We give them a future.

In the process, we also may restore humans to themselves. There is perhaps no animal in the Northwoods capable of engendering greater wonder than the black bear. Patiently and doggedly attentive to the smallest detail of the world in which it lives, this great hulk of an animal survives one ant, one tiny berry, one caterpillar, one nut at a time. By observing bears as they truly are—complex beings exquisitely calibrated to the circumstances of their home ground—we might capture some of the things we came to the woods to find: the pleasure of wondering at the magnificent workings of a world that is not of our making. As the environmental writer Paul Shepard observes, bears are the "ideogram of men in the wilderness." He writes:

> Like us, the bear stands upright on the soles of his feet, his eyes nearly on a frontal plane. The bear moves his forelimbs freely in their shoulder sockets, sits on his tail end, one leg folded like an adolescent slouched at the table, worries with moans and sighs, courts with demonstratable affection, . . . snores in his sleep, spanks his children, is avid for sweets and has a moody, gruff, morose side. . . . [He is] wily, strong, agile; independent in ways that we humans left behind when we took up residence in the city.

In the mythologies of many northern American Indians, bears were considered the most powerful and most intelligent animal in nature's pantheon and were emulated for their cunning, resourcefulness, and ability to survive. Photograph copyright Stan Tekiela / NatureSmartImages.com.

SUGGESTIONS FOR FURTHER READING

Alerstam, Thomas. "Animal Behaviour: The Lobsters Navigate." *Nature* 421 (2003): 27–28.

Boles, Larry C., and Kenneth J. Lohmann. "True Navigation and Magnetic Maps in Spiny Lobsters." *Nature* 421 (2003): 60–63.

Eder, Stephan H. K., Hervé Cadiou, Airina Muhamad, Peter A. McNaughton, Joseph L. Kirschvink, and Michael Winklhofer. "Magnetic Characterization of Isolated Candidate Vertebrate Magnetoreceptor Cells." *Proceedings of the National Academy of Sciences* 109 (2013): 12022–27.

Fair, Jeff. *The Great American Bear.* Minocqua, Wis.: Northword Press, 1994.

Henricksson, John. *A Wild Neighborhood.* Minneapolis: University of Minnesota Press, 1997.

Kimchi, Tali, Ariane S. Etienne, and Joseph Terkel. "A Subterranean Animal Uses the Magnetic Compass for Path Integration." *Proceedings of the National Academy of Sciences* 101 (2004): 1105–9.

Mora, Cordula V., Michael Davison, J. Martin Wild, and Michael M. Walker. "Magnetoreception and Its Trigeminal Mediation in the Homing Pigeon." *Nature* 432 (2004): 508–11.

Rogers, Lynn L. "Effects of Mast and Berry Crop Failures on Survival, Growth, and Reproductive Success of Black Bears." *Transactions of North American Wildlife Resources Conference* 41 (1976): 431–38.

———. "Home, Sweet-Smelling Home." *Natural History* 98, no. 9 (1989): 61–66.

Rogers, Lynn L., and Edward L. Lindquist. "Black Bears and the Oak Resource in Northeastern Minnesota." In *The Oak Resource in the Upper Midwest: Implications for Management,* ed. Steven B. Laursen and Joyce F. deBoe, Proceedings of a conference held June 3–6, 1991, St. Mary's College, Winona, Minn.

Waterman, Talbot H. *Animal Navigation.* New York: Scientific American Books, 1989.

Star–Nosed Moles
and a Life
Down Under

When it comes to their physical design, some animals look as if they have had the benefit of multiple revisions. Not so the star-nosed mole.

These tiny wetland residents could not appear more hastily constructed. It is as if they were assembled as an afterthought using the mismatched, leftover pieces of other animals. If you want to find them in the great file cabinet of Life, look under "E" for Evolutionary Work in Progress. Indeed, it seems that only time—and lots of it—will be able to smooth out their ungainly proportions, abrupt transitions, and peculiar add-ons.

Shapeless, like a baked potato, the palm-sized mole is not without its redeeming features. It sports a coat of soft and lustrous black fur, for example. Aesthetically speaking, however, it is all downhill from there. The animal's eyes are squinty, as if it had diverted all of its excess energy to growing other body parts. Take the mole's paws, for starters. One part bird claw to two parts catcher's mitt, they are outsized and scaly, ending in a set of curved, pearly-pink nails that are as long as they are tough.

Then there is the matter of the animal's tail. Highly irregular in its thickness, the tail is constricted at the base, swollen along its midsection, and tapered at the end. In winter the bulge becomes even more pronounced, as if the tail were a snake that had swallowed an egg.

Powerful paws and a magnificent nasal appendage are just some of the star-nosed mole's amazing adaptations to a secretive life underground and underwater. Courtesy of Kenneth C. Catania.

But when it comes to strange features, nothing can top the animal's nose (no pun intended). Like the other six members of its family in the United States, the star-nosed mole has a pointy, slightly upturned snout. But this tiny underground dweller goes to great lengths to differentiate itself from its plain-Jane relatives. Sprouting from the tip is a frilly structure of twenty-two fleshy tentacles that are arranged in a circle around the animal's nostrils. Uneven in length, they range from one-quarter to one-half inch long. When engaged, the highly mobile tentacles sway like an anemone caught up in the ocean's currents.

Long known to science, this spectacular schnoz has driven many observers to abandon objectivity and raid the realms of poetry in their struggle to find adequate descriptive language. "A medusa-like rosette," exclaimed mole experts Martyn Gorman and David Stone in one of their research papers. "Fresh bits of sirloin being extruded through a meat grinder," countered science writer Natalie Angier in a 2010 article on ugly animals in the *New York Times*.

Uncovering the intricate workings of these nasal appendages, however, has proven far more challenging. As laboratory subjects go, moles are fairly high-maintenance animals. They require large containers of soil and have voracious appetites. A single mole consumes up to twenty worms per day in captivity. In the wild, they spend most of their lives underground, threading through burrows that often lead to streams and ponds. Although star-nosed moles range from southeastern Canada into the eastern United States, with populations as far south as Georgia, they are seldom seen except perhaps by the occasional hiker who stumbles across a dead mole that has been dropped by a predator at the side of a country road or a pond shoreline, says biologist Evan Hazard.

Still, researchers have been able to study these secretive animals in enough detail to learn that the animal's odd assortment of physical attributes outfits it perfectly for the challenges of life in the netherworld. It turns out that far from being an evolutionary rough draft, star-nosed moles have had millions of years to perfect their adaptations to life underground. Moles (like humans) trace their ancestry to the first mammal, a tiny shrewlike creature that hunted insects, spiders, earthworms, and other invertebrates as the steps of dinosaurs thundered about its head. During the Cretaceous period about 135 million years ago, small insectivorous mammals belonging to the order Insectivora underwent an evolutionary flowering. The resulting assemblage included such common species as moles, shrews, and hedgehogs, among others. The dinosaurs came and went, but some 350 species of insectivores survived to the present day.

Over geologic eons, star-nosed moles may not have perfected harmony of design so much as excellence of engineering. The tail, for example, performs double duty. When held erect, it trails along the ceiling of the mole's underground tunnels, much the way human hands might feel their way down a dark corridor. During winter, the tail stores fat like a portable pantry. When prey becomes scarce, the moles simply draw on their reserves to tide them over to better times.

The massive front paws perform multiple functions as well. When used like powerful shovels, they can excavate tunnels through wetland soils with remarkable speed. Scientists observed one mole, for example, that tunneled 235 feet in a single night. Not bad for an animal that only grows up to seven inches and weighs in at an average of two ounces.

Switch from land to water and the powerful front paws become amphibious-assault

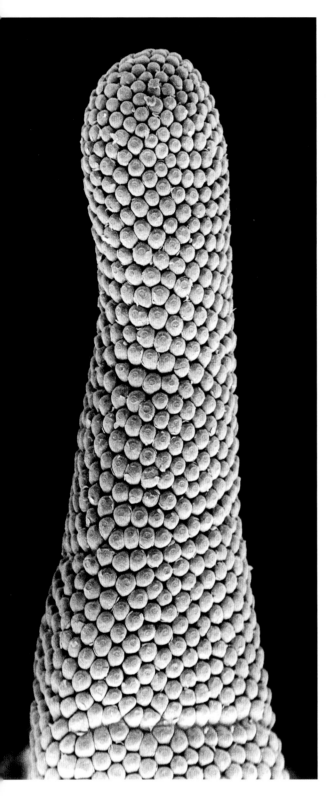

gear. Moles are formidable aquatic, as well as terrestrial, predators. Indeed, like beavers, many moles maintain burrows that open under water. Not surprisingly, the star-nosed mole is an expert swimmer and has even been spotted paddling under the ice in winter. The broad, flat paws, like a pair of flippers, help propel the animals through the water to intercept aquatic staples such as crayfish, insects, and even frogs and fish.

But the evolutionary coup de grace is the mole's anemone-like nose. Contrary to what you might think, this elaborate appendage is not used to enhance the sense of smell—although the animal's powers of olfaction have recently earned it a place in the zoological history books. In 2006, Vanderbilt University scientist Ken Catania published research in the prestigious science journal *Nature* showing that star-nosed moles and a species of water shrew *(Sorex palustris)* can smell their prey while submerged, becoming the first mammals known to use underwater olfaction. In the study, Catania released scent trails into the water. A mesh screen prevented the moles from contacting the odors with their noses, thereby ruling out the use of the nasal feature as an olfactory tool. Instead, the moles were able to smell by a far more fantastical method: they rapidly exhaled and inhaled bubbles, between five and ten times per second. The exhaled bubbles picked up odor molecules, which the animals sensed when they were inhaled.

So if star-nosed moles do not use their elaborate nasal design for smelling, why then have they gone to such great lengths where their sniffers are concerned? Microscopic analysis of the nose surface has revealed a honeycomb pattern of tightly packed pimple-like structures, known as papillae. Named Eimer's organs in honor of the German scientist Theodor Eimer, who in 1871 discovered similar features on the plain noses of European moles, each papilla contains a core of epidermal cells that function as sensory receptors. Although it measures only about 0.6 inches from tip to tip at its widest point, the entire nasal structure of star-nosed moles contains more than twenty-five thousand Eimer's organs.

In a 1993 issue of the *Journal of Mammalogy,* a trio of researchers argued that the animal's nose is used to detect the electrical fields created by the sweat and mucous of its prey. Catania, on the other hand, suggests that such a bundle of

The nose tentacle of a star-nosed mole *(left)* is tightly packed with sensory receptors known as Eimer's organs. The entire nasal structure *(opposite)* contains more than twenty-five thousand Eimer's organs, making it one of the most highly sensitive touch organs among animals. Courtesy of Kenneth C. Catania.

Eimer's organs increases the animal's sensitivity of touch. In 1996 he and neuroscientist Jon Kaas published research showing that more than one hundred thousand nerve fibers link the mole's nasal appendage to the animal's brain, nearly six times the number of neural connections between the human brain and hand. According to the researchers, the mole's nose "is clearly a major source of information about the mole's external environment and may be one of the most sensitive and highly developed touch organs among mammals."

Catania and Kaas determined how important the sense of touch was to moles by measuring how much space on the brain's hard drive was devoted to processing information from the animal's nasal organ. Their research showed that a larger-than-average portion of the brain's cortex was taken up by the somatosensory area, that is, the part of the brain that is responsible for registering and ordering the blitz of stimuli relayed by the sense of touch. Indeed, tests revealed that neurons in a large part of the cortex lit up like a switchboard when the nose was stimulated.

But the most astounding discovery was this: the star pattern of the nasal tentacles is represented, in a kind of mirror image, in the actual physical structure of the brain.

Microscopic cross sections of brain tissue from the star-nosed mole *(below)* reveal a point-by-point correspondence between neural pathways and the animal's nasal tentacles. Courtesy of Kenneth C. Catania.

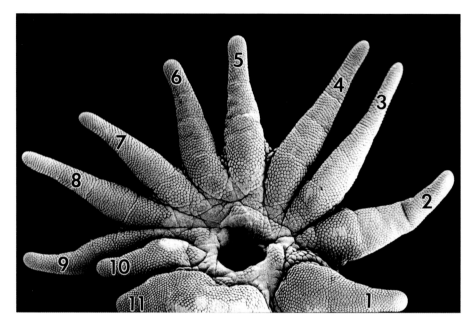

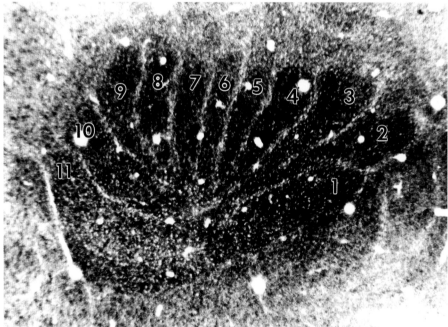

Cross sections of brain tissue reveal a point-by-point correspondence between neural pathways and nasal tentacles.

Information travels on this neural interstate at lightning speeds, as Catania and Kaas discovered after observing moles on the prowl for food. When searching out prey, the tentacles are extended in a forward position, where they constantly "finger" the world in their path. When the mole lifts its nose, the tentacles reflexively shift backward. When the nose is lowered, they project forward again. Each second they touch a different area—not once but ten times or more, thereby receiving an extremely refined tactile impression of all that they encounter. In this way, they stay in almost instantaneous contact with their world. In less than a second moles are able to carry out a complex sequence of actions. In what amounts to no more than a blink of an eye, moles are able to distinguish a worm from a tree root and snap it up faster than a hungry youngster can slurp a spaghetti noodle.

Star-nosed moles have about twelve times the number of Eimer's organs as other mole species. If its nasal frill confers such a predatory advantage, then why haven't other moles developed a similar snout? The reason for the difference may have something to do with the varying soil conditions in which moles live, the researchers say. The star-nosed mole's nasal organ is an extremely delicate tool. Its bright reddish-pink color indicates that the many blood vessels on which its sensory functions depend lie very close to the surface. Furthermore, to maximize its sensitivity, the skin covering the nose is thin, only one-twentieth as thick as human skin. The development of such a tender organ, Catania says, is possible only in the loose, mucky soils around swamps and streams. Indeed, the noses of moles that inhabit drier environments are by necessity far more blunt instruments. Subject to greater abrasion and desiccation, they are protected by a tougher, thicker skin that, not surprisingly, contains fewer Eimer's organs. "Why don't other moles have this star?" Catania asks. "Well, the fact is that no other mole *could* have it."

If evolutionary success is defined as having the precise keys for unlocking the doors to survival, then the star-nosed mole can be considered a master locksmith. As for harmony of design, when you have attained such a high degree of fitness between form and function, who needs it? ❧

SUGGESTIONS FOR FURTHER READING

Angier, Natalie. "A Masterpiece of Nature? Yuck!" *New York Times,* 9 August 2010.

Catania, Kenneth C. "Underwater 'Sniffing' by Semi-Aquatic Mammals." *Nature* 444 (2006): 1024–25.

Catania, Kenneth C., and Jon H. Kaas. "The Unusual Nose and Brain of the Star-Nosed Mole." *BioScience* 46, no. 8 (1996): 578–86.

Gorman, Martyn L., and R. David Stone. *The Natural History of Moles.* Ithaca, N.Y.: Comstock Publishing Associates, 1990.

Gould, Edwin, William McShea, and Theodore Grand. "Function of the Star in the Star-Nosed Mole, *Condylura christata.*" *Journal of Mammalogy* 74 (1993): 108–16.

Hazard, Evan. *The Mammals of Minnesota.* Minneapolis: University of Minnesota Press, 1982.

Moriarty, John J. "Itty-Bitty Bug Biters." *Minnesota Volunteer,* January/February 1998: 31–41.

Rankin, Bill. "Star of the Swamp." *National Wildlife,* December/January 1997: 32–33.

NEARSHORE

N 1855 THE GERMAN ETHNOGRAPHER JOHANN GEORG KOHL SPENT FOUR MONTHS among the Ojibwe in the vicinity of the Apostle Islands. In his popular 1860 classic, *Kitchi-Gami: Life among the Lake Superior Ojibway,* Kohl recounts how he had been moved by the story of an Indian girl who walked for miles through the forest to Superior's shores carrying her ailing father on her back "because the old man wished to see the lake once more ere he died." The Lake Superior Ojibwe, Kohl was told, "are attached to the Kitchi-Gami as the French Swiss to their Lake of Geneva."

This devotion to Superior would come as no surprise to today's lake lovers. In *The North Shore Experience,* a rare survey of North Shore visitors conducted by Minnesota Sea Grant in summer 1981, 80 percent of respondents listed "watching the lake" as one of their leisure pursuits, making it the most popular in the researchers' poll of thirty-nine activities. When people were asked to list their reasons for visiting the shore, the response "(to) enjoy the scenery" consistently received the highest rating.

Research only confirms what the devoted visitor already knows—that Superior's

The nearshore includes the terrestrial portion of the Minnesota North Shore watershed that is closest to Lake Superior. Here, steep, lake-facing slopes of birch, aspen, or mixed hardwood and conifer forests abruptly end in lakeside cliffs and beaches or transition into broad terraces of forests on fine-grained soils. North Shore rivers and streams complete their journey to Lake Superior through narrow canyons with cascading falls carved through billion-year-old bedrock.

edge is a powerful place. And the lake's cold waters, rock-bound coast, and fickle shifts in weather have helped to keep it that way. Absent is the buzz of jet skis and motorboats that has fouled the air and shattered the peace of many smaller inland lakes. And the smell of balsam fir is a far more common aroma than that of coconut tanning lotion. "The coolness of the [North Shore's] waters and the ruggedness of the adjoining terrain," write Timothy B. Knopp and Uel Bland in *The North Shore Experience*, "have discouraged the kind of recreational development typical of our ocean beaches; it does not invite participation in swimming, sunbathing or other indulgences characteristic of warmer climes."

On wind-still days in early summer it is possible to lean back into a hollow of rock along the lake and listen, without interruption, to a song sparrow calling from a thicket of wild rose or waves sipping at the shore. Undimmed by the glow of city lights, the night sky startles with its darkness and the brilliant cut of innumerable stars. From their shore-edge vantage point, many urban dwellers have glimpsed for the first time the Milky Way, the path of faint light that, according to the beliefs of the native Ojibwe, is a celestial highway trod by souls departing into the afterlife.

We seem to be drawn to Lake Superior for the same reason that we visit the great cathedrals of Europe or climb mountain peaks: to measure our lives against the infinite and find the part of ourselves that is unbounded by the bricks and mortar of flesh and blood. The shoreline of the North Shore provides an ecological backdrop that is well suited to such meditational experiences. According to Knopp and Bland, it "appeals to those seeking a place that stimulates the mind and challenges the spirit. The very harshness of the landscape can serve to magnify the significance of life."

Fire, Ice, and Flowing Water: The Natural Forces That Shaped the Nearshore

Compared with the other terrestrial components of the North Shore watershed (the headwaters and North Shore Highlands), the shore edge and adjoining lake terrace are the most familiar to visitors. Not only do they contain extensive segments of the shore's only major thoroughfare—Highway 61—but they also host the greatest concentration of lodgings and scenic tourist attractions, including soaring overlooks of Lake Superior and the roaring waterfalls of North Shore rivers.

The character of this landscape's distinctive ecology was forged in the deep shadows of time many thousands, even millions, of years ago. The lake terrace portion of the watershed skirts the base of the ancient hills of the North Shore Highlands, where it forms a series of intermittent, forested tablelands. Occupying the old lake plain and wave-cut beach terraces of Superior's glacial ancestors, the lake terrace extends as far as five miles inland. Along Highway 61, stretches of it can be seen most easily along the road between Duluth and the Split Rock River, Lutsen and the Cascade River, and between Colville and Paradise Beach.

Numerous streams cut through the lake terrace, some of which have gouged spectacular gorges. Ribbons of gravel and cobble that once served as the beaches of Superior's ancestral lakes thread across the land. At intervals, great promontories of rock rise up, some of them continuing to the lake edge to form the North Shore's most recognizable headlands, such as Palisade Head, Silver Creek Cliff, Lafayette Bluff, and Shovel Point in Lake County and Carlton Peak and Hat Point peninsula in Cook County.

The Kadunce River, like many North Shore streams, gouges a deep, serpentine gorge through the lake terrace before flowing into Lake Superior through a deep cobble beach. Photograph courtesy of Tony Ernst.

The boundaries of the lake terrace end where the rocky shoreline begins. Scattered here and there along the lake edge are communities of low-lying plants adapted to the exposed and often severe conditions. Most of them hunker in the crevices of rock, disappearing altogether at the point where the turbulent ebb and flow of the lake prevents even a tenacious crust of lichen from taking root.

For generations, this rocky rim has attracted urban-stressed tourists who would rather do nothing more than spend a morning lounging on a lakeside porch or pitching stones into the lake. Little do many of them realize, however, that the very ground beneath their tranquil retreats once was a scene of unimaginable violence.

About 1.1 billion years ago, in the Middle Proterozoic era, parts of mid-North

Great promontories of rock tower over Lake Superior, forming some of the North Shore's most recognizable headlands. Some of them, such as Hat Point in Grand Portage, were formed when magma moved upward through Earth's crust along massive fissures and then cooled and solidified. Photograph by Ken Harmon.

America began to turn inside out, transforming the region into a great field of poisonous fumes where magma as hot as 1,832 degrees Fahrenheit squirted through crevices or poured out like rancid pancake batter onto the surface. Back then, the North Shore was "a very hazardous environment," writes John Green, the preeminent authority on North Shore geology.

During this time, hot, buoyant rock deep within Earth began to slowly migrate toward the surface. The center of this rising column of molten rock was located under the present-day basin of Lake Superior. As it pushed upward, it began to melt and stretch thin Earth's crust until it tore open a great horseshoe-shaped gash, known as the Midcontinent Rift. Measuring some 1,375 miles long, the rift curved northeastward from present-day Kansas to near Thunder Bay, Ontario, before bending southward through Michigan and into Ohio. Over the next twenty-five million years, hundreds of great lava flows poured out of the fractured ground. According to one estimate, an astonishing 312,000 cubic miles of lava—one hundred times the current volume of Lake Superior—gushed out onto the planet's surface.

The area underlying what is now the basin of Lake Superior became ground zero for magma eruption. So massive and prolonged were the rift era's volcanic events that geologists can trace most of the rock found in the lake terrace and shore edge today to this tumultuous time.

Geologists divide the rocks created during this period into two categories: extrusive and intrusive. Extrusive rocks were formed by magma that was squeezed, or extruded, onto Earth's surface. In the Lake Superior region, extrusive rocks are primarily of basaltic composition; that is, they are rich in calcium, magnesium, and iron, which gives them their dark color.

Flow after flow oozed up onto the ground, the viscous material spreading out over an area measuring hundreds of square miles. On the North Shore, according to Green, successive flows of basalt accumulated to depths of up to six miles.

The wave-tumbled stones of basalt, a dark, bluish rock rich in calcium, magnesium, and iron, are the most common rocks on the North Shore's cobble beaches. Photograph by Gary Alan Nelson.

Originally these "flood basalts" were horizontal. In time, however, the center of the rift subsided under the weight of the solidified magma, canting the edges at an average angle of ten degrees. Today, visitors to the Superior shoreline can identify them by their telltale tilting toward the lake. Among the places where evidence of ancient volcanic events can be most easily seen are at Artist Point in Grand Marais and along Superior's shores in the Temperance River State Park.

Among the most unusual of the North Shore volcanics, Green observes, is a type of rock known as rhyolite, which was created when basalt magmas located deep within Earth actually melted some of its older, crustal rock. The fusion formed a new magma of granitic composition, which erupted and cooled on the surface as rhyolite. One of these flows is especially well exposed at Palisade Head and Shovel Point. Here, vertical columnar joints that formed as the flow cooled are dramatically displayed in sheer cliffs that soar to 180 feet above Lake Superior.

Intrusive rocks, on the other hand, were formed in places where the rising magma was prevented from breaching the surface, forcing the molten rock to solidify below Earth's crust. Some of these rocks, such as diabase, crystallized close to the surface. Another type of rock, the coarse-grained and especially erosion-resistant anorthosite, is believed to have been formed near the base of the planet's crust some twenty to twenty-five miles below Earth's surface. Rising magma later dislodged great chunks of these profundal rocks and floated them up near the surface, where they became trapped in the cooling diabase.

Intrusive rock formations, such as this overlook on Bean Lake, form the hilly topography north of Beaver Bay, Silver Bay, and Palisade Head. Photograph by Gary Alan Nelson.

The chemical and structural differences between extrusive and intrusive rocks played a major role in shaping the character of today's North Shore. Trapped beneath the surface, intrusive rocks cooled more slowly. This long curing, Green explains, allowed the formation of larger crystals in the rock as well as fewer and more widely spaced fractures, features that make them resistant to erosion.

Basalts that flowed onto Earth's surface, on the other hand, cooled at a faster rate, preventing large crystal development. Depending on the thickness of the lava flow, this dense, fine-grained rock may be riddled with numerous closely spaced joints and fractures. Basalts are prone to crumbling when exposed to natural weathering or stress from flowing streams, waves, and ice, particularly the Ice Age glaciers, which plowed up the region in twenty separate episodes over a period of two million years. The dramatic promontories of rock that are found on the North Shore today withstood the grinding of the ice sheets, while the surrounding softer rocks, among them the flood basalts, were whittled down to the gentle contours of the modern lake terrace.

Evidence of this differential erosion is obvious along the coast. Many of the shore's most spectacular overlooks, for example, rest on great outcrops of intrusive rock. Split Rock Lighthouse perches on a single monolithic block of anorthosite. Carlton Peak and several prominent outcrops within Tettegouche State Park are examples of anorthosite encased in a matrix of diabase. Yet another intrusive—the dark-colored diabase—forms the major mineral constituent of Hat Point, Silver Creek Cliff, and Lafayette Bluff.

About 12,000 BP, water would join fire and ice in helping to shape profound

Rhyolite beach stones are typically reddish in color (due in part to traces of iron) but may range from pale salmon to buff. Poor in calcium, magnesium, and iron, they are rich in potassium and silica. Photograph by Chel Anderson.

Among the most unusual of the North Shore volcanic rocks, rhyolite was created when basalt magmas located deep within Earth actually melted some of its older crustal rock. Vertical columnar joints that formed as the flow cooled are displayed in sheer cliffs that soar to 180 feet above Lake Superior at Palisade Head and Shovel Point. Photograph by Gary Alan Nelson.

Extrusive rocks like basalt quickly cooled at the surface of lava flows, resulting in the formation of small crystals that made the rock more susceptible to fracturing and erosion by glaciers, natural weathering, and stream flows. The hammer in the photograph of this lakeshore cliff marks the contact between two flows. The more erodible bottom flow is topped by a more erosion-resistant flow. Photograph by John C. Green.

topographical change in the nearshore zone. During this period of rapid warming, which ended the Ice Age, the ice sheet beat a rapid retreat from the deep basin of Lake Superior. Meltwaters streamed off the glacier and ponded along its margins. Over the next several thousand years, a series of postglacial lakes, which rose and fell in identifiable stages, would fill the portions of the basin that had been vacated by the ice.

The first precursor to modern-day Lake Superior was

Intrusive rocks, such as the anorthosite of Carlton Peak above Tofte, formed when magma cooled more slowly while trapped beneath the surface. Anorthosite's large crystals and fewer, more widely spaced fractures made it especially resistant to erosion, even by the Wisconsin glacier. Photograph by Chel Anderson.

Glacial Lake Duluth. Confined by higher land on one end and the retreating ice sheet on the other, Glacial Lake Duluth began to form around 12,000 BP. The fledgling lake attained its fullest extent around 11,000 BP, roughly encompassing the western third of today's lake. The earliest and deepest of Superior's ancestors, Glacial Lake Duluth featured shorelines that were nearly twice as high as those now found around Superior. During its heyday, all of the present-day communities along the North Shore would have been completely submerged under its waters. Drowned too would have been its many state parks.

With Superior's modern outlet, the St. Marys River near Sault Ste. Marie, still buried under thick ice, Glacial Lake Duluth drained to the west into the St. Croix River via Wisconsin's Brule River. But as the ice receded across the basin, lower outlets were uncovered along the lake's southern shore. Geologists have identified several of these abandoned channels, including one that crossed the Upper Peninsula and funneled Superior's waters into ancestral Lake Michigan.

As the glacial waters first breached the barriers of ice or sedimentary debris that blocked each new outlet, catastrophic floodwaters surged through the gaps, dramatically draining the lake basin until the waters eventually reached equilibrium flows. For millennia, Lake Superior was like a yo-yo dieter on a long-term weight-loss program. The waters of Glacial Lake Minong, the successor to Lake Duluth, measured 450 feet above sea level. By 9,500 BP, the Lake Houghton stage, water levels had dropped even farther, to 375 feet above sea level. By about 6,000 BP, the Lake Nipissing stage, water levels had risen to 605 feet above sea level, 3 feet above the modern average of 602 feet, which became stabilized some 2,200 BP.

Each lake stage left its signature on the land. As early as the 1890s, scientists began to comb Superior's coast for evidence of this geologic past. They discovered, for example, abandoned shorelines that were located high above the modern lake edge. Clues such as wave-cut cliffs and old beach terraces allowed scientists to reconstruct the waistline of Lake Superior as it expanded and contracted before finally attaining its present size. They named the glacial lakes in honor of the locations where the shorelines of each proglacial lake stage were most distinctly carved on the land.

The water levels of Glacial Lake Duluth, which averaged 500 feet higher than today's Lake Superior, left some of the most prominent marks on the basin's rim. The city of Duluth's popular Skyline Drive, perched some 540 feet above the surface of present-day Lake Superior, more or less follows the highest beach ridge that was formed by the waves of Glacial Lake Duluth. Radiocarbon dated to about 12,000 BP, this strandline extends ninety miles along the Minnesota shore.

Lower beaches that were created by the wave action of later glacial lake stages reach as far northeast as the Pigeon River. Composed variously of gravel, rock shingle, and cobblestones, many of these abandoned beaches are plainly visible today along the Minnesota North Shore. Between the town of Schroeder and the Temperance River, for example, Highway 61 follows a former shoreline of Glacial Lake Minong. Measuring 670 feet in elevation—68 feet above the surface of today's lake—the beach dates to 10,000 BP. Just west of the Cascade River, County Road 97 (otherwise known as Cascade Beach Road) trails another prehistoric beach of Lake Minong. And travelers driving eastward on Highway 61 from the Devil Track River to just west of the Brule River traverse a pebble beach terrace that was created by waves lapping at the shore of Glacial Lake

Superior's historic lake levels changed several times following the retreat of the Wisconsin ice sheet. Among the precursors to the present-day lake were Glacial Lake Duluth *(top)*, Glacial Lake Minong-Houghton *(center)*, and Glacial Lake Nipissing *(bottom)*.

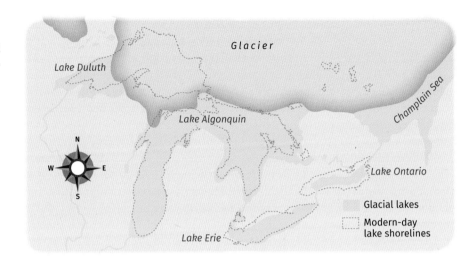

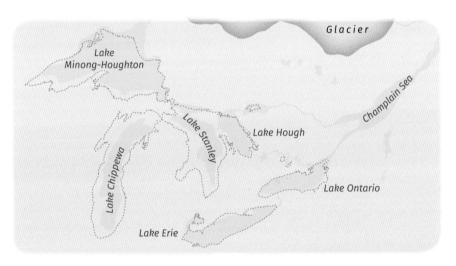

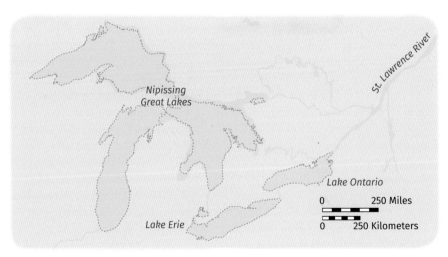

Nipissing some 6,000 years ago. This beach marks the final proglacial lake level of Superior before it reached its modern level.

Visible too along the lakeshore are erosional features known as wave-cut cliffs, whose abrupt rock faces stand in stark contrast to the gentle inclines of bedrock ledges. The formation of these sheer rock walls depends on two geological conditions: the joint structure of the rock and the angle at which storm waves attack the cliff face. Bedrock shore ledges, for example, are composed of lava flows piled up like pancakes on top of one another. The dominant fractures in the rock tilt toward the lake. As a top layer erodes, the layer beneath it becomes exposed. "Where a gently-tilted lava flow contact is the major plane of weakness," Green writes, "waves will strip off the overlying material and leave a wide, open ledge." But wave-cut cliffs, which are riddled with vertical columnar joints, are like stacked blocks. A full frontal assault by waves, Green says, "will take blocks away column by column, forming, and maintaining a vertical cliff." The most dramatic wave-cut cliffs along the North Shore include Silver Creek Cliff, Palisade Head, and Shovel Point. These sheer faces were created over the past several thousand years by modern Lake Superior.

The glacial lakes left other calling cards on the land as well. The lakes served as catchments for sediments that were eroded from the newly exposed land or shed from the melting glacier. Fine sediments such as silt, sand, and clay were carried into the deeper, less turbulent waters, where they settled out to the bottom. North Shore geologists can chart the reaches of these former lakes, most notably Glacial Lake Duluth, which covered the widest area, by surveying the fine sediments they left behind on the lake terrace and among the hills of the North Shore Highlands. A soils map of Lake and Cook Counties shows a sheet composed primarily of red clay some one to one hundred feet deep roughly paralleling the Lake Superior shoreline. In St. Louis and southwestern Lake Counties, this clay belt is fairly continuous and typically extends about five miles

Stranded far above present-day Lake Superior, this ancient cobble beach marks the former shoreline of a precursor to Lake Superior. Photograph by Virginia Danfelt.

235

East of Grand Marais, Highway 61 follows a gravel beach and wave-cut cliff of the Nipissing stage of Lake Superior. This beach line marks the final proglacial lake level of Superior before it dropped to its modern level. Photograph by Chel Anderson.

inland. Farther east along the shore, the clay belt narrows, reaching between one to five miles inland. Frequently interrupting this tongue of clay are bedrock hills and intermittent sand and gravel deposits.

According to geologist Green, localized patches of pebbles, cobbles, or even boulders are mixed into the lake terrace's clay matrix. Such anomalies, he observes, probably occurred when sediment-laden icebergs floated into the lake and melted, dropping their loads. Other deposits could have resulted from the slumping of nearshore sediments or the deltas of debris that formed at the mouths of glacial rivers.

Nature Giveth—and Taketh Away: Erosion along the Shore Edge

Watch a storm wave throw its weight against a cliff without even so much as a shudder from the rock, and you might believe that the North Shore's hard coastal rind has outlasted even time itself. With plants whose lineages can be traced to the Ice Age and rocks that comprise some of the oldest on Earth, the shore edge seems immutable.

And in some respects, it is. In time, the enormous magma flows that poured through the Midcontinent Rift effectively sealed this great crustal fissure. "Although the entire Lake Superior area was an extremely active volcanic area for millions of years in Late Precambrian times," Green writes, "no volcanic activity has occurred here in the last billion years." Neither do earthquakes, he adds, "occur with sufficient frequency or intensity to be of concern."

Yet research has shown that Minnesota's coastline is not as stable as it seems. Interspersed among the vertical headlands and gently sloping ledges of rock that dominate the shoreline are bluffs and beaches composed of highly erodible materials. The Minnesota Board of Water and Soil Resources (MBWSR), an agency that administers several programs designed to protect the state's soil and water, estimates that the North Shore has some sixty miles of unstable clay embankments, most of them located between Duluth and Silver Bay. Just as vulnerable are the shorelines of sand and gravel that commonly occur north of Grand Marais. And contrary to what you might think, "not all bedrock is absolutely stable," say the authors of a 1989 study published in the *Journal of Great Lakes Research*. "Some of it is highly fractured and shows signs of collapse." The authors report that about 25 percent of bedrock bluffs along the Minnesota shore are vulnerable to battering waves.

These coastal features find a measure of protection from the sheltering arms of rocky points that protrude into the lake and help to break the kinetic energy of pounding waves. They also benefit from the pattern of prevailing northwesterly winds that push damaging storm waves to the southern shore of the lake.

But the Minnesota shore edge is nonetheless at risk from erosion, some of it severe, especially when Superior's water levels are above normal. Clay bluffs, even vegetated ones, Green points out, are very unstable. Vulnerability to slumping increases during periods of heavy rainfall and during the spring, when soils are saturated with snowmelt and surface runoff. As waves undercut the toes of bluffs, weakening their structural support, whole sections of hillsides can collapse into the water.

Erosion threats to the North Shore also increase during Superior's infamous nor'easter storms. The geographic position of the Minnesota coast leaves it especially

Not all of Minnesota's Lake Superior coastline is ancient bedrock. Postglacial lakes deposited fine sediments of clay soils, which are now exposed along stretches of shoreline. The North Shore has an estimated sixty miles of clay embankments that are highly erodible and unstable even when they are vegetated. Courtesy of the Minnesota Lake Superior Coastal Program.

exposed to assaults from the east and northeast since winds from these directions are able to pick up steam over a 310-mile stretch of open water. These long fetches can produce waves more than twenty feet high.

Coastal erosion has been well documented throughout the lower Great Lakes, but its effects on Lake Superior's shorelines have been largely overlooked due to sparse settlement, say Carol A. Johnston and James Sales, authors of a 1994 study that identified erosion-hazard areas along the Minnesota North Shore. To date, studies on Lake Superior have primarily focused on severe erosion-prone areas such as the red-clay bluffs on the Wisconsin shoreline. These bluffs contribute 58 percent of the five tons of fine-grained sediments that are flushed into the lake each year.

While the predominance of bedrock along the Minnesota North Shore has slowed the pace of erosion, Johnston and Sales observe, the northern coast nonetheless has lost its share of lake frontage. Just how much land is claimed by Superior each year? In their study Johnston and Sales compared aerial photographs of the Minnesota coastline taken in 1939 with images taken in 1975. The researchers measured the loss of land and then divided it by the number of years between photographs to determine the annual erosion rate. They also compared the 1975 photos with ones taken in 1988–89 to gauge the toll that record-high lake levels in the 1980s had taken on the shoreline. The researchers expected—and found—a high rate of erosion along clay banks. They discovered, however, that sand and gravel beaches were just as prone to shoreline loss. Between 1939 and 1975, the average loss of the lake edge in highly erodible areas was 0.46 feet per year. Some areas lost as much as 1.1 feet per year. Not surprisingly, nearly all the detectable erosion measured between 1975 and 1988–89 occurred in these areas. Some shorelines experienced a whopping annual loss of 4.5 feet of shoreline during years of high water levels.

Nearshore Plant Communities

Superior is not content to simply gnaw at the North Shore's coastal edge. The lake's presence is felt far inland, governing everything from air temperatures and the direction and velocity of the wind to the amount of precipitation along its shores. "The physical structure and living communities of the land along the lake's edge are as much a function of the lake's ecosystem as the fish in its depths," observe the authors of the 1997 report *The Land by the Lakes: Nearshore Terrestrial Ecosystems.*

Superior's effects on the coastal zone did not escape the notice of early visitors. C. W. Hall, a member of the state's Geological and National History Survey team, reported in 1880 that Superior "is so large and deep as to exert an almost oceanic effect on the climate near its shores, such as imparting a certain moisture to the air and diminishing the liability to sudden changes in its temperature; the temperature is through each season quite equable and very low."

Such accounts offer forest ecologists important clues about the presettlement composition of the nearshore forest. Unfortunately, the historic record tells an incomplete story. To piece together a picture of the past, forest ecologists rely, in part, on information compiled by the land surveyors of the Government Land Office during the latter part of the nineteenth century. Although detailed in many respects, these reports do not draw a complete vegetational picture. Surveyors divided the landscape into one-square-mile sections. In their written descriptions of each section, they focused much of their attention on identifying the individual trees that were used to mark survey boundaries. Information about other landscape features, such as topography, soils, and hydrology, was abbreviated. Their brief notes on woody vegetation largely singled out dominant tree species and shrubs, especially commercially valuable ones, while ignoring herbaceous (nonwoody) species.

To help flesh out this sketchy picture, today's scientists take on the role of ecological detectives. They combine historic information with layers of data gathered from intact native plant communities that have escaped human-caused disturbances such as logging, agriculture, grazing, and an unnatural frequency of fire.

Using these tools, ecologists have been able to draw several conclusions about the lake terrace's plant communities. They point out that terrace forests would have comprised many of the same species that are found in the headwaters forest, with the exception of red pine and jack pine. (Because they favor the drier, well-drained soils found farther inland, these species would not have been abundant.) But the heightened humidity, cool temperatures, and richer, cooler, and moister soils along the lakeshore and terrace would have imposed a markedly different disturbance regime, conferring a greater resistance to the infernos that frequently swept headwaters' forests. Studies by Minnesota forest ecologist Lee Frelich and scientists from the U.S. Forest Service (USFS) estimate that catastrophic disturbances (forest-leveling winds and severe fires that kill overstory and understory trees as well as the seedbed in the soil) were relatively rare along the Superior lakeshore, occurring every 150 to 1,000 years.

During the extended lulls between conflagrations, nearshore forests developed old-growth conditions, among them a high degree of structural diversity. The forest floor, for example, would have been littered with downed tree trunks and branches in various stages of decay. The vertical structure of the forest would have been equally complex

Nearshore forests have lost much of their structural and species diversity due in large part to the catastrophic wildfires in the wake of Euro-American settlement and herbivory by white-tailed deer. As aging paper birch trees die, they often leave behind little more than bluejoint grass in the understory. Photograph by Chel Anderson.

239

with living trees and shrubs of many different ages and heights interspersed with the occasional standing dead snag. And a careful examination of the canopy overhead would have revealed a highly variable mosaic of species that corresponded to the great heterogeneity of conditions on the ground, including fine-scale differences in soil conditions and microtopography.

These forests would have also harbored a greater number of large, long-lived trees such as white pine, white spruce, white cedar, yellow birch, and heart-leaved birch. Adding to their primeval quality was the abundance of lichens and mosses. Some scientists speculate that the abeyance of fire may have allowed mosses to form deep blankets on the ground, much as they do on undisturbed islands such as Susie Island and a few remnant locations on the mainland shore. These conjectures seem to be corroborated by early visitors to the area, such as botanist Benedict Juni, a member of the Geological and Natural History Survey of Minnesota. On his 1879 reconnaissance of the Minnesota North Shore, Juni commented on the abundance of lichens and mosses, observing that "there is scarcely a rock or tree that is not partly or entirely covered by the different species. The ground is often covered by a layer nearly a foot in thickness, which conceals the angular fragments of rock beneath. In many places there is no soil except what is entangled in the mosses and lichens. . . . The trees in proximity to the lake, and especially on the side facing it, are draped with a moss-like lichen (*Usnea longissima*?). This hangs in long tufts of about a foot and a half from trunk and branches."

The long-lived conifer-rich forests that dominated the nearshore zone provided an important complement to neighboring forests in the North Shore watershed. For example, while many communities in the North Shore Highlands also commonly attained old-growth status, their uplands tended to be dominated by sugar maple and other deciduous species, which created the conditions for distinctly different suites of plants and animals.

The mature, conifer-dominated forests near the lake also counterbalanced the greater prevalence of younger boreal woodlands found farther inland in the headwaters region. Swept more frequently by catastrophic fire, which reset the successional clock, a large proportion of the headwaters' forest communities harbored species that thrived in a landscape dominated by large patches of primarily younger forests. The nearshore forests, on the other hand, provided many more refugia for plants and animals that flourished in stable, undisturbed conditions, including some rare species of lichen that require consistently high levels of humidity. At the same time, they served as a kind of species bank for neighboring fire-strafed forest communities, supplying them with recolonizing seeds, particularly those that could easily be picked up and carried by the wind and quickly germinated in bare mineral soils.

Investigations by plant ecologists and land-use historians also suggest that forest communities in the lake terrace would have exhibited some markedly different compositional patterns from those we see today. White pine, for example, is tolerant of clay soils and would have thrived in the fine sediments laid down by glacial lakes along the North Shore. These rich, moist soils, along with the infrequent occurrence of catastrophic fire, created prime conditions for the establishment and persistence of white pine–dominated forest communities. According to survey records, these forests were particularly prominent on the extensive tracts of clay that underlie the soils from Duluth to Illgen City. Northeastward of Illgen City, such broad plains of clay were interrupted by till-layered slopes dominated by sugar maple, and by hills whose cover of shallow, well-drained soils on bedrock supported forests with a greater susceptibility to fire. Under these conditions, white pine–dominated communities were probably far less common. Instead, this stately conifer likely grew in small patches or as individuals in mixed hardwood-conifer or maple-dominated northern hardwood forests.

The great forests of white pine, many of which towered some 170 feet or more above the forest floor, attained a grandeur the likes of which the North Shore may never see again. Today, remnants of these awe-inspiring forests still exist in a privately owned fifteen-hundred-acre reserve along the Encampment River and in an area southwest of Beaver Bay in Lake County.

These trees are valuable not only for their beauty and the many ecological services they offer to wildlife but also for their genetic reserves. Scientists speculate that white pine entered Minnesota from the eastern United States some seven thousand years ago via two different routes. Some trees are thought to have migrated along the North Shore of Lake Superior before spreading west and north throughout Minnesota. Other trees may have come via a more southerly route through Wisconsin and Michigan before migrating north along the western end of Lake Superior. The oldest white pine trees that today grace the shores of Lake Superior may be the genetic descendants of these pioneers from the east. As such, they form a unique component in the gene pool of white pine in northern Minnesota.

While the lake terrace forests may not have been as susceptible to the conflagrations that regularly visited the headwaters, they were not immune to fire. The persistence of white pine in expansive tracts of lake terrace forest suggests that cool, ground-hugging fires regularly swept through some areas. Ecologists point out that white pine–dominated forests require periodic fires to prevent the buildup of fuels that support

Remnants of the awe-inspiring white pine forests that once towered above the forest floor still exist in a privately owned fifteen-hundred-acre reserve near the Encampment River.

catastrophic wildfires. Such hot blazes often destroy mature trees. But the species has evolved characteristics that give older trees the ability to survive low-intensity burns, including heat-resistant bark and an absence of low-growing branches. At the same time, cool-burning flames work to ensure the future survival of pine in the forest. Not only do these surface fires strip the forest floor down to mineral soils, a condition that favors white pine regeneration, but by clearing out competing species, they increase the odds of success for new pine seedlings.

A far more common agent of change throughout the lake terrace, however, was windthrow. The scale of these disturbances varied, from the uprooting or snapping of a few trees to a flattening of whole patches of mature trees, such as resulted from the blow-downs that occurred in a stand of large aspen trees in 1992 at Butterwort Cliffs Scientific and Natural Area east of Cascade River State Park and in a grove of old-growth white pine near the Encampment River in 1993.

Openings in the tree canopy—whether localized gaps or extensive tracts—allowed many different tree species to gain a foothold over multiple points in time. As a result, windfalls helped to create and maintain the mosaic of intricate diversity that character- izes nearshore forests.

Where Land Meets Water: Plant Communities on the Edge

While many plant species, such as aged white pine, garnered attention for their commer- cial value, other plants generated excitement among researchers simply for their rarity, beauty, and scientific interest. As early as 1878 scientists paced the basalt in search of rare plants, especially arctic-alpine species. These holdovers from glacial times, which became separated in some cases by more than six hundred miles from their main ranges in the Arctic tundra or in the cordilleran, or mountainous, regions of western North America, sparked heated debates in botanical circles. Research has revealed that sites particu- larly rich in disjunct plant communities are concentrated along the lake's northern shore from Duluth to Sault Ste. Marie and along the coastlines of the lake's northerly islands, including Isle Royale. On the South Shore, disjuncts are primarily concentrated on the Keweenaw Peninsula. Of the 340 species of flowering plants and ferns whose main North American range lies in the Canadian Arctic Archipelago, 46 have been documented in the Lake Superior region to date. (The Minnesota North Shore hosts 23 disjunct species.) (See "Hay Pickers and Grass Gatherers: Botanical Exploration along the Lakeshore.")

To this day, surveys continue to turn up new county and state records for shore-edge species, including mosses, grasses, and liverworts. Scientists have documented unique and rare native plant communities in the nearshore area, including a type of wetland known as a rich fen near Hovland. On Shovel Point and a few other scattered locations, plant ecologists have also located remnant spruce-fir woodlands. Especially abundant in northern mosses and lichens, these relic ecosystems provide important refugia for boreal species in a place where two great vegetational systems—the boreal and temperate forests—meet and mingle.

New information continues to be unearthed about animals as well as plants. In 1990 researchers discovered that eroded fractures in the bedrock of North Shore cliffs serve as rare winter hibernacula for at least four species of bats: big brown, little brown, northern myotis, and tricolored (also known as eastern pipistrelle).

But research into the ecological relationships among shore-edge species has not kept pace with the cataloging of individual species. Take, for starters, disjunct plants. Although their existence has been recorded in the annals of botany for more than a century, little is known about even some of their most fundamental lifeways. After thousands of years along the rocky shores of Lake Superior, have arctic-alpine disjuncts evolved survival strategies that differ significantly from their cousins in alpine meadows or the Arctic tundra? Are any of these differences encoded in their genes? What is the relationship between the disjuncts and shore-edge animals, particularly pollinators?

Scientists have begun to lay some of the groundwork for a better understanding of these ecological workings. U.S. and Canadian agencies have devised an ecosystem classification scheme for shoreline communities in the Great Lakes. Their data reveal great intricacy and complexity in what appears on the surface to be undifferentiated stretches of rocky habitat.

Bedrock beach makes up more than 75 percent of the Minnesota North Shore coastline. To the untutored eye, these gently tilted shelves of basalt are little more than scoured rock with a random scattering of opportunistic plants. But a closer look at the shore edge has revealed a predictable range of habitats between the lake-terrace forest

Low, gently sloping bedrock beaches facing great fetches of open lake are exposed to the most punishing storm waves and ice scouring. Except for lichens, the rock can be devoid of soil and vegetation for up to one hundred feet from the water's edge. Courtesy of Chel Anderson.

and the water. The character and location of these habitats are largely dependent on underlying bedrock topography, that is, the contours of the terrain and the surface texture of the rock.

Also governing their character and location are the type and frequency of disturbance by water and ice. For example, where the gently sloping shoreline faces great fetches of open lake and is therefore exposed to the most punishing storm waves and ice scouring, the rock can be devoid of soil and vegetation for up to one hundred feet from the water's edge. On the other hand, where topography helps to limit the incidence of ice scour or dissipates the intensity of waves before they hit the shore—cases in which the bedrock shore presents a more vertical face to the elements or is protected by an offshore island or reef—plants may gain a toehold closer to the lake.

Just beyond the scour zone in locations that are somewhat protected from the full brunt of the waves and ice is a habitat type known as the low wet bedrock beach. Pounded by storm waves or caked with winter ice, this lip of beach hosts only tough, low-growing plants such as mosses in the cracks of rock and a few lichens.

Farther from the water, where bedrock ledges and rocky outcrops rise up like shields against the battering of the waves lies what researchers call the intermediate bedrock beach. Intermediate/wet beaches are regularly bathed by lake spray, which enables them to support lichens, low shrubs, and herbs, particularly in protected pockets where the rock surface is rough and eroded into a matrix of cracks and ledges. Plants that grow in the intermediate/wet beaches include bird's-eye primrose, harebell, and Kalm's lobelia. The odds of success for plants rooted in the shallow soils of the rock shore may be boosted by some species of lichens whose algae are capable of fixing nitrogen, thereby providing nutrients that otherwise would be scarce.

As their name suggests, intermediate/dry bedrock beaches are dry more often than not and typically feature rocks with a smoother surface. Lacking an abundance of nooks and crannies for capturing the windblown organic materials that provide rooting substrate, plants are sparse with the exception of lichens. Here, crustose lichens cover between 50 and 90 percent of the surface of the rock like a plush velvet tapestry. This weather-beaten fringe supports a lichen flora of nearly one hundred species. Some of these species are slow growing (expanding only a fraction of an inch each year) and may be extremely old. Indeed, the limited investigations into the life history of lichens suggest that individual plants may live for hundreds of years or more.

Because they lie at some distance from the dynamism of waves and ice and are highly fractured with crevices, hollows, and cracks, high/dry bedrock beaches support an even greater variety of plants and a wider array of lichen species dominated by foliose (leafy) and fruticose (branched) lichens. These lichens are joined by a variety of mosses, herbs such as upland white aster, hairy goldenrod, and flat-topped aster, and woody plants including shrubby cinquefoil, sweet gale, and ninebark.

Visitors traversing these highly fractured beaches occasionally may encounter an especially species-rich habitat known as a perched meadow. Tuft-forming grasses including tufted hair grass and spike trisetum grass along with sedges such as tufted bulrush and lenticular sedge give these miniature plant communities a meadow-like appearance. Located in and around hollows of rock that contain seasonal pools, perched meadows pack a surprising number of plant species into a very compact space. In addition to grasses and sedges, they support herbs such as butterwort and small false

Perched meadows form in and around seasonal bedrock pools. Protection from waves and ice has allowed these communities to develop considerable diversity and pack a surprising number of plant species into a very compact space. Photograph by Carol Reschke.

Bedrock glades are found at the very top of the bedrock beach. Taking root on thin soil over bedrock, these communities include a scattering of trees and shrubs and patchy turf composed of sedges and dense lichen cover. Photograph by Carol Reschke.

asphodel, which share rocky niches with woody plants such as mountain alder, shrubby cinquefoil, sweet gale, and ninebark. So ecologically rich and diverse are these miniature plant communities that on a reconnaissance of the lake's eastern shoreline in 1938, Canadian botanist R. C. Hosie counted more than thirty-one species of plants clustered around a pool that measured no more than eight feet across and contained a mere two feet of water. Data collected by the Minnesota Biological Survey in 2001 revealed a similar number of vascular plants associated with perched meadows along the Minnesota coast.

At the very top of the bedrock beach, closest to the terrace forest, is a transition zone known as a bedrock glade. Taking root on thin soils is a plant community that includes a scattering of trees and shrubs as well as a layer of turf composed of sedges that covers the bedrock like a grassy toupee. Commonly found in the bedrock glade are shrubs such as round-leaved juneberry, bearberry, bush honeysuckle, juniper, and creeping juniper, and trees such as balsam fir, paper birch, black spruce, white spruce, and white cedar. During winter, these species often are coated with frozen wave spray and fog, making twigs and branches vulnerable to pruning by strong winds and heavy ice buildup. Consequently, these species often take a stunted, bonsai-like form known as krumholz ("crooked wood"), which is indicative of harsh conditions.

Whether these habitats support animals that are as unusual or rare as many of their plants is largely unknown. The authors of *The Land by the Lakes,* which examined bedrock habitats around the Great Lakes, observed that "although there is a possibility the animals inhabiting bedrock beaches, glades, and cobble beaches are unique because of the special habitat requirements, little is known about the fauna of these communities."

What scientists do know is that the shore edge provides, according to the report, a "distinctive environment for wildlife, in many ways different from the adjacent inland areas. This coastal areas has a more moderate climate and unusual physical structures, such as sand spits, islands, or bluffs, which meet the needs of a diverse range of wildlife species." Splash pools near the lake, for example, provide breeding ponds and nurseries for many aquatic insects and amphibians. In addition, their hollows supply surrounding plant communities with a ready source of water, vegetating what otherwise might be a relatively barren rock field. In the spring, migrating shorebirds linger along the coastal edge, using it as a staging area while the ice on inland waterways thaws. Migrating songbirds and butterflies, such as monarchs, concentrate on peninsulas or island chains to wait out stormy weather or replenish vital fat reserves. Hawks and other raptors rely on updrafts created by shore bluffs to conserve precious energy on their migratory journeys.

The human pressures on the nearshore terrestrial zone—from stepped-up development and recreational activities to the effects of global warming—are not likely to lessen in the foreseeable future. If we are to become good stewards and thoughtful inhabitants of this rich and fragile place, a better understanding of its ecological workings will become more important than ever.

From Javelins to Jacuzzis: Human Changes on the Land

Reconstructing the prehistoric human occupation of northeastern Minnesota has proven to be as challenging—and highly selective—as compiling a picture of its vegetational past.

To date, the majority of prehistoric sites that have been identified in the Arrowhead region lie inside the Boundary Waters Canoe Area Wilderness (BWCAW), says Walt Okstad, a historian of the Superior National Forest (SNF). Does the concentration of sites in the BWCAW mean that ancient people used this region more intensively than other parts of northeastern Minnesota? Not necessarily. Aboriginal groups were highly mobile, Okstad says, leaving widely scattered evidence of their passing. Many promising places are off limits to exploration because they exist on private land or became submerged under water several thousand years ago when a climatic swing to wetter conditions caused lake levels to rise. Other parts of the region feature thin soils on bedrock, conditions that are not conducive to the accumulation of human debris.

Uncovering archaeological sites, Okstad points out, is primarily a function of where scientists look and why. Excavations tend to be located on public land because regulations often require a series of scientific evaluations, including archaeological surveys, to be conducted prior to development, such as the construction of campsites in the BWCAW. Because a good campsite today was probably a good campsite thousands of years ago, the odds are good that ancient artifacts will be found.

But even sites rich in artifacts offer only the sketchiest of clues. "Archaeologists only work with what they can prove," Okstad says. And that means relying on objects that have survived the ravages of time. In the case of the earliest people to inhabit northeastern Minnesota, this means studying bits of burned bone, stone tools, and lithic scatter, that is, flakes that are left over from the process of working rock into weapon tips sharp enough to penetrate the tough hides of large animals.

Other scraps of daily life, such as clothing, storage bags, most food remains, housing materials, and tools fashioned from bone and antler, have disappeared into the mulch of history. Made of organic matter, they decomposed rapidly in the northland's acidic soils. Dave Radford, a Minnesota Historical Society archaeologist who works in Minnesota's state parks, also points out that other forces, both severe conditions and human-caused disturbance, may have claimed their share of objects. Along the length of the Minnesota coastline, between Lake Superior and Highway 61, scientists have identified relatively

Archeological evidence indicates that humans have occupied Minnesota's Lake Superior watershed for thousands of years. This adze, made from Knife Lake siltstone, was created by Paleo-Indian people more than eleven thousand years ago. Photograph courtesy of USDA Forest Service / Superior National Forest Service.

few archaeological sites. And it is not because researchers have not looked. Over the past three decades, Radford says, highway reconstruction and state park developments have triggered mandatory archaeological surveys prior to any digging. To date, these surveys have turned up less than a dozen lithic scatter sites left behind by ancient hunters. Radford suspects that the material remains of prehistoric people in many coastal sites may have been obliterated by wind, waves, and erosion over time. Today, backhoes are exacting a toll on potentially promising sites, such as the abandoned beach ridges of Superior's ancestral lakes, as these sand and gravel lodes are mined for road and housing construction.

Still, archaeologists have collected enough evidence to determine that humans occupied the North Shore watershed as far back as 10,000 to 11,000 BP. They theorize that Paleo-Indian people entered the region along an ice-free corridor that trended southwest to northeast between the Superior and Rainy lobes of the Wisconsin ice sheet. Living within the shadow of a mile-high glacier whose meltwaters flooded the lowlands, these early people were forced onto the higher, drier ground of moraines, eskers, and glacial-lake beaches. Large projectile points that would have been attached to wooden spears have been discovered along these dry gravel ridges and hilltops in the SNF, where Paleo-Indians once pursued their primary prey: migrating herds of barren ground caribou.

Living "off the hoof" kept hunting bands small. According to Gordon Peters, retired SNF archaeologist, Paleo-Indian population densities within the boundaries of what is now the national forest likely averaged about one person per sixty-two square miles. Although few in number, these early people covered a lot of ground. Okstad observes that projectile points uncovered in Minnesota can be traced to rock deposits located as far away as North Dakota, northern Wisconsin, and Hudson Bay.

As the region's ecology evolved, so did the technology of these early people. Indeed, Peters proposes that "changes in culture are really nothing more than a response to a changing environment." Around 7,000 BP, he observes, the beginning of what is known as the Archaic period, the northern forest made a transition from a tundra-boreal forest to a warmer, drier coniferous forest in which red, white, and jack pine flourished. The lifeways of native people adapted accordingly. As the soggy tundra became dry land and lakeshores assumed more defined boundaries, aboriginal people took to the water, felling the massive trunks of red and white pine and working them with axes, wedges, and adzes to create canoes.

Native people may have developed such watercraft to better exploit a new resource—fish. During the Late Archaic period (3000–1000 BC), writes anthropologist Charles E. Cleland, the prehistoric people of the upper Great Lakes region began to look to the water for food. Gear recovered from the mouths of rivers, such as copper gaff hooks and harpoon heads, indicates that they stalked sturgeon and other river-spawning fish. This fishing economy grew more varied over time. A few centuries before the birth of Christ, Cleland writes, prehistoric people added nets to their arsenal of fishing technology, as indicated by the stone sinkers that have been found in lakes and rivers. According to Ojibwe legend, as recorded by the Jesuit missionary Pierre de Charlevoix in his 1761 *Journal of a Voyage to North America,* the mythological trickster figure Nanabozho taught the Ojibwe to fish with nets after observing spiders capture their prey with webs.

Such implements undoubtedly served prehistoric people well since, according to Okstad, there was little further technological innovation in the archaeological record for thousands of years. Indeed, up until 500 BC, the opening of what is known as the Initial Woodland period, native people continued to use dugout canoes and atlatl technology. But around 500 BC, a dramatic change surfaced in the archaeological record—clay pottery.

Given that clay had been abundant and readily available for thousands of years after the ice sheet's retreat, why did northland Indians suddenly utilize this raw material in the manufacture of pottery? To answer this question, Okstad again points to changing environmental conditions. Paleobotanists have demonstrated that around 2,500 to 3,000 BP, the climate swung back into a cooler, wetter mode. Water levels in lakes and rivers began to rise. Analyses of pollen cores show a major increase in the pollen of grasses, which scientists attribute to a surge in the growth of wild rice.

A kind of caloric insurance, wild rice served as a reliable source of sustenance when other foods were scarce. As such, it likely revolutionized the prehistoric subsistence economy. Archaeologists such as Peters believe that its abundance contributed to a swelling in the populations of native people. It is no surprise that this food staple was given the name *mano' min,* or "good fruit," or that the September moon became known as Man-o-min-e-geez-is, the "moon of the wild rice." Writing in 1900 of the Indians of the northern lakes country of Minnesota and Wisconsin, the ethnographer Albert Ernest Jenks observed, "It is believed that more geographic names have been derived from wild

Wild rice became a dietary staple for native people when it began to proliferate under a cooler, wetter climate approximately twenty-five hundred to three thousand years ago. A change in the climate also allowed paper birch to reinvade the forest, enabling prehistoric Indians to replace their heavy dugout canoes of pine with a lighter-weight birch bark. Seth Eastman, *Gathering Wild Rice,* 1851–57. Courtesy of the James Ford Bell Library, University of Minnesota.

rice in this relatively small section of North America than from any other natural vegetal product throughout the entire continent."

The benefits of wild rice were twofold: not only was this foodstuff abundant, but also surplus harvests could be stored for later consumption. To preserve wild rice, however, the grains had to be parched. The need for more durable cooking utensils that could withstand roasting fires may have compelled prehistoric Indians to turn to pottery.

The centrality of wild rice to the lives of native people may also have given rise to another technological innovation, Okstad says. The climatic shift to cool, wet conditions allowed boreal species, such as birch, to reinvade the Lake Superior region. Taking advantage of the changed species composition of the forest, prehistoric Indians replaced their heavy dugout canoes of pine with birch-bark canoes around AD 1000. The greater responsiveness of these lightweight canoes undoubtedly helped their occupants to more easily navigate the thickly vegetated stands of wild rice beds during the fall harvest.

Around this same time, as northern Indians supplanted dugouts with birch-bark canoes, they relinquished the atlatl in favor of a new technology with which northern Indians have subsequently become identified—the bow and arrow.

But according to some anthropologists, viewing early Lake Superior people primarily as hunters rather than anglers constitutes a case of mistaken identity. Prehistoric people in the upper Great Lakes region, Cleland notes, moved "from hunters to

The ecological conditions of the Northwoods required the Ojibwe to be highly nomadic in their use of Minnesota's lakes and forests, as seen in this 1846 painting of a seasonal Ojibwe encampment by Paul Kane. Reproduced with permission of the Royal Ontario Museum; copyright ROM Images (912.1.9).

generalized fishermen to specialized fishermen." He asserts that the centrality of fishing to Ojibwe culture has been largely overlooked in the scientific literature. Instead, archaeologists and anthropologists have chosen to focus on the role of native people in the fur trade, amplifying the scientists' "own cultural predisposition to cast these fishermen in the roles of hunters, warriors, and fur traders," Cleland says.

The organization of the Ojibwe calendar around fishing would have a large impact on the kind of social organization that Euro-Americans encountered when they first entered the Lake Superior region. The rapids of Sault Ste. Marie, among other locales, drew large crowds for the spring and fall spawning runs of fish such as sturgeon, lake trout, and whitefish. Yields were substantial, Cleland writes, requiring a "large labor force to clean fish, gather firewood, build smoking racks, sustain fires, turn the smoking fish, and pack the preserved fish." These jobs were primarily carried out by women.

According to anthropologist Thor Conway, places rich in spawning fish have attracted people for millennia. After conducting archaeological surveys of Whitefish Island in the St. Marys River near Sault Ste. Marie, for example, Conway discovered that the site served as a fishery for Ojibwe and other tribal groups for some two thousand years.

But unlike many agrarian-based tribal groups to the south, where fertile soils and long growing seasons promoted the development of sedentary societies, the native people of the north country did not establish permanent villages. The ecological conditions in which they lived forced them to remain highly nomadic. Theresa Schenck, an American Indian ethnohistorian, points out that the Ojibwe political and social organization evolved out of what is known as the seasonal round—a widely scattered series of camps that exploited the seasonal bounty of the lakes and forest. "The smallest unit of Ojibwa society, the patrilineal band, often split into individual families to spend the winter in the forest where they had hopes of finding game," Schenck observes in her 1997 book *The Voice of the Crane Echoes Afar: The Sociopolitical Organization of the Lake Superior Ojibwa, 1640–1855*. "In the early spring, when the sap began to flow in the maple trees, the band came together to make sugar. Later, spring fishing would be done at the shores of a lake. Early summer was the time when many bands gathered together, usually at a common fishing ground, to renew alliances through marriages and religious rites. Bands dispersed again to gather wild rice at the end of the summer, and then went to their fall fishing and hunting grounds. This nomadic life was the source of much frustration on the part of the Jesuit missionaries, who believed that a sedentary life was a prerequisite for conversion to Christianity."

Many scholars posit that the fur trade was responsible for irreparably disrupting the traditional subsistence patterns of native people. But anthropologist Cleland disagrees. "When considering the subsistence economies of historic Great Lakes Indian tribes," he writes, "one's thinking may be so shaped by the drama of fur-trade commerce that one sometimes forgets that Indians continued to eat after the arrival of Europeans." Strong evidence suggests "that the agricultural, hunting, wild rice, and fishing economies of Woodland Indian peoples were not abandoned or substantially changed even in the face of the immense social trauma of Euro-American contact," Cleland adds. "In fact, the knowledge and skill developed by prehistoric and protohistoric Indians for life in the Great Lakes forests were continued into the [twentieth] century."

Schenck agrees, arguing that confining tribal people to reservations, not engagement in the fur trade, was the single most important cause of the dissolution of traditional tribal structure. "The Ojibwa were one of the last Northern Algonquian groups to be contacted by the French," Schenck observes. "Living in areas at first not coveted by Europeans, they were able to retain their traditional way of life far longer than many other Algonquian peoples. . . . The Ojibwa of Lake Superior resisted enforced settlement for more than two centuries. . . . As long as the Ojibwa resisted all efforts by missionaries and government officials to assign them to permanent residences, they were able to maintain their sociopolitical organization intact."

But with the advent of industrial-scale mining along Superior's southern shore, the geographical boundaries around the Ojibwe way of life began to tighten. In the early nineteenth century, for example, prospectors located troves of copper and iron in Michigan's Upper Peninsula. By 1847 a series of mining booms had so transformed the landscape that the famous north-country priest Frederic Baraga hardly recognized the wild and lonely land he had entered as a new missionary only twelve years earlier. "I was dumbfounded at the fast spreading of civilization on these shores of Lake Superior," he exclaimed upon arriving in his former territory on the bustling Keweenaw Peninsula after a long absence. "I found in many places neat houses with nice, carpeted rooms. In one house there was even a piano on which a young American woman played very skillfully."

As lands along Superior's southern shore were coveted by logging and mining interests, the federal government negotiated treaties in 1825 and then again in 1837. In these agreements the Ojibwe ceded large blocks of land in Michigan and Wisconsin in exchange for reservation lands and hunting and fishing rights.

As lands along Superior's shores were coveted by logging and mining interests, the federal government negotiated treaties in which the Ojibwe ceded large blocks of land in Michigan and Wisconsin in exchange for reservation lands and hunting and fishing rights. Shown here are lands ceded by the Ojibwe in the Lake Superior basin between 1825 and 1854.

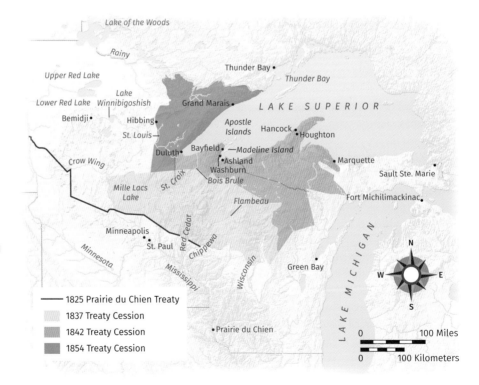

Only after other parts of the Lake Superior shoreline had been developed did Euro-Americans set their sights on the north. Although occupation of the North Shore by whites lagged decades behind that of Superior's more accessible southern shore, the fate of both regions was closely linked. Scientists of the time theorized that these mineral-rich formations continued under Lake Superior and emerged on the lake's northern shore. Their speculations fanned copper fever, which spread across the lake. But before prospectors could file any claims, the northern frontier too had to be formally wrested from its native occupants. Bowing to repeated pressure from mineral seekers, politicians, and industrialists, the U.S. government negotiated a treaty with the Ojibwe Indians in 1854 that opened Minnesota's Arrowhead region and additional parts of Wisconsin to white settlement. The initial response by claimants was swift. By 1856 post offices had already been established at the French River, Beaver Bay, Grand Marais, and Grand Portage. When the Methodist minister James Peet arrived on the North Shore that same year, he exclaimed in his diary that for thirty miles along the coastline "nearly all the land is claimed and a shanty built every half mile."

Meanwhile, the La Pointe Treaty of 1854 established reservations on both ends of the North Shore, at Fond du Lac and Grand Portage. The changes brought by this new pattern of imposed confinement, as well as the ecological transformation of the North Shore watershed by Euro-American newcomers, would change the character of the land and its people as dramatically as the retreat of the ice sheet thousands of years before.

Timber, Trout, and Tourism

"Ever since human beings have been in northeastern Minnesota," observes archaeologist Peters, "they've had an impact on the ecosystem." Tree-ring studies suggest, for example, that prehistoric people may have set fires to clear large openings in the forest to promote the growth of young vegetation such as aspen sprouts. This tender browse would have been especially attractive to their primary prey—moose. Scientists also suspect that native people routinely torched sandy scrub areas to make way for sun-loving blueberry shrubs. And according to Peters, the rapid increase of wild rice in the pollen record some 2,500 to 3,000 years ago quite possibly reflects aboriginal conservation measures. By embedding a portion of the kernels from each year's harvest in clay balls and tossing them overboard, a practice that was commonly witnessed during the historic period, early peoples may have helped to expand the range and increase yields of existing wild rice beds.

To some extent, Peters adds, humans have always manipulated the environment to obtain products from the land. In prehistoric times, however, populations of native people in the region had been relatively low, and their tools for effecting widespread ecological changes were limited. As a result, human-caused disturbances, such as fires or the exploitation of trees for fuel and fiber, rarely—if ever—exceeded the natural boundaries of variability that govern northland ecosystems.

But the arrival of European newcomers to the region beginning in the seventeenth century raised the level of resource extraction from subsistence use by a relatively small group of people to an industrial scale in which raw materials fed the consumer appetites of entire nations. Suddenly, northern ecosystems underwent changes that defied long-established boundaries of variability. Take, for example, the fur trade. In the late eighteenth century, Grand Portage became one of the busiest settlements west of the Appalachian Mountains. Despite the fact that fur traders rarely ventured far from this northern outpost and the shores of a few well-traveled water routes, their activities had a profound effect—both directly and indirectly—on the land and its inhabitants. By encouraging the unrestrained trafficking in beaver hides—in the process nearly extirpating the beaver from its northland haunts—the fur trade eliminated backwaters created by beaver dams. Scientists conjecture that the number of countless communities of plants and animals that depended on these still-water habitats plummeted. Decades—and even centuries in some cases—may have passed before they became reestablished on the land. (See also "Headwaters.")

As voyageurs pushed into the continental interior, they caused other changes that ranged far beyond their footsteps. For example, early historical accounts chronicle dense clouds of smoke from rampaging fires that were sparked by careless campfires. Other blazes were set intentionally. In his journal of a trip to the Canadian North Shore of Lake Superior in 1848 John Cabot writes that his voyageur guides routinely set fire to trees along the lakeshore, in one case torching a lichen-draped spruce in order to signal to their companions in a separate party of boats. Some blazes were ignited out of sheer boredom. Cabot recounts how one such "recreational" fire "unfortunately ran along the ridge of the beach, and, in spite of [the voyageurs'] utmost exertions, marched with a broad front into the woods. It was an exciting spectacle, the eagerness of the flames to seize upon each fresh tree, winding round it like serpents, crackling and rushing furiously through its branches to the top, until every fragment of dry bark, lichen, &c., was consumed. . . . When we left, the fire was in full progress and was probably stayed only by a swamp beyond."

But by far the greatest disruption to North Shore ecosystems came with the era of Euro-American settlement in the mid-nineteenth century. What brought people to a place that more often excited fear, rather than promise, as it did in David Thompson, who coasted along the North Shore in a fragile birch-bark canoe in 1822 and noted in his diary that he found "an Iron bound Shore of Rock with very few places to camp"?

Copper fever. In 1848, Joseph G. Norwood conducted the first U.S. geological survey of the Minnesota North Shore, testing his hypothesis that a vein of copper ran beneath the waters of Lake Superior, connecting Michigan's mineral-rich Upper Peninsula with the North Shore. After traveling up the coast by canoe from Fond du Lac to Grand Portage, Norwood kindled the hopes of prospectors by noting signs of copper in several streams.

But no claims could legally be filed because the land along the shore was in possession of the Lake Superior Ojibwe. Several failed attempts by the U.S. government to gain title to the land did not dissuade illegal trespassers, however. At the mouth of the French River, for example, Norwood recorded the existence of a cabin that was built in 1846 by a mining company agent hoping to stake a preemption claim for his employer.

Illegal incursions on Indian lands only increased as scores of other get-rich-quick schemers flooded towns on the western end of the lake. Writing of his visit to Superior, Wisconsin, in September 1854, Robert B. McLean, who in 1856 became Beaver Bay's first postmaster, recalls, "What conversation I heard around me all turned toward copper and copper claims. There were rumors of great masses of pure copper and large veins full of copper that could be traced for long distances, but they were all on the North Shore and that was Indian territory, white men were not allowed to enter." But that did not stop McLean and his comrades. As he left his hotel on September 15, a fellow prospector advised him to pack a bedroll and an ax. "We are going to sneak over onto the North Shore," he confided to McLean, "and try and find where those masses of copper and that big vein that we have heard so much about are." That night, they set sail in secret for the North Shore, where they carried out mineral explorations as far as the Knife River.

On September, 30, 1854, the government finally succeeded in pressuring the Ojibwe to sign a treaty ceding a great triangle of land that included the Minnesota North Shore. Even before the treaty was officially ratified by the U.S. Congress, a wave of prospectors staked their mineral claims, largely in the vicinity of the Knife and French Rivers. And to compound their hoped-for mineral profits, many prospectors and their investors also became land speculators, platting towns in places where they suspected great quantities of ore would someday be discovered. Within a few short years, paper towns—and a few actual settlements—dotted the coastline. By 1858, writes historian Glenn R. Sandvik, twenty-three town sites were officially registered, and nearly the same number of other towns were platted but not yet registered. Included in these heady days of urban planning was the establishment of Lake and Cook Counties on March 1, 1856.

But dreams of copper riches sputtered and died along the Minnesota coast, unlike those along Superior's southern shore, where copper veins spawned millionaires overnight. Most towns never got off the drawing boards. The economic panic and depression that beset the nation in 1857, followed by the turmoil of the Civil War, dampened

Copper fever set off a wave of Euro-American settlement along the North Shore in the nineteenth century. This 1870 painting by June Lowry depicts the early days of Beaver Bay, including the schooner *Charley,* which for decades transported passengers and lumber to points all across the lake. Courtesy of the Lake County Historical Society.

mineral and land speculation. Hopes were temporarily revived in 1865 when Minnesota State Geologist Henry Eames set off a stampede for gold on Lake Vermilion and another copper rush on the North Shore. Having traveled as far as the Temperance River, Eames claimed to have found evidence of copper along several rivers, most notably on the Knife, Stewart, and Split Rock Rivers. Although Eames admitted that his explorations provided only a glancing profile of the region's mineral potential, he wrote, "Enough has been demonstrated to convince even the incredulous, of the vast deposits within its limits. The stimulus given by the discoveries made is already being felt, and the next season will doubtless witness the labors of hundreds of hardy explorers and miners along the Minnesota Shore of Lake Superior."

Precious metals were not the only minerals to excite the imaginations of hopeful millionaires. The discovery of corundum along the Canadian North Shore at the turn of the nineteenth century sent a group of Two Harbors businessmen to take a closer look at the rock deposits bordering Crystal Bay and on Carlton Peak. The rock on the American side turned out to be anorthosite, however, and not the superhard corundum that was so valuable as an abrasive in industrial processes. All was not lost. The second group of investors to take over mineral explorations and mining operations at Crystal Bay and Carlton Peak formed the nucleus of what would become one of the state's most successful corporations, Minnesota Mining and Manufacturing Company, otherwise known as 3M.

Although these various schemes had little staying power, they nonetheless left their mark. To expose the rocky substrate of their claims, prospectors regularly set fire to the forest, introducing a powerful disturbance that was rare in many nearshore plant communities before their arrival. C. W. Hall, a member of a team of scientists from the state's Geological and Natural History Survey, described their actions in a report following a visit to northeastern Minnesota in 1879. Hall decried the actions of the prospectors, calling them a "lawless set of 'explorers' who own nothing and feel no responsibility." Within a few years, Hall continued, the prospectors "have destroyed more timber and burned down to the rock more of what could be made fine grazing land than probably all the gold and silver they will ever find can pay for. . . . Very often when an 'explorer' finds a vein of quartz or calcite with a sprinkling of galenite or blende, crossing one of the river gorges so common in that part of the State, he takes its direction and coolly proceeds to set fires along its course, that when the leaves and soil are burned off he may examine it in various places across the country, and, if there be a trace of any metal, proceed to mark out mining locations." Today, those with a compass and a willingness to bushwhack through tangled forests still can stumble across their test shafts blasted into the bedrock.

But some North Shore communities would discover assets nearly as valuable as mineral deposits—safe, accessible harbors in a treacherous rock-bound coast. To the northwest of Lake County, in what is known as the Mesabi Range, lay one of the world's greatest deposits of iron ore. As early as 1884, rail cars traveling the Duluth and Iron Range Railroad began hauling ore to Two Harbors, where the railroad built shipping facilities. Two Harbors became a bustling port town. The first shipment to leave its docks in 1884 included some sixty-two thousand tons of iron destined for the nation's burgeoning steel mills.

But large, navigable harbors were in short supply and failed to provide the region

In the early years of logging, operators used the larger rivers on the North Shore, such as the Pigeon River, to drive logs to Lake Superior, where they were rafted to shore-edge mills. Photograph by Carl Henrickson/Campbell and Shiels, 1924. Courtesy of the Kathryn A. Martin Library, University of Minnesota Duluth, Archives and Special Collections.

with a sufficiently large industrial base. Instead, officials steered economic development to one resource that the North Shore possessed in abundance—forests. By the late 1850s, three sawmills had opened in Lake County. One of them—a mill operated by the German immigrants Christian Wieland and his four brothers at the mouth of the Beaver River—survived the ups and downs of the North Shore's fledgling lumber market for twenty-five years before closing in 1884. In its heyday, the lakeside mill was one of the most prolific producers in the region, churning out twenty-two thousand board feet of lumber per day.

Early mills, such as the Wielands' operation, supplied wood primarily for a regional market. After the town of Marquette, Michigan, was nearly leveled by a blaze in June 1868, writes Jessie C. Davis in a history of Beaver Bay, residents stood in line to unload the Beaver Bay lumber that was delivered by the Wielands' schooner, *Charley*. So great was the demand that the schooner crisscrossed the lake for months delivering building supplies. That same year, miners struck silver on a small island off the Canadian coast near Thunder Bay. The Wieland mill provided building material for everything from new housing and offices to structural headers for the mine. But a far more bustling market lay closer to home. In 1871 the Lake Superior and Mississippi Railroad was completed from St. Paul to Duluth, opening a long-hoped-for trade channel between the upper Mississippi River to ports throughout the Great Lakes and along the Atlantic Seaboard.

Along with the shipments of prairie wheat came boom times to the cities of Duluth and Superior. As Davis points out, "Many of the early homes built in Duluth were made of Beaver Bay lumber."

To feed their sawmills, the Wieland brothers and other early operators cut trees from their own backyards—literally. According to forester John Fritzen, "In the early days immense stands of virgin white pine were to be found all along the North Shore and the pioneer sawmills had little difficulty in obtaining their raw material close by." The Wielands, for example, cut most of their timber from the banks of the Beaver River and nearby Beaver Bay and used oxen to haul logs down to their lakeside mill.

But around 1892 commercial logging by out-of-state outfits began to eclipse the pace and scale of such family-owned enterprises. As farmsteads and the great prairie cities of Chicago and St. Louis flourished on the treeless plains, the demand for building and railroad lumber grew. Having logged out much of the white pine stands in Michigan and Wisconsin by 1890, timber corporations set their eyes on the wealth of straight white pine that grew in the North Shore's nearshore forest and farther inland.

Watershed by watershed, Fritzen writes, logging outfits worked their way up the shore, cutting white pine for lumber and lath, and cedar for shingles along the Gooseberry River in 1903, the Beaver River in 1906, and the Manitou and Temperance Rivers in 1909. They exploited the more easily accessible trees along waterways first, in some cases, damming rivers and dynamiting waterfalls to ease the course of logs floating down to Superior during river drives. When rivers proved unnavigable for log drives, companies sometimes built bypasses on land, such as the road constructed by the Lesure Lumber Company in 1895. In winter, sleighs skidded down the icy road from the inland reaches of the French River watershed to the mouth of the French River, slowed only by workers who threw sand and hay in their tracks. In more elaborate cases, rail spurs started at the mouths of North Shore rivers and worked their way inland, such as the ten-mile corridor up the Split Rock River that was built by the Merrill and Ring Company in 1899. Holding ponds were constructed at the mouths of these and many other rivers, where logs could be collected into rafts and then towed across the lake to mills at Ashland, Wisconsin; Baraga, Michigan; and Duluth.

With the advent of railroad building, however, such practices would become obsolete. "Prior to 1900," writes historian J. C. Ryan, "most logs were moved from the woods by draying, skidding, or sleigh hauling them to the rivers and lakes, and driving or rafting them to the mills. This type of logging operation reached back from the shores of the lakes and streams only about six to ten miles and when all timber within an area was logged off some method of transportation had to be found to move what remained. To meet this problem logging railroads were constructed. Some were built by railroad companies for the purpose of getting the log-hauling business, as well as that of carrying supplies to the small logging towns that sprang up with the industry, and others were built by logging companies to reach their own logging operations."

So indispensable had the railroads become in transporting timber from the far inland reaches of the North Shore watershed, Fritzen writes, that by 1901 some seventy-five miles of tracks crisscrossed the North Shore watershed. By 1905, that number jumped exponentially. One of the most active was the Duluth and Northern Minnesota Railway, constructed by the big Michigan firm Alger, Smith and Company. The railway began operations in 1899 in the Knife River valley. By the time it was fully built out, its tracks extended from the Knife River to as far as Cascade Lake.

North Shore riverways often contained serious obstacles for log drives, such as the High Falls on the Pigeon River (*below, right*). Engineers circumvented this obstacle by diverting dammed river water into a flume that carried logs safely downstream. Courtesy of the Cook County Historical Society, PH-B129.

At full build-out, the Alger, Smith and Company line, shown here in 1899 at Knife River in Lake County, extended as far as Cascade Lake in Cook County. Courtesy of the Minnesota Historical Society.

Around the turn of the twentieth century, timber companies began constructing an elaborate network of railroads to transport logs from the far inland reaches of Minnesota's Lake Superior watershed. By the 1920s, nearly all the valuable timber supplies had been exhausted, and the lines were dismantled. Courtesy of the Kathryn A. Martin Library, University of Minnesota Duluth, Archives and Special Collections.

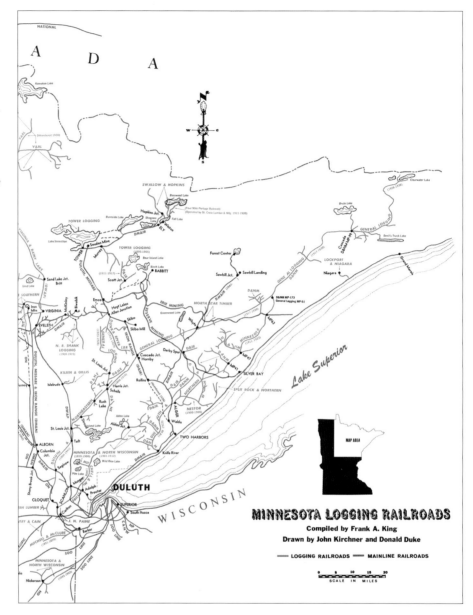

The cutting would continue without pause from 1892 to 1912 until nearly all the valuable timber had been cut and hauled away. As loggers cleared the forest, land speculators, in cooperation with state and federal governments, advertised the cutover hectares to hopeful farmers. From 1900 to 1910, writes Thorvald Schantz-Hansen of the University of Minnesota's Forest Experiment Station in Cloquet, the number of farms in Lake County boomed, largely due to demand for local produce from logging camps and the availability of transport by local rail lines. But by 1920, as supplies of timber began to dwindle, so did local markets and transit lines. With the forests denuded and burned, the Duluth and Northern Minnesota Railway was sold and then dismantled in 1924. New

summer-home residents provided a temporary demand for local food products. With the onset of an agricultural depression in 1930, however, the bottom dropped out, and the number of farms in Lake County plummeted.

Among those who weathered the boom-and-bust cycles of agricultural settlement and timber and mineral exploitation were the families of commercial fishermen, who reclaimed abandoned homesteads along the North Shore coast in the 1880s. Most of these families emigrated from Sweden and Norway.

Many immigrants found a mainstay for their incomes in commercial fishing. Eager to enlist workers to supply fish for its Lake Superior operations, the Booth Fisheries Company of Duluth offered easy credit terms on fishing outfits—everything from boats to nets. The ranks of fishermen swelled after Norwegian immigrants flooded the fishing market in the 1880s. According to historian Newell Searle, 37 commercial fishermen operated along the North Shore in 1879. In a newspaper account of his trip up the shore in 1890, Duluth resident Burt Fesler counted no fewer than 106 fishermen from Duluth to Isle Royale. By 1917, Searle writes, the number had jumped to 273.

Fishing families diversified their modes of livelihood by supplementing commercial fishing with subsistence hunting and agriculture in the warmer seasons. During winter, they logged the North Shore forests, supplying railroad ties, poles, and lumber

The livelihoods of many Swedish and Norwegian immigrants to the North Shore consisted of subsistence hunting, agriculture, and logging, which supplemented a more mainstay income from commercial fishing. Courtesy of the Kathryn A. Martin Library, University of Minnesota Duluth, Archives and Special Collections.

for the mines, houses, and railroads that were booming on the Mesabi Range and in Two Harbors and Duluth.

But their lives were not immune to fallout from the egregious cut-and-run logging practices of big corporations, which devastated Superior's nearshore forests around the turn of the twentieth century. Prior to Euro-American settlement, stand-replacing conflagrations were rare in lake-terrace forests. But as sparks from logging trains set fire to the great piles of woody debris left behind by the lumberjacks, catastrophic fires occurred with alarming frequency. In her book *Lake Superior,* historian Grace Lee Nute notes that in the late nineteenth and early twentieth centuries, fires strafed most parts of the North Shore watershed at regular intervals.

On the North Shore, where fires typically swept the landscape from west to east or southwest to northeast, human settlements were frequently threatened. For some residents, Lake Superior was their only refuge. On September 9, 1908, for example, a blaze that dogged Grand Marais for two weeks finally came within two miles of the town's borders, forcing many of its six hundred residents to camp out on the beach. The fire was one of many that roared from the tinder-dry headwaters, which had not seen rain in two months. Conditions were so perilous that a naval gunboat had been sent to pick up burned-out settlers along the entire shoreline. Smoke from the fire joined with smoke from other fires in the Lake States, Canada, and northeastern United States, writes fire historian Stephen Pyne, to cover places as far away as "New York City in a 'heavy gray pall.'"

The U.S. Forest Service declared the 1908 fire season to be "one of the worst in the last quarter century." Little did the agency know that the record would be bested a mere two years later. On May 11, 1910, conditions once again were so dire that the North Shore made front-page news in the *New York Times*. One telegraph operator managed to type out a brief message describing the general panic before narrowly escaping the approaching inferno himself. "The entire northeastern portion of Minnesota's forest region is smoldering and smoking," the newspaper reports exclaimed. "Wild animals are rushing to the lake shore before the fires." A telegraphed message from Grand Marais warned that "forest fires were bearing down on that village, and that it seemed doomed."

A follow-up account published on May 13 described in horrific detail the mayhem caused by the blazes:

> Settlers in the vicinity of Grand Marais, having lost their homes and everything else except the clothes on their backs, began arriving in that village to-day, according to reports by wireless telegraph.
>
> One group arrived early to-day, after they had spent the night in the bed of a stream. Part of the time they were compelled to submerge themselves to keep the flames from burning the clothes from their bodies. In this party were five children, who were so severely burned that they had to be cared for in a hospital.
>
> At Gooseberry River and along Beaver Bay the flames are eating their way fiercely through the woods. From Park Bay to Tofte, and as far back as the eye can see timber is on fire. From Good Harbor to Grand Marais fires are burning at intervals.

Grand Marais once again escaped the flames, but the village of Tofte was not so lucky. The fire had burned much of Tofte to the ground, including its newly erected

A postlogging wildfire on May 11, 1910, swept so close to Grand Marais that a telegraph operator issued a message that the tiny settlement "seemed doomed." Grand Marais escaped the flames, but much of Tofte was burned to the ground. Courtesy of the J. Henry Eliasen Collection, Cook County Historical Society, PH-A175/6.

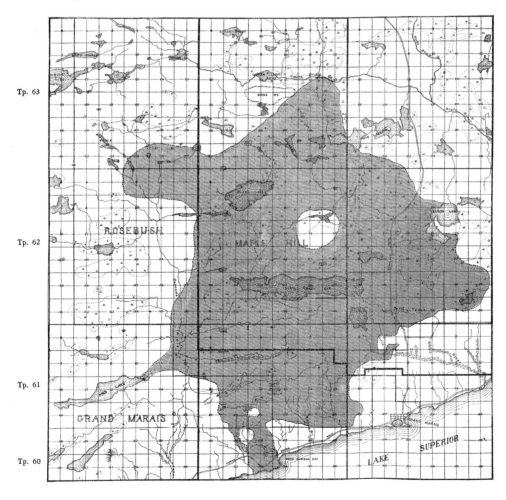

The year 1910 was a record-setting year for wildfires across the United States, including the Minnesota watershed of Lake Superior. This map shows the extent of one large fire in Cook County.

church, town hall, and sawmill. Photographs of the town taken in subsequent years show the tiny rebuilt outpost against a ridge so denuded of trees that Carlton Peak, about one mile to the northwest, is clearly visible in the background. The devastation of the nearshore forest by the flames was extensive enough that even as late as the 1930s, Nute writes, "The land beside the scenic coast drive was practically a continuous forest of blackened stumps."

The tragedy of laying waste to the upper Midwest's great pine forests was compounded by the unscrupulous sale of agricultural pipe dreams that drew thousands of hapless, impoverished settlers to the northland's marginal soils. The devastating history of the cutover lands can best be read in county tax records. Schantz-Hansen writes that by 1934 in Lake County, for example, some 21 percent of the land was tax delinquent—and the number of tax-forfeited acres was growing. After shearing the land of its valuable timber, many logging companies simply defaulted on their tax obligations. Having lost local markets for their produce and already struggling against marginal growing conditions, cash-strapped farmers abandoned their land, transferring the ballooning tax burden to more solvent residents, many of whom succumbed under the pressure in a domino-like fashion. (See also "Highlands.")

Faced with an environmental, social, and economic catastrophe of monumental proportions—not to mention an outraged public—state and federal governments proposed sweeping changes to land-use policies. In Lake County, as in many regions throughout the northern cutover region, officials came to the conclusion that the economic uses to which the land was best adapted were recreation and forestry. In a 1927 article, George S. Wehrwein, a professor of land economics at the University of Wisconsin, affirmed the changing priorities for the use of rural northern lands: "The Lake States are beginning to realize what a financial asset their scenery is. The old notion of converting all land into farms is giving way. The thousands of lakes, the rocky lands, the forests are attracting the dwellers of the cities that cluster at the foot of the lakes and the inhabitants of the flat but fertile prairies."

The marketing of nature's assets was nothing new to the North Shore. As early as the mid-1800s, anglers from around the world undertook difficult, and often dangerous, journeys to Lake Superior in hopes of hooking the trophy-sized brook trout made famous by such books as *Superior Fishing,* an 1865 account of sports travels by Robert Barnwell Roosevelt, uncle of President Theodore Roosevelt.

So popular did the brook-trout fishery become that, in his 1880 report, Professor Hall of Minnesota's Geological and Natural History Survey accurately predicted that angling pressures could imperil the fish's future. "This beautiful and universally admired species inhabits, in great numbers, the many small rivers flowing into Lake Superior," Hall writes. "These streams, in fact, have become one of the most famous fishing grounds on the continent. That they may continue so, they must be protected. Those within the State of Minnesota are visited annually by large numbers of amateur fishermen, who go in parties, and thus make most enjoyable vacation excursions. A boatman and a cook are engaged at Duluth or some other accessible point, who load into a sail-boat a store of provisions and other essentials to comfort and pleasure, and then take the excursionists to the best trout streams around the lake. One stream after another is visited. A camp is pitched beside each where it empties into the lake. Then, for several days, perhaps a week, the river banks are lined with the creeping, stealthy forms of the fishermen, throwing every temptation the ingenuity of man can devise before the eyes of the wary trout." (See also "Healing a Watershed for Coaster Brook Trout.")

Not all the new forms of tourism, however, required a hardy constitution and a bedroll. Touring the sights from the luxury of a rail car or a ship's berth had become a wildly popular pastime among fashionable Americans, as the writer Edward Bellamy observed in 1874: "I would call attention to the fact that sentimental love of the beautiful and the sublime in nature, the charm which mountains, sea, and landscape so potently exercise upon the modern mind through a subtle sense of sympathy, is a comparatively modern and recent growth of the human mind. . . . It is almost within a century, in fact, that this susceptibility of the soul seems to have been developed."

Smitten by the new craze for rugged, picturesque landscapes and anxious to avoid vacation watering holes along the East Coast that had opened their doors to vacationers of lower incomes and social status, wealthy Easterners packed rail cars bound for natural wonders in the far-flung reaches of the continent. By the 1880s, writes historian Earl Pomeroy, it had become so easy to reach remote locations in safety and comfort that whole families boarded trains to destinations that only a few years before would have required months of arduous, dangerous trekking. According to *Outing and the Wheelman* magazine, "As they have acquired wealth with the increase of years, the American people have become nomadic in their habits."

Steamship operators began to advertise Lake Superior cruises to tourists in the mid-nineteenth century. Michigan photographer B. F. Childs, seen here painting a sign for his gallery circa 1870s, was among the earliest purveyors of images of area tourist attractions. Collection of Jack Deo, Superior View Gallery.

Stream fishing, as seen here in the mouth of the Baptism River circa 1890, was among the earliest recreational activities to attract visitors to the North Shore. Photograph by Northwestern Photo Company. Courtesy of the Minnesota Historical Society.

TO SEEKERS OF HEALTH AND PLEASURE.

Grand Pleasure Excursion for the Season of 1874

—FROM—

BUFFALO, ERIE, CLEVELAND AND DETROIT,

TO DULUTH AND ST. PAUL,

PASSING THROUGH

LAKES HURON AND SUPERIOR.

To Continue during the Summer Months.

A Daily Line of STEAMERS will run from **Buffalo, Erie, &c.,** to **Saut Ste. Marie, Marquette** and **Duluth,**—Connecting with Cars on the **Lake Superior & Mississippi Railroad,** running to St. Paul, Minn.

From **St. Paul** Steamers run Daily on the Mississippi River, during the season of Navigation, to **La Crosse, Prairie du Chien, Dubuque** and **St. Louis,**—Connecting with the Lines of Railroad running to **Milwaukee, Chicago** and **Detroit,**—thus furnishing a ROUND TRIP of over *two thousand miles,* by land and water, through one of the most healthy and interesting regions on the Continent.

DULUTH TO BISMARCK, DAKOTA,

VIA

NORTHERN PACIFIC RAILROAD

This new and HEALTH-RESTORING LINE OF TRAVEL, by means of steamers on the UPPER LAKES OF AMERICA affords an extended EXCURSION of 1,650 miles from BUFFALO TO BISMARCK, Dakota—connecting with Steamers on the Red River of the North, and on the Upper Missouri, extending for 1,200 miles. further to FORT BENTON, Montana—forming altogether the

GRANDEST EXCURSION IN THE WORLD.

In 1855, the canal at Sault Ste. Marie opened, allowing seamless passage for tourist cruises from the lower Great Lakes into Lake Superior. Advertisements, such as this 1874 broadside "Sailing on the Great Lakes and Rivers of America," attracted wealthy East Coast tourists to Superior's remote reaches.

No destination seemed too far, including the Lake Superior region. In 1855, the canal at Sault Ste. Marie opened, allowing the seamless passage of steamers from the lower Great Lakes into Lake Superior. Little more than a decade later, an illustrated Lake Superior travelogue appeared in the May 1867 issue of *Harper's New Monthly Magazine,* accompanied by an etching of the geological formations of the Pictured Rocks lakeshore on the cover. Such accounts whetted the appetites of the traveling public, and soon people were gazing at the shores of Lake Superior from the crowded rails of a steamship. By 1872 a tourism guide for "seekers of health and pleasure" in the Lake Superior region noted no fewer than six different rail and steamer routes leading to the lake. Referred to in the guidebook as an "Inland Sea," Superior was extolled as "one of the grandest and most instructive features of the American Continent, in a physical point of view. Here Nature has been lavish in her gifts to man—affording pure and sparkling waters— a healthy and invigorating climate—useful and precious metals, and various kinds of fish of a delightful flavor, all abounding in this favored region; where Pictured Rocks, mountains, water-falls, islands, bays and varied land and water scenery greet the eye of the observing traveller."

Grand accommodations catered to the upscale tourist trade, such as the Hotel Chequamegon, built by the Wisconsin Central Railroad in 1877 on the site currently occupied by a newer hotel of the same name in Ashland, Wisconsin. Among the people on its exclusive guest list were John D. Rockefeller, William Cullen Bryant, and Marshall Field.

At the same time, commercial fishing families began to capitalize on the increasing traffic by renting out rooms and small cabins to travelers of more modest means. Joining the anglers and sightseers were hunters as well as health seekers who suffered from asthma, hay fever, or tuberculosis. Before the advent of effective medications, many of those afflicted by respiratory ailments found relief along Superior's coast.

And then—as now—the shores of inland lakes were advertised as a way to escape the cities' stifling heat. Before the widespread use of air conditioning, such getaways provided the only sure antidote to the Midwest's hot, humid summers. For some visitors, lakeside living was such a restorative that they returned year after year to the same destination. Take, for example, Lionel Chapman of Kansas City, who, according to one newspaper account, was brought to Minnesota in a refrigerator train car in 1924 after suffering a sunstroke. "Restored to health and vigor by the cool breezes from the big lake in the Land of Naniboujou," the article observed, Chapman resolved to spend his remaining summers on the Minnesota shore of Lake Superior.

In the aftermath of the cutover, however, tourism began to be shaped by some markedly new trends. Chief among them, writes Paul S. Sutter in his book *Driven Wild,* was the unprecedented activism of state and federal governments in promoting tourism and building a national tourism infrastructure—with public lands as its centerpiece. Indeed, within the first two decades of the twentieth century, two major government agencies were established that would chart the course for the future of U.S. lands held in the public trust. In 1916 the National Park Service (NPS) was created. In 1905, the USFS was officially formed after the nation faced the prospect of a timber famine that threatened to cripple its economic development as well as undermine its national security. The creation of the national forest system, writes former USFS chief Mike Dombeck, "was in response to the cut-and-run era of timber harvests that left the United States with

80 million acres of denuded forests known as clear-cuts, mostly in the East and upper Midwest." Forest lands were reserved not only to restore the nation's timber supplies but—just as important—to protect the nation's water supplies. Early on, Dombeck observes, the architects of our public forest system recognized that "water is perhaps the most important forest product."

The national movement to establish public forests reached into the far corners of the country, including the Minnesota Arrowhead region. As early as 1902, some 500,000 acres of forest, most of it tax-forfeited land, were set aside in northeastern Minnesota. This reserve became the nucleus of the SNF, which by the time President Theodore Roosevelt officially added it to the national forest system in 1909 had grown to approximately 1.1 million acres. (Today, the SNF comprises nearly 3 million acres.) The state followed suit in 1911, creating its own Minnesota Forest Service. Like the SNF, the state forest system, historian Searles observes, was created expressly "to protect and manage cutover lands."

Not all residents in neighboring communities welcomed the idea of taking land off the tax roles, particularly those in counties whose soaring tax-forfeiture rates left them desperate for economic development. But many analysts, including Schantz-Hansen in his 1934 assessment of the cutover lands, argued that government-managed timber reserves could be coupled with a lucrative new use—recreation—as a strategy for economic recovery. As Barrington Moore of the Council on National Parks, Forests, and Wild Life emphasized at the 1924 National Conference on Outdoor Recreation, organized by President Calvin Coolidge, "Forest recreation is a forest product," on par with grazing, timber production, and stream protection.

Furthermore, unlike the disastrous policy of agricultural development, these land uses would work with, rather than against, the physical character of the land. "Probably no other resource possessed by Lake County presents such possibilities for immediate returns as the recreational values," Schantz-Hansen writes. "The shore line of Lake Superior, the inland lakes, the attractive streams, all are potential sources of income. The development of the recreational values is not incompatible with the use of the areas for forest purposes. In fact, the two uses supplement each other. Without an adequate stand of trees there is a marked decline in the recreational possibilities of any area." The development of land by seasonal residents, Schantz-Hansen argued, could help put tax-strapped counties back on the road to economic solvency. "Summer homes," he observed, "by increasing the taxable value of the land, and tourists can do much to bridge the gap between the present condition of cut-over lands and their rehabilitation by natural or artificial means."

Schantz-Hansen's prescription was echoed throughout the north country. In his economic analysis of similar problems in the cutover lands of northern Wisconsin, land economist Wehrwein concurred: "Even though recreational lands do not occupy a large area, they are very important from the standpoint of taxation. In the first place they are the most valuable land in the county, with the exception of land used for urban purposes. . . . In the second place, recreational land is highly improved. . . . Not only are recreational lands a large part of the tax base of these counties but they have been comparatively free from delinquency."

The coupling of forests and recreation as a strategy for economic recovery in cutover country was an idea whose time had come. The convergence of several important trends in American society made such new land-use proposals feasible: average Americans were suddenly presented with the means and the motives to explore the great outdoors. Perhaps nowhere are these changes best summarized than in the 1924 Proceedings of the National Conference on Outdoor Recreation. In a series of spirited presentations and discussions, conference attendees highlighted several recurring themes. Addressing the issue of the federal role in recreational planning, Joseph Hyde Pratt of North Carolina pointed to recent national legislation that established the eight-hour workday, allowing American workers greater leisure time. The average citizen could not be trusted, however, to put this newly available free time to best use. "Our labor leaders and legislators are gradually shortening the hours of work and lengthening those available for recreation," he observed. Pratt spoke for the majority of attendees when he added, "It becomes, therefore, most essential that wise provision should be made for this leisure time."

The productive use of leisure time, as Pratt and others pointed out, had serious implications for personal mental health. Noting a rise in the number of hospitalizations for "mental disease," Pratt observed that "recreation and supervised play is psychologically necessary in order to retain a normal equilibrium in the midst of the deadening monotony and excessive strain of the life of many of our people to-day, both in the city and in the country."

In the eyes of many conference participants, the nation's energy was being zapped by the stressful and artificial conditions of modern life. Some even suggested that Americans were draining the capital of vigor that had been banked by their pioneer ancestors, thereby losing the very qualities that made America a great nation. Preserving wild areas, where citizens could make contact with these pioneer conditions, was seen as nothing less than helping to preserve the environment in which the national character could continue to be forged, as Moore of the Council on National Parks, Forests, and Wild Life stressed in his address to the group in 1924:

> It is the consensus of opinion that to-day we in America are living largely upon the inherited mental and physical vigor of our ancestors; that the stout frames and stouter hearts bred by a long battle against the primeval forests and the enemies it contained constitute the greatest heritage passed down to us by pilgrim and cavalier, but that unless we again consciously and universally turn the minds and hearts of our people toward a life in the open, that American vigor or "pep" will soon reach its crest. Some 50 years ago that great apostle of park and playground, Frederick Law Olmsted, stated that the average human nervous system could stand only 3–4 generations of city life with its lights and noise, poor air, and lack of exercise; that there was a reason why the majority of leaders in every walk of life came from the farm, and that in physical and mental inheritance as well as in financial it was frequently "three generations from shirt sleeves to shirt sleeves"; that the decline of great families was frequently due to the "running out" of mental and physical vigor occasioned by the continued drain of city life upon the nervous and physical vigor of succeeding generations. If that were true in the early [1870s], how much more potent is his argument to-day with the speed of our life in the cities multiplied manyfold—automobile, airplane and jazz. Recreation in the great outdoors is a superb antidote for many of the social and economic ills of to-day. Fortunately, American civilization has not drifted too far.

By the 1920s, many Americans had time and dollars to explore the great outdoors. State and county governments began to develop recreation as a strategy for economic recovery in cutover country such as the Arrowhead region. Courtesy of David A. Lanegran.

As Moore intimated, outdoor recreation was essential not simply to personal health but to the commonwealth. Major William A. Welch of Palisades Interstate Park cited the experience of a fellow officer during World War I, who "found a striking contrast in the self-reliance, resourcefulness, and adaptability to new conditions between units made up of city boys and units composed of boys from the rural counties of northern Maine."

At risk also was America's domestic security. Young people with too much time on their hands were seen as a menace to the public peace. Welch declared that "unemployed leisure time is a source of great danger to the moral welfare of youth. There is much truth in the saying, 'A busy boy is usually a happy and good boy.'" Governor Martin G. Brumbaugh of Pennsylvania echoed this sentiment: "Idleness and loafing are alike the enemies of good government. These are the seed beds of much lawlessness, unrest, disorder, and class hatred—the great source of unstable government. Industry and wholesome, constructive recreation are the effective antidotes for many national ills."

But far more than simply offering Americans healthy distraction, outdoor recreation was viewed as the very training ground for good citizenship. Echoing Welch's sentiment—"we will not have a virile citizenship unless a virile people"—Fulton Oursler proclaimed, "It is an idle truism to repeat that without a strong body no man can realize to the full his God-given mental powers, nor can his spirit soar and in soaring carry with it the uplifted soul of a nation. . . . We have come to realize that true citizenship lies in the full development of the individual, and that it is a part of the duty of the state to provide readily accessible opportunity for that development, so that the trinity of man—his body, his brain, and his soul—may grow together in a great, divine, and useful strength."

Americans received such encouragement to explore the outdoors at a time when they had unprecedented opportunities to do so. According to historian Sutter, after Henry Ford's first Model T rolled off the assembly line in the midteens, car ownership grew exponentially. In 1913, more than one million cars were registered in the United States. By 1922 that number had increased tenfold and then more than doubled to twenty-three million by 1929.

In the early 1920s, Sutter writes, at least half the nation's automobiles—some five million cars—were used by Americans to indulge in what had become a popular pastime: "tramping and camping." At the National Conference on Outdoor Recreation, F. V. Colville of the American Automobile Association estimated that in 1924 American motorists logged some six billion miles in recreational excursions. "The whole world of outdoor recreation is at the command of the motorist," he exclaimed. "One end at the road is at his doorstep. At the other end is the place of his desires."

Advocates of the new motor-borne tourism urged government—at all levels—to build an infrastructure for the traveling public. According to Sutter, the state parks movement, for example, first gained traction in the early 1920s largely due to the efforts of the National Conference on State Parks. The organization's motto, "A State Park Every Hundred Miles," summed up its goal of establishing a system of state parks that would serve as stepping-stones to the more far-flung national parks of the western United States.

To link this great open-space system, outdoor recreation advocates exhorted the federal government not only to invest in safe and serviceable highway infrastructure but also to exploit the recreational potential of the open road. They stressed highway-beautification strategies such as preserving forested rights-of-way, removing billboard advertising, posting the names of waterways on bridges, and installing roadside monuments marking historic events. One conference attendee, identified only as Dr. Johnson, addressed the subject of national highways and national parks, observing that "a road is more than a track, just as a home is more than a house, and instead of our road-building activities concerning themselves almost exclusively with provision for transport, more thought [ought] be given to the right of way and the landscape as seen and enjoyed by the motorist." In part, driving had become an end in itself. "Think of the pleasure derived from motoring," he rhapsodized, "the sense of mastery of physical conditions, the chat and laughter of the company, the landscape as it unfolds on either side, ever varying and presenting a series of objects to delight the eye, instruct the mind, enlarge the vision, and above all to lift one out of narrowness and provincialism into the largeness and fullness of all that gives meaning to the word America."

Going Up to the Cabin: Recreational Development in the Nearshore

When David Naymark and a companion set out in a horse-drawn buggy from Duluth to Grand Marais in summer 1911, touring the shore by land had little of the kind of joie de vivre extolled by Dr. Johnson. In a reminiscence of the journey, Naymark recalls that he and his friend, afflicted with what he called the "adventure bug," decided to travel a wagon road that had been built from Duluth to the Pigeon River in the 1880s. The men found the road largely deserted since most summer traffic among the North Shore settlements took place by boat, whether it was hauling barrels of salted fish, transporting residents in medical emergencies, or delivering mail. They quickly discovered the reason why.

The pair made good time at the outset, traveling the twenty-eight-mile corduroy road from Duluth to Two Harbors in a single day. Not far beyond Two Harbors, however, the dirt road all but disappeared. Naymark and his companion could barely make out its faint outline in the overgrown right-of-way. Averaging a mere five miles per day, the men spent the greater part of their waking hours cutting and clearing windfalls out of the path or placing tree branches in front of the wagon's wheels so that they would not sink into patches of soft soil. They picked their way down steep river gorges and waded into the water to test its depths. They guided their horse and cart through mudholes while mosquitoes swarmed their bodies "like bees," Naymark wrote. They spent their nights on soiled mattresses in logging camps, hurrying into the sauna the following morning to rid themselves of bedbugs and lice. By the time they reached Grand Marais, the adventuring duo took one look at their skinny, bedraggled horse and boarded the steamer *America* for the return trip to Duluth.

A mere fourteen years later, what had been little more than a brushed-over pathway through the woods became a pleasure route for roadsters. In 1925 an improved Highway 61 (known then as the Babcock Scenic Highway) was completed to the Canadian border, offering motorists sweeping vistas of Lake Superior and roaring river waterfalls as well as more intimate twists and turns through forested glens. By 1926 the Minnesota Arrowhead Association had issued tourist brochures that depicted a Northwoods bustling with automobile-based activities. In one advertisement, tour buses and Model Ts whizz effortlessly along a sleek, modern roadway, with Split Rock Lighthouse and an offshore tour

HIGHWAY No.1, NEAR LUTSEN, NORTH SHORE, LAKE SUPERIOR.   107553

boat beckoning in the background. In another, riders with cameras in hand disembark a cherry-red taxi to photograph blue, spruce-framed mountains rising on the horizon as a passenger train streaks past. Never mind that the scene looked more like Glacier National Park than the North Shore—the message was clear: accessible to modern transport, the North Shore's natural spectacles could be taken in with speed, comfort, and safety.

And accommodations sprang up to meet the growing demand. Norwegian immigrant Emil Edison is credited with starting the shore's first cabin-style resort. Initially, Edison rented out his three tiny lakeside cottages to executives overseeing the construction of area railroads. Capitalizing on the burgeoning tourist trade, Edison opened for business in 1910 as the Star Harbor Resort. In time, he expanded his operation to include twenty-four cabins. (When the originals were removed in 1990 to make way for more luxurious structures, one of them was rescued and now resides in the collection of the Minnesota Historical Society.)

Other North Shore families followed suit, and a vacation destination, known as the Scandinavian Riviera, was born. As Minnesota historian Mark Haidet points out, the growth of recreation along the shore was part of a larger statewide trend. The period from 1917 to 1930, he writes, represents "the real flowering of the [Minnesota tourism] industry." In 1930, estimates put the number of resorts in Minnesota at thirteen hundred.

Development along the shore was brisk. According to a brochure issued by the owners of Bob's Cabins located in Larsmont just east of Duluth, during the 1920s more than 150 rental cabins were built in the Larsmont area alone. Accommodations for middle-class families were simple. The original structures, the brochure points out, "had roll roofing, varnished siding, bare stud interior walls, bare wood floors, and outdoor privies. Each cabin was furnished with one or two iron bedframes . . . a wood stove, and a washstand with a pitcher. Later, with electricity, came the shower house, with water pumped from the lake. Each cabin then got one light bulb hanging from the ceiling. The rent was $4 to $6 a night—$1 extra if you wanted sheets or a higher wattage bulb!"

The original prospectus for the Naniboujou Club, which was issued in 1928, featured such "high-end" amenities as formal lawns, tennis courts, a golf course, bathhouses, and piers for steamships. The market collapse in October 1929 forced developers to scale back the project. Courtesy of Naniboujou Lodge.

The eye-popping iconography of the dining room at Naniboujou Lodge is drawn from Cree Indian mythology. Courtesy of Naniboujou Lodge.

For more well-appointed lodgings, travelers could join the Naniboujou Club, which was officially dedicated near Grand Portage in July 1929. The prospectus for the club, which counted the likes of Ring Lardner, Babe Ruth, and Jack Dempsey among its members, featured renderings of a palatial lodge surrounded by formal lawns, tennis courts, a golf course, bathhouses, and piers for steamships that would ferry vacationers across the Great Lakes. Following the crash of 1929, which occurred just three months later, the developers were forced to scale back the project. Nonetheless, the lodge retains a kind of fantastical flair. It is best known for a communal great hall with eye-popping iconography, drawn from Cree Indian mythology, that is rendered in kaleidoscopic paints and patterns.

Many well-to-do vacationers simply opted to build getaways of their own. In 1921, a group of influential Twin Citians purchased some fifteen hundred acres for an Adirondack-style camp along the shores of Lake Superior near Two Harbors. The private compound, known as the Encampment Forest Association, still exists today in the shade of one of only a few significant remnants of white pine–dominated forest along the North Shore.

But the Minnesota tradition of "going up to the lake" was fully born when middle-class vacationers began to construct their own accommodations—in some cases, with the help of state and federal governments. In 1915, the U.S. Congress authorized the secretary of agriculture to lease suitable lands within the national forests for "summer homes, hotels, stores or other structures needed for recreation or public convenience." For an annual fee of $10 to $25 and a lease period of thirty years, homeowners could build summer cabins on up to five acres of public land. Benton MacKaye of the USFS observed, "Uncle Sam has gone into the real estate business, but with public service as the object and not profits."

"To every citizen of the country is given the opportunity to own a summer home built on forest land," proclaimed USFS landscape architect Arthur H. Carhart in a 1920 article in *Good Housekeeping* magazine. "If only these heat-cursed city dwellers had knowledge of the forests," Carhart wrote, "many of them could escape to their own vacation lands and be free from the heat demons. They could walk or rest in cool shade, or

The designs in the 1928 plan book *The Real Log Cabin* by Minneapolis architect Chilson Darragh Aldrich reflected the growing popularity in the 1920s of rustic, back-to-nature getaways.

with fly and tackle stalk gamy trout. Inviting trails would lead them to tree-bordered pools, or a climb would take them to a point where a kingdom of crags would be spread at their feet. But best of all, they could live in their own little cabins amid stately trees."

The appeal of this back-to-nature sentiment is evident in the enduring popularity of *The Real Log Cabin,* a kind of plan book of log designs. Written by Minneapolis architect Chilson Darragh Aldrich, the book was published by Macmillan in 1928 and kept in print for two decades. During his career, Aldrich, who built a series of log buildings for himself and his wife on the shores of Lake Superior in Hovland, designed log structures in thirty states and in Canada. He summed up the lure of rustic architecture: "It is being increasingly borne in upon the members of the present generation that they have lost much by not being pioneers. Much in the stuff that goes to make character. Much of the downright exhilarating fun. With increasing frequency, therefore, one finds that the genuine American is going back to the good old ancestral custom—bootlegging his pioneer kick as it were—and building himself a log cabin away from the honks of man."

But building a cabin was not the only way to escape the "honks of man." For the price of a canvas tent, nature lovers of more moderate means too could take up residence under the trees, thanks to an army of workers known as the Civilian Conservation Corps. The young men enrolled in this Depression-era public works program developed many tourist facilities in the public parks and forests along the North Shore, including campgrounds, outhouses, picnic shelters, administration buildings, and hiking trails. Among their work assignments, writes Minnesota historian Nute, were cleanup operations in the forests along Highway 61. They removed the charred remains of the North Shore's great fires, she writes, "for the benefit of tourists."

If They Come, You Must Build It: The Hidden Costs of Recreational Infrastructure

Already in 1931 the Minnesota conservationist Ernest Oberholtzer was expressing fears that without proper planning, the tourism industry could prove just as damaging to north-country ecosystems as the waves of other economic activity that preceded it. "We are at a transition stage in our growth," he wrote. "The older exploitation of the public domain has failed. We have had mining booms, logging booms, drainage booms. Only wreckage and blasted hopes remain. Mushroom communities have been built up and public facilities provided at high cost, only to become a public liability when the boom is over. We are likely soon in the same way to have a tourist boom, which for lack of wise direction may be as evanescent and do as much damage as any of the other booms. Only if we direct the tourist growth wisely, supplying the higher satisfactions it wants, will it prove the boundless, permanent force which our economic life invites and requires. Otherwise, we are due for ill-advised roads, cheapened landscapes, and the flight of our tourist trade to other states and provinces less favored by nature but more advanced in public planning. Everything has been exploited helter-skelter for maximum immediate benefit. We need now to work for permanency."

Oberholtzer's words have proven to be surprisingly prescient, say the critics of the current building boom along the Minnesota coast. The North Shore, they contend, stands at a crossroads. To date, the shore has ranked near the top on a number of ecological indicators. According to the Environmental Protection Agency's 1997 report *Land by the Lakes,* compared to the other Great Lakes, the coastal region of Lake Superior is the only one to receive consistently high marks on a range of ecological indicators, ranking "good" when it comes to "retention of shoreline species/communities; retention of natural shoreline processes (unarmored shoreline); representation of biodiversity in lakeshore parks and protected areas." Superior gets mixed, but improving, ratings on "gains in biodiversity investment areas." Overall, according to the report, "Lake Superior is rated highly for three out of the four indicators since its lightly developed shoreline and extensive parks system have kept its nearshore terrestrial ecosystems in generally excellent condition."

Whether the Minnesota North Shore—the most developed part of the shoreline—will maintain such high standards depends on the willingness of residents to work for the permanency that Oberholtzer so passionately urged. Land-use critics worry that economic forces and demographic trends may threaten to overwhelm even the best-laid growth plans. That is because attracting tourists has become big business. According to the state's Explore Minnesota tourism office, in 2009 the leisure and hospitality industry generated $48.5 million in gross sales in Cook County alone. In addition, tourism-related jobs comprised 51 percent of the total private-sector jobs, making the county the most tourist-dependent of any in the state.

But the industry's phenomenal success is taking a toll on north-country communities and ecosystems. One of the biggest causes of environmental stress is the sheer number of people who now visit the shore. Exact numbers are hard to come by, but many agencies look to the number of park visitors as an indication of overall tourist visitation trends. According to the Minnesota Department of Natural Resources Division of Parks and Trails, in 2009 some 1.3 million people visited the Tettegouche, Gooseberry Falls, and Split Rock Lighthouse State Parks. During some seasons, park facilities can be jammed beyond capacity. In autumn 2004, for example, Gooseberry Falls State Park tallied an astounding 10,000 to 11,000 leaf peepers each weekend during the peak fall-color period.

Changed too are the patterns of tourist travel. In the early years of tourism along the shore, visitation was largely confined to the summer months. Most resorts opened on Memorial Day and closed on Labor Day. The winter season was short as well. Depending on the amount of snowfall, the ski area at Lutsen opened around Christmas and closed in early March.

Today, the shore is thronged almost year-round, attracting nearly as many visitors in February and March as the peak periods in July and August. One of the biggest catalysts for this change is the ski area at Lutsen. The operator, Lutsen Mountains, generally opens at least a portion of its runs in mid-November and keeps slopes skiable until mid-April. According to a 2011 article in the *Cook County News-Herald,* the extended ski season is made possible by artificial snowmaking, which draws as much as 108 million gallons of water over the season from the Poplar River. Lutsen Mountains was permitted to tap the river until 2014, when the completion of a pipeline would allow the ski-hill operators to draw water for snowmaking from Lake Superior. The lake's waters will also be used to irrigate the nearby public golf course and to provide vacation homes and area businesses with a reliable alternative to the scarce groundwater they had used as a water source. The State of Minnesota will pick up the lion's share of the $5 million price tag for the pipeline, and the remainder will be matched by local and regional funding.

A driver of tourism's economic engine in Cook County, the ski area has had synergistic effects in many North Shore communities. A multimillion dollar expansion of the ski hill in the 1990s, for example, prompted an explosion in the construction of motel and condo units to serve the brisk winter traffic. According to the report "North Shore Land Issues: The Real Costs of Growth," published by the Minnesota Pollution Control Agency (MPCA) in 2000, the shore saw the construction of fifty new lodging establishments between 1990 and 1995. The stepped-up pace of development persisted into the latter part of the decade. If the amount of snowfall and duration of snowpack continue to diminish in other parts of the state, as predicted under many climate change scenarios, greater numbers of other outdoor winter enthusiasts, such as snowmobilers and cross-country skiers, will travel to the shore in search of reliable powder.

But the pressure on North Shore ecosystems is not due simply to the boom in numbers of visitors and the facilities built to accommodate them. Today's developments have far larger ecological footprints than those in the past, primarily due to changing

In the early years, the North Shore tourism season traditionally opened on Memorial Day and closed on Labor Day. Expanded developments, such as those of the Lutsen Mountains ski area, are part of a trend toward year-round visitation. Photograph by Virginia Danfelt.

The Star Harbor Resort, shown here circa 1920, is credited with being the North Shore's first cabin-style resort. Courtesy of the Minnesota Historical Society.

The Star Harbor's replacement, the Grand Superior Lodge, reflects a trend toward greater density and luxury amenities in modern recreational developments. Copyright 2009 Minnesota Department of Natural Resources.

visitor expectations. Far from settling for such "luxuries" as fresh sheets and a higher wattage bulb, many visitors now expect—and routinely receive—such amenities as outdoor and in-room hot tubs, dishwashers, and high-speed Internet access. (See also "Where Has All the Sewage Gone? Development and Water Quality.") No longer content with just fishing, swimming, and boating, the top recreational pursuits of vacationers in the past, today's visitors patronize resorts that offer heated swimming pools, full-service restaurants, golf and sailing lessons, and children's programming.

Resorts with far more modest ambitions have been unable or unwilling to keep up with these demands. As a result, many of the no-frills resorts dating to the state's historic heyday of tourism have largely been swallowed up by lakeside condominium resorts, a trend that Minnesota architect Robert Roscoe bemoaned in a 1998 issue of *Architecture Minnesota*. During the 1970s, 1980s, and 1990s, Minnesota lost 40 percent of its small mom-and-pop resorts. These getaways, which once were "far more numerous than the state's 10,000 lakes," Roscoe observed, have been "replaced by large-scale resorts that are mini-suburbs of densely packed time-share condos and large, expansive recreation buildings that line up along strips of asphalt parking lots. The mom-and-pop places, we are told, lack the 'amenities' the new resort complexes offer. . . . Northern Minnesota itself used to be amenity enough."

Indeed, so rare have these accommodations become statewide that the early twentieth-century mom-and-pop resorts of Cass County were included in the Preservation Alliance of Minnesota's 1998 list of the state's ten most endangered historic buildings. Their reason for such an unorthodox nomination, according to the alliance, was this: "the rustic resorts that represent the Minnesota tradition of going up north to the lake are fast disappearing."

Star Harbor Resort, now Grand Superior Lodge, reflects the changing times. Once little more than a smattering of twenty-four tiny cabins that housed a maximum of 128 visitors, the complex of hotel rooms, condos, and lake homes today can accommodate 250 people on a mere 10.9-acre site.

At the same time, many visitors, primarily baby boomers, are sinking deeper roots into the shore by building vacation and retirement homes of their own. According to the MPCA report, from 1990 to 2000, fifteen out of seventeen North Shore municipalities saw population increases, some of them as much as 35 percent. To maximize the dollars invested in their properties, owners are spending more time in these homes year-round. To offset costs, many others offer their property as vacation rentals during unoccupied times of the year. And like resort accommodations, modern residences often feature all the comforts of home—and then some. As the MPCA report points out, modern "cabins" resemble "showcase homes in urban subdivisions."

The greater the square footage, the bigger the downsides for surrounding ecosystems. Clearing a forested lakeshore for a year-round, suburban-style house with a typical expanse of lawn to the water's edge greatly increases the amount of sedimentation and phosphorus that reaches the lake, both of which can stimulate the growth of noxious algae.

Removing native ground cover and exposing underlying soils to the elements is not the only cause of these spikes in pollution. Impervious surfaces such as rooftops, driveways, patios, and lawns send huge amounts of storm runoff pouring across the land, increasing its velocity and decreasing the areas in which it can slowly soak into the

Coastal development in North Shore towns like Tofte (*above*) and Two Harbors (*below*) can have adverse effects on nearby waters. Impervious surfaces, such as parking lots and rooftops, increase the temperature and velocity of stormwater while decreasing the areas in which it can slowly soak into the ground. Without filtration, water laden with sediments, chemical pollutants, and nutrients flows directly into Lake Superior. Copyright 2009 Minnesota Department of Natural Resources.

ground. Even light traffic from lawn mowers, golf carts, and footfalls can compact soils, decreasing their ability to absorb rainfall. The ground sheds water, like the proverbial duck's back, carrying sediments, chemical pollutants, and nutrients into surrounding sensitive lakes and rivers.

Just as worrisome is the thermal pollution in storm flows. Roof tiles, asphalt, and cement typically store more solar radiation than the ground, heating up the water that courses across their surface. Runoff entering streams from urbanized areas has ranged as much as 10.8 degrees Fahrenheit warmer than storm flows from undeveloped watersheds.

Playing on the Edge: The Impact of New Recreational Trends in the Nearshore

An economic engine in its own right is the growing interest in physical fitness and outdoor exploration, which has brought a whole new set of recreationists to the shore edge. With only 4 percent of the Minnesota coast in public hands, mostly in state parks, these new users are exerting enormous pressure on the North Shore's public shorelines. For example, interest in touring the North Shore by boat—be it by kayak or cabin cruiser—has grown and with it the demand for the construction of waterfront facilities. In 1993 the Minnesota legislature designated the Lake Superior Water Trail, a 150-mile route from the St. Louis Bay to Pigeon Point. Designed with kayakers in mind, the route seeks to establish land access from the lake at three- to five-mile intervals. Access points may range from something as simple as a launch site to developed primitive campgrounds with latrines.

Far more ambitious was a proposed $40 million publicly financed plan ($77.4 million in 2014 dollars) to develop a series of safe harbors and full-service marinas along the North Shore. In 1989, the Legislative Commission on Minnesota Resources (LCMR) funneled $100,000 through the Minnesota DNR to the North Shore Management Board to study the feasibility of developing harbors along the North Shore. At the time, only three sites along the shore offered services and shelter to boaters. Public outcry over the proposal tabled plans for harbor development along undeveloped stretches of coastline such as Horseshoe Bay in Hovland and Sugarloaf Cove in Schroeder. "Many people viewed the plan as a referendum on the North Shore's future," wrote Greg Breining in a 1993 article for the *Minnesota Volunteer*. "Should government spend money to draw even more tourists to the shore? Or should the public draw a line against further development to protect the shoreline's character?"

Critics worried that their special scenic or natural features would be destroyed by the development that accompanies harbor construction, infrastructure that can range from breakwaters and docks in the lake to launch ramps, parking lots, restrooms, gas pumps, laundry facilities, and fish-cleaning stations. Instead of carving up new shoreline, funds have been spent to upgrade facilities to varying degrees at existing sites.

Recreational infrastructure has been built for landlubbers as well. The Gitchi-Gami State Trail accommodates bicyclists and pedestrians with a paved, ten-foot-wide trail, which, once completed, will stretch eighty-six miles from Two Harbors to Grand Marais.

And for extreme-sports enthusiasts, the shore's sheer cliff faces are proving to be a popular draw—and not just during the summer months. Using crampons and ice picks, outdoor enthusiasts inch their way up frozen waterfalls and ice caves in the zero hours

Interest in physical fitness and outdoor exploration are driving new recreational developments in the nearshore landscape, such as the Gitchi-Gami State Trail, which accommodates bicyclists and pedestrians with a paved, ten-foot-wide trail. Photograph by Virginia Danfelt.

The use of the cliff top as a tourist lookout and a staging area for rock climbers led to a dieback of vegetation along Shovel Point *(above)*. An innovative restoration plan, which called for the replanting of native vegetation and the installation of signage, boardwalks, platforms, ropes, and permanent anchors, has helped to meet the needs of recreationists *(below)* while improving the ecological integrity of the site *(opposite)*. *(above)* Copyright Minnesota Department of Natural Resources–Philip Leversedge; *(below and opposite)* photographs by Chel Anderson.

of winter. According to Canadian ice-climbing expert Shaun Parent, Lake Superior is ranked as North America's third-best ice-climbing destination.

The growing sports-equipment industry, says Dave Olfelt, a resource specialist for Tettegouche State Park, is a driving force in much recreational development. Park officials routinely field phone calls from even low-tech sports enthusiasts, such as snowshoers, who want to see trails developed for their sport. "People can buy all kinds of toys at REI, and they want to use them," he points out. "Everybody wants a piece of the parks. And everybody's got a legitimate claim. But nobody wants to believe that they will be the culprit in the demise of special habitats."

And such activities are not without their impacts. For example, the Cliff Ecology Research Group has studied cliff faces on the Niagara Escarpment near Toronto and found that tree density along popular warm-weather climbing routes was cut by half when compared to undisturbed sites. Existing trees in these areas were also ten times more likely to be injured than those in unclimbed areas. Cliff fauna appear to be negatively impacted as well. A study published in the 2003 issue of *Conservation Biology* showed that the number of land snails along popular climbing routes on the escarpment plummeted by 80 percent as compared to undisturbed sections of cliffs. Researchers speculated that the compaction and dislodging of soil by climbers may be largely to blame for the declines.

Several studies of a popular climb along the cliffs of Shovel Point in Tettegouche State Park have revealed cause for similar concerns. These great walls of rock are home to several plants that have been listed as endangered, threatened, or of special concern. In addition, a number of regionally rare and endangered animals make their homes or seek seasonal shelter in these towering outcrops, including peregrine falcons, which were reintroduced in the 1980s following their near extirpation by the widespread use of DDT.

In 1995 Hamline University biologist Michael A. Farris published the results of research showing that vegetation along the cliff face of Shovel Point was sparser than vegetation in unclimbed areas. Although he could not conclusively pinpoint the cause for the decline of some plant species, Farris cautioned that the technical climbing could put some populations, such as the cliff's fragile umbilicate lichens, at risk. The edges of these foliose, or "leafy" lichens, are fragile. And because they are attached to the rock at a single point, they can be easily dislodged.

A matter of equal, if not greater concern, however, was the growing dieback of vegetation along the cliff top that climbers used as a staging area. "The top of Shovel Point . . . [has] suffered the most damage," Farris observed, mainly severe erosion caused by the loss of plants and compaction of soils. If left unchecked, he warned, "further erosion will cause the loss of more trees in these areas . . . [and] could change the amount of runoff reaching the cliff face."

Searching for ways to head off damage to this sensitive landscape, in 1996 park manager Phil Leversedge consulted with colleagues from the Minnesota DNR and University of Minnesota Duluth. The trio developed an innovative plan for managing recreational activities on the site as well as restoring some of its native plants. Park personnel collected seeds from nearby grasses and wildflowers and germinated them in greenhouses. Seedlings then were planted in damaged or denuded areas. Signage, boardwalks, and ropes were erected to funnel visitors away from revegetated plots.

Managers also confined the footprint of climbers by constructing a staging platform and installing signage that directs climbers to tie ropes on fixed anchors in the rock instead of around the trunks of cliff-edge trees. Citing research showing that "educating the visitor on the value of natural resource rehabilitation and management leads toward acceptance and compliance of managed recreational use," the researchers also established a citizens' advisory committee, which gathered input from rock climbers and others at the outset of the project. To date, climber compliance is high, and replanted vegetation is flourishing.

For park managers, who are charged with drafting management and restoration policy for public lands, predicting the popularity of recreational pursuits has proven to be a special challenge. When construction began on the Superior Hiking Trail in 1987, for example, Olfelt and his colleagues were taking bets on how quickly the trail would be brushed over through lack of use. But by 2000 the footpath, which now runs 205 miles along the ridge overlooking Superior, had become so popular that *Backpacker* magazine named it one of the ten best hiking trails in the United States.

Motorized traffic has skyrocketed as well. According to DNR estimates, between 1986 and 1992–93, trail use by snowmobilers jumped 900 percent, reflecting a 28 percent increase in the number of snowmobile registrations during roughly the same time.

The mandate to manage public lands for multiple purposes has resulted in contentious debates among recreational users with widely different expectations about outdoor experiences. Examining attitudes toward winter trail use by snowmobilers and cross-country skiers, a survey of Cook County citizens published in 2003 by Minnesota Sea Grant found that "residents feel that although snowmobilers may spend more money than x-c skiers, they create costs, particularly environmental costs, as well as significant unwanted social impacts. Residents are less willing to see growth in snowmobiler use of local trails and a significant minority would wish to see fewer snowmobilers."

The debate over motorized trail use sharpened in the 1990s with demands by ATV enthusiasts to convert the 146-mile-long North Shore State Trail to off-road use. Currently, motorized use of the trail, which runs from Duluth to Grand Marais, is restricted to winter use when the trail is snow covered. Citing trail development for cyclists, ATV advocates contend that they should be afforded the same opportunities for trail development. Critics charge that traffic from ATVs, as well as a whole array of other motorized sports vehicles collectively known as off-road vehicles (ORVs), have created tens of thousands of deep ruts in public lands across the nation. So widespread has this damage become that in some places these deep grooves now alter the flow of water across the land, resulting in degraded water quality, erosion, and damaged habitats for plants and animals.

These and other outdoor-products markets are growing so quickly that park planners are hard-pressed to anticipate the effects of new recreational activities in land-management plans. The 1997 plan for Tettegouche, for example, prohibits the use of personal watercraft on the park's interior lakes. There may come a day, Olfelt says, when park planners could add the Baptism River to the list of waters that are off-limits to jet skis. "We're not looking too far down the road," he says. "They're a credible threat. Who knows what else is out there?"

But even more passive forms of recreation, such as quiet communing by the lake, do damage to soils and vegetation. Foot traffic on popular Artist Point, for example, has created a spiderweb of informal trails in the small forest that abuts the rocky shore, compacting the soil and setting up conditions for serious erosion. Visitors who venture out onto the rocks for the stunning view of the Sawtooth Mountains have worn away the fragile carpet of rock-loving lichens. Some lichens are such slow growers—expanding at a fraction of an inch each year—that even if left undisturbed, centuries may pass before they attain their former abundance. Gone too from many rock crevices is anchoring vegetation for the rare and delicate plants that fired the imaginations of botanical explorers at the turn of the twentieth century, including butterwort and small false asphodel. According to the report *Land by the Lakes,* "Arctic-alpine disjunct communities on the rocky shores and cliffs along the north shore of Lake Superior are generally protected from disturbance because they are inaccessible and the climate is harsh, though the area is attractive to tourists. Recently, however, second-home development

The increase in North Shore development has accelerated the demand for infrastructure to service new technologies. Some developments, such as these communications towers in Grand Marais, may conflict with the ecological needs of nonhuman residents and the aesthetic desires of people. Photograph by Chel Anderson.

has begun to encroach. Recreational use (marina development) and trampling of vegetation also has the potential for significant vegetative impact."

Servicing the increasing numbers of tourists and residents has placed other, less obvious burdens on the environment. Modern-day shore lovers may look for a place to get away from the "honks of man," as Aldrich so vividly put it. But they do not necessarily want to leave behind the comforts and conveniences of home. With the increase in North Shore development has come the demand to construct greater capacity for new and rapidly changing technologies, such as computers and cell phones. Already, local battles have been waged over plans to erect radio and cell phone towers. Residents complain that the metal structures mar pristine ridgetop vistas. They also point out that communications structures can be lethal to migrating birds. Conservative estimates from the U.S. Fish and Wildlife Service say that between four and five million birds in the United States are killed each year in collisions with these towers. Not only do they add to the landscape's hazards and visual clutter, but their lights also contribute to the growing problem of stellar pollution, which in some heavily developed areas of the shore has begun to dim the starry brilliance of the night sky.

Greater numbers of tourists and new residents also present budgetary, as well as environmental, challenges to local communities. Take, for example, services as basic as waste disposal. According to the MPCA, each year the average Minnesotan produces one ton of solid waste. The growing mounds of garbage can overwhelm the budgets of many small communities. Unable to locate an affordable and acceptable location for a new landfill, Cook County closed its overflowing dumpsite in 2000 and began paying haulers to truck waste across the state to distant disposal sites. The sustainability of this short-term solution remains questionable as fuel costs rise and the siting of landfills in such distant locations becomes more restrictive.

One of the most obvious—and costliest—accommodations to the rise in tourism and second-home development along the North Shore is the renovation of Highway 61. The narrow, two-lane highway with its stomach-lurching dips and curves once was adequate for servicing traffic that was far sparser, slower-moving, and more seasonal than today. Use of the highway has changed dramatically, however. In 2000, the Federal Highway Administration designated Highway 61 as one of only fifteen scenic highways known as All-American Roads. Today, as in the past, motorized travel along the shore is advertised as a destination in itself. But both the speed and volume of traffic have increased substantially from the time when carefree vacationers toured the shore in open-air Model Ts. Resident commuters sprinting to and from work and eighteen-wheelers barreling to warehouses in Duluth share the roadway with scenery-loving vacationers in lumbering RVs. The result: an exponential rise in white-knuckle encounters.

Faced with demands to increase the roadway's safety and efficiency, in the 1990s the Minnesota Department of Transportation (MNDOT) began work on a decades-long plan for upgrading Highway 61 to meet modern traffic standards. In general, plans called for rebuilding the entire roadway to the Canadian border with twelve-foot driving lanes, generous paved shoulders, and passing lanes at regular intervals. By 2010, MNDOT had reconstructed portions of the highway and completed such major projects as bypassing dangerous curves with two traffic tunnels.

But because of the highway's rugged character and remote location, rehabilitation costs have been staggering. Building a quarter-mile tunnel through Silver Creek Cliff alone, one of two such tunnel projects, carried a price tag of $23 million. In the process, workers removed some 497,000 cubic yards—about 4,200 semitrailer loads—of rock and soil, some of which were used to build breakwaters and offshore spawning reefs for lake trout.

Renovating the roadbed also has tapped public coffers. A 4-mile leg of highway built to ease congestion between the Onion River and County Road 34 cost $12.2 million. The reconstruction of 3.4 miles between the Split Rock River and Chapin's Curve was completed in 2011 to the tune of $14.2 million. Due to funding shortages, some improvements have been temporarily tabled, and highway funds allocated to maintaining the existing road.

Not only has roadway funding strained state budgets, but it has also threatened to bankrupt county coffers. The problem of financing the spiraling costs of county-road maintenance has become particularly acute. The Cook County map is spiderwebbed with roads that once were little used and poorly maintained. But now, as growing vacation-home development has pushed into the rural areas of the county, demands for public upkeep of roads that provide access to these remote properties continue.

The construction of roads has taken a toll on north-country ecosystems as well. In the early years of highway construction along the shore, road builders routinely mined sand, gravel, and cobble from nearby beaches and the mouths of rivers and streams. Geologist John Green points out that the gravel bar at the mouth of the Gooseberry River was heavily plundered for building materials used in Duluth roadways. Decades later, natural forces still have not rebuilt the deposit to its former volume.

Today, the materials for roadbeds, as well as building and septic system construction, are excavated from deposits that lie farther inland. These operations are located on remnants of the region's ancient glacial past, such as eskers and the ancient beaches of Lake Superior's ancestral lakes. As the pace of development speeds up, so too the destruction of these relict Ice Age features, not to mention the plant and animal communities that have colonized their reaches. Geologist Green points to a pit near the Brule River that he says is "gradually consuming" what once was an ancient beach or river-delta deposit.

The amount of aggregate—a term that describes mixtures including sand, gravel, and crushed rock—used in road building is staggering. According to the 1998 report *Minnesota's Aggregate Resources: Road to the 21st Century,* it takes some 20,000 tons of aggregate to build one mile of four-lane highway, and 120 tons to build a new home. Minnesotans consume 10.5 tons per person each year to "maintain and construct roads, develop infrastructure, support building and construction projects and for use in industrial applications," say the authors of the report.

Since transportation accounts for most of the costs of aggregate, siting mines close to locations in which the materials will be used is crucial. According to the report, adding twenty miles to the hauling distance doubles the cost. Transporting materials thirty-four miles is considered the maximal economically feasible distance. As fuel prices rise, these distances may become even more restrictive.

But the demand for materials is expected to increase just as supplies become less readily available. Already, the report points out, Cook and Lake Counties are experiencing a scarcity of aggregate materials as those in easily accessible pits become tapped out.

And the opening of new mines has become increasingly contentious—for many reasons. Gravel operations do not make good neighbors. They pollute the air with diesel fumes and dust, which coats the surface of area plants. The application of calcium chloride, a chemical that helps reduce the dispersion of fine sediment particles from the access roads, has been implicated in the deaths of amphibians such as salamanders that come into contact with it while crossing treated roadways. Trucks and rock-crushing machines create noise that can travel up to one mile from the site of operation. Slow-moving vehicles pulling out into busy thoroughfares pose traffic hazards. Disturbed gravel pits have become hot spots for invasive species even though recent ordinances and policies require that abandoned pits on public land be revegetated. And such sites can result in long-term water-quality problems when rock that is fractured in the mining process enables contaminated surface-water flows to mix with groundwater.

But North Shore road building and maintenance have far more subtle ecological effects than just the polluted air and hollowed-out hillsides of gravel operations. Roadways are primary migration routes for exotic species, a phenomenon that was noted along the North Shore as early as 1891. On a plant-collecting trip to northern Wisconsin and the Minnesota North Shore, botanist L. S. Cheney recorded a number of introduced plants, most notably tall buttercup, that occurred along even some of the most rudimentary footpaths on the North Shore and its inland reaches. "It was observed at all old fishing stations along Lake Superior," he wrote. "At Grand Portage it literally covered the land occupied by the village, and also the adjoining fields—perhaps fifteen acres in all. From this village the plant has been carried westward along the canoe route to the portage between North Lake and South Lake, a distance of sixty miles. It was observed at most of the portages between the two points, usually at the landings. This distribution was evidently effected either by travellers carrying the flowers or seed, intentionally or otherwise, or by the seed adhering to the hair of dogs or other animals."

Over time, the construction of new roads, as well as the exponential growth in the traffic they carry, has only multiplied opportunities for the spread of noxious, nonnative plants. In their 1953 study of Cook County plants, botanists Fred K. Butters and Ernst C. Abbe observed, "The weed flora is extensive especially along roadsides. . . . The more certain snow cover in the winter, along with the cool summers, may well explain the success of many north European weeds in Cook County."

Disturbed soils around roads provide prime conditions for the establishment and growth of nonnative invasive species that would not have been able to gain a foothold in intact ecosystems. Included in the list of problem plants are bird's-foot trefoil, white sweet clover, yellow sweet clover, smooth brome, common St. John's-wort, spotted knapweed, purple loosestrife, Queen Anne's lace, and common reed or phragmites, a

Mined aggregates, such as sand, gravel, and crushed rock, are essential materials in the construction of roadbeds, buildings, and septic systems. The affordability of future supplies is a concern in Cook and Lake Counties as existing pits are tapped out and the siting of new ones becomes increasingly contentious. Photograph by Chel Anderson.

noxious grass whose dense clumps inhibit the growth and germination of fellow plants and whose water-guzzling habits may lower water tables in the wetlands that it colonizes. Phragmites can thrive in wet ditches along roadsides. These migration corridors enable this exotic species to spread far and wide. While phragmites reproduce primarily by cloning, the seeds of many other exotic plants are borne aloft by the wind, transported by right-of-way mowing equipment, or stirred up by passing cars. Unfortunately, some of them, such as showy lupine, have been deliberately planted by well-meaning residents.

Allowing greater access to remote areas via roads creates opportunities for people to inadvertently carry seeds into areas that are otherwise buffered from invasion. Often the spread of nonnative invasives is deliberate, however, much to the dismay of ecologists and conservationists, who cite abundant evidence that exotic species imperil native ecosystems. For example, while they may contain a greater percentage of local native species than in the past, the seed mixes sown on newly bared slopes by state and local road authorities and their contractors routinely include those of quick-growing nonnatives, as well as some inappropriate native species. And the matting of low-grade hay that is used to stem the flow of stormwater runoff in roadside ditches also has come under greater scrutiny. This imported hay, grown in distant locales, contains the seeds of weedy species and has been implicated in the spread of ragweed along the Gitchi-Gami bike trail. Concerns about these and other infestations led officials in Cook County in 2006 to begin discussions about the passage of an ordinance requiring that all hay used in roadway operations be locally grown.

Scientists have cataloged a host of other environmental ill effects caused by roads. In a 2000 issue of *Conservation Biology,* Stephen C. Trombulak and Christopher A. Frissell write that roads have the potential to alter the ecological conditions of huge swaths of adjoining land. The amount of land taken up by Highway 61 alone from Duluth to the Canadian border is conservatively estimated at 2,078 acres. This thoroughfare creates an extensive and permanent open edge that in the forests of northeastern Minnesota rarely, if ever, occurs in nature.

These open, sun-drenched rights-of-way can also reduce humidity levels in adjacent forests. This condition is a factor in the dieback syndrome that has killed scores of birch trees along many sections of Highway 61. For animals, such as amphibians, that rely on cool, moist environments, the warmer and drier conditions can pose thermal barriers, serving as de facto fences that cut the animals off from the seasonal pools in which they reproduce and the woodlands that provide cover.

Animals that do venture the crossing expose themselves to a greater chance of predation, including run-ins with passing cars. Studies have shown a direct correlation between vertebrate mortality and road width and traffic volume. Along the North Shore, such run-ins can be especially dangerous to humans in the case of white-tailed deer that migrate to the shore during the winter to escape deeper snows farther inland and to find easy meals at backyard feeders.

For some animals, roadways may prove to be too formidable to cross. Research has shown that many animals, including black bears and timber wolves, shy away altogether from areas with high road densities. On the other hand, some species, such as snakes in search of basking sites, are lured to the warmth of asphalt. Also attracted to these heat islands are many species of insects, which, in turn, entice feeding birds. Busy arterials can become death traps.

The amount of land taken up by Highway 61 from Duluth to the Canadian border is conservatively estimated at 2,078 acres. The highway and adjacent utility corridors create an extensive and permanent open edge, which rarely, if ever, occurs naturally in the forests of northeastern Minnesota. Photograph by Paul Stafford. Copyright Explore Minnesota Tourism.

But roads have even more subtle, far-reaching, and long-lasting impacts. Scientists studying the effects of roadway noise on breeding birds in highway corridors discovered that up to a point traffic noise had no impact on the density of breeding birds in highway-corridor habitats. But when decibels exceeded certain thresholds—limits that varied from species to species—the number of breeding birds plummeted. In his studies of a fifteen-mile stretch of Route 2 near Cambridge, Massachusetts, Harvard road ecologist Richard T. T. Forman noted that common birds, such as chickadees, herons, and egrets, are unfazed by the din of passing traffic. But sensitive grassland species, including bobolinks and meadowlarks, vacate habitats adjacent to busy thoroughfares.

Even though breeding males in some species turn up the volume of their singing around noisy traffic, researchers hypothesize that the sounds of traffic can be loud enough to drown out their calls, leading to declines in reproduction. Others suggest that birds may abandon highway habitats due to stress. Laboratory experiments with vertebrates have shown that chronic noise causes numerous physical symptoms associated with stress, including higher blood pressure and levels of stress hormones, faster heart rates, reduced body size, and enlarged adrenal glands. In the case of the grassland birds that he has studied, Forman suggests that traffic sounds can drown out calls of alarm. "When there are eggs on the nest and a cat shows up, or a snake, or a hawk," he notes, "the adult male or female makes an alarm click or call, and the adults freeze and so are not seen. Those alarms are similarly critical when baby birds are fledged and on the ground. If the traffic noise is loud enough, the birds can't hear the alarms." Birds with low-pitched calls are at a special disadvantage.

Roads exert other long-lasting impacts on adjacent environments. Trombulak and Frissell note that travel corridors dramatically increase compaction in adjacent soils. They point to studies which found that the soils in these areas are two hundred times more compact than in undisturbed sites. The damage to the soils may persist long after roads are decommissioned.

Highways also radically alter the chemical composition of adjoining lands. Scientists have tallied elevated levels of five types of chemicals in roadside environments, write Trombulak and Frissell, including excess nutrients such as nitrogen, which is released during the fuel combustion process, and ozone, which can cause respiratory problems in animals as well as humans. Roadsides also are polluted with heavy metals. Elevated levels in the tissues of plants and animals frequently are found many hundreds of feet from the roadbed. Also lodged within the soils are pollutants from gasoline, motor oil, roadway sealants, rubber tires, soot, and auto exhaust. Research conducted by scientists from the U.S. Geological Survey in 2003 showed that these common roadway by-products are a major source of suspected carcinogens known as polycyclic aromatic hydrocarbons (PAHS).

By covering the land with impervious surfaces and channeling runoff into ditches, roads alter the natural pattern of runoff across the land and serve as conduits for routing sediments and contaminants into lakes and streams. According to Trombulak and Frissell, sediments laced with heavy metals, salt, organic molecules, ozone, and nutrients course into streams. Of concern is not only vegetation flanking the roadbed, which can accumulate high levels of toxic pollutants and transfer them to the animals that ingest these plant tissues, but also aquatic systems, even those that may lie at considerable distance from the roadway.

Salt of the Earth—and Water Quality

Since Highway 61 closely shadows the lakeshore and crosses numerous tributaries, these effects are especially worrisome; contaminants are easily washed into nearshore waters, the zone that fosters the greatest biological productivity in Lake Superior. Yet, most automobile-related pollutants have received little scrutiny along the North Shore, including monitoring of one of the most common chemicals related to motorized travel—road salt. Research in other northern regions of the United States suggests that applications of road salt, which is freely dispensed in liquid and crystallized form to combat freezing road conditions, may be wreaking ecological havoc.

According to the Salt Institute, road salt has been used effectively since the 1930s to help keep ice and snow from bonding to pavement. But it was not until the 1960s that salt use became widespread in order to keep roadways safe and passable during winter storm conditions. A review conducted by the National Academy of Sciences' Transportation Research Board in 1991 concluded that deicing salts cut the accident rate on U.S. highways by 88 percent.

But along with the exponential increase in paved surfaces has come a dramatic rise in the application of salt. Reports published in the 2005 *Proceedings of the National Academy of Sciences (PNAS)* note that the sales of highway salt in 1940 amounted to 164,000 tons. Today, that number has skyrocketed hundredfold. Of the 39.7 million tons of rock salt mined in the United States each year, 19.8 million tons are applied to paved roadways, most of them in the Northeast and Midwest.

The Salt Institute maintains that "when salt is used properly, it does not present environmental harm." But research suggests that this is not the case. A 2005 study published in the *Wildlife Society Bulletin,* for example, implicates salt in the demise of some species of birds, in particular, a group known as cardueline finches. These birds, which include crossbills, grosbeaks, and pine siskins, vacate boreal regions in the winter when supplies of conifer seeds run low. They migrate to more southerly latitudes, including the Lake Superior region, in search of food.

Primarily seed eaters, these birds frequent roadsides, where they pick up the grit that helps them to grind coarse foods in their gizzards. But sand and clay are not the only items on the menu. The diets of herbivorous and granivorous birds (those that survive on plants and seeds) tend to be deficient in salt. Many birds are attracted to salt-covered roadways to stock up on this badly needed supplement.

Instead, they may be getting a lethal overdose. Terrestrial birds are especially sensitive since their systems are poorly equipped to eliminate excess salt. Some birds die outright from ingesting too much salt (an acute poisoning known as salt toxicosis), particularly during cold winters when the water they need to mitigate the effects of toxicosis is frozen. Even at nonlethal levels, the overconsumption of salt causes disorientation and other problems, making birds vulnerable to run-ins with moving vehicles. The roadkill of pine siskins, for example, has become so common a phenomenon in places such as Mount Revelstoke National Park, British Columbia, that local residents call them "grill birds," referring to the frequency with which they are smashed on the grills of moving vehicles. The birds are especially vulnerable, say scientists Pierre Mineau and Lorna J. Brownlee, not on "busy multi-lane highways servicing a large metropolitan area" but on smaller roads that offer vegetative cover, such as many North Shore roads, including Highway 61.

Excess salt poses problems for animals in aquatic as well as terrestrial environments when salt dissolves in rain or snowmelt and trickles into neighboring soils and streams. "The effects on aquatic biota of temporary surges of salt that often accompany runoff from roads to surface and groundwaters have received little study," Trombulak and Frissell point out.

Research findings suggest that temporary pulses of salt that drain from roadways each winter often do not become diluted in the environment but are accumulating to harmful levels in terrestrial and aquatic ecosystems. A 2014 study in the PNAS showed, for the first time, that these concentrations can alter the course of animal development. A sampling of roadside plants in Minnesota revealed that some species stored sodium in concentrations that were many orders of magnitude greater than those that were located 330 feet or more from roadbeds. Salt in roadside milkweeds, for example, measured thirty times higher; northern pin oaks registered levels that were fifty times higher. Both plants are the favored foods of caterpillars, including monarch butterflies. In laboratory experiments, the adults of monarch caterpillars that were raised on the salt-laden plants developed pronounced physical differences including bigger thoracic muscles in males and larger eyes in females, both of which could confer survival advantages in the wild. However, the study's author, Emilie Snell-Rood of the University of Minnesota, cautioned against drawing the conclusion that road salt is good for butterflies since follow-up research on the ecology of these salt-boosted butterflies has yet to be conducted in the wild.

While some roadway salts may accumulate near their point of application, others can travel great distances. Studies of Kampoosa Bog in western Massachusetts, for example, reveal elevated levels of sodium and chloride that were carried in rain and snowmelt some two thousand feet from the Massachusetts Turnpike, which borders the wetland. "Inputs of road salt are now enormous and water moves that salt around locally and regionally," say the authors of the PNAS report "From Icy Roads to Salty Streams." In areas around Baltimore where more than 40 percent of the ground is covered in impervious surfaces such as buildings, parking lots, and roads, streams and rivers already have exceeded salt limits recommended for sensitive freshwater species.

A major study published in the same issue points out that concentrations of chloride as low as 0.03 ounces per gallon can damage roadside vegetation. Plants suffer from salt exposure in several ways. Brine sprayed by passing vehicles coats the needles of conifers, causing them to drop prematurely. Deciduous trees are injured when salt burns sensitive buds. Continued salt buildup can interfere with plants' absorption of vital nutrients such as potassium, calcium, and magnesium. And it can disrupt osmosis, a physical process that maintains water balance. Not only does excess salt lock water in the soil through osmotic tension, but in some cases it actually draws water out of plants into the ground.

In time, these altered conditions can lead to a change in the composition of whole plant communities. Some plants, especially invasive ones such as phragmites, are tolerant of higher salt levels. The salting of freshwaters has been blamed, in part, for the spread of phragmites in northern wetlands, where they grow in dense mats at the expense of native species.

Even large bodies of water are not immune to the effects of ongoing inputs of salt. Since the 1850s, according to a 1993 article published in the *Journal of Great Lakes Research,* the salinity of many nearshore waters in the heavily developed watersheds of the lower Great Lakes has increased threefold. The culprits? Road salt and industrial discharges. So salty have these waters become that some species of brackish-water algae—hitchhikers into the Great Lakes in the ballast water of commercial ships—have been able to adapt to life in these altered aquatic habitats.

Authors of the research published in *PNAS* drew a connection between the salinization of inland waters and the amount of impervious surfaces that covered a watershed. At the same time, however, they "observed strong increases in the baseline concentration of chloride in rural watersheds with low density of roadways." Examination of seven streams and rivers in Maryland, New York, and New Hampshire over periods ranging from twenty to forty years has revealed that chloride concentrations in some water bodies have as much as doubled since the 1960s. What surprised researchers was that "chloride concentrations in the rural streams did not return to baseline levels in summer, even when no salt was being applied. One reason is that salt concentrations build up over many years and remain high in the soil and groundwater. Groundwater seeping into streams often keeps water flowing during the driest periods, typically in summer. If the groundwater is salty, the stream will be salty," the researchers write, adding that "once groundwater becomes salty, it typically will take decades to centuries for the salts to disappear, even when road salting ends."

Research published in other journals corroborates these findings. A study reported in the 2003 issue of *Environmental Pollution* found that wells and springs in six of the fourteen counties that make up the Mohawk River basin in New York State contained a mean value of salt of 0.02 ounces per gallon, more than three times that recommended for the drinking water of people on a low-sodium diet. Researchers pointed out that less than 6 percent of the watershed is urban. In the *PNAS* study, scientists obtained seasonal salinity readings from streams in the rural watersheds of New Hampshire's White Mountains that were similar to those found in the Hudson River estuary.

Such rises in salinity are not likely to peak anytime soon. Cheap, plentiful, reliable, and easy to store and apply, salt far outcompetes its chemical competitors. In a 1991 study by the Transportation Research Board (TRP), commissioned by the U.S. Congress, researchers compared the efficacy and the total costs of using calcium magnesium acetate (CMA) versus salt. The study calculated such costs as damage to motor vehicles and transportation infrastructure, which is conservatively estimated to carry an annual price tag of $2 billion to $4 billion. Although CMA has a more benign impact on both the built and natural environment, its cost—some twenty to thirty times the price of salt—was expected to increase winter road-maintenance budgets by a factor of five.

According to the *PNAS* researchers, "If the construction of roadways and parking lots were to continue to increase at its current rate, there likely would be large changes in baseline salinity across many northern regions of the United States and in other urbanizing areas throughout the world." Rural areas will not be exempt. If present rates of salinization continue, by the next century, the authors add, "baseline chloride concentrations in many rural streams [will become] toxic to sensitive freshwater life and not potable for human consumption."

"No one is suggesting that society should instantly ban rock salt use," say the authors of the *PNAS* report. "Nonetheless . . . there are real, long-term consequences to its use, particularly for freshwater systems and soils. Understanding which environments are more likely to transfer salt from roads, streams, and groundwater could help managers identify sensitive species and highway segments that need alternative methods of deicing. More generally, a prudent step would be to adopt a 'less is more' policy, reducing the amounts of salt applied and considering alternatives where economically feasible. As is so often the case today, society is left to balance a discrete, positive benefit (safer roads) with more dilute environmental costs that build over decades and take decades to recover."

Building on the Edge

At first glance, the North Shore appears to be a solid wall of rock. But as Carol A. Johnston and James Sales point out in a 1994 study of the shoreline, "More than half of the coastal zone is composed of erodible glacial and post-glacial deposits in the form of glacial till, sand, gravel, clays, and silts." Placing heavy loads such as houses on these soils has increased the occurrence of some kinds of erosion. The zoning ordinances of Lake and Cook Counties have attempted to circumvent problems by mandating that structures be set back no less than forty feet from the normal high-water mark. Unfortunately, many structures along the shore were built before such safeguards were put in place. A survey conducted in the late 1980s revealed that existing properties had an average setback of only fifty-five feet from the lake edge. According to the authors of an article in the *Journal of Great Lakes Research,* "Most of the properties on this edge of water were developed during low water periods when shore users were lulled into false security by the relatively stable appearance of the shoreline."

The results of this research have given policy makers some guidelines for steering development away from hazardous areas. But caution is sometimes difficult to exercise in the North Shore's hot real-estate climate. "Its relative proximity to the thriving Minneapolis/St. Paul urban area," Johnston and Sales observe, "has made the Minnesota north shore increasingly popular for resort and second-home development."

Some property owners have taken the job of preventing shoreline erosion into their own hands by installing a variety of engineered solutions, such as riprap and concrete sea walls. The question remains, Who will pick up the tab for the more than $30 million that experts estimate it will cost to install erosion-hazard infrastructure along vulnerable stretches of the North Shore—taxpayers or lakeside property owners? Along the Minnesota coast, nearly all of these projects have been subsidized by taxpayers through state and federal grants. "Coastal property owners who have chosen to live on the edge of water have largely been attracted by the amenities of the shoreline," say the authors of the *Journal of Great Lakes Research* article, and should be willing to shoulder the risks of living in potentially hazardous environments. "If human ability to affect Great Lakes water levels is limited, then damage to shoreline and shore properties should be considered as an inherent operating cost of living on the shoreline," they say.

The authors point out, however, that erosion control is voluntary. Armoring the shoreline is expensive and effective only if property owners have the economic means and exercise the vigilance to continually maintain and occasionally replace control structures. In the meantime, sediments from problem properties continue to wash into the lake. And this pollution has concerned those charged with maintaining Superior's high ecological quality. Because of Lake Superior's great depth, researchers point out, the centers of the lake's biological activity lie in a relatively narrow band of nearshore waters, where excessive sediment loads can disrupt critical ecological processes. Furthermore,

clay particles, which are highly mobile in water, naturally bond with pollutants such as PCBs and carry toxins via currents into the far reaches of the lake.

Recognizing that soil erosion is both a public and private problem, the MBWSR has solicited grants to help municipalities and private property owners along the North Shore stem the flow of sediments into the lake. The Lake County Soil and Water Conservation District (SWCD), an agency overseen by the MBWSR, has made cost-share monies available to mediate erosion from the highly vulnerable clay bluffs in southwestern Lake County. In an innovative intra-agency partnership, MNDOT donated waste rock from the blasting of the Lafayette Bluff and Silver Creek Cliff tunnels in the early 1990s to nearby property owners. In one project, more than twelve thousand cubic yards—about a hundred semitrailer loads—of the material was used to riprap about 1,255 feet of shoreline on Lake Superior near Silver Cliff. The cost of the project topped $20,000, half of which was supplied by the SWCD.

In some cases, public agencies have collaborated with private organizations to help mitigate erosion problems. Anglers scrambling down a steep clay embankment to fish the mouth of the Knife River, for example, had so destabilized the soils that whole chunks of Lake Superior shoreline were sloughed away by heavy storms. To help safeguard the bank from further trampling, in the 1990s the MBWSR matched a $4,119 contribution from the Lake Superior Steelhead Association to construct a one-hundred-foot timber stairway.

More than 50 percent of the North Shore's coastal zone is composed of erodible soils. Rock revetment, shown here at the bottom left of a shoreline slope, is an effective—but expensive—method for protecting developments such as lakeside properties and Highway 61 from erosion. Courtesy of the Minnesota Lake Superior Coastal Program.

Such protective measures are costly, however, and only a few organisms, such as game fish, have attracted organized constituencies that are able to match private dollars with public investments.

Upstream/Downstream: A River in Trouble

Soil erosion is not just a lake-edge phenomenon. As coastal real estate becomes ever more scarce and pricey, development has moved inland. Most of this development has occurred piecemeal, with motels and houses climbing the slopes of the terrace's hills to catch a Lake Superior view. In some cases, incremental development—claiming five, ten, twenty acres at a time—has reshaped entire watersheds to serve the desires of the traveling public. Despite the building boom, say the authors of the 2003 MPCA study, "Surprisingly few data have been collected for the North Shore streams over the past 30 years in the systematic manner needed to define present conditions and detect water quality trends over time." The MPCA embarked on a plan to remedy this oversight. By 2013, the agency had initiated biological monitoring of streams throughout the state, including many North Shore streams.

Unfortunately, for some streams the results have been ominous. Take, for example, the Poplar River, which feeds into the lake at Lutsen. In its 2003 report, the MPCA added the final three-mile leg of the Poplar River to its list of impaired waters after the river fell short of meeting standards set forth by the federal Clean Water Act. (Once a water body is added to the impaired list, the state is given thirteen years to pinpoint the sources of degradation and make a good-faith effort toward cleanup.) Readers of the report were not surprised by the other three Lake Superior streams that made the list—Amity Creek and the Talmadge and French Rivers—since they course through the greater Duluth metropolitan area. But the listing of the Poplar River was puzzling since the river originates in the BWCAW and its 114-square-mile watershed lies 92 miles from Duluth, the nearest big city.

MPCA officials set up two monitoring sites, one of which sampled water from a location some three miles upstream from the river's mouth and another that took water-quality measurements in the lower reaches of the river near Lake Superior. Measurements taken on the upper river compared favorably to those taken from pristine North Shore waterways, such as the Brule River. But where the river approached Lake Superior, the waters were the color of a chocolate milkshake. Not surprisingly, tests showed that suspended solids (a measure of sediment erosion) had increased sixfold. The Poplar's waters also were chemically altered. Near the mouth of the river, phosphorus levels had doubled. Elevated too were concentrations of chloride, which increased twofold in spring snowmelt. Most worrisome, however, were the high levels of mercury, which routinely exceeded State of Minnesota water-quality standards.

The degradation of the lower Poplar River had already taken a biological toll. Although steelhead still venture up the river during the spring spawning season, fisheries biologists have measured no natural reproduction in the stream since 1994. They believe that the clean, well-oxygenated gravel on which trout eggs depend have become choked with silt. And without clean riffles, streams also produce fewer insects that feed juvenile fish.

Where do the sediments and chemical pollution originate? As in most compromised

The Poplar River largely runs clear in the upper reaches of its watershed through the North Shore's headwaters and highlands, where forests are unbroken by few developments. Courtesy of the South St. Louis County Soil and Water Conservation District.

The nearshore reaches of the Poplar River run through the most intensely developed recreational complex on the North Shore. Since 2003 public–private partnerships have begun to address problems such as large flows of eroded sediments into the river. Courtesy of the South St. Louis County Soil and Water Conservation District.

watersheds, researchers point out, the sources are numerous and their effects cumulative. But this much is clear: The waters of the Poplar flow clean and clear as they enter the most intensely developed recreational complex on the North Shore, which includes the Lutsen Mountains ski area, the Superior National Golf Course, a range of vacation lodgings including condominiums, houses, and resorts along with several wastewater lagoons that were built to service them. The waters that emerge downstream of the development are turbid and polluted.

Scientists believe that this human alteration of the landscape appears to be aggravating already-sensitive natural conditions within the Poplar River watershed. For example, although researchers did not pinpoint the exact cause for the high mercury levels, they strongly suspect that the clearing of forests exposed atmospheric mercury that had fallen to earth and become attached to soil particles. These contaminated soil particles then were carried into the river via stormwater runoff. Increases in the volume and velocity of these stormwater flows have altered not only the land but also the coursing of the river. According to the report, "The river makes several sharp turns, causing bank erosion, as it winds through its valley. Some of this bank erosion could be termed 'natural,' although the water volumes and runoff rates are increased due to human activity (land use change) in the valley, such as gully erosion, exposed clay soils, open areas (i.e., removal of the trees), impervious surfaces (roads, roof tops, and parking lots) and treated wastewater."

In 2005, two years after the MPCA added the river to the impaired waters list, landowners along the lower Poplar River in Lutsen formed the Poplar River Management Board (PRMB). Participants included not only property owners but also representatives from the MPCA, the Minnesota DNR, the Cook County SWCD, and scientific and engineering experts. According to the PRMB's website, the group's goal is "to protect and improve the water quality of this segment of the river" and to ensure its removal "from the impaired list as soon as possible."

Since then, relying on data from scientific research, the PRMB and other interested stakeholders have identified the most significant sources of sediment in the watershed and implemented $1.7 million in sediment-reduction projects paid for with state and private funds. Their efforts paid off. In February 2013, the MPCA provided updated data on sediment trends in the Poplar River. The agency estimated that sediment loads from 2002 to 2006 averaged about 1,000 tons per year. Average loads fell for the years 2009 through 2011 to about 660 tons per year, a reduction of about 35 percent. Additional projects are planned.

Maintaining good water quality, say the authors of the MPCA report, is not only essential to the healthy functioning of ecosystems and the organisms they support but also to the region's economic vitality. "As travel and tourism are heavily dependent upon the quality of the recreational experience," they write, "a primary Minnesota 'product' will be to maintain the quality of our North Shore recreational areas so as to compete with premium recreational areas of the region (Wisconsin, Michigan, Canada, and western states). As was shown for northern Minnesota lakes, recreationists will seek alternative bodies of water, or reduce their level of activities, in response to perceived water pollution. . . . A recent study found that for lakes in north central Minnesota, water quality has a positive relationship with property prices, and management of the quality of lakes is important to maintaining the natural and economic assets of the region."

The Perpetual Full Moon: Light Pollution

In November 2005 Cook County citizens gathered at the first of a series of town hall–style meetings to discuss the future of the tiny town of Grand Marais. Residents voiced predictable concerns: lack of affordable housing, runaway harbor-front development, preserving open space, maintaining the area's outstanding water quality. But on Mike Bauer's wish list was an item that likely would have caught many of his other fellow citizens by surprise—the preservation of the night sky. "No one else can see the northern lights as well as we can," the Grand Marais resident told a reporter for the *Cook County News-Herald.*

Bauer's comment was only a slight exaggeration. In a 2001 issue of the *Monthly Notices of the Royal Astronomical Society,* the Italian astronomer Pierantonio Cinzano

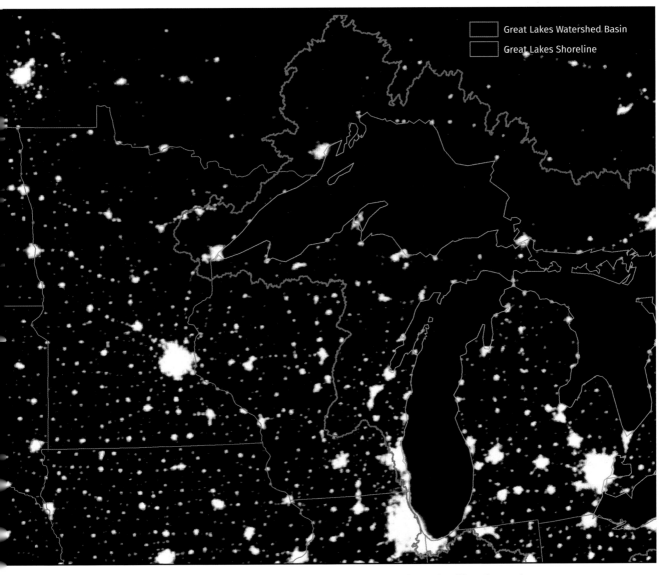

This night sky map of the western Great Lakes shows that the Lake Superior basin still retains much of its original nocturnal wildness, but even small towns such as Two Harbors and Grand Marais emit enough light pollution to reduce visibility to only two hundred to three hundred stars. Courtesy of the Natural Resources Research Institute GIS Lab, University of Minnesota Duluth.

and his colleagues published the first world atlas of artificial night-sky brightness. According to the research, more than one-half of the people living in the European Union and more than two-thirds of those in the continental United States can no longer see the Milky Way with the naked eye. The team's maps revealed that few regions of the continental United States have escaped the spotlights. Of the five Great Lakes, for example, only Huron and Superior still retain some of their original nocturnal wildness. The edges of Lakes Ontario, Erie, and Michigan are so brightly lit by artificial lights that the outlines of their shores are clearly legible in nighttime satellite images.

Scientists refer to lights that bleed into the night sky and blot out the stars as astronomical pollution. Before the advent of widespread electrical illumination, the average star gazer could pick out some twenty-five hundred celestial bodies in the night sky. This view, writes Gary Andrew Poole in a 2003 article in the *New York Times,* guided people on their journeys, "inspired them in art and science and caused them to ask themselves about their place in the universe." Today, residents gazing into the skies over a suburb or a small town, such as Two Harbors and Grand Marais, that emit a moderate amount of light pollution may see only two hundred to three hundred visible stars. Light pollution in big cities can cut that number down to only a few dozen.

Perhaps most worrisome is that you do not need to be standing under a streetlamp or staring into your neighbor's backyard floodlights to experience a diminished overhead view. Researchers have found that the domes of diffused light that hover above major metropolitan areas, a phenomenon known as sky glow, can illuminate the heavens as far as 188 miles away. Sky glow is created by light reflecting off moisture and dust in the air.

But the problem of light pollution is troubling not simply for its dimming of the human imagination but also for the disruptions it causes in the biology and behavior of the animals that share our nocturnal landscape. "Most species depend on light and dark for some portion of their daily or seasonal life cycle," write Catherine Rich and Travis Longcore of the Urban Wildlands Group and editors of the 2006 book *Ecological Consequences of Artificial Night Lighting.*

The scientific term for the alteration of natural patterns of light and darkness is ecological light pollution. Some of the most famous examples of animals affected by this kind of contamination include sea turtle hatchlings, species that have been hard hit by ocean-front development. Instead of heading into the sea after emerging from their sand-beach nests, the tiny turtles are lured landward by electrical lights, where many end up in the jaws of hungry predators or die of exposure.

Equally well publicized are collisions between birds and towers, largely because their spectacular death tolls often make headline news. According to the Toronto-based organization Fatal Light Awareness Program (FLAP), an estimated one hundred million birds die each year after colliding with human-made structures ranging from monuments, broadcast towers, buildings, and smokestacks to oil platforms and sea-going vessels. Especially treacherous are the concrete canyons created by high-rises in large cities throughout the world. Confusing big city lights with the constellations that birds use as navigational aids, whole flocks of night-flying migrants can be lured into the maze of urban towers, where they crash into windows and die or become disoriented by the bright lights and reflected glass. Eventually, many succumb to exhaustion, making them easy pickings for predators. During migration season, FLAP volunteers routinely comb the sidewalks of downtown Toronto at dawn looking for dead, stunned, or injured birds. According to FLAP, some 450 species are affected by these collisions, among them threatened or endangered species such as the cerulean warbler and Henslow's sparrow. Each day during migration season, the body count can top one thousand birds. How many more injured or disoriented birds have disappeared into a predator's mouth or under a street-sweeper's broom is anybody's guess. (See also "The Leading Edge: North Shore Bird Migration.")

The dangers are increased when misty, foggy weather forces the birds to fly at lower altitudes, conditions that are common during spring and fall migrations. Some of the largest recorded downings of birds occur during nights like these. In 1954 a storm that raged for two nights at Warner Robins Air Force Base in Georgia claimed the lives of fifty thousand birds after they mistook runway lights for stars and crashed into the ground. In 1981 more than ten thousand birds collided with floodlit smokestacks at a generating plant near Kingston, Ontario.

Especially tragic are episodes that decimate whole populations of a single species. One of the largest bird kills in Kansas history, for example, occurred on a snowy, foggy January night in 1998 when an estimated five thousand to ten thousand lapland longspurs slammed into the guy wires of radio transmission towers. Subsequent research has shown that towers with red lights, rather than white ones, not only attract greater numbers of birds but also lead to more erratic flight patterns. The reason: "Birds' magnetic compasses seem to break down in red light," says Sidney Gauthreaux, a bird researcher at South Carolina's Clemson University.

But scientists are only now discovering that increased illumination is having negative effects on a whole range of animals. And as human development continues to encroach on rural and wilderness areas, refuges for night-loving animals are shrinking. "The most noticeable effects," say Longcore and Rich, "will occur in those areas where lights are close to natural habitats. This may be in wilderness where summer getaways are built, along the expanding front of suburbanization, near the wetlands and estuaries that are often the last open spaces in cities, or on the open ocean, where cruise ships, squid boats, and oil derricks light the night."

Of the animals studied, reptiles and amphibians appear to be especially sensitive to environmental lighting conditions. Many species of snakes, frogs, and salamanders naturally take their cues from lunar cycles, restricting their activities during full moons and stepping up foraging activities in periods of darkness. Artificial lighting can send signals that interfere with such basic life functions as feeding and breeding. For instance, laboratory research has shown that lights severely curtailed the ability of gray treefrogs to detect prey. To avoid being eaten themselves, some frog species play possum in brightly lit areas, remaining motionless long after lights have gone out. Researchers have documented that not only do many species of frogs eat less under these artificial conditions, but they also reduce or halt their calling during mating season. As a result, many frog populations have plummeted or disappeared in areas where electrical illumination never allows darkness to fall or where patches of light may be as effective as fences in keeping animals from using parts of the landscape they need in order to survive.

Researchers caution that it does not take lights with the equivalent brightness of a Kmart parking lot to disrupt such essential functions. Squirrel treefrogs have been shown to decrease their activity when light levels exceed 0.001 lux (a lux is a unit of illumination; one lux is the equivalent of that shed by dim interior lights). Even soft lighting can alter behavior, as ecologists Sharon Wise and Bryant Buchanan discovered in the wilds of Virginia. The scientists found that red-backed salamanders, which normally emerge from beneath the leaf litter to hunt about an hour after dusk, delayed foraging for an additional hour after a string of tiny holiday lights was activated. According to Wise, "Artificial night lighting has the potential to shorten foraging periods, limit food intake, and depress rates of growth, reproduction, and survival."

In some ecosystems, the intrusion of lights may have subtle but far-reaching effects. In 2000 limnologist Marianne Moore and her colleagues published research demonstrating that light pollution may have serious consequences for aquatic food webs. The scientists studied lakes located in lighted conditions ranging from inner-city lakes in Boston to remote, rural bodies of water. Moore points out that these bodies of water are highly susceptible to the effects of artificial light encroachment since they often are not shaded by trees or buildings. Suburban lakes, for example, register artificial light intensities that are five to thirty times greater than that of the full moon.

To study the effects of illumination on these ecosystems, Moore and her team focused on the behavior of a common group of zooplankton known as *Daphnia*. Under natural conditions, *Daphnia* frequent deep waters during daylight hours, migrating up into the water column under the cover of darkness to graze on algae. Lights from roads, lake homes, and boathouses, however, illuminate these upper waters, discouraging the extent to which these tiny animals are willing to venture into areas where they are more visible to predators. The distance proved significant. Moore's team found that zooplankton migrated 6.5 to 10 feet higher into the water column under natural darkened conditions than zooplankton in lighted waters. "If *Daphnia* or other zooplankton do not migrate to the surface of the wetland to forage on algae because light levels are too high," Moore observes, "then the whole aquatic food chain is in jeopardy." In time, writes Ben Harder in an article in *Conservation in Practice,* "algae populations could explode in response to reduced predation, and those blooms would deplete dissolved oxygen critical to fish, crowd out other photosynthesizers, and cast unwanted daytime shade on submerged aquatic vegetation that provides habitat for juvenile fish."

But not all nocturnal animals are light averse. Some species exhibit an attraction to artificial light known as "flight-to-light" behavior. Declines in certain moth populations, for example, have been blamed on the death lure of artificial light. Not only are these moths exposed to greater predation, but the flight-to-light behavior can disrupt feeding, mating, egg laying, and numerous other functions.

Some animals have evolved to take advantage of what ecologists call the "night light niche." Studies of orb-weaving spiders along bridges in Austria show that far more spiders weave their webs along well-lit sections than in darkened ones, where night-flying insects are less abundant. Some faster-flying bat species are able to feast on insects that are attracted to lights. Urban crows are thought to favor roosts near lights to avoid predation by owls.

But such disruptions in predator-prey relationships can stack the deck for some species while reducing the odds for others. "Many of the effects of artificial light may resonate up and down food chains, dragging whole ecosystems into imbalance," writes Harder. "And by modifying the playing field on which nocturnal organisms develop, interact, and reproduce, artificial light may sculpt not only their individual lives but also the biological evolution of their species." For slower-moving bats, the advantages of

Research has shown that light pollution from waterfront developments can seriously disrupt aquatic food webs. Photograph by Jarrod Lopiccolo.

having a bonanza of food concentrated under streetlights may be offset by an increased risk of predation by owls. Crows may be able to outwit owls by establishing resting sites in well-lit trees, but increases in their numbers can wreak havoc for other species. Crows, observe Longcore and Rich, "are aggressive, and artificially increased population levels can be detrimental to other native bird species."

According to the International Dark-Sky Association, many of the problems caused by light pollution can be solved without posing hazards or inconvenience to people. In homes, businesses, and communities along the North Shore, for example, residents can take simple steps to protect the darkness of the night sky and see immediate results— unlike other threats to their environment and quality of life. The association estimates that 30 percent of artificial light in the United States is wasted. Pointed skyward, it washes out the stars while dangerously blinding or disorienting people with glare. These problems can be avoided by using properly designed light fixtures that direct light to the ground, where it is most effective.

The ability to gaze up into the heavens and see a sky drenched in stars, the authors warn, is not something that people in rural or remote regions can continue to take for granted. "The fading away of the night sky is an issue not only in cities, but also in the countryside and in developing waterfront communities," warn David Liebl and Robert Korth, authors of "Sensible Shoreland Lighting," a pamphlet published by the University of Wisconsin–Extension, Madison.

Safeguarding the opportunity to gaze into a night sky blazing with stars is nothing less than preserving the wholeness of being fully human. "We human beings lose something of ourselves when we can no longer look up and see our place in the universe," writes David Crawford of the International Dark-Sky Association in Tucson. "It is like never again hearing the laughter of children; we give up part of what we are." ✎

SUGGESTIONS FOR FURTHER READING

Ad Hoc Aggregate Committee. *Minnesota's Aggregate Resources: Road to the 21st Century.* Aggregate Resources Task Force, 1998.

Agassiz, Louis. *Lake Superior: Its Physical Character, Vegetation, Animals Compared with Those of Other and Similar Regions.* With a Narrative of the Tour by J. Elliot Cabot. Boston: Gould, Kendall and Lincoln, 1850.

Aldrich, Chilson D. *The Real Log Cabin.* Minneapolis: Nodin Press, 1994.

Anderson, Jesse, Mark Evenson, Tom Estabrooks, and Bruce Wilson. *An Assessment of Representative Lake Superior Basin Tributaries 2002.* St. Paul: Minnesota Pollution Control Agency, 2003.

Avise, John C., and Robert L. Crawford. "A Matter of Lights and Death." *Natural History* 90 (1981): 6–14.

Backes, David. "Wilderness Visions: Arthur Carhart's 1922 Proposal for the Quetico-Superior Wilderness." *Forest and Conservation History* 35 (1991): 128–37.

Bishop, Hugh E. "Restoring Rustic Dreams." *Lake Superior Magazine,* June-July 2001: 35–39.

Breining, Greg. "Storm over Safe Harbors." *Minnesota Volunteer,* January/February 1993: 4–15.

Bruland, Kenneth W., and Minoru Koide. "Lead-210 and Pollen Geochronologies on Lake Superior Sediments." *Quaternary Research* 5 (1975): 89–98.

Butters, Fred K., and Ernst C. Abbe. "A Floristic Study of Cook County, Northeastern Minnesota." *Rhodora* 55, no. 650 (1953): 21–85.

Carhart, Arthur H. "A Forest Home for Everyone." *Good Housekeeping,* June 1920: 39–120.

———. *The National Forests*. New York: Alfred A. Knopf, 1959.

———. *Timber in Your Life*. Philadelphia: Lippincott, 1955.

Cheney, L. S. "A Contribution to the Flora of the Lake Superior Region." *Transactions of the Wisconsin Academy of Sciences, Arts and Letters* 9 (1893): 233–54.

Cinzano, Pierantonio, Fabio Falchi, and Christopher D. Elvidge. "The First World Atlas of the Artificial Night Sky Brightness." *Monthly Notices of the Royal Astronomical Society* 328 (2001): 689–707.

Cleland, Charles. E. "Indians in a Changing Environment." In *The Great Lakes Forest: An Environmental and Social History,* ed. Susan L. Flader. Minneapolis: University of Minnesota Press, 1983.

———. "The Inland Shore Fishery of the Northern Great Lakes: Its Development and Importance in Prehistory." *American Antiquity* 47 (1982): 761–84.

Committee on Land Utilization. "Land Utilization in Minnesota: A State Program for the Cut-over Lands." Minneapolis: University of Minnesota Press, 1934.

Davis, Jessie C. "Beaver Bay Original North Shore Village." Duluth: St. Louis County Historical Society, 1968.

Disturnell, John. *Lake Superior Guide, Giving a Description of All the Objects of Interest and Places of Resort on This Great Inland Sea; with an Account of the Iron, Copper and Silver Mines; Also, Commercial Statistics in Regard to the Product of the Mines, Fisheries, &c. with a Township Map including the Lake Superior Region and Northern Minnesota.* Philadelphia: J. Disturnell, 1872.

Dodson, Stanley. "Predicting Diel Vertical Migration of Zooplankton." *Limnology and Oceanography* 35 (1990): 1195–1200.

Dombeck, Mike. "The Forgotten Forest Product: Water." *New York Times,* 3 January 2003.

Eames, Henry H. *Report of the State Geologist on the Metaliferous Region Bordering on Lake Superior.* St. Paul: Press Printing Company, 1866.

Edgerton, Angelique D. "The Structure of Relict Arctic Plant Communities along the North Shore of Lake Superior." M.S. thesis, University of Minnesota, May 2013.

Farris, Michael A. *The Effects of Rock Climbing on the Cliff Flora of Three Minnesota State Parks.* St. Paul: Minnesota Department of Natural Resources, 1995.

Fesler, Bert. *The North Shore in 1890.* Annual Meeting of the Tri-County Historical Assembly, Grand Portage, August 23, 1930.

Frelich, L. E. "Range of Natural Variability in Forest Structure for the Northern Superior Uplands." Prepared for the Minnesota Forest Resources Council and the National Forest in Minnesota. 13 pages. 1999.

Fritzen, John. *History of North Shore Lumbering.* Duluth: St. Louis County Historical Society, 1968.

Godwin, Kevin S., Sasha D. Hafner, and Matthew F. Buff. "Long-Term Trends in Sodium and Chloride in the Mohawk River, New York: The Effect of Fifty Years of Road-Salt Application." *Environmental Pollution* 124 (2003): 273–81.

Goldman, Jason G. "Road Salt Changes Urban Ecosystems in Big Ways." *Conservation,* 11 June 2014. http://conservationmagazine.org/2014/06/road-salt-changes-urban-ecosystems-in-big-ways/.

Gorenzel, W. Paul, and Terrell P. Salmon. "Characteristics of American Crow Urban Roosts in California." *Journal of Wildlife Management* 59 (1995): 638–45.

Green, John C. "Beaches in the Making: Geology in Action." *Natural Superior,* Late Summer/Fall 1999: 6–9.

———. *Geology on Display: Geology and Scenery of Minnesota's North Shore State Parks.* St. Paul: Minnesota Department of Natural Resources, 1996.

Green, John C., Mark A. Jirsa, and Carol M. Moss. *Environmental Geology of the North Shore.* St. Paul: Minnesota Geological Survey, 1977.

Harder, Ben. "Degraded Darkness." *Conservation in Practice* 5 (2005): 21–27.

Henderson, Troy. *Lake Superior Country: 19th Century Travel and Tourism.* Chicago: Arcadia Publishing, 2002.

Hosie, R. C. "Botanical Investigations in Batchwana Bay Region, Lake Superior." Bulletin No. 88. Ottawa: National Museum of Canada, 1938.

Hummel, Joan. "A Cabin by the Lake." *Minnesota Volunteer,* May/June 1993: 8–17.

Jackson, Robert B., and Esteban G. Jobbagy. "From Icy Roads to Salty Streams." *Proceedings of the National Academy of Sciences* 102 (2005): 14487–88.

Johnston, Basil H. *By Canoe and Moccasin.* Lakefield, Ont.: Waapoone Publishing and Promotion, 1992.

Johnston, Carol A., Brian Allen, John Bonde, Jim Sales, and Paul Meysembourg. *Land Use and Water Resources in the Minnesota North Shore Drainage Basin.* Technical Report NRRI/TR-91/07. Duluth: Natural Resources Research Institute, 1991.

Johnston, Carol A., and James Sales. "Using GIS to Predict Erosion Hazard along Lake Superior." *Journal of the Urban and Regional Information Systems Association* 6 (1994): 57–62.

Kaushal, Sujay S., Peter M. Groffman, Gene E. Likens, Kenneth T. Belt, William P. Stack, Victoria R. Kelly, Lawrence E. Band, and Gary T. Fisher. "Increased Salinization of Fresh Water in the Northeastern U.S." *Proceedings of the National Academy of Sciences* 102 (2000): 13517–20.

Kingman, Joseph R. *History of Encampment Forest Association.* Two Harbors, Minn.: Encampment Forest Association, 1946.

Kneipp, L. F. "Recreational Use of the National Forests." *Journal of Forestry* 28 (1930): 618–25.

Knopp, Timothy B., and Uel Bland. *The North Shore Experience.* Research Report No. 8. St. Paul: Minnesota Sea Grant College Program, University of Minnesota, 1983.

Kohl, Johann Georg. *Kitchi-Gami: Life among the Lake Superior Ojibway.* London: Chapman and Hall, 1860. Reprint, St. Paul: Minnesota Historical Society Press, 1985.

Liebl, David S., and Robert Korth. "Sensible Shoreland Lighting: Preserving the Beauty of the Night." Madison: University of Wisconsin-Extension, 2000.

Longcore, Travis, and Catherine Rich. "Ecological Light Pollution." *Frontiers in Ecology and the Environment* 2 (2004): 191–98.

MacKaye, Benton. "Recreational Possibilities of Public Forests." *Journal of the New York State Forestry Association* 3 (1916): 4–32.

Mills, Edward L., Joseph H. Leach, James T. Carlton, and Carol L. Secor. "Exotic Species in the Great Lakes: A History of Biotic Crises and Anthropogenic Introductions. *Journal of Great Lakes Research* 19 (1993): 1–54.

Mineau, Pierre, and Lorna J. Brownlee. "Road Salts and Birds: An Assessment of the Risk with Particular Emphasis on Winter Finch Mortality." *Wildlife Society Bulletin* 33 (2005): 835–41.

Minnesota Coastal Zone. *North Shore Data Atlas.* St. Paul: Minnesota Coastal Zone Program, 1978.

Minnesota Pollution Control Agency. "North Shore Land Use Issues: The Real Costs of Growth." WQ/Lake Superior Basin #2.03. Duluth, Minn.: Minnesota Pollution Control Agency, 2000.

Moore, Marianne V., Stephanie M. Pierce, Hannah M. Walsh, Siri K. Kvalvik, and Julie D. Lim. "Urban Light Pollution Alters the Diel Vertical Migration of *Daphnia.*" *Proceedings of the International Association of Limnology* 27 (2000): 1–4.

Morton, Ron, and Carl Gawboy. *Talking Rocks: Geology and 10,000 Years of Native American Tradition in the Lake Superior Region.* Duluth: Pfeifer-Hamilton Publishers, 2000.

National Conference on Outdoor Education. *Proceedings of the National Conference on Outdoor Education.* Washington, D.C.: Government Printing Office, 1924.

———. *Proceedings of the National Conference on Outdoor Education.* Washington, D.C.: Government Printing Office, 1926.

North Shore Management Board. "North Shore Harbors Plan: Recreational Boating Harbors Plan for the North Shore." Duluth: North Shore Management Board, 1991.

Nute, Grace Lee. *Lake Superior.* New York: Bobbs-Merrill Company, 1944.

Oberholtzer, Ernest C. "Conservation and the Economic Situation in Minnesota." *Minnesota Municipalities,* December 1931: 2–7.

Perkins, Sid. "Paved Paradise?" *Science News* 166 (2004): 152–53.

Pomeroy, Earl. *In Search of the Golden West: The Tourist in Western America.* Lincoln: University of Nebraska Press, 1957.

Possis, Ann. "Making Snow and Making History on the Hill." *Cook County News-Herald,* 29 November 2002, 1A–8A.

Pyne, Stephen. *Year of the Fires: The Story of the Great Fires of 1910*. New York: Viking, 2001.

Rasid, Harun, Robert S. Dilley, Dale Baker, and Peder Otterson. "Coping with the Effects of High Water Levels on Property Hazards: North Shore of Lake Superior. *Journal of Great Lakes Research* 15, no. 2 (1989): 205–16.

Reed, Christopher. "Driving Birds Away." *Harvard Magazine,* May/June 2005: 11–12.

Reid, Ron, and Karen Holland. *The Land by the Lakes: Nearshore Terrestrial Ecosystems.* 905-R-97-015c. Chicago: U.S. Environmental Protection Agency, 1997.

Reijnen, Rein, Ruud Foppen, and Henk Meeuwsen. "The Effects of Traffic on the Density of Breeding Birds in Dutch Agricultural Grasslands." *Biological Conservation* 75 (1996): 255–60.

Rich, Catherine, and Travis Longcore, eds. *Ecological Consequences of Artificial Night Lighting*. Washington, D.C.: Island Press, 2006.

Richburg, Julie A., William A. Patterson, and Frank Lowenstein. "Effects of Road Salt and *Phragmites australis* Invasion on the Vegetation of a Western Massachusetts Calcareous Lake-Basin Fen." *Wetlands* 21 (2001): 247–55.

Roscoe, Robert. "Mom and Pop Resorts Northern Minnesota." *Architecture Minnesota,* July/August 1998: 7, 46.

Ryan, J. C. "Minnesota Logging Railroads." *Minnesota History* 27, no. 4 (December 1946): 300–308.

Sandvik, Glenn N. "Land Office Buchanan: Emporium of the North Shore.' *Minnesota History* 52 (Fall 1991): 279–88.

Schantz-Hansen, Thorvald. *The Cut-Over Lands of Lake County*. Bulletin 304. St. Paul: Agricultural Experiment Station, University of Minnesota, 1934.

Schenck, Theresa M. *The Voice of the Crane Echoes Afar: The Sociopolitical Organization of the Lake Superior Ojibwa, 1640–1855*. New York: Garland Publishing, 1997.

Schwartz, George M. *A Guidebook to Minnesota Trunk Highway No. 1*. Bulletin No. 20, Minnesota Geological Survey. Minneapolis: University of Minnesota, 1925.

Shippee, Lester Burrell. "The First Railroad between the Mississippi and Lake Superior." *Mississippi Valley Historical Review* 5 (1918): 121–42.

Smith, Lloyd L., Jr., and John B. Moyle. "A Biological Survey and Fishery Management Plan for the Streams of the Lake Superior North Shore Watershed." Technical Bulletin No. 1. St. Paul: Minnesota Department of Conservation, Division of Fish and Game, 1944.

Snell-Rood, Emilie C., Anne Espeset, Christopher J. Boser, William A. White, and Rhea Smykalski. "Anthropogenic Changes in Sodium Affect Neural and Muscle Development in Butterflies." *Proceedings of the National Academy of Sciences* 111, no. 28 (2014): 10221–26. http://www.pnas.org/content/111/28/10221.

Summer Resorts Northern Iowa and Minnesota. Chicago: Rand, McNally and Co., 1883.

Sutter, Paul S. *Driven Wild: How the Fight against Automobiles Launched the Modern Wilderness Movement*. Seattle: University of Washington Press, 2002.

Transportation Research Board. *Highway Deicing: Comparing Salt and Calcium Magnesium Acetate*. Special Report 235. Washington, D.C.: National Research Council, 1991.

Trombulak, Stephen C., and Christopher A. Frissell. "Review of Ecological Effects of Roads on Terrestrial and Aquatic Communities." *Conservation Biology* 14 (2000): 18–30.

Upgren, Arthur R. "Night Blindness." *Amicus Journal,* Winter 1996: 22–25.

Wehrwein, George S. "A Social and Economic Program for the Sub-Marginal Areas of the Lake States." *Journal of Forestry* 29 (1931): 915–24.

———. "Some Problems of Recreational Land." *Journal of Land and Public Utility Economics* 3, no. 2 (1927): 163–72.

Wehrwein, George S., and Kenneth H. Parsons. "Recreation as a Land Use." *Wisconsin Agricultural Experiment Station Bulletin* 422 (1932): 3–31.

Wolff, Julius F., Jr. "Some Vanished Settlements of the Arrowhead Country." *Minnesota History,* Spring 1955: 177–84.

INTERNET RESOURCES

The Butterfly Effect, https://www.cbs.umn.edu/blogs/cbs-connect/snell-rood-road-salt

Fatal Light Awareness Program, www.flap.org

International Dark-Sky Association, www.darksky.org

Lake Superior-Poplar River Water District, www.poplarriverboard.com/pdf/Water.District.Fact Sheet.pdf

MPCA Stream Monitoring, http://www.pca.state.mn.us/index.php/water/water-monitoring-and-reporting/biological-monitoring/biological-monitoring-of-water-in-minnesota.html

NRRI Coastal GIS, A Coastal Atlas for the North Shore of Lake Superior, http://www.nrri.umn.edu/coastalgis

Poplar River Management Board, http://www.poplarriverboard.com/

Salt Institute, www.saltinstitute.org

Urban Wildlands Group, www.urbanwildlands.org

What's in a Name?

One who understands the origin of the place names of any region
knows its history.

William E. Culkin, *North Shore Place Names* (1931)

Wᴇɴ ᴛʜᴇ Fʀᴇɴᴄʜ ғɪʀsᴛ ᴇxᴘʟᴏʀᴇᴅ ᴛʜᴇ sʜᴏʀᴇs ᴏғ Lᴀᴋᴇ Sᴜᴘᴇʀɪᴏʀ ɪɴ ᴛʜᴇ seventeenth century, they discovered that the lakes, rivers, harbors, and landforms had already been carefully observed and richly named by the region's native inhabitants. If the Indian names were easy to remember and pronounce, the French recorded them on their maps. Scholars point out that in the mid-seventeenth century, an important period of exploration and map making, fewer than one hundred Frenchmen occupied the Lake Superior basin. According to historian Alan H. Hartley, these Europeans "wielded a toponymic [i.e., place-naming] power far out of proportion to their numbers: it is fortunate for our place-name heritage that they were accurate and sympathetic recorders of Ojibway names."

Today the lake remains ringed with Ojibwe names or their French versions, some of which were later translated into English. Luckily, successors to the French were not only interested in the history of these names but also dedicated to preserving this legacy for future generations. When members of the Minnesota Geological Survey team, for example, mapped the Arrowhead region in the late nineteenth century, they consulted numerous historical accounts and Indian-language studies, including an 1853 Ojibwe dictionary compiled by Lake Superior missionary Bishop Frederic Baraga. "Much care was taken to secure correctly the Ojibway names of the streams and lakes," writes Warren Upham, author of the definitive 1920 volume *Minnesota Geographic Names*. "Their translations were commonly used in that survey, as also by the earlier explorers and fur traders, government surveyors, and lumbermen."

As a result, we now possess a written record—albeit an episodic one—of the relationship between the region's original people and the land. "Place names," writes Virgil V. Vogel, author of *Indian Names on Wisconsin's Map,* "are cultural artifacts which tell us as much about how people lived, as do the relics dug from the ground." From them we can learn something of native people's "cosmic views, their values, their understanding of their place in the nature, and their ways of life," he adds.

Chances are that if a name refers to a natural feature or phenomenon, it has an Indian origin since native people did not use personal names to mark places. Most native place-names exhibit a faithfulness to the details of the natural world, either as an expression of the site's usefulness, its special beauty, or its spiritual aura. The act of naming was no casual matter. According to the White Earth missionary the Reverend J. A. Gilfillan, who published a list of Indian-derived place-names in 1885, "The Ojibway Indian is a very close observer, a name either of a person, or a place with him always *means something,* and is never a mere arbitrary designation as with us, but expresses the *real essence of the thing,* or its dominating idea as it appears to him."

Indeed, keen observation, writes ethnographer Theresa S. Smith, was a matter of life and death. The lifestyle of northern Indians, she writes, "was not only physically but intellectually demanding. Traditional Ojibwe people observed the natural world with great care and precision because an accurate understanding of one's environment was essential to one's very survival. These people were neither vague nor romantic in their descriptions of the world, and their complex understanding of natural phenomena is reflected in their language."

The Rosebush River, for example, takes its name from the Ojibwe word *Oginekan,* or rose berries. Packed with nutrients (three rose hips are said to have as much vitamin C as an orange), the fruit of wild rose bushes was prized for its health-giving properties.

So too other parts of the plant. The Ojibwe scraped the inner layer of the shrub's root bark and then soaked the shavings in water to create medicinal eyedrops.

The Indian word *Gamanazadika-zibi,* "place-of-poplar river," which gave rise to the name of the Poplar River in Lutsen, was based on a large stand of balsam poplars that grew near the river. According to the early-twentieth-century ethnologist Frances Densmore, the tree's roots, blossoms, and plump, resinous springtime buds were important ingredients in a concoction used to treat heart ailments. The Ojibwe also mixed the buds with bear grease to create a poultice that was applied externally to heal sprains.

For some plants and animals, place-names provide one of the few records of their passing. The Pigeon River, from the Ojibwe *Omini-zibi,* referred to the now-extinct passenger pigeons that once appeared in great numbers to nest each spring near its shores. The birds were last seen on the shore around 1900. By the early 1940s woodland caribou had disappeared from northern Minnesota, leaving behind place-names such as Caribou River and Caribou Lake.

Unfortunately, not all place-names were recorded with a deep appreciation or respect for Indian culture. Nothing generated more misunderstandings than references to the supernatural. The name of the Manitou River, for example, was derived from the Ojibwe words *Manidowish* or *Manidobimadga-zibi,* meaning "spirit or ghost." Historians theorize that the ethereal mists created by the falls of the Manitou River may have inspired the spiritual reference. Sometimes, however, whites simply substituted the word *devil* for *spirit,* as in the case of the Devil Track River, which has its origin in the Ojibwe name *Manido bimadigako-wini-zibi,* meaning "The Spirits (or God) Walking Place on the Ice River." Citing Wisconsin, which has no fewer than thirty-three place-names with the word *devil* in them, Vogel observes:

> The word has been applied to places by whites because of a misunderstanding of the names given to them by the Indians. There is, in fact, in native cosmography, no single evil spirit comparable to the Devil of the Christians. There are many spirits of all shades from good to bad, called Manito by the Algonquians and Wakan by the Siouans. As Ruth Landes has explained it, "In the world-view provided by Ojibwa religion and magic, there is neither stick nor stone that is not animate and charged with potential hostility to man. . . ."
>
> Europeans simplified these words so that we have innumerable names on the map resulting from their translations. Whites commonly thought of Indians as "devil worshippers" because of their propitiation of evil spirits, and so Indian words meaning "spirit" or "sacred" or "mysterious" were simply put down as "devil."

After the North Shore was opened to white settlement in 1854, many new names began to appear, adding layers to its rich toponymic legacy. Unlike the native people before them, these newcomers often chose human rather than nature-based names. Some honored military heroes or the region's early explorers. Others acknowledged the workers and settlers of the region. The lake landings from which logs were gathered into rafts to be floated to sawmills across the lake, for example, routinely bear the names of logging camp foreman, such as King's Landing near Beaver Bay and Kennedy's Landing near Little Marais.

As Vogel points out, place-names also chart the migration of people. Hovland was named by one early settler for his hometown in Norway. A compatriot, Hans Engelson, named the town of Tofte for the village he left behind in the Bergen district of Norway. Swedish immigrants chose the name Lutsen to commemorate the site of the German battlefield in which the Swedish King Gustavus Adolphus was killed in 1632.

Whether they refer to people, faraway battles, or living organisms, place-names often are regarded as disembodied artifacts of history. In many cases, however, they provide a moving record of human experience in response to a particular piece of earthly real estate. Place-names are emotional and cognitive, as well as geographic, markers that commemorate mourning for the death of a friend or the loss of an overseas home-land, offer practical prescriptions for survival in difficult places, or express gratitude for nature's beauty and bounty. The camp-side readings of George Eliot's novel *Daniel Deronda* must have given the members of the Minnesota Geological Survey group in 1880 such great pleasure after a hard day's work in the field that they named the bay on which they were camped Deronda Bay. Kimball Creek was named in honor of Charles D. Kimball, a member of an earlier geological reconnaissance, who on the night of August

WOODLAND CARIBOU

By the early 1940s woodland caribou had disappeared from northern Minnesota, leaving behind place-names such as Caribou River and Caribou Lake. Gilbert Boese, *Woodland Caribou*, 1940 (watercolor). Courtesy of the Minnesota Historical Society.

8, 1864, inexplicably jumped from his team's boat and fled into the woods. The next morning his distraught colleagues found his clothes and shoes on the beach and the footprints of his bare feet leading into Lake Superior, from which his body was later recovered. Profound feelings of loss and tenderness are captured in their decision to name the creek after their drowned colleague: "As a slight tribute to his memory we leave his name to the mountain stream where an inscrutable Providence separated him from his companions and the world."

What's in a name? It can bring to life something as simple as the pleasure of human companionship, good food, or the beauty of nightfall along the North Shore, as it does in an account written in 1865 by state geologist August H. Hanchett as he watched a night-fishing party of Ojibwe at Wauswaugoning Bay (from the word *Wasswewini-wik-wed* or "making-a-light-by-torches"). As darkness descended, Hanchett writes, suddenly a "dozen or more torchlights burst to view under the shade of the black spruces crowning the bluffs of the east shore of the bay; they moved in line a dozen yards from each other in a westerly direction, presently they began to dart hither and thither, the lights reflected down in the clear water gave them the appearance of so many fiery comets, as each native captured a fish his shout and the shrill laugh of the boys and girls in the canoes echoed and multiplied until the bay, cliffs, crags and forest seemed a merry pandemonium; of course sleep was out of the question, but we were amply remunerated with a breakfast of the most delicious white fish [an] epicure could covet."

SUGGESTIONS FOR FURTHER READING

Culkin, William E. *North Shore Place Names.* St. Paul: Scott-Mitchell Publishing Co., 1931.

Densmore, Frances. "Uses of Plants by the Chippewa Indians." In *Forty-fourth Annual Report of the Bureau of American Ethnology to the Secretary of the Smithsonian Institution, 1926–1927.* Washington, D.C.: Government Printing Office, 1928.

Fritzen, John. *Historic Sites and Place Names of Minnesota's North Shore.* Duluth: St. Louis County Historical Society, 1974.

Gilfillan, J. A. Rev. "Minnesota Geographical Names Derived from the Chippewa Language." In *The Geological and Natural History Survey of Minnesota.* Fifteenth Annual Report for the Year 1886. St. Paul: Pioneer Press Company, 1887.

Hartley, Alan H. "The Expansion of Ojibway and French Place-Names into the Lake Superior Region in the Seventeenth Century." *Names* 28 (1980): 43–68.

Peters, Bernard C. "The Origin and Meaning of Chippewa Place Names along the Lake Superior Shoreline between Grand Island and Point Abbaye." *Names* 32 (1984): 234–51.

Smith, Theresa S. *The Island of the Anishnaabeg: Thunderers and Water Monsters in the Traditional Ojibwe Life-World.* Moscow: University of Idaho Press, 1995.

Upham, Warren. *Minnesota Place Names: A Geographical Encyclopedia.* 3rd ed. St. Paul: Minnesota Historical Society Press, 2001.

Vogel, Virgil J. *Indian Names on Wisconsin's Map.* Madison: University of Wisconsin Press, 1991.

WHERE HAS ALL
THE SEWAGE GONE?

Development
and Water Quality

I N 1846 THE TRAVEL WRITER CHARLES LANMAN VENTURED FROM HIS HOME in Washington, D.C., to the Mississippi River, paddling north to its headwaters before heading east on the St. Louis River into Lake Superior. Near present-day Duluth, he recorded in his journal that he experienced "the power of the Omnipotent" while reflecting on the "uninhabited wilderness in every direction around me."

If Lanman were to visit the mouth of the St. Louis River today, he would no longer recognize the land he described in such rhapsodic detail little more than 160 years ago. Cluttering the primordial view would be houses and factories, highways and shipyards. Remarkably, however, if he pushed his birch-bark canoe into the lake and peered over the edge into Superior's crystalline depths, he would thrill to the same sense of wonder. Kayakers who ply Superior's nearshore waters today could easily describe their own experience using the words from Lanman's diary entry of August 1846: "In passing along its rocky shores in my frail canoe I have often been alarmed at the sight of a sunken boulder, which I fancied must be near the top, and on further investigation have found myself to be upwards of twenty feet from the danger of a concussion." So intrigued was Lanman by the water's clarity that he had "frequently lowered a white rag to the depth of one hundred feet, and been able to discern its every fold and stain."

Of the five Great Lakes, Lake Superior is by far the clearest. Its waters are also the cleanest. With the exception of such pollution hot spots as Duluth, Ashland, and Thunder Bay, the Lake Superior watershed lacks the concentration of people, industry, and agriculture that lines the shores of the more heavily developed lower Great Lakes. In 1984 the Minnesota Pollution Control Agency (MPCA) recognized Superior's "exceptional recreational, cultural, aesthetic, [and] scientific resources" by declaring it an Outstanding Resource Value Water (ORVW).

This formal acknowledgment of Superior's notable qualities has brought greater scrutiny to practices along the Minnesota shore of the lake. ORVW regulations, for example, ban or strictly control a "new or expanded discharge of any sewage, industrial waste or other waste . . . unless there is not a prudent and feasible alternative to the discharge."

Pollution hawks say that protections like these are vital to the continued health of the lake but warn that they are just the beginning. The control of point-source discharges—those that spew out of the tail end of a pipe from a factory or sewage-treatment plant, for example—has helped to keep many harmful pollutants out of the lake. But Superior is far from being protected against increasing levels of contamination. Regulators are just beginning to tackle pollution from what they call nonpoint sources. Dangerous toxins are carried to the lake via the atmosphere hundreds, sometimes thousands, of miles from their place of origin and deposited into Superior via dust, rain, and snowfall.

Not all contaminants, however, can be traced to upwind coal-fired utilities or pesticides from southern cotton fields. State and regional authorities are attempting to stem the flow of more localized pollutants. High levels of roadway salt contained in snowmelt runoff, for example, can be toxic to aquatic plants and animals. (See also "Nearshore.") Sediments that course into the lake from road cuts and construction sites contain nutrients that stimulate excessive algal growth, leading to a reduction in water transparency. Excessive sediment loads also are implicated in the smothering of fish-spawning habitat and the clogging of fish gills. They can further harm fish and other aquatic organisms by accelerating the loss of oxygen from the lake both in the summer and in winter under the ice. (See also "Healing a Watershed for Coaster Brook Trout.")

Among the most troubling is the runoff that oozes from the North Shore's many failing septic systems, one flush of the toilet at a time. In 1991, the North Shore Management Board, a group that represents local governments on the management of the Minnesota coastline of Lake Superior, used aerial photography to survey thirty-two miles of Superior shoreline from the Lester River in Duluth to Encampment Island located just east of Two Harbors. Researchers isolated wet spots in 410 sites that they suspected were signs of seeping septic systems. An on-ground inspection of 231 sites by officials from the Western Lake Superior Sanitary District (WLSSD) revealed that 70 percent of septic systems were failing to treat wastewater adequately. Most of the systems were antiquated and in need of replacement. Of special concern to the investigators, however, were the malfunctions of expensive, new systems that discharged partially treated sewage beneath the soil or onto its surface.

Data gathered on this stretch of coastline are representative of septic failure rates shore-wide. According to WLSSD estimates, more than half of all North Shore systems are not up to snuff; that is, they release wastewater before disease-causing organisms, nutrients, and chemicals are fully removed. Problems range from systems that seep partially treated effluent year-round under the surface of the soil to systems that fail only when the soil is saturated with heavy rainfall or snowmelt, causing them to send pulses of polluted runoff across the land and into surface waters.

Rich Axler, a limnologist at the University of Minnesota's Natural Resources Research Institute (NRRI), Duluth, who studies the aquatic ecosystems of oligotrophic, or nutrient-poor, lakes such as Superior, points out that failing septic systems are of concern for two reasons. Partially treated wastewater contains pathogens that pose a threat to human health. They include disease-causing microorganisms such as viruses (enteroviruses and hepatitis, among others), bacteria (salmonella and shigella), and parasites (cryptosporidium and giardia). These organisms can be carried into groundwater, where they may contaminate the wells that people depend on for drinking water. Humans can also become infected by coming into contact with wastewater that has ponded in backyards due to malfunctioning septic systems.

When pathogens flow into Superior, they can become concentrated in areas with poor water circulation, such as harbors, which are heavily used for recreation. They also endanger the health of people in households and institutions that rely on unfiltered water from the lake for their drinking supplies.

Discharges from failing septic systems also contain high levels of nutrients that can upset the chemical and biological composition of aquatic ecosystems. Lakes, for example, receive minerals, such as phosphorus and nitrogen, in runoff. These substances, which fertilize the lake's plant life—most notably, microscopic algae and rooted aquatic plants—can make their way into surface waters either in dissolved form or by adhering to soil particles that are flushed into lakes during spring snowmelt or storm surges.

Oligotrophic lakes, such as Lake Superior, have evolved in relatively unfertile conditions. (*Oligotrophic* comes from the Greek *oligos,* meaning "small," and *trophe,* meaning "nutrients.") Superior's oligotrophy is due to several factors. The annual nutrient load that Superior receives from the surrounding land is relatively meager compared to its vast volume. That is because Superior drains a small land area relative to its size. Its catchment basin measures 49,300 square miles, while the lake covers 31,280 square miles. (By comparison, the Mississippi River drains an area of 1.2 million square miles.)

Furthermore, the glaciers plowed away much of the region's topsoil and largely scoured the basin of softer, calcium-rich rocks such as limestone. As a result, much of the land is covered by thin, nutrient-poor soils or underlain by hard, crystalline bedrock that resists erosional forces such as wind, ice, and rain. Unlike the lower lakes, a large percentage of the basin also remains forested, a land cover that serves to hold soils in place and to tie up nutrients in plants and litter. Because of these natural conditions, Lake Superior's algal community is nutrient starved, especially for phosphorus.

Organisms in Superior and other lakes with low productivity farther inland have adapted over thousands of years to thrive in these relatively austere conditions. The algae that live in them, for example, are opportunists that have evolved the ability to quickly grow whenever nutrients, especially phosphorus and nitrogen, are present. Herein lies the problem. Inadequately treated effluent contains high levels of phosphorus and nitrogen that result naturally from the microbial breakdown of organic materials such as human waste. When septic systems are poorly sited—say, on thin soils, on impervious clays, on land with high water tables, or too close to shorelines—the effluent can flow into lakes, inundating them with fertilizing compounds. When the volume of these nutrients exceeds a lake's capacity to assimilate them, they overstimulate the growth of sensitive plants such as algae, turning crystalline waters a murky green or brown or filling them with floating scums of blue-green algae during ice-free parts of the year.

Nuisance algae do more than just make life miserable for swimmers and boaters, however. When these booming algal populations die, they are decomposed by bacteria, fungi, and other microorganisms that in the process of degradation deplete the water of the oxygen needed by fish and other aquatic life forms.

In time, the nutrient-enriched conditions lead to permanent changes that cascade throughout the lake's biological community. Populations of zooplankton and benthic organisms (the invertebrates that live in and on the lake bottom) decline in response to changes in the lake's plant community and the loss of oxygen from deeper waters and bottom sediments, conditions that become especially acute during summer when lakes are stratified. The reduction in food and oxygen eliminates species such as lake trout and walleye, which require cold water and high oxygen levels to survive. These chemical and biological disruptions leave lakes more susceptible to invasions of exotic plant and fish species, degrading not only their ecological integrity but also their recreational and aesthetic value. Left unchecked, the overfertilization can accelerate the eutrophication process, the natural chemical, physical, and biological changes that lakes undergo as they age.

Skeptics admit that nutrient overloads have led to serious problems in many of the state's inland lakes and in the lower Great Lakes, particularly Lake Erie. (See also "Lake Superior.") But they argue that Lake Superior is large enough to assimilate the septic runoff of even a heavily developed coastline such as the North Shore. Tim Kennedy, who oversaw the regulation of on-site septic systems in Cook County as director of planning and zoning from 1977 to 1998, said he frequently encountered this perception, especially when homeowners were faced with the prospect of installing a costly new system for the purpose of environmental protection. "People comment, how could we possibly pollute Lake Superior with a sewer system? It's like putting a drop in the bucket," he observes.

Nuisance algae, often associated with nutrient-laden runoff from poorly sited septic systems, can disrupt the areas of greatest biological productivity in the lake—in bays or around the mouths of rivers and streams. Photograph by Craig Blacklock.

It would be a drop in the bucket, Axler says, "if all the contaminants got dumped into the middle of the lake and stirred up with a giant paddle, but this is not the way pollutant loads work." Most septic discharges are carried to Superior via groundwater flows and streams that enter the lake at the shore. Therefore, septic pollution can disrupt chemical and biological processes where the lake is perhaps most vulnerable—in bays or around the mouths of rivers and streams, centers of the greatest biological productivity and human use.

Furthermore, these contaminants can be trapped in nearshore areas during critical times of the year when deepwater organisms, such as lake trout, visit shallow waters to feed or breed. In the spring, for example, the relatively warm waters of tributaries that flow into the lake along the shore are prevented from mixing with colder offshore waters by a vertical temperature barrier known as a thermal bar. During this time, offshore temperatures normally range from 32 to 39 degrees Fahrenheit, while nearshore temperatures may reach 50 degrees Fahrenheit. The warmer temperatures and flush of nutrients in the spring runoff fuel biological productivity in these nearshore zones. (Anglers, for example, locate this fertile seam of life by looking for the distinct gradient between the turbid, green- or brown-colored water of the nearshore zone and the clear, blue waters of the open lake.) Of concern to scientists such as Axler is the fact that the lake's thermal bar could trap and concentrate stream-borne contaminants in the productive nearshore zone and circulate them via currents along the lake's shoreline.

Polluted inputs into these areas close to shore, Axler says, do not necessarily remain there forever. By the 1960s, excessive phosphorus loadings into Lake Erie, for example, had turned the lake into a bowl of thick algal soup. Fish kills due to oxygen depletion were common. Studies in the early 1970s showed that in the large ecosystems of the Great Lakes, eutrophication first develops in nearshore regions before eventually moving out into the open lake.

When the North Shore was sparsely settled, Axler says, septic runoff probably had little impact on the lake's nearshore ecosystem. But now that densities have increased, he and other researchers—and some longtime lakeshore dwellers—are concerned that nutrients from septic systems in concert with other factors, such as the increased runoff of lawn fertilizers and soil erosion from stepped-up development, could degrade coastal water quality. "There are some well-done scientific studies, such as those conducted on the once-pristine Lake Tahoe in the Sierra Nevada Mountains," Axler says, "that show that the noxious growth of attached algae—the slimy stuff that grows on the rocks in some areas and not in others—is linked to nutrients in water seeping into the lake from below or from surface and tributary runoff. In Lake Superior this kind of seepage will likely have no real impact on the lake water as a whole. The nearshore zone, however, can be affected. This is where you get your drinking water in a lot of cases, it's where people recreate and where biological resources such as spawning fish are concentrated."

Building a Better Sewage Trap

Concerns about treating the sewage of burgeoning human populations are not new on the North Shore. According to Jeff Crosby, environmental health specialist for the St. Louis County Health Department, the area's first on-site sewage-treatment system—the outhouse—was a safe and effective way in most circumstances to contain wastes and

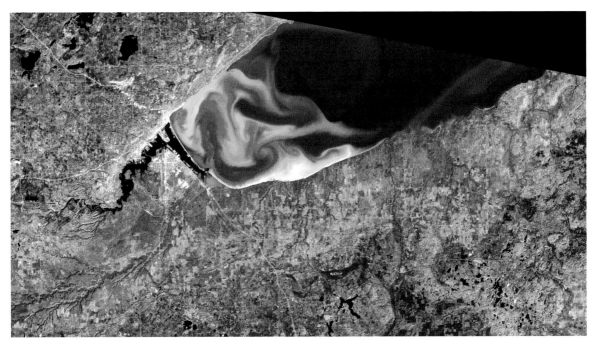

expose them to organisms that killed problem-causing microbes and degraded organic materials.

Problems arose, however, with the advent of indoor plumbing. Once it was mixed with water, sewage could travel far from its place of origin into rivers and lakes—the source of drinking water for many people. As population densities grew, so did the amount of raw sewage—and related outbreaks of such diseases such as typhoid fever, which is caused by ingesting food or water contaminated with the typhoid bacillus. According to limnologist Wayland R. Swain, "Early in this century, all of the major centers of population [in the Great Lakes] seemed to be adversely affected by diseases directly associated with sewage pollution." In 1896, for example, Duluth set a city record for typhoid infections—2,020 cases, about one in every twenty people.

The city instituted chlorine treatment for drinking water in 1912 and virtually eliminated the threat from this highly infectious disease by killing the typhoid pathogen in wastewater. The remedial step was part of a larger binational initiative to rid Great Lakes waters of disease-causing organisms from untreated municipal sewage, particularly in the heavily polluted waterways that connected the Great Lakes. Just three years earlier the United States and Canada signed the 1909 Boundary Waters Treaty, which addressed the public health threats from contaminated water. In the treaty, each country asserted its right to use the waters of the Great Lakes. At the same time, the United States and Canada recognized their shared environmental responsibility and asserted that "boundary waters and waters flowing across the boundary shall not be polluted on either side to the injury of health or property on the other."

By 1940 the rate of typhoid-related mortalities in the Great Lakes had plummeted, largely due to the construction of sewage-treatment plants and filtration facilities for drinking water. Still, many rural households remained outside the reach of centralized

Septic pollution can be trapped in bays, depleting oxygen levels during critical times of the year when deepwater organisms, such as lake trout, visit shallow waters to feed or breed. Landsat image provided by Dr. Peter T. Wolter, Iowa State University.

municipal systems. The sewage strategies they developed, Crosby says, initially focused on shunting waste out of sight. Some people buried old cars or perforated drums, for example, and ran septic effluent into them where it could slowly seep into the ground. In time, however, public health officials recognized the need not only to divert sewage away from homes and yards but also to provide treatment that could kill problem pathogens.

Ironically, these early attempts at wastewater disposal provided a crude model for traditional septic designs that are still in use today. These systems treat waste in two stages. Household sewage first flows into a watertight tank, where solids are separated. Heavier material sinks to the bottom of the tank, while smaller particles float to the surface. Here, anaerobic bacteria (those that live in the absence of oxygen) begin the important process of decomposition.

Pretreatment in the septic tank (ideally for a period of about thirty-six hours) is not sufficient, however, to kill all disease-causing microbes. Septic wastes need to undergo a second level of biodegradation. Each time a toilet is flushed or a kitchen sink is drained, new wastewater enters the tank, forcing the partially treated water into perforated pipes that fan out into a drain field of gravel or soil. As wastewater seeps into the drain field, aerobic bacteria (which are far more efficient workers than their anaerobic cousins) finish the job of sewage treatment. If the system is designed properly, the soil itself also acts as a fine filter, removing most organisms before the wastewater reaches the water table.

This final process of water "polishing" is both simple and ingenious. The aerobic bacteria secrete a sticky film that enables them to adhere to the soil or rock. Eventually they form what is known as a biomat that lines the bottom and walls of the drain field. In systems that are working properly, the effluent pools in the bottom of the drain field, known as the saturated zone, and then slowly percolates into surrounding soils, called the unsaturated zone, for treatment. The biomat serves as a kind of mucousy filter that dramatically slows the flow of wastewater so that it trickles into, rather than inundates, the surrounding soils.

This trickle-down effect is vital to the successful removal of pathogens. Slowing the movement of water maximizes sewage treatment by lengthening the contact that bacteria-rich soils have with the effluent in the unsaturated zone. Here, the negatively charged soil particles capture the positively charged pathogens in the effluent. In time, the captive pathogens die due to changes in temperature, a lack of moisture or food, or from the attack of other bacteria, viruses, or microscopic animals.

The biomat also serves a second important function. By slowing the flow of water, it helps to keep the surrounding soils in which final treatment takes place from becoming saturated, that is, losing the air pockets, or soil pores, that aerobic bacteria need to survive. When these soils are flooded, the pores become compacted, and the bacteria die of oxygen starvation.

State of Minnesota septic standards require that effluent pass through three feet of dry (unsaturated) soils to be considered treated. According to a publication from the MPCA, "Maintaining an unsaturated zone surrounding the trenches is the single most important factor in preventing transmission of pathogens." "Weepy" drain fields—trenches that are soggy around the edges—indicate that the function of the biomat has become impaired and that effluent is escaping to the surface before being thoroughly treated.

Crosby says that contrary to public perception, on-site septic systems can provide wastewater treatment that results in a higher level of water quality than that of most municipal treatment plants. He points out, for example, that during periodic heavy rainfalls when municipal plants are overwhelmed by surges of wastewater and storm runoff, operators sometimes are forced to open the floodgates. These surges, known as infiltration, can flush two million gallons of raw sewage into the lake in a single episode.

Even under ideal conditions, municipal plants do not provide problem-free treatment. The process produces polluting chlorine disinfectants and leftover sludge known as biosolids. Furthermore, the water they discharge is not completely cleansed of contaminants but contains levels of pollutants within allowable limits set by state and federal authorities.

Treatment plants can also pollute the environment in less overt ways. Axler, who has studied the aquatic system of Lake Tahoe (an even more oligotrophic system than Superior), says that the municipal sewage system that was installed near the lake in the 1960s has not completely stemmed the flow of nutrient-rich septic runoff into the sensitive lake. Sewer pipes are made to shift slightly to avoid cracking. These shifts cause minor leaks, known as exfiltration, that allow effluent to seep into the ground in small, but continuous, doses. The excess nutrient loading has led to huge increases in the growth of periphyton, the slimy algae that grows on rocks and piers. Far less obvious to Tahoe dwellers, however, is another indicator of lake degradation: declining water transparency. Since 1968 Secchi disk readings have registered the dramatic loss of sixty-two feet in water clarity.

On-site systems have another important advantage over large, centralized treatment facilities. Crosby and Axler point out that compared to the cost of constructing additional plants, on-site septic systems provide a more affordable treatment alternative in sparsely populated rural areas. In 1972, alarmed by the growing pollution of the Great Lakes, the United States and Canada signed the Great Lakes Water Quality Agreement. It was accompanied in the United States by the Clean Water Act, which, among other provisions, mandated—and subsidized—new sewage-treatment plants to control excessive loadings of nutrients into the Great Lakes. From the 1960s through the mid-1980s, the United States and Canada shelled out a combined $8 billion for the building of new plants. But local jurisdictions can no longer count on the federal government for footing the bill for municipal sewage-treatment facilities.

When they are well-designed and fully functioning, on-site septic systems can circumvent the need for hugely expensive infrastructure, Crosby says, assimilating dangerous microbes and other pollutants almost completely into the environment at a fraction of the cost. "It's actually a step backwards," Crosby says, "to pipe wastewater to a plant when you can get this level of treatment near the source."

On-Site but Not out of Mind: Failing Septic Systems

So, given its virtues, why has on-site wastewater treatment prompted so many concerns? Why, for example, did a citizens' task force charged with developing a water-management plan for Lake County in 1991–92 single out failing septic systems, along with airborne pollutants, as the top two threats to water quality in the county?

To understand many of the North Shore's septic problems, you have to go back to the glaciers, which scraped the land down to bedrock in some areas while pock-marking the terrain with shallow watery basins in other places. Many of today's septic woes result also from the legacy of ancient glacial lakes whose bottom sediments blanket huge sections of the nearshore area, particularly in Lake County. Wayne Seidel, a conservation technician for the Cook and Lake County Soil and Water Conservation Districts, points out that 85 percent of Lake County residents—nearly eleven thousand people—live along the coast within five to six miles of Lake Superior. The area lies in a belt of heavy clay soils, a substrate "that's inherently unsuited to traditional septic treatment systems," Seidel says. Clay, which is composed of fine, densely packed grains that form a largely impervious substrate, prevents wastewater from the slow percolation it requires for treatment. Clay soils also become more easily saturated than more porous sandy or loamy soils. Once they are saturated, say, by heavy rainfall or snowmelt, clay soils remain wet for a longer period of time, creating an oxygen-poor environment that kills most beneficial microbes. Wet clay soils also repel excess water rather than absorb it, causing effluent to pool on the soil surface. The contaminated water then can be readily swept into waterways by rainstorms or snowmelt.

Many site conditions in Cook County are equally inhospitable. Property owners often find bedrock beneath a few precious inches of soil or high water tables. As the

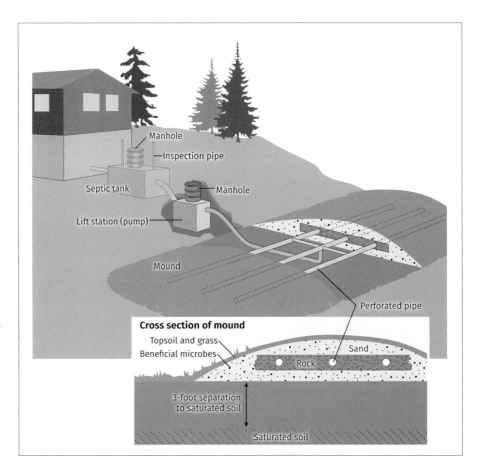

In the late 1970s and early 1980s, researchers at the University of Wisconsin developed a new septic technology, known as the mound system, which treats sewage by passing effluent through a constructed mound of clean sand or crushed rock.

pace of real estate activity continues to skyrocket, septic problems surface with greater frequency. Cook County sanitarian LeRoy Halberg observes that most lots with adequate site conditions for accommodating traditional septic systems are either occupied or are prohibitively expensive for new property seekers on limited budgets. As a result, "we're getting greater pressure to develop the lots that have poorer soils. Many of them have been left over for a reason: Where could you put the sewage systems? But people still want to move to Cook County. That hasn't stopped."

To help owners overcome the natural limitations of their land, in the late 1970s and early 1980s researchers at the University of Wisconsin developed a new septic technology known as the mound system. In sites that lacked the three feet of dry soil necessary for the adequate treatment of effluent, a mound of drain rock, sand, or soil was constructed on the surface of the ground. The size of each mound varied depending on such factors as estimated water usage and soil type.

The mound system quickly became one of the most widely used alternatives to traditional septic designs on the North Shore. However, in the 1991 study of septic systems by the North Shore Management Board, investigators found that 50 percent of mound systems they inspected were failing, even new systems. "If mound systems are designed correctly, installed to the right specifications, if they are maintained and operated correctly by the landowner, they'll function well for twenty-five years," Seidel says. "Some nitrates aren't being collected by the system but, by and large, they're doing a good job of treatment. But these are all big ifs."

In the rush to implement the new technology, many mound systems are not installed correctly. Construction equipment driven over drain fields compacts materials, crushing the air pockets that sustain aerobic bacteria. Sometimes the wrong material is used for drain fields. Mound systems require clean sand or crushed rock, that is, materials without silt or clay, which can become saturated and block the slow percolation of effluent.

Sometimes mounds are poorly sited because owners did not want to install systems in places where they might obstruct scenic views or interfere with other uses. To maximize results, mound systems must be located away from woody vegetation such as trees, whose roots can infiltrate drain fields. They also require sunny sites to sustain the cover of shallow-rooted vegetation, such as turf, that quickens the rate of moisture evaporation.

In most cases of failure, the design of the mound was too small to adequately handle waste loads, a problem that many regulators often blame on the inexperience of increasing numbers of urban property owners, who know little about the function and maintenance of septic systems. Accustomed to the flush-and-forget-it attitude of urban wastewater systems, many users simply overburden the ability of on-site septic systems to treat waste. Mound systems function correctly only if people use water sparingly. A system that is designed to treat 400 gallons a day, for example, works best when water use is limited to 100 gallons a day. But few households practice such stringent conservation, largely because users are unaware of how much water they actually consume. Figures compiled by the Minnesota Extension Service estimate that the average family of four uses 260 gallons of water per day, 40 percent of which is used to flush toilets. "A lot of homeowners don't realize how much water they consume," says Barb McCarthy, a soils scientist at the NRRI, who is working on developing alternative technological solutions to wastewater problems in northern Minnesota. "Most say they don't use much water but then do six to seven loads of laundry on a Saturday, which probably translates to about 300 gallons of water."

The irony, Axler says, is that in the land of ten thousand lakes, including along the shore of the world's largest lake, "we need to practice substantial and sometimes radical water conservation."

Solving Septic Problems through Technology and the Law

Regulatory and technological steps have been taken to address pollution caused by septic systems. In 1994 the State of Minnesota passed a law known as the Individual Sewage Treatment Act. It mandated that local sewage ordinances conform to statewide standards and called for the training and state licensing of septic contractors. The law also stipulated that in the sale or transfer of property, owners must disclose the condition and location of septic systems and bring their systems into compliance before the property changes hands.

Critics charge that the law does not go far enough. It only targets new homes or homeowners seeking to renovate their properties, leaving many existing polluting systems in place. Furthermore, new conventional systems, which can cost individual homeowners anywhere from $8,000 to $15,000, and resort owners $100,000 or more, are prohibitively expensive for many people. And without a strong educational program that informs users about water conservation and maintenance, even new systems can fail.

Left out of the picture altogether are older homes and cabins built on lots that are simply too small, too rocky, or too wet to install new conventional sewage systems. Built up decades before setback ordinances and septic regulations were instituted, the north shore of Devil Track Lake in Cook County, for example, became dotted with a series of old cabins whose failing septic systems contributed significant pollution to the lake. The tiny lot sizes, typical of many older recreational developments, precluded the siting of compliant new systems. In an innovative land swap with the state in 1998, the Trust for Public Land (TPL), a national organization that protects land as open space, traded lakeshore that it owned elsewhere in the county for state land across the road from the troubled parcels. The TPL then donated this land to Cook County, which, in turn, made parcels available to the Devil Track property owners. The trade gave owners the option to purchase the additional land needed to site compliant septic systems.

Such opportunities are rare, however. To help solve tough septic problems, NRRI researchers have tested new technologies for several hard-to-fit sites around the Arrowhead region. In a demonstration project on Grand Lake, located about fifty miles north of Duluth, McCarthy and Axler teamed up with sanitarian Crosby to remedy the septic woes of a group of ten homes that were constructed in the 1930s on a boggy, low-lying shoreline of the lake. Some residents had piped sewage into buried drums, using drain fields located between the homes as a catchment for secondary treatment. The wet, boggy soils, however, did not provide thorough sewage treatment. Furthermore, without adequate drainage, the effluent frequently pooled in residents' backyards, where it killed grass, leaving an unsightly black crust. To make matters worse, residents were plagued by a distasteful odor that Crosby describes as "like that of a hog barn." In desperation, one resident installed an expensive mound system only to have the system's heavy gravels sink into the underlying ground layer of soft peat.

McCarthy, Axler, and Crosby secured funding to hire two consulting engineers to help design a collective system for the ten households. Incorporating the spruce-forested peatland located behind the cabins, the team selected a new septic technology that uses constructed wetlands to imitate the water-cleansing processes of natural ones. On a rise of land located in the middle of the wetland, the team dug two cells. Pipes from the cluster of homes carried wastewater into the first cell, pumping it into the system just below the surface of a lined gravel bed planted with native cattails. The roots of the plants not only supply oxygen to root-dwelling aerobic bacteria that break down waste, but they also serve as a stable substrate for the growth of the bacterial biomat. In addition, the network of roots within the gravel filters larger particles. The plants work a double shift by also taking up some of the nitrogen and phosphorus in the wastewater. The pretreated effluent then moves into an unlined second cell, where it undergoes additional treatment before slowly dispersing into the surrounding soils for further breakdown. On average, the process takes about ten days.

The proper functioning of constructed wetlands, like conventional systems, is based on a simple operating philosophy: the slower the treatment, the cleaner the water. As a result, the team strictly limited water usage to about one

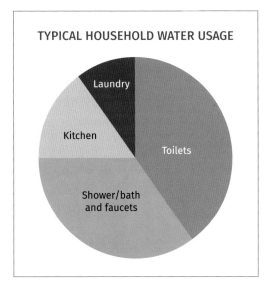

TYPICAL HOUSEHOLD WATER USAGE

A typical household of four uses 260 gallons of water each day. The proper functioning of on-site septic systems, however, depends on strict water conservation in ordinary household uses ranging from showers and toilet flushing to laundry.

TYPICAL WATER USE (IN GALLONS)

Action	Typical Use	Conservative Use	Ultra-Conservative Use
Toilet flushing	6 (old standard)	1.5–3 (low flow)	Composting toilet
Tub bath	30 (1/2 filled)	15 (1/4 filled)	Sponge bath
Shower	10 min: 50 (5 gal/per minute) 3 min: 15 (5 gal/per minute)	25 (2.5 gal/min) 7.5 (2.5 gal/min)	Camper style (3 gal)
Laundry (full load)	Top loading: 50–60 (older models) Front loading: 33 (older models)	40 (newer models) 17–28 (newer models)	Laundromat
Dish washing	Machine: 12–15 (old-reg cycle) Hand: 16 (faucet rinse)	6–9 (new-reg cycle) 6 (basin rinse)	
Teeth brushing	2 (faucet running)	1/8 (wet brush, brief rinse)	
Hand washing	2 (faucet running)	1 (basin; brief rinse)	
Shaving	3–5 (faucet running)	1 (basin; brief rinse)	

hundred gallons per day per household to avoid flushing the system with too much water. And to help residents monitor their consumption, they installed meters in each home.

For Every Solution a New Problem

In the future, McCarthy predicts that multihousehold collector systems like these will replace mound systems on problematic sites. In rural areas such as the North Shore, they may solve more than just ecological problems. The hefty price tag for individual on-site septic systems is currently a barrier to building badly needed housing for low- and moderate-income residents. Multiplying the number of system users will help to lower the cost of septic-system construction, say advocates for affordable housing.

Alternative technologies, however, bring with them a whole array of other potential problems. Until now the zoning regulations for septic systems have been used as a de facto control on development. "Sewage treatment is one of the limiting factors for development in Cook County," Kennedy says. "If you can't provide sewage treatment, you're limited with what you're able to do on your piece of property."

New septic-system technologies allow greater development densities that bring with them a host of cultural and ecological changes. Cook County zoning ordinances for properties fronting Lake Superior, for example, currently mandate lot sizes with a two-hundred-foot minimum width. The regulation is driven by the constraints of traditional septic designs—it ensures that adequate space is provided for septic installation as well as for the safe separation of sewage treatment from drinking-water wells. New technologies could dramatically reduce the space needed for disinfection, allowing for a greater density of development that threatens to further degrade the shore's wild character in a wave of new home and resort construction. In fact, these technologies could serve as an incentive for property owners to subdivide their large lots (which normally measure one acre or more in size), particularly under the pressure of rising taxes. "Heavy pro-development people see these new technologies as a cure-all," Axler points out. "They reason that if we can treat the water, then we can put in more homes and more businesses and derive more taxes."

Land-use decisions, critics caution, do not take place in a vacuum but are part of a larger ecological matrix. This interconnectedness of human and natural systems is perhaps nowhere more evident than in technologies designed to mitigate human wastes. Stepping up the pace of development without recognizing the region's finite natural resources and limited ecological carrying capacity could be disastrous for many terrestrial and aquatic ecosystems. Axler points out that while new technologies will safeguard the public from enteric diseases, they still will only partially remove algae-stimulating nutrients, discharging enough nitrogen and phosphorus to degrade streams and lakes.

Greater development also poses additional threats to water quality. "If you increase the density of development along the shore through collective septic systems and other alternatives, you may solve one water quality issue but create many others," says David Abazs, former water-plan coordinator for Lake County. More development, he says, increases fertilizer runoff, loss of shoreline vegetation, soil compaction, and sediment runoff from construction sites, road cuts, and residential and commercial properties. According to the MPCA, sediment losses can range anywhere from 5 to 120 tons per acre from one construction site alone, as high as a hundred times the amount of soil that runs off a comparable parcel of agricultural land.

Crosby is also concerned that in a rush to develop properties with unfavorable site conditions many new technologies could be implemented before the bugs have been adequately worked out for their functioning in northeastern Minnesota's demanding environmental conditions. While initial reports are positive, most of the new nontraditional septic technologies have been field-tested only for less than a decade. "Sometimes in the stampede to solve a perceived problem," he says, "you actually end up creating more problems than you started with. The North Shore needs to take a good look at that."

The question is, do North Shore residents have a coherent vision for the future of the region that can withstand new challenges and resist the kind of pell-mell development that has ruined the natural beauty and ecological health in other wild parts of the United States? Kennedy believes that even with expanded sewage-treatment capacities, residents will not be tempted to amend current zoning regulations that keep Cook County from becoming like other popular recreational destinations in the state where the lakes are ringed with dwellings and their waters choked with pleasure boats. "Just because we develop new technologies," Kennedy says, "doesn't mean we want to change the area."

Karen Evens, former water-plan coordinator for Cook County, is less certain about the region's preparedness to successfully meet new challenges. Already, she says, new technologies such as sand filters and drip irrigation have allowed development to occur on lands that previously were not eligible, because they lacked the appropriate site conditions for traditional mound and trench systems.

Even more worrisome, she says, is the North Shore's checkered track record for anticipating and coping with change. In the course of one generation, the North Shore landscape has been transformed beyond recognition. The pace of development is not likely to let up anytime soon. Cook County is one of the three fastest-growing regions in Minnesota, itself the fastest-growing state in the Midwest. "People in the 1950s," Evens observes, "would have said that today's development wasn't possible." ❧

SUGGESTIONS FOR FURTHER READING

Bangay, Garth E. "Great Lakes Development and Related Water Quality Problems." *Proceedings of the International Association of Theoretical and Applied Limnology* 21 (1981): 1640–49.

Minnesota State Board of Health, Minnesota Commissioner of Game and Fish, and the Wisconsin State Board of Health (1928–1929). "Investigation of the Pollution of the St. Louis River below the Junction of the Little Swan, of St. Louis Bay, and Superior Bay, and of Lake Superior Adjacent to the Cities of Duluth and Superior." St. Paul, 1929.

Moen, Sharon. "Spotlight Hits Unsavory Subject." *Seiche,* June 2000: 1–9.

Moore, Anne Perry. "Up North and Personal: Sprawl in Northern Minnesota." *Minnesota Environment* 1 (Fall 2000): 5–7.

Perich, Shawn. "Wastewater Woes." *Minnesota Volunteer* 58 (January/February 1995): 42–48.

Rice, Jason. "Flush with Ideas." *Lake Superior Magazine,* June/July 2004: 38–43.

Swain, Wayland R. "The Great Lakes: An Example of International Cooperation to Control Lake Pollution." *GeoJournal* 5, no. 5 (1981): 447–56.

———. "Great Lakes Research: Past, Present, and Future." *Journal of Great Lakes Research* 10, no. 2 (1984): 99–105.

INTERNET RESOURCES

Great Lakes Information Network, http://www.great-lakes.net

Minnesota Department of Natural Resources, http://www.dnr.state.mn.us

Minnesota Pollution Control Agency, http://www.pca.state.mn.us

Minnesota Sea Grant, http://www.seagrant.umn.edu

Minnesota Shoreland Management Resource Guide, http://www.shorelandmanagement.org

National Small Flows Clearinghouse, http://www.nesc.wvu.edu/wastewater.cfm

Shoreline Alterations: Natural Buffers and Lakescaping, www.dnr.state.mn.us/publications/waters/shoreline_alteration.html

University of Minnesota Extension Service, Onsite Treatment Program, http://www.septic.umn.edu/

Water on the Web, http://wow.nrri.umn.edu

Botanical Exploration
along the Lakeshore

Two or three days previously the Chief had noticed, among the passengers, a gentleman out for his holidays on a botanical excursion to Thunder Bay, and, won by his enthusiasm, had engaged him to accompany the expedition. At whatever point the steamer touched, the first man on the shore was the Botanist, scrambling over the rocks or diving into the woods, vasculum in hand, stuffing it full of mosses, ferns, lichens, liverworts, sedges, grasses and flowers, till recalled by the whistle that the captain always obligingly sounded for him. Of course such an enthusiast became known to all on board, especially the sailors, who designated him as "the man that gathers grass" or, more briefly, "the hay picker" or "haymaker." They regarded him, because of his scientific failing, with the respectful tolerance with which fools in the East are regarded, and would wait an extra minute for him to help him on board if the steamer were to cast loose from the pier before he could scramble up the side. . . .

At Michipicoten Island some of the passengers went off with the Botanist to collect ferns and mosses. He led them on a rare chase over rocks and through woods, being always on the lookout for the places that promised the rarest kinds, quite indifferent to the toil or danger. Scrambling, puffing, rubbing their shins against the rocks, and half breaking their necks, they toiled painfully after him, only to find him on his knees before something of beauty that seemed to them little different from what they had passed by with indifference thousands of times. But if they could not honestly admire the moss or believe that it was worth going through so much to get so little, they admired the enthusiasm, and it proved so infectious that before many days almost every one of the passengers was smitten with "the grass mania," or "hay fever," and had begun to form a collection.

Reverend George M. Grant, *Ocean to Ocean* (1873)

IN SUMMER 1848 LOUIS AGASSIZ, THE RENOWNED HARVARD SCIENTIST, LED A group of students and professional naturalists on an expedition along the Canadian North Shore of Lake Superior. Although Euro-American travelers had retraced the route from Sault Ste. Marie to the Kaministiquia River countless times since the French first plied Superior's waters in the mid-seventeenth century, no one had surveyed its plant and animal life in great detail. Before 1810, botanist Edward G. Voss points out, travelers into the upper Great Lakes "were more interested in furs, or in the souls of Indians, than they were in study of the natural history of the region." But the nation had changed in two hundred years. So keen was the country's interest in botany, zoology, and geology that a narrative of the group's journey, published in 1850, enjoyed brisk sales, even though it described a region on the remote northern frontier that most nineteenth-century readers would have found hard to imagine, much less visit.

By the time Agassiz toured Lake Superior, he had made a name for himself as a geologist and ichthyologist. His book *Lake Superior: Its Physical Characters, Vegetation, and Animals* would add botany to his long list of accomplishments. When he explored the crevices of rocky outcrops between the forest and the lake, Agassiz was surprised to find tiny plants that bore a striking resemblance to those that grew above the tree line in his native Swiss Alps. It proved to be a startling revelation. The shoreline plants are known as arctic-alpine disjuncts, so called because they are separated hundreds of miles (and in some cases more than a thousand miles) from their main range in the Arctic tundra or in the mountainous regions of western North America.

Agassiz's botanizing whet the appetites of his fellow scientists. Not long after he published his findings, expeditions in both the United States and Canada began to document the occurrences of disjuncts at points all around the lake. Given the difficulties and dangers of travel at the time, the number of collecting activities in the Lake Superior basin during the nineteenth century is nothing short of remarkable. In Minnesota, for example, a gravel road connecting Duluth to the Canadian border was not completed until the early 1920s. Until then, survey teams were forced to travel by boat up the North Shore's treacherous coastline. Nonetheless, as early as 1878 Benedict Juni filed the first, albeit brief, report on the plants of the North Shore for the Minnesota Geological and Natural History Survey. The Juni expedition was followed in summer 1879 by a small group that included Thomas Roberts, a University of Minnesota student who would later become an accomplished ornithologist and the first director of the Museum of Natural History at the University of Minnesota in Minneapolis. Like Juni, Roberts collected plants only episodically. According to his account, the scientists traveled by boat from Duluth to the mouth of the Devil Track River but "landed only here and there along the shore." Still, Roberts made important additions to Juni's list, including a record of one arctic-alpine disjunct, the butterwort, which he noted thriving on the rocky shoreline around splash pools and mossy crevices.

These fledgling state-sponsored surveys drew particular attention to the finger of wind-raked rock known as Artist Point near the Grand Marais harbor. Rich in disjuncts (five species have been identified) and easily accessible by boat, the "Point," as it became known among the growing circle of Superior botanists, was especially popular. Hunched with magnifying glass in hand, many a plant hunter would slowly pace the ancient basalt looking for tiny blooms nodding from its crevices or a rosette of leaves stitched into pincushions of moss. Defying the ravages of waves and ice to this day are such rare prizes as butterworts, whose pinwheels of pale green leaves cling like glistening star-fish to the wet faces of the bedrock shore. In May and June they can easily be spotted from a distance by their elegantly spurred purple flowers, each with a white spot at the mouth. Less conspicuous but even more unusual are the tiny white bottle-brush blooms of alpine bistort. In the Great Lakes alpine bistort is, for the most part, confined to the shores of Lake Superior near the Canadian border. Anchored by the mats of other vegetation in bedrock cracks, the plant occupies microhabitats that often measure no more than ten square feet in size. In fact, so limited is the plant's documented occurrence here that the combined size of these scattered niches on the Minnesota and Ontario shores adds up to less than 2.5 acres.

Not surprisingly, the opening of the region to the automobile stepped up the pace and scope of botanical exploration. When Roberts surveyed the Point in 1879, Grand Marais was a "'settlement' of but one house," write botanists Frederick K. Butters and his colleague Ernst C. Abbe. Within decades, however, the construction of highways and tourist lodgings around Lake Superior would enable scientists to cover more terri-tory—and in greater safety and comfort. In summer 1924, for example, Butters helped to conduct the first extensive botanical survey in Cook County. "It is no mere coincidence," he recalls in a 1953 summary of the county's flora, "that this was the era of the Model T Ford and that old Highway No. 1 [the precursor to today's Highway 61] from Duluth to the Canadian border at the Pigeon River had been opened but a few years previously." Indeed, road building in Canada paved the way to new reports of disjunct plant commu-nities as late as the 1960s following the completion of the final leg of the Trans-Canada Highway in 1960.

Research has revealed that sites particularly rich in disjunct plant communities are concentrated along the lake's northern shore from Duluth to Sault Ste. Marie as well as along the coastlines of the lake's northerly islands, including Isle Royale. On the South Shore, disjuncts are primarily concentrated on the Keweenaw Peninsula. Of the 340 species of flowering plants and ferns whose main North American range lies in the Canadian Arctic Archipelago, forty-six have been documented in the Lake Superior region to date. (The Minnesota North Shore hosts twenty-three disjunct species.)

Despite the fact that the shore edge remains one of the least studied ecosystems within the Lake Superior basin (largely because the species it harbors have little commer-cial value), new plants are occasionally added to the list. In 1995–96 three mosses were discovered in a survey of the Sugarloaf Cove Scientific and Natural Area near Schroeder.

Harboring a rare and diverse flora, Grand Marais Harbor and Artist Point (in the distance) became a popular destination for botanists beginning in the late 1800s. Courtesy of the Cook County Historical Society, A522.

New species are occasionally added to the list of arctic-alpine plants that find refuge along the lakeshore. Hoary whitlow grass, *Draba cana*, seen here in the center of the image, was first documented on a Minnesota cliff in 1999. Photograph by Donald Cameron.

New records for the North Shore, as well as for the state of Minnesota, have also been registered by the Minnesota Biological Survey (MBS), a county-by-county inventory of the state's native species of plants and animals and native plant communities that was begun by the Minnesota Department of Natural Resources in 1987. Since 1999 a thorough MBS examination of the shoreline in Cook County has turned up one previously undocumented arctic-alpine species: hoary whitlow grass.

Even after more than a century of botanical exploration, close studies of the shoreline continue to reveal new species of minute plants that have clung to life on Superior's rocky edge for thousands of years. Each discovery is momentous. Disjunct plants are descendants of botanical pioneers that bear humble witness to a past that lies beyond human memory or record, to the wasting of towering ice sheets, and to the birth of nothing less than the lake itself, the rivers that once cut down bedrock to feed it, and the immense forests that took root one tree at a time on bare glacial soils.

Disjuncts: The Pioneer Fringe

Arctic-alpine disjuncts became Lake Superior's claim to botanical fame, but the notoriety would not come without its share of scientific controversy. Indeed, the intensity of the debate often seemed far out of proportion to the disjuncts' actual numbers and diminutive size. What intrigued scientists, in particular, was this: if the last glacier to occupy the region had essentially sterilized the land, uncovering barren ground as it retreated, then how had the arctic-alpine plants come to occupy their shoreline habitats?

Early twentieth-century scientists developed multiple hypotheses that attempted to account for the spotty occurrence of arctic-alpine species in temperate latitudes,

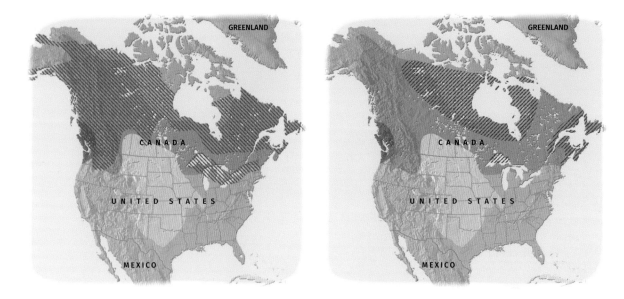

Butterwort
(*Pinguicula vulgaris*)

Knotty Pearlwort
(*Sagina nodosa*)

Tundra

Coniferous forest

Temperate rainforest

Deciduous forest

Grassland

Desert

Populations of arctic-alpine disjuncts around Lake Superior, such as butterwort *(left)* and knotty pearlwort *(right)*, are separated by hundreds of miles (and in some cases more than a thousand miles) from their main range in the Arctic tundra or in the mountainous regions of western North America.

including those in the Lake Superior basin. Some postulated that arctic-alpine disjuncts were relative newcomers to the Superior shoreline, their seeds deposited by moving water or animals such as birds. Others suggested that they were products of seeds that traveled the strong prevailing westerly-northwesterly winds and sprouted when they landed in propitious shore-edge habitats, a phenomenon known as jump dispersal.

Such origins were possible, skeptics admitted, but highly improbable. The odds against seeds first getting past the vast windbreak of boreal forest that has separated the Arctic tundra from the Superior shoreline for thousands of years and then landing in what amounts to a mere ribbon of suitable habitat are enormous. Furthermore, if the seeds had been dispersed by the wind, why were arctic-alpine plant communities not more widespread in the intervening area between their disjunct locations and the Arctic?

Still others posited that disjuncts survived the ebb and flow of the ice sheets in southerly refuges such as the unglaciated pockets of Wisconsin, Illinois, Iowa, and Minnesota known as the Driftless Area. According to these scientists, they began to disperse from these botanical asylums when the glaciers retreated, but a warming of the climate, which began about 11,000 years ago, created unsuitable conditions for their survival. Some species, however, were able to maintain footholds in a few scattered locales hospitable to cold-loving plants such as mountaintops and the exposed shores of Lake Superior. But today's plant geographers have largely discounted this idea. They base their judgment on a technique that is widely used for determining the distribution of plants and animals in the past; that is, they map their present distributions. Scientists point out that most of the northern species that persist in the Driftless Area today are of boreal, rather than Arctic, origin. The paucity of such species suggests that driftless zones never were major havens for arctic plants.

One of the most controversial explanations for the occurrence of disjunct plants was supplied by the renowned botanist Merritt L. Fernald. Pointing to the widely isolated distribution of arctic-alpine species in lower latitudes, from the Rocky Mountains in Alberta and Colorado to mountaintops in New Hampshire and Maine, Fernald argued that these populations survived the coming and going of the ice sheets on glacial features known as nunataks—mountaintops that were tall enough to keep their peaks above the ice during glaciation. Fernald's hypothesis became known as the "Persistence theory." In 1934 he visited the Keweenaw Peninsula on Superior's southern shore and documented not only arctic disjuncts on bluffs along the lake but also alpine species whose nearest relatives could be found in North America's Rocky Mountains or along the West Coast. One year later he published an account of his findings, suggesting that West Bluff near Eagle Harbor, Michigan, once was a nunatak. Arctic-alpine species around Lake Superior, he claimed, survived localized extinction by the glaciers by clinging to elevated refugia like West Bluff. From there they subsequently spread to other points around the lake. According to Fernald, these plants failed to gain ground outside of these outposts because their longtime isolation on nunataks had led to biological senescence, that is, a loss of genetic variability within their populations.

Fernald's rationale was roundly debated. Proponents pointed to what they claimed were other unglaciated peaks on the Slate Islands along Superior's Canadian North Shore and on Bruce Peninsula in Lake Huron. One of the biggest blows to Fernald's argument, however, came in a 1937 paper published in *Science* by a geologist who had visited the Michigan sites that Fernald claimed were Pleistocene nunataks. The article's author found abundant evidence that the bluffs had indeed been glaciated. Present were striations near the peaks that had been etched into the rock by the restless ice sheet and several different types of glacial erratics (rocks that were not of local origin but clearly had been carried over long distances and deposited on the site by moving ice).

The nunatak hypothesis for Lake Superior disjuncts has largely fallen by the wayside since scientists now generally agree that the entire Superior basin was submerged under glacial ice prior to twelve thousand years ago. In recent years, sophisticated field techniques have allowed plant geographers to construct more plausible—although no less hotly contested—scenarios of vegetational change in the wake of the receding glacier. In 1953 botanist Abbe of the University of Minnesota proffered a third explanation, which has gained the widest acceptance to date. Analyses of pollen samples seem to bear out his theory. Phytogeographers suggest that about eleven thousand years ago a belt of discontinuous tundra-like habitat, no wider than about sixty miles at any given point, formed along the southern margin of the retreating ice sheet, an unglaciated region known as the periglacial zone. The fossil record shows that such a zone existed throughout much of northeastern Minnesota and western Ontario as well as along the newly deglaciated eastern coastline of Lake Superior until about ten thousand years ago.

Scientists theorize that as the ice sheet receded, the prevailing winds carried the seeds of arctic plants from ice-free refugia in northwestern Canada and surrounding regions to the periglacial zone. They were joined by those of arctic herbs that trace their origins to the eastern coast of North America; and others, to the driftless region of the Midwest.

The seeds of arctic plants were tailor-made for long-distance travel. Small and lightweight, they could be picked up easily by the wind and skittered across rafts of glazed ice that lay scattered across the landscape in the glacier's retreat. Others were lofted high into the atmosphere, into what is known as the aeolian sphere, and deposited far from their parent communities.

Some might consider the life that awaited the new migrants as the human equivalent of Siberian exile. Tracts of available soils often lay among great fingers of still-melting ice. In other cases, the plants colonized coarse gravel debris that blanketed blocks of ice and sometimes the glacier itself. Even during the height of summer, the plants had cold, damp feet and an unrelenting wind around their heads.

But the raw terrain contained an essential ingredient that arctic plants required in abundance—freedom from competitors, particularly trees that could shade them out or slow their dispersal by serving as windbreaks for their airborne seeds. In places where soils were thin and all but the top inch of ground perpetually frozen, trees would have been discouraged from taking root. Although clusters of trees pocketed the landscape, they undoubtedly were low-growing, stunted, and tilted at odd angles from the effects of frost heave, much like those in spruce forests near Arctic latitudes today. Proximity to the frigid waters of Glacial Lake Duluth as well as to the main body of the glacier and its katabatic winds (the unremitting blasts of cold air that roared off the slopes of the ice sheet) resulted in cold temperatures and cool, cloudy summers that would have further inhibited the wholesale establishment of trees in the periglacial zone.

The cold, wet conditions, while forbidding for trees, would have been ideal for pioneering tundra herbs, including butterwort and alpine bistort, which in modern times have been documented growing in the shadows of glaciers in British Columbia and on Ellesmere Island. Indeed, analysis of fossil seeds taken from southern St. Louis County show that the arctic plant mountain avens, for example, was once widespread in late-glacial times.

The plants followed the retreat of the glaciers by migrating along the shorelines of proglacial lakes and the deep gorges of abandoned spillways that were cut by glacial rivers. But the question remains: If these plants once covered a far greater geographic area, why did they become restricted to the small, isolated populations found along the lake edge today?

Botanists point out that arctic herbs thrived in conditions that were too harsh for most other plants. So when these conditions changed, so did their prospects for survival. The rapidly warming climate that began about 11,000 years ago enabled the boreal forest on the southern edge of the periglacial zone to quickly extend its range far to the north. In the race to the north, the trees managed to stay one step ahead of the arctic herbs.

But some botanists hypothesize that it was more than just the stiff competition from trees that kept disjuncts from following suit. According to David R. Given and James H. Soper, the acidic soils derived from the granite of the Canadian Shield may have slowed or prevented the tiny arctic plants from colonizing terrain laid bare by the glacier's northward retreat. Along the shoreline of Lake Superior, they note, disjuncts are not abundant on granitic rock. On the other hand, the boreal forest, composed mainly of more temperate species such as spruce, jack pine, birch, balsam fir, balsam poplar, alder, and willows, thrives in acidic soils and tolerates water-logged conditions, adaptations that may have enabled it to quickly colonize the newly exposed ground. The forest was so successful in exploiting this new niche that its expansion was abruptly halted only at the sixty-fifth parallel. Here the severe environmental conditions of Arctic tundra, similar to those that existed on the margins of the glacier, have prevented the establishment of taller vegetation such as shrubs and trees.

For the most part, the boreal forest that separates the Lake Superior region from the tundra is little changed from the one that overran the region in postglacial times. Today, much of the forest is black spruce muskeg—a mossy, boggy tangle of spruce trees that rapidly regenerates itself following disturbance. In what is known as the "wet blanket" effect, the forest forms a living barrier to the migration of plant species such as disjuncts that cannot tolerate its soggy, acidic conditions.

Suited to less acidic conditions and unable to grow in the dim light of the boreal forest, the arctic-alpine plants that once were broadly distributed in the tundra-like communities along the margins of the glacial ice became confined to the only habitat that retained the conditions to which they were adapted—the exposed cliffs and foggy shorelines of Lake Superior. Along the hard, rocky rind of the lake, the "remnants of a sort of pioneer fringe," as Butters and Abbe refer to disjuncts, flourished where most other plants could not even hope to survive.

Disjuncts: Stressed Out—and Never Happier

In Minnesota arctic-alpine disjuncts occupy an extremely narrow strip of exposed shore-line—at most only about sixty-five feet wide—between the wave-splash zone of Superior and the lake-edge forest. According to botanists Given and Soper, their persistence on Superior's shores over the past thousands of years has depended on two key conditions: stress and disturbance. The shore edge offers both—in spades. "The best developed assemblages [of arctic-alpine species]," they point out, "are generally found on rugged shores of hard rock, exposed to wave action and inclement weather and in close prox-imity to deep water."

But disjuncts occupy a knife edge between survival and annihilation since these outposts of vegetation can be devastated by the very forces on which they depend. Take something as simple as waves. Proximity to spray from Superior's surf is critical to the replenishment of moisture in plants exposed to drying winds and the full glare of the sun. During major storms, however, waves that can reach heights of nearly twenty-five feet crash on the shore and flush plants out of even the most protected cracks. In winter, breakers routinely pluck out chunks of ice that form in the crevices of the rock and uproot the plants embedded in them.

Paradoxically, the very disturbances that prove most destructive to disjunct plants are essential to their long-term survival. Storms, for example, can thrust floating sheets of ice far up onto the land in a phenomenon known as ice push. The invading floes clip off tall or unprotected vegetation like a blunt mower, keeping the shoreline free of trees and shrubs that could transform the exposed terrain on which low-growing disjuncts depend.

And the action of waves and ice help to weather the rocky shoreline and create habitat niches that are favored by disjuncts. Vast stretches of the Superior coast are composed of diabase outcrops that naturally break down into the shallow basins and steplike ledges and crevices that are especially hospitable to disjuncts. Indeed, the greatest number of species are found in places with the most varied topographical features such as ledges, fissures, and bedrock pitted with splash pools. Botanist Douglas R. Lindsay suggests that the absence of most disjuncts from the shorelines of the other Great Lakes may be due, in some cases, to their limestone substrate, which not only fractures differently but also dries out more quickly.

Adding to the stressful—but ultimately beneficial—conditions of the shore edge are

Like a mower, shore ice clips trees and woody shrubs, creating the open habitat niches that are favored by arctic-alpine disjuncts. NOAA Great Lakes Environmental Research Laboratory.

cold temperatures. Agassiz was the first to observe that disjuncts routinely are found in proximity to Superior's cold waters. Indeed, research has shown that disjuncts are largely missing from segments of the shoreline where the midsummer surface temperatures of adjoining waters exceed 59 degrees Fahrenheit. (The fact that these temperatures rise to 64 degrees Fahrenheit in some of the lower Great Lakes may be another reason why disjuncts are less abundant along their shorelines.) Helping to keep temperatures low and humidity levels high, particularly during the summer, are frequent fogs, which ebb and flow along the shore like waves, rarely extending inland more than 165 feet. Also contributing to the chilly conditions along the northern shore are frequent upsurges of deep, cold water, known as upwellings, that occur when persistent winds or seiches strip away upper layers of warmer water.

But benefiting from the freebies of natural forces that carve prime habitats and discourage or weed out competitors is only part of the disjuncts' survival story. What allows them to thrive in a place where many plants would struggle or die? For starters, they are able to draw on a wide repertoire of specific responses to its particular challenges. Take, for example, chilly temperatures. In cold regions, plants that send up tall stems waste valuable carbohydrates as well as run the risk of exposing flowers to frigid blasts that could delay or prevent the development of seeds. Disjuncts survive by hunkering down close to the ground where wind speeds are slower. The tallest of the disjuncts, such as northern paintbrush and alpine bilberry, measure no more than eighteen inches.

The plant form of many disjuncts is not only small but also compact. Many of them assume the protective structures of leaf rosettes. Others have adopted cushion-like forms. Serving an aerodynamic function much like the curve of an automobile wind-shield, the rounded surface of these low, densely branched mounds deflects cold, desic-cating winds. This allows higher temperatures and humidity levels to be maintained within the more vulnerable inner chambers of the plant.

By growing small and low to the ground, disjuncts are able to take advantage of the most hospitable habitat that the lake edge has to offer. The rocky shoreline, researchers discovered, is riddled with small thermal pockets. Indeed, microhabitats located within this often damp, chilly, and windswept terrain, Lindsay observes, provide "conditions for plant growth [that] are considerably better than one might expect." In studies conducted on an assemblage of arctic-alpine plants near Thunder Bay, Ontario, Lindsay discovered that air temperatures within two to four inches above the surface of the shoreline rock were significantly higher than those measured at a distance of three feet due to the radia-tion of solar heat from the rock.

Plants multiply this thermal advantage by taking up residence in rocky clefts or around splash pools. The dark-colored diabase commonly found along the shore heats up quickly in the sun. This high level of thermal conductivity results in rapid heat transfer to crevices, where plants tend to congregate to take advantage of precious pockets of soil and protection from the elements. Small pockets of stagnant air over these sheltered niches help to minimize heat loss. Lindsay's records show that air temperatures in crev-ices protected from the wind averaged 10 degrees Fahrenheit higher than over exposed zones. And nighttime heat loss was less dramatic for the soils in crevices located near splash pools than for those surrounded by rock largely due to the high capacity of water to store heat and then slowly release it.

Arguably, no other disjunct is better suited to survival on the North Shore than

The diminutive butter-wort favors the cool, moist crevices of bedrock shore habitats that are sheltered from intense wave action. Photograph by Virginia Danfelt.

343

the diminutive butterwort, the botanical equivalent of the Swiss Army knife when it comes to possessing mechanisms for meeting the myriad challenges of life on the shore edge. True to their Arctic origins, butterworts are compact, fibrous-rooted perennials that grow low to the ground, so low, in fact, that their leaves seem to clutch the rock or cushions of crevice mosses with an almost muscular grip. The tiny butterwort bloom, as elegant and elaborate as any fist-sized lady's slipper orchid, tops a stem that measures a mere one to four inches.

On an evolutionary tally, however, the butterwort would score twice as many points for invention than for beauty. On Superior's exposed rock shore, the soil and nutrients needed for plant growth often are in short supply since decaying plant materials are at a premium and the shore edge receives few inputs from the surrounding land. Disjuncts survive on mere thimblefuls of soil that have accumulated in rocky cracks and crevices.

To solve this food-shortage problem, butterworts resort to carnivory, using their leaves as a kind of flypaper. The brilliant sheen of the butterwort's basal leaves is caused by sunlight reflecting off countless tiny stalks topped with pill-shaped glands that cover the leaf surface. Each gland secretes a drop of clear fluid, from which the butterwort derives its name. The Latin root of the plant's genus—*Pinguicula*—*pinguis,* or fat, refers to the oily texture of this fluid.

When insects alight on the leaves, they become caught in the leaf's gummy secretions. But this is no ordinary flypaper. Before they suffocate, the struggling victims stimulate the production of digestive enzymes that are secreted from smaller, more numerous sessile glands on the leaf surface. At the same time the leaf margins slowly roll inwards, causing the inner surface of the leaf to flex and form a dish that fills with the enzyme-laced fluid. In this digestive pool, the insects' soft parts are efficiently broken down. The resulting liquid is absorbed into the plants' system through the sessile glands, providing the plants with nutrients, especially nitrogen, that their environments are sorely lacking.

The butterwort is well equipped for the challenges of life on the shore edge. Small insects trapped and digested on the butterwort's gummy leaf surface provide the plants with the nutrients, especially nitrogen, that their environments lack. Photograph by Skip Moody / Dembinsky Photo Associates. Copyright 2013; all rights reserved.

Some arctic plants, such as butterworts, have also evolved other backup strategies to compensate for the deficiencies of their environment. Take reproduction. Instead of relying solely on sexual reproduction, butterworts produce a tight, resting bud that begins to form in the center of the plant's pinwheel of leaves in late summer. The dormant resting bud may be dislodged and scattered by wind, rain, frost action, ice, or animal activity. By winter's end, if the bud has landed in a favorable site, leaves and roots will sprout from the hibernaculum, and a new plant is born.

As if to leave nothing to chance, butterworts have yet another ploy in their arsenal of reproductive strategies. The plant's leaf axils form smaller winter buds known as gemmae. Like the resting winter buds, the gemmae also can be dispersed through disturbance. If they beat the odds and come to rest in a propitious location, the gemmae, like seeds, can sprout and form minute new plants in the spring.

An Uncertain Future on the Edge

For the most part, scientists can offer only conjectures about the lifeways of Lake Superior disjuncts based on extrapolations from studies conducted on their relatives in the Arctic tundra. Still, there are major environmental differences that prevent scientists from drawing easy comparisons. For example, Lake Superior exerts a moderating influence on the nearshore climate, lengthening the growing season by stalling the onset of autumn frosts and reducing the risks of late-spring or early-summer frosts. In addition, North Shore disjuncts may be subject to higher temperatures during the growing season than their Arctic counterparts. And due to the linear, patchy nature of their microhabitats, they may also have developed different strategies for wind or insect pollination. In fact, some disjunct species may have evolved such pronounced adaptations to their Superior habitats that their genetic makeups and ecological functions may diverge sufficiently from their Arctic relatives that the disjuncts could be considered a separate subspecies.

These speculations are among the many knowledge gaps that plague scientists concerned about the future survival of disjuncts on Superior's shores. Equally troubling is the lack of in situ research and comprehensive monitoring. Many botanists fear that entire disjunct communities may disappear long before they even are discovered and documented. The exploding gull populations in the Lake Superior basin, for example, pose a serious threat to disjuncts as nesting birds colonize new reaches on the shorelines of the lake's islands and mainland. Studies of both island and mainland shore-edge plant assemblages have shown that pecking displays by these densely nesting birds can uproot plants over a wide area. Furthermore, their droppings can transform native plant communities by altering nutrient regimes as well as introducing invasive weedy species from the mainland. (See also "Nature or Nuisance? Gulls in the Great Lakes.")

Gulls are not the only ones to invade the disjuncts' limited habitat in ever-greater numbers. Property owners, for example, often are unaware that when they purchase lakeside property, they buy not only a view of Lake Superior but sometimes habitat for rare species. Erosion of sediments from development and runoff laced with fertilizers and herbicides can damage disjunct communities.

The trampling of tiny plants is also a problem, particularly on state and federal land. Only 4 percent of the Minnesota shoreline is in the public domain, mostly in state parks.

Exploding gull populations in the Lake Superior basin pose a serious threat to rare disjunct plants when nesting birds disturb vegetation and alter nutrient regimes with their droppings. Photograph by U.S. Fish and Wildlife Service.

A growing number of recreational interests are competing for access to these limited parcels. Interest in touring the North Shore by boat—be it by kayak or cabin cruiser—has grown and with it the demand for the construction of waterfront campsites and additional harbors that would shelter boaters from Superior's dangerous and unpredictable moods. For those who want to tour the shore on solid ground, a lakeside bike trail now brings them closer to the edge. In some places, plants that once found refuge on vertical cliff faces are now regularly disturbed by the activities of rock climbers.

But soils and vegetation have suffered from even the most passive shore-edge recreation—communing with the lake. Foot traffic on popular Artist Point, for example, has created a spider web of informal trails in the small forest that abuts the rocky shore, compacting the soil and setting up conditions for devastating erosion. Visitors who venture out onto the rocks for the stunning view of the Sawtooth Mountains have worn away the fragile carpet of rock-loving lichens. Some are such slow growers—expanding at a fraction of an inch each year—that even if left undisturbed, centuries may pass before they attain their former abundance.

And what about the fate of the Point's disjuncts, tiny plants that have fired the imaginations of botanical explorers for more than a century? To date, no one has documented the localized extinction of any species. Much anecdotal evidence exists, however, to suggest that the populations of some species, including small false asphodel and butterwort, have suffered substantial declines.

Highly specialized plants such as Superior's arctic-alpine disjuncts also stand on the front lines of global environmental change and, like canaries in a mine, may be the first to signal vegetational shifts in the region. The thinning of the ozone layer may pose special hazards for plants such as disjuncts that grow fully exposed to the sun. In 1998 scientists at Utah State University published results of a study in which thirty-four species of plants were exposed to high doses of ultraviolet-B (UV-B) radiation. Such levels are similar to those which penetrate Earth's atmosphere when its protective ozone layer is thinned by anthropogenic chemicals. Researchers found that intense exposures of UV-B radiation slowed the development of pollen in nineteen of the species tested.

The biggest question mark for the future of these plants may be the effects of global warming. According to some scenarios, rising carbon dioxide emissions in the atmosphere could lead to climate changes that would dramatically lower water levels in Lake Superior. Would disjuncts be able to migrate quickly enough to new lake-edge habitats, or would they be outcompeted by trees, shrubs, and forbs before they had a chance to establish new territories? Will the diminishing ice cover predicted for the Great Lakes lessen the impact of ice push and confer a competitive edge to these other species? With mean lake temperatures expected to rise, would the shore edge provide a warmer, more hospitable environment for taller vegetation? If disjuncts are outcompeted or unable to evolve adaptations quickly enough to stave off extinction, what would happen to the suites of insects, birds, and small mammals that may have come to depend on them?

And what are the implications of their disappearance for humans who depend on disjuncts not for material sustenance but spiritual replenishment? Many North Shore lovers visit the lake edge to experience what writer Bill McKibben describes as a "delicious sense of smallness" in the face of an immense universe. Disjuncts foster this sense of smallness by testifying to some of the most ancient events on Earth and to the powerful and dynamic forces that continue to shape today's landscape. If they disappear, will our sense of wonder be diminished by their absence? ❧

SUGGESTIONS FOR FURTHER READING

Bergquist, Stanard G. "Relic Flora in Relation to Glaciation in the Keweenaw Peninsula of Michigan." *Science* 86, no. 2220 (1937): 53–55.

Butters, Fred K., and Ernst C. Abbe. "A Floristic Study of Cook County, Northeastern Minnesota." *Rhodora* 55, no. 650 (1953): 21–85.

Davis, Margaret B. "Phytogeography and Palynology of Northeastern United States." In *The Quaternary History of the United States*, ed. H. E. Wright Jr. and David G. Frey. Princeton, N.J.: Princeton University Press, 1965.

Edgerton, Angelique D. "The Structure of Relict Arctic Plant Communities along the North Shore of Lake Superior." Master's thesis, University of Minnesota, 2013.

Fernald, M. L. "Critical Plants of the Upper Great Lakes Region of Ontario and Michigan." *Rhodora* 37, no. 438 (1935): 197–341.

Given, David R., and James H. Soper. *The Arctic-Alpine Element of the Vascular Flora at Lake Superior.* Publications in Botany 10. Ottawa: National Museums of Canada, 1981.

Juni, Benedict. "The Plants of the North Shore of Lake Superior." In *The Geological and Natural History Survey of Minnesota: The Seventh Annual Report for the Year 1878*. Minneapolis: Johnson, Smith and Harrison, 1879.

Lindsay, D. R. "Migration and Persistence of Certain Arctic-Alpine Plants in the Lake Superior Region of Ontario." *Lakehead University Review* 1 (1968): 59–69.

Pielou, E. C. *After the Ice Age: The Return of Life to Glaciated North America*. Chicago: University of Chicago Press, 1991.

Roberts, Thomas S. "Plants of the North Shore of Lake Superior, Minnesota." In *The Geological and Natural History Survey of Minnesota: The Eighth Annual Report for the Year 1879*. St. Paul: Pioneer Press Co., 1880.

Soper, James H. "Botanical Observations along the Lake Superior Route." *Proceedings of the Royal Canadian Institute* 10 (1963): 12–24.

Soper, James H., and Paul F. Maycock. "A Community of Arctic-Alpine Plants on the East Shore of Lake Superior." *Canadian Journal of Botany* 41 (1963): 183–98.

Voss, Edward G. *Botanical Beachcombers and Explorers: Pioneers of the 19th Century in the Upper Great Lakes*. Vol. 13. Contributions from the University of Michigan Herbarium. Ann Arbor: University of Michigan Herbarium, 1978.

Wessels, Tom. *The Granite Landscape: A Natural History of American's Mountain Domes, from Acadia to Yosemite*. Woodstock, Vt.: Countryman Press, 2001.

INTERNET RESOURCES

Minnesota Department of Natural Resources, Minnesota Biological Survey, www.dnr.state .mn.us/eco/mcbs

BETWEEN A ROCK
AND A LAKE

Life on the
Cliff Edge

The North Shore is composed primarily of vertical headlands interspersed with gravely beaches and sloping bedrock shores. Little is known about the ecology of cliff habitats, such as Palisade Head, in part because these vertical ecosystems are difficult to access. Copyright John and Ann Mahan.

Despite the dangers and difficulties of navigating its dangerous coastline, the North Shore lays claim to a long history of scientific investigation. Reports of botanical rarities along the lake edge have been published since the late nineteenth century. Geologists have probed the rocky shoreline not just for valuable minerals but also to advance academic inquiry into such subjects as the glacial history of the region. As early as 1893, Andrew C. Lawson published a report of his travels from Sault Ste. Marie to Duluth in which he identified the wave-cut cliffs and former beaches of the proglacial lakes that preceded Superior's modern-day form.

Not all aspects of the shore have been so thoroughly scrutinized, however. To this day the ecology of the plants and animals that inhabit the sheer rock outcrops has largely remained a mystery. When it comes to these vertical habitats, such knowledge gaps are not unusual, say Douglas Larson, Peter Kelly, and Uta Matthes of the Cliff Ecology Research Group at the University of Guelph, Ontario. Until recently cliffs were regarded as "hostile, lifeless, and seemingly impossible to sample," the researchers point out. "Indeed they do not even show up on the most detailed aerial photographs." Even such postcard staples as Gibraltar and Dover, whose images are recognized the world over, they say, "have attracted no scientific inquiry."

Since the 1980s, the Cliff Ecology Research Group has helped to overhaul the ecological reputation of cliffs with studies in the eastern United States and in France, Germany, England, and Wales. The group is best known in both scientific circles and among the general public for its focus on the Niagara Escarpment, the limestone edge of an ancient seabed that slices across much of the upper Great Lakes region. Today most of the escarpment lies buried under a thick blanket of glacial till, but one hundred miles of exposed rock wall has broken the surface in southern Ontario, where a stretch of it overlooks downtown Toronto. Intermittent outcrops, like hyphens across the landscape, also appear in Wisconsin and Michigan.

When the scientists dangled from climbing ropes to view the escarpment up close, they discovered a complex assemblage of microhabitats. Even though soils were thin or nonexistent and environmental conditions swung from one extreme to another, they supported a great diversity of plants and animals. These tiny pieces made up the vast ecological puzzle of a cliff face. Its intricate fit of ledges and niches formed runways for small mammals, platforms for nesting birds and basking snakes, and root holds for clinging plants. Even the

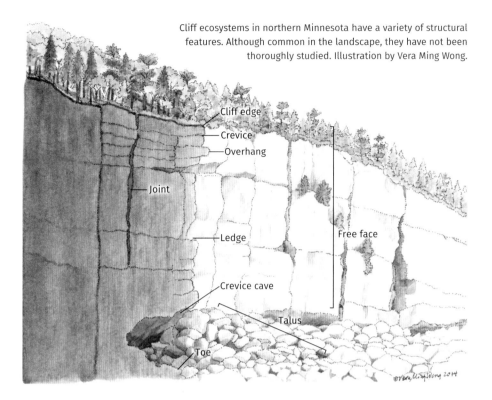

Cliff ecosystems in northern Minnesota have a variety of structural features. Although common in the landscape, they have not been thoroughly studied. Illustration by Vera Ming Wong.

surface of the rock face itself was a rich tapestry of many different species of lichen. Yet it was easy to understand why scientists ignored such ecological richness and complexity for so long—the details of cliff faces are difficult to see beyond distances of more than fifty feet.

Since the 1980s, the cliff ecology group has introduced startling new revelations that have revolutionized the ways in which scientists think about this vertical world. Take, for example, the northern white cedar trees that form the overstory of the escarpment's vertical woodland. In 1993 dendrochronologists cored a group of live cedar trees. Such sampling was no easy feat since their roots, trunks, and branches were grossly contorted in a gravity-defying attempt to cling to the cliff face. Much to their surprise, the tree-ring counts dated individuals that were up to 1,654 years old. Cores taken from a dead tree lying at the bottom of a cliff near Toronto, Ontario, revealed that it was 1,900 years old when it died. Researchers suddenly realized that the cliff-face forests of the Niagara Escarpment were old-growth forests, in fact, the oldest and least disturbed forests in eastern North America.

Yet compared to the towering conifers of old-growth forests in the Pacific Northwest, these ancient trees are tiny. Even the Methuselah of trees that were sampled measured only 10 feet tall and less than 12 inches in diameter. That is because cliff-dwelling trees are painfully slow growers, expanding less than 0.04 inches in diameter each year. The group has documented one 25-year-old tree that, at 4 inches high and 0.3 inches in diameter, was a mere twig. "It's the same everywhere," says Douglas Larson, director of the research group. "Real old, real tiny trees."

Research conducted in the 1980s showed that in some parts of the Great Lakes cedar trees reach grand old ages of a thousand years or more. Photograph by Virginia Danfelt.

Cryptoendolithic organisms—tiny algae, fungi, and bacteria that live within rocks—may provide cliff-edge trees with the small doses of fertilizer they need to survive. Photograph by Douglas W. Larson.

Microscopic root fungi are thought to play a vital role in helping cliff plants access scarce water and nutrients. They are distributed throughout the cliff ecosystem via fungal spores in the feces of small animals, such as this rock vole. Copyright Minnesota Department of Natural Resources—Kelly L. Pharis.

The researchers compared the cliff-face cedars with those that grew in swamp forests farther inland. Slow growth, longevity, and twisted forms characterized the cliff dwellers. The differences were due to environmental conditions, however, not to divergent genetic profiles. In fact, when cliff-face seedlings were transplanted to more benevolent inland conditions, they prospered like their cousins.

That led to another nagging question: How did the gnarled and weathered cedar trees survive, much less reach such grand old ages, under the cliffs' extreme conditions of temperature and moisture, particularly when many of their roots were anchored in thin soils or nothing but solid rock? Contrary to first impressions, the researchers discovered that the trees were not deficient in nutrients or water. The reason: their roots were heavily colonized by mycorrhizae, microscopic fungi that are known to boost a tree's access to such nutrients as phosphorus. In fact, mycorrhizae covered about 80 percent of the surface of tree roots. (Scientists suspect that the fungal spores of mycorrhizae are distributed throughout the cliff ecosystem in the feces of small animals.) These microbial helpers enabled the cedar trees to grow more extensive roots, which, in turn, gave them added surface area for the greater absorption of water.

As it turns out, the cedar trees may benefit from another group of microscopic aides. One of the most startling outcomes of the cliff research was the discovery of cryptoendolithic organisms (*crypto* meaning "hidden"; *endo* meaning "in"; *lithic* meaning "rocks"). Researchers uncovered a rich array of tiny algae, fungi, and bacteria that had taken refuge inside the rock to escape the extreme fluctuations of temperature, moisture, pH, and wind that buffet its surface. Many of the algae species subsequently were identified as cyanobacteria, nitrogen-fixing organisms that are capable of taking nitrogen from the air and making it available to plants. Researchers theorize that these organisms may provide rock-clutching roots with the small doses of fertilizer that trees need to survive.

These adaptations have helped to create an entire ecosystem of astounding stability that has changed little since the melting of the last ice sheet. Examination of dead woody debris, for example, has shown that the density of cedar trees, their growth rates, and their form and structure have remained essentially the same over the past 3,400 years. The trees are accompanied by a distinctive complement of species that scientists theorize first became assembled during the peak of the last glacial advance when plants retreated to the limestone cliffs of Tennessee, Kentucky, and the Carolinas and then moved north as the ice receded. "This refugial status of cliff ecosystems," say Larson, Matthes, and Kelly, "suggests that considerably greater biodiversity may be present on cliffs than is currently recognized."

Whether such ecological mechanisms exist among organisms on the cliffs of volcanic rock along Superior's northern shore is anyone's guess since, as Larson observes, no one has studied them in great detail. But if the work of the cliff ecology group is any indication, astounding discoveries are likely to be had simply for the looking. ❧

SUGGESTIONS FOR FURTHER READING

Larson, Douglas W., Uta Matthes-Sears, and Peter E. Kelly. "The Cliff Ecosystem of the Niagara Escarpment. In *Savannas, Barrens, and Rock Outcrop Plant Communities of North America*, ed. Roger C. Anderson, James S. Fralish, and Jerry M. Baskin. Cambridge: Cambridge University Press, 1999.

Larson, Douglas W., Uta Matthes, and Peter E. Kelly. *Cliff Ecology: Pattern and Process in Cliff Ecosystems*. Cambridge: Cambridge University Press, 2000.

THE LEADING EDGE

North Shore
Bird Migration

IN THE EARLY MORNING HOURS OF AUGUST 14, 1826, A CANOE PARTY ACCOMPAnying Thomas McKenney, an official with the U.S. Indian Department, broke camp on the South Shore of Lake Superior to make an eighteen-mile crossing of Fishing Bay. No sooner did they push out into the lake than a southerly breeze had "freshened into a blow." The men put down their paddles and rigged a sail, taking advantage of the wind at their backs.

Not all of the creatures on the lake, however, found a friend in the wind. At six o'clock that evening, McKenney's party noticed a lone pigeon, undoubtedly a passenger pigeon, beating its way into the gusts toward their canoes. When it alighted on the sail, McKenney ordered the men to lower the sheet. "It was too feeble to fly," he wrote in a later account of his travels. "Its heart beat as if it would break." After feasting on cracker crumbs and Lake Superior water (McKenney filled his mouth with lake water so that the bird could sip it directly from his own lips), the pigeon revived, even living long enough to accompany McKenney back to his home in Washington, D.C.

The pigeon had found a rare asylum. Without McKenney's intervention, it would have been fodder for the frying pan of the hungry paddlers since, as he observed of his companions, "The disposition of our voyageurs [was] to kill and eat whatever fell in their way." And out on the open lake, the bird would have no doubt dropped of exhaustion into the water. In fact, McKenney speculates that the pigeon was probably the sole survivor of a flock crossing the lake from Canada. "Thousands of them," he writes, "perish in crossing every season."

McKenney provides one of the earliest written accounts of the dangers that Lake Superior poses to land birds, particularly migratory ones. Unfortunately, more than 150 years later, scientists still have little detailed information on how birds negotiate the

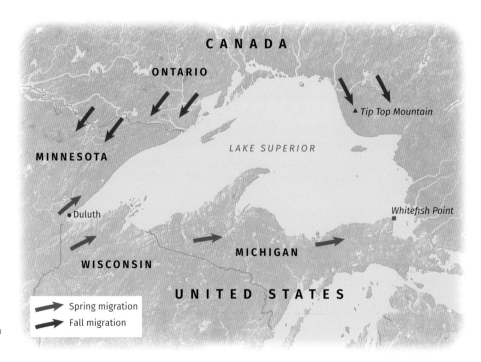

Many birds migrate along routes that minimize or avoid the hazards of open water over Lake Superior.

hazards of lake crossing. Indeed, anecdotal sightings like McKenney's and others from the historical record are commonly used to supplement patchy research.

We are only just beginning to understand how perilous—and complicated—the Lake Superior crossing can be. Not surprisingly, some birds simply avoid it altogether. Day-flying migrants en route to and from their breeding grounds in northern Minnesota and central Canada navigate partly by sight and typically follow the lake's shoreline like prudent canoeists in fragile crafts. Keeping away from open water is especially critical for birds such as raptors that conserve energy by gliding on thermal updrafts. (Thermals are great columns of hot air that are created when the sun heats the ground's surface and the air molecules above it, causing them to expand and rise. Spanning up to five miles in diameter, thermals disappear over Superior's cold expanse.)

Some species, such as passerines (perching birds that include songbirds), however, commonly migrate at night. The cooler, moister air helps the birds to better retain water while maximizing the loss of body heat. The darkness also serves as a cover from predators. And flying by night allows insect-eating birds to feed during the day.

After studying the autumn bird migration on Isle Royale in 1905, ornithologist Max Minor Peet noted that major waves of migratory birds often coincided with periods of high atmospheric pressure when skies were clear, temperatures were cold, and northwesterly crosswinds helped to speed the crossing of the birds. He concluded that passerines seemed to minimize migratory risks by heading out over the lake during these optimal climatic conditions.

This strategy is not without its downsides, however, especially when it comes to crossing big bodies of water like the Great Lakes. Among the trickiest is the weather on the lake, which can be as fickle and unpredictable for birds as it is for people. The sudden switch to a low pressure system ushers in clouds, fog, rain, and headwinds, as it did in the early hours of September 7, 2005, when gusts of up to forty miles per hour were implicated in a mass downing of birds along the North Shore. Residents reported isolated slicks of dead birds from Schroeder to the Brule River. These contrary conditions bring birds down to earth, where they drop of exhaustion into the lake or must navigate a collision course with human-built obstacles. Even the sparsely developed landscape on Isle Royale at the turn of the century posed numerous hazards for birds. In his 1908 report for the Michigan State Biological Survey, Peet noted the "fatal attraction of the lighted windows of resorts and the light-houses." The keeper of the lighthouse in Siskiwit Bay, he wrote, "reported that hundreds of birds lost their lives every spring and fall at his light alone. It was mainly on cloudy nights that the birds struck the lighted windows and the lantern, but some were killed on other nights."

Migrating birds are not safe from such lethal collisions even out on the open lake. Take, for example, what happens when commercial shipping lanes intersect with flight paths. In a 1964 article for *Audubon* magazine, Officer P. T. Perkins, a thirty-one-year veteran of commercial shipping on the Great Lakes and an avid bird watcher, recounted his experience of the autumn bird migration on Lake Superior. On the night of August, 20, 1961, Perkins headed up to the pilothouse to assume the midnight watch. "I could tell from the sounds and the birds fluttering on deck that a huge migration was in progress," Perkins writes. "We were sailing eastward on Lake Superior from Devil's Island in the Apostle group. The fog had changed to a misty drizzle. When I aimed the searchlight upward, the beam revealed heavy flights of passing birds. Others fluttered around the ship's lights by the dozen. Normally, the pilot house is kept dark at night, but on the few occasions that a light was necessary, the birds would strike, or flutter against, the windows. At times there were so many on the bridge deck that the lookout had to be careful not to step on them."

The peak of the migration continued throughout the night and finally subsided at about three in the morning. When his shift ended at four o'clock, Perkins took a flashlight tour of the deck. "The toll was heavy," he recalls. "Few people realize the force with which these ounces of flesh and feathers hit an object. I dispatched the hopelessly injured and placed those that might recover in a sheltered room."

Such mass casualties are only amplified when migrating flocks are caught out over the open lake during violent weather. Peet noted that following autumn storms, particularly those from the southeast, birds in great numbers were blown onto Isle Royale, their energy so spent that many could be picked up and easily handled. Isle residents and commercial fishermen reported that in the aftermath of such storms, Washington Harbor was strewn with the floating bodies of dead birds that had been washed in from the surrounding lake waters. "The loss as shown by those collected at the harbor," Peet observed, "could be but a slight proportion of the vast numbers which must have perished in the open lake."

But optimal climatic conditions provide no guarantee of safety. Migrating birds must contend with another routine danger: fatigue. Even the fifteen-mile flight from Superior's northern coast to Isle Royale during fair weather took a heavy toll on some segments of the bird population. After each major migratory wave, Peet noted great numbers of dead or exhausted birds, the majority of which were new fledges that had not yet built sufficient fat reserves to sustain them through the rigors of flight. Indeed, studies in 1993–95 of migrating birds on Whitefish Peninsula, located some fifty miles west of Sault Ste. Marie, showed that birds on their southbound fall migration carried less fat than birds that were headed north in spring. "Since a land crossing does not stress migrants to the extent that a water crossing does," says Jeannette Morss, director of the Whitefish Point Bird Observatory, "it is not surprising that spring migrants arriving on the Lake Superior shoreline are carrying more fat than fall migrants." As a result, to replenish energy reserves lost on their strenuous water crossing of the lake, fall migrants lingered longer on Whitefish Peninsula and gained more weight there than did spring migrants.

Both spring and fall migrations, however, seemed to share one characteristic: Whether en route to summer breeding territories or to their wintering grounds, weary, hungry birds eagerly seek out prominent landfalls that offer a place to rest and refuel. Indeed, researchers have identified a series of islands or peninsulas that jut out into the lake as great collectors of migrating birds. Sites of major bird concentrations have been documented in the vicinity of Whitefish Point (where more than 320 bird species have been observed during migration), on Thunder Cape on the southern tip of Ontario's Sibley Peninsula, on Isle Royale, and on the Keweenaw Peninsula, which is centrally located along the lake's southern shore. In the western part of Lake Superior, Outer Island, the most northerly and easterly of the chain of twenty-one islands that makes up the Apostle Islands National Lakeshore, has been recognized as a major migratory stopover. Recent surveys have shown that more than 200 species of birds, up to 98 percent of them passerines, use a twenty-acre sand spit on the south end of the island as a respite station.

Yet even on these isolated outposts, migrants cannot let down their guard. Some hawks appear to throw caution to the wind and follow migrating flocks of passerines out over the open lake. During the fall migration on Isle Royale, Peet observed sharp-shinned and sparrow hawks "creating great havoc" among sparrows and warblers. Peet concluded that the hawk migration appeared "to have been intimately connected with the migration of the smaller birds upon which they preyed, and seems to give at least one instance of bird migration being influenced by the food supply."

Nearly a century later, scientists observed a similar stalking phenomenon in the

Fall migration for hawks, such as this merlin, may be closely timed with the migration of the smaller birds on which they prey. Photograph by Phil Seu.

Apostle Islands as well. According to Geoffrey Smith, author of a 1998 bird survey report on Outer Island, "Long-term monitoring has indicated that the island appears to be regionally, and possibly nationally, significant for falcon migration," particularly for merlins and peregrines. The pandemonium that results when predator and prey share close quarters is aptly described by writer Michael Van Stappen, who took part in island surveys: "Loud squawks often came from the pines where merlins were chasing down flickers. Sometimes peregrine falcons came racing by after both merlins and flickers, terrorizing birds, then breaking off the chase at the last minute. I watched one peregrine at a great height repeatedly dive or 'stoop' like a stunt pilot on flocks of flickers trying to gain altitude as they left the island. The scene was a wild circus the likes of which I had never before seen."

Nonetheless, certain birds may prefer the threat of avian predators to the dangers of open-lake crossings. It appears as if some warblers undertake the Lake Superior portion of their journey during the day, says Duluth ornithologist Laura Erickson. After monitoring the fall migration of warblers along the Minnesota North Shore of Lake Superior from 1988 through 1990, Erickson observed some curious anomalies. A large number of warblers deviated from their normal pattern of nocturnal migration to travel the North Shore segment during daylight hours. Further research needs to be conducted to determine just what percentage of the flock opts for this alternate course and whether their choice confers any survival advantages. Nonetheless, Erickson says, the preliminary evidence suggests that for some birds at least, the lake "elicits a change in migratory behavior." "Lake Superior, a cold, deep lake with few islands, presents a significant hazard to warblers. When the wind is likely to carry nocturnal migrants, which are influenced by tail winds, over the water, the normal advantages of nighttime flight are in part offset by the advantage of flying by day, when the shoreline can be negotiated visually."

That much of our knowledge about migrating birds in the Great Lakes has not advanced beyond theories and hunches only underscores the need for more rigorous research. Even something as fundamental as identifying and mapping migratory pathways is more critical than ever. Known obstacles to birds, including communications towers and wind turbines for energy production, are sprouting throughout the Great Lakes basin with little understanding of their effects on migrating flocks.

On the North Shore, studies of bird migration patterns are struggling to keep pace with both the siting and construction of wind turbines for generating electricity. Interest has grown exponentially since recent research revealed the shore's great promise in wind energy. In 2008, Mike Mageau, director of the Environmental Studies program at the University of Minnesota Duluth, began a series of studies to assess wind-energy prospects along the North Shore. State maps had rated this potential as poor, but Mageau discovered just the opposite. Mageau and his students installed anemometers on towers along the ridge that parallels the lakeshore. At one hundred feet high, the researchers clocked average wind speeds of fifteen to twenty miles per hour. These velocities match those measured at Buffalo Ridge in southwestern Minnesota, a site that is considered the best for wind power in the state. Locations for the strongest winds along the shore center on high points that are in close proximity to Lake Superior.

These are precisely the places favored by migrating birds. Ground-breaking work in 2008–10 by Anna Peterson, a Ph.D. candidate at the University of Minnesota, is making the case for considering the Minnesota North Shore as one of North America's major flyways. Millions of birds, including warblers, cedar waxwings, northern flickers, hawks, and owls, migrate along these ridges and the Lake Superior shoreline on their way to and from their northern breeding grounds.

Constructing new structures in their pathways simply compounds the damage caused by existing hazards—such as shipping lanes and roadways. Take the October migration of 1990, the year Erickson has dubbed the North Shore's Great Highway 61 Massacre. Unseasonably cool weather had driven migrating warblers close to the ground in search of insects. Unfortunately, some of the best foraging was on the roadway's warm asphalt. Here the birds encountered the first of many deadly obstacles en route to their southerly wintering grounds—a steady stream of tourists in search of the shore's spectacular fall colors. The bodies of dead and dying warblers caught in the heavy traffic littered the entire length of the highway. Erickson and her colleagues hurried to get a body count, but the dangers posed by the high-speed traffic prevented them from compiling even a reasonable estimate. "It was heartbreaking," she recalls. "It's one thing when a peregrine falcon snatches up a bird—we know that its life energy will sustain the peregrine. When the killer is a Ford Falcon, the death is entirely wasted."

SUGGESTIONS FOR FURTHER READING

Bardon, Karl J. "Fall Diurnal Migration of Passerines and Other Non-Raptors at Hawk Ridge and Lake Superior, Duluth (2007–2011)." *The Loon* 84 (2012): 8–19.

Erickson, Laura L. "Daytime Warbler Migration in Fall along Lake Superior's North Shore." For the Birds. http://www.lauraerickson.com/bird/Species/Warblers/DaytimeWarblerMigration Paper.html.

Johansen, W. R., Jr., W. C. Scharf, and F. M. Danek. "Migration Pathways, Habitat Associations, Stopover Times, and Mass Changes of Migrant Landbirds on the Whitefish Peninsula, Chippewa Co., Michigan." Unpublished report for the Whitefish Point Bird Observatory, Paradise, Mich., 1995.

Kelleher, Bob. "North Shore Attracts Wind Power, Migrating Birds." Minnesota Public Radio website, 29 September 2010. http://minnesota.publicradio.org/display/web/2010/09/29/north-shore-wind-power-birds/.

Peet, Max Minor. "The Fall Migration of Birds at Washington Harbor, Isle Royale, in 1905." In *An Ecological Survey of Isle Royale, Lake Superior*, ed. Charles C. Adams. Michigan State Biological Survey Report. Lansing, Mich.: Wynkoop Hallenbeck Crawford Co., 1909.

Perkins, J. P. "A Ship's Officer Finds 17 Flyways over the Great Lakes." Part I. *Audubon* 66, no. 5 (1964): 294–99.

Smith, Geoffrey. "Migratory Bird Survey: Apostle Islands National Lakeshore." A report to the Apostle Islands National Lakeshore, Bayfield, Wis., 1998.

Stewart, Scott. "Scoping Out Lakeside Hot Spots." *Lake Superior Magazine,* August/September 2001: 20–23.

Van Stappen, Michael. *Northern Passages: Reflections from Lake Superior Country.* Madison, Wis.: Prairie Oak Press, 1998.

INTERNET RESOURCES

Hawk Ridge Bird Observatory, www.hawkridge.org

LAKE SUPERIOR

The Lake Superior that we see today is a sculptural tour de force created by two different artists: fire and ice. Although these masters were equally powerful, their artistry could not have been more different. Think of the glaciers as a twentieth-century sculptor who added his own touches to one of Michelangelo's half-finished marbles. The ice sheets were never able to completely obliterate the legacy of volcanic action that shaped Superior's basic contours. Nonetheless, they would carve distinctive patterns of their own.

In the beginning, however, fire ruled. About 1.1 billion years ago, a horseshoe-shaped gash, known as the Midcontinent Rift, opened in Earth's crust. The rift stretched northeastward from present-day Kansas to near Thunder Bay, Ontario, before bending southward through Michigan and into Ohio. Over the next 100 million years, wave after wave of liquefied rock poured out of the opening. According to one estimate, an astonishing 312,000 cubic miles of lava—one hundred times the current volume of Lake Superior—gushed out onto the planet's surface.

The epicenter of magma production lay under Lake Superior. In his book *Geology on Display: Geology and Scenery of Minnesota's North Shore Parks,* John Green uses the words of a fellow geologist to describe the sulfurous setting that once existed in what is now the Superior basin: "The eruptive scene is unimaginable! A mini-ocean of eerily fuming basaltic lava. A region the size of Maine engulfed in the acrid fumes of volcanic gases. The searing heat of a blast furnace across the whole of it. Literally, hell on Earth."

The eruptions formed a sea of magma that spread over thousands of square miles and accumulated up to five miles deep. In time, it cooled and hardened into a dense and heavy basalt. The displacement of so much lava from the planet's interior caused the surface of the land to slowly sink along the axis of the rift. The subsidence created a large, shallow basin that would become the rough outline of Lake Superior.

Most of the basin now lies under water, but visitors can find evidence of these ancient volcanic fireworks along the Lake Superior shoreline. At Artist Point in Grand Marais, for example, the congealed flows form great sheets of basalt that gently tilt in the direction of the ancient rift.

Incredibly, after this period of enormous upheaval in which the earth was literally turned inside out, the Lake Superior basin would remain relatively stable for the next one billion years. For much of its life, the lake was a dry catchment for boulders that tumbled over its rim and eroded rocks, gravels, and silt that were deposited by the surrounding rivers.

The calm, however, would not last. About two million years ago Earth's mild climate began to fluctuate. Long cool periods, which lasted between sixty thousand and ninety thousand years, caused great mantles of snow to form in the northern reaches of North America and Eurasia. Over time, more snow accumulated than melted. The compacted snow formed mountains of ice, some as high as two miles.

Although monumental, these great ice sheets were anything but static. The warmth of the earth and the heat generated by the friction between the land and ice kept the bottom layers of the glaciers melted. The pressure of mounting snow in the center of the ice masses caused the edges to bulge. At the height of North America's last glaciation, which ended some 191,000 years ago, the Laurentide ice sheet flowed across Canada and the northern third of the United States as far south as the Ohio valley. With chisel-sharp rocks embedded on their undersides, glaciers filed away hillsides. Plucking great blocks of rock the size of a jumbo SUV from present-day northern Minnesota, the roving ice

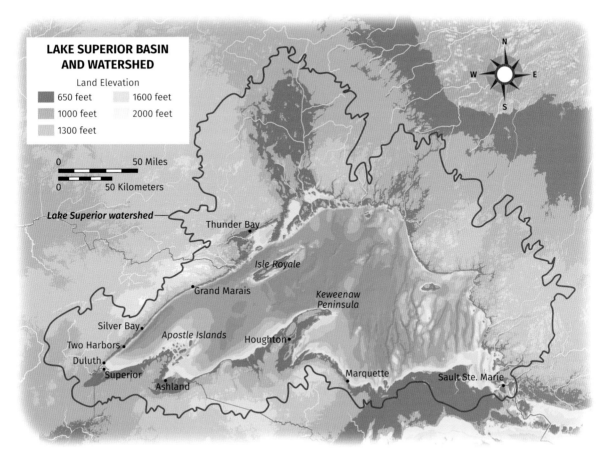

LAKE SUPERIOR BASIN AND WATERSHED

Land Elevation

- 650 feet
- 1000 feet
- 1300 feet
- 1600 feet
- 2000 feet

0   50 Miles

0   50 Kilometers

Lake Superior watershed

Thunder Bay

Isle Royale

Grand Marais

Keweenaw Peninsula

Silver Bay

Apostle Islands

Houghton

Two Harbors

Duluth

Superior

Ashland

Marquette

Sault Ste. Marie

Water, not land, makes up most of Lake Superior's 49,300-square-mile watershed. As a result, relatively few nutrients flow into the lake to fuel the growth of phytoplankton. Superior is the most oligotrophic, or nutrient-poor, of the Great Lakes, a condition that determines the lifeways of all the plants and animals that inhabit its cold, clear recesses.

Approximately 1.1 billion years ago, a horseshoe-shaped gash, known as the Midcontinent Rift, opened in Earth's crust. From this rift successive flows of lava spread across the land to become the bedrock of the Lake Superior basin. South of Lake Superior these rocks are mostly covered by younger sedimentary layers.

dropped them on the Kansas prairies, where they lie stranded today.

Earth's long glaciations were interrupted by short, comparatively warm climatic intervals, known as interglacial periods, in which the glaciers retreated, exposing an abraded land. Their frigid meltwaters ponded in the ice-gouged earth. Torrential rivers sliced a path through thick piles of gravel and rock that had been bulldozed by the glaciers' frontal plows. In time, however, many of these features would be obliterated by the grinding mass of yet another mantle of ice.

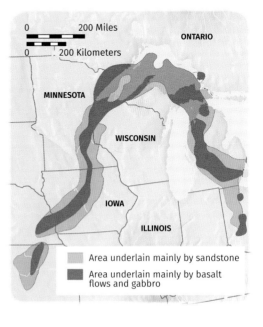

0   200 Miles

0   200 Kilometers

ONTARIO

MINNESOTA

WISCONSIN

IOWA

ILLINOIS

- Area underlain mainly by sandstone
- Area underlain mainly by basalt flows and gabbro

Three of the postglacial lakes stages in the Great Lakes region. Among the precursors occupying the Lake Superior basin were Lake Duluth *(top)*, Lake Minong-Houghton *(center)*, and Nipissing Great Lakes *(bottom)*. Lake levels changed as different outlets were breached, and the crust rebounded following the retreat of the continental ice sheet and its great weight. By 3,000 years ago the Lake Superior we know today was born.

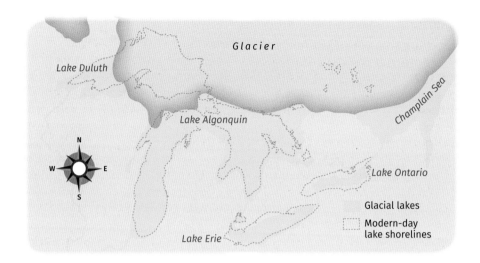

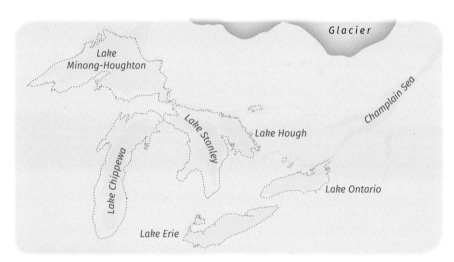

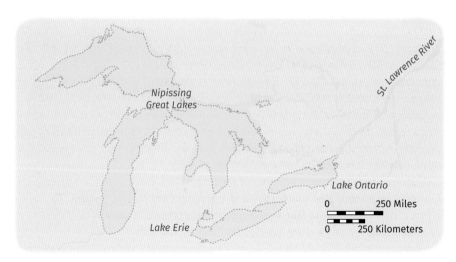

In the past two million years, Earth has undergone about twenty glaciations. Reconstructing this restless cycle of freezing and thawing in the Lake Superior region is difficult since each episode largely wiped out evidence of prior activity. Most of the glacial features of present-day Lake Superior had emerged from beneath the ice by about 11,500 years ago, when the last glacier in the region, known as the Wisconsin ice sheet, had retreated to the northeast corner of the basin. (See also "Highlands.") In its wake, meltwater poured into the volcanic depression that had been scoured and deepened by massive lobes of ice. Glacial Lake Duluth, confined to what is now the western third of Lake Superior, was the first part of the catchment to fill with water. Measuring five hundred feet higher than today's lake, it would have swamped all but the highest reaches of present-day Duluth. The summit of the glacial lake's ancient beaches can still be traced along Skyline Drive, which towers over the city at an elevation of about eleven hundred feet.

By 9,500 years ago, Superior was completely free of ice. Lake levels, however, continued to rise and fall as various outlets for the basin's water were alternately blocked and opened. The shorelines of these precursors to Lake Superior are a common geologic feature along many stretches of the North Shore's coastline. (See also "Nearshore.") About 3,000 years later, water levels stabilized, and the Lake Superior that we know today was finally born.

Living Richly on Meager Means: Life in the Lake

The meltwaters that gushed from the retreating glacier into the basins of the Great Lakes were clear as glass and bone-chillingly cold. Scientists classify these waters as oligotrophic; that is, they contain few fertilizing compounds. Having low fertility, the Great Lakes could not support vast aquatic pastures of microscopic plant life, limiting the number of animals that took up residence within their icy expanse.

The fertility of Superior's waters has changed relatively little since glacial times—for reasons that are both natural and cultural. The hard, crystalline bedrock that dominates the Superior basin is resistant to weathering and therefore supplies few nutrients to the water, unlike the basins of the lower Great Lakes where the presence of softer, more erodible sedimentary rock, such as limestone, imparts a greater array of dissolved minerals, most notably calcium carbonate. In fact, scientists compare the low chemical content of Superior's waters to that of rainwater.

The glaciers also stripped most of the topsoil off this volcanic bedrock, removing another potential source of nutrients. Eroded sediments supply not only calcium but also silica, phosphorus, and nitrogen. These elements are essential to the growth of phytoplankton, the free-floating, photosynthesizing algae and bacteria that form the base of Superior's food chain. Together they determine the amount of life the lake can support.

The contours of the Lake Superior basin further limit the amount of nutrient-laden materials—soil particles and plant and animal detritus—that can flow into the lake from the surrounding land. Steep coastal ridges hug much of the lake's shoreline. As a result, water, instead of land, makes up most of Superior's 49,300-square-mile watershed. Of this enormous expanse, the lake occupies 31,700 square miles, or 64 percent. The ratio of land area drained relative to lake area is 1.5 to 1. By contrast, in most of the world's large lakes (excluding the other Great Lakes), this ratio is at least 6 to 1.

Dampening productivity too are lake temperatures as bracing as the waters that first streamed from the melting glacier. Superior is the deepest—and the coldest—of the Great Lakes. Although summer temperatures of surface waters along protected stretches of the lake's southern shoreline and around the Apostle Islands can reach 70 degrees Fahrenheit, the annual average temperature of waters in the open lake hovers around a numbing 40 degrees Fahrenheit. Surface water temperatures along Minnesota's North Shore rarely climb above 50 degrees Fahrenheit.

The region's thin soils and colder temperatures discouraged the kind of heavy agricultural and industrial settlement that developed along the lower Great Lakes. Take the city of Chicago, for example, which started in 1818 as a small cluster of twelve families. By 1900 the settlement had reinvented itself as a booming prairie metropolis with some 2 million inhabitants. Other regions in the Great Lakes would undergo similar population surges. From 1900 to 1960, the population in the Lake Erie basin mushroomed from 3 million to 10.1 million people. By contrast, during the same period, the population in the Superior basin grew from a mere 400,000 to nearly 800,000 people. And unlike its neighbors to the south, the region still continues to host more trees than people. Despite extensive logging around the turn of the century, more than 75 percent of Superior's watershed remains forested, a land cover that retains soils more effectively than the intensive agriculture practiced along the shores of the lower lakes.

Because there is little algal growth and few suspended sediments, dissolved minerals, and organic materials to block the penetration of sunlight, visibility, particularly in the waters of the open lake, can extend beyond depths of seventy-five feet. This transparency both thrilled—and unnerved—early visitors to the lake. In his 1827 book *Tour to the Lakes,* Thomas McKenney, commissioner of the U.S. Department of Indian Affairs, exclaimed how from his canoe "the rocks at the bottom [of Lake Superior] were seen as distinctly at the depth of twenty feet, as if they had been on the beach. Nothing can be more pellucid than this water; and nothing sweeter to drink."

Superior's oligotrophy does not mean, however, that the lake is a biologically simple system. In fact, its low productivity may have prompted organisms to develop more intricate survival strategies than those in more provident locales. Only recently have researchers begun to appreciate the complexity of the lake's fundamental ecosystem processes.

Most of the basic research into the lifeways of Lake Superior has centered around the needs of fish, especially lake trout, largely because of their economic value in sport and commercial fisheries. Proving especially helpful have been the keen observations of generations of commercial fishermen. They have long known, for example, that Superior's top piscatorial predator is very discriminating about its habitat. Before their numbers began to crash in the late 1940s and 1950s, lake trout had begun to differentiate themselves to an astonishing degree. Some populations chose to spawn on the submerged peaks of offshore mountains, while others were faithful to certain rivers or nearshore reefs. To prevent overlap among species, they spawned at different times. In some populations, these behavioral differences manifested themselves morphologically—fishermen were able to distinguish among different stocks of lake trout by their distinctive physical features, such as the shape or coloration of a fish's body and fins. (See also "A Mansion of Many Rooms: The Return of Lake Trout to Superior.")

As early as 1871 scientists sought to better understand the food web that supported these commercially desirable species. That year members of the U.S. Lake Survey embarked on a cruise to collect basic data about the lake. Among their activities was a dredging of the lake floor. Although limited in scope, these samplings yielded an amazingly comprehensive list of the creatures that occupied the benthos (the community of organisms that make their home either in or on the sediments of the lake floor). Yet, "despite this auspicious beginning, nearly a century elapsed before any significant advances were made on the benthic biology of the largest freshwater lake on earth," write David G. Cook and Murray G. Johnson of the Canada Centre for Inland Waters. It was not until commercial fish stocks nosedived in the 1940s and 1950s that investigations into the lake's benthic habitats would be renewed. (See also "Searching High and Low: The Science of Lake Superior Exploration.")

When referring to benthic habitats, researchers distinguish between the bottom of the lake in relatively shallow waters and the lake floor in the profundal zone (waters more than 230 feet deep). Each has its own distinctive ecology—and challenges for science.

In terms of species diversity, the nearshore benthos is far more varied than that in deep water. Here, the lake bottom can sustain between twenty and thirty different invertebrates (animals without backbones, such as clams, insect larvae, and aquatic worms) versus the meager catalog of animals that occupy the profundal zone, including *Diporeia* (amphipods or scuds), mysids (opossum shrimp), pisidiids (tiny clams), oligochaetes (freshwater worms), and chironomids (midge larvae).

The numbers reflect key differences between these two lake environments. Fed by streams and warmed by the sun, nearshore waters have relatively higher temperatures. Warmer water, proximity to stream-borne nutrients, and deeper light penetration contribute to more productive conditions for phytoplankton, such as algae, the staff of life for a wide range of lake-dwelling animals. Streams also wash insects and other aquatic animals into the lake, where they either take up residence or supplement the

Lake Superior's unique ecology makes it an evolutionary hot spot for boreal freshwater snails. A subspecies of *Helisoma anceps*, shown here, and several other snail species occur only in the Lake Superior and Hudson Bay watersheds. Photograph by T. Travis Brown.

diets of their fellow creatures. Not surprisingly, the organisms of the nearshore benthos can be concentrated in numbers as high as 930 per square foot.

The nearshore also offers a wider variety of habitat, from stretches of cobble, sand, and clay to gravel, boulders, and clean bedrock. (See also "Searching High and Low: The Science of Lake Superior Exploration.") And these environments are subject to a greater array of physical disturbances, which can themselves drive greater species richness. Malacologist Arthur H. Clarke notes that this "unique ecology" has served as an incubator for fostering species richness in gastropods. In 1969, he published a list of eight subspecies of boreal freshwater snails that have developed in postglacial times in the mostly vacant waters left behind by the retreating Wisconsin ice sheet. Seven of them occur only in the Lake Superior and Hudson Bay watersheds, making them, he writes, "the most active recent site for freshwater gastropod evolution in boreal eastern North America."

The shifting environment, while a boon to some species, has been a bane to research scientists. Its changeable nature makes it especially difficult to study. Despite the relative abundance of organisms and their proximity to land, the nearshore remains one of the least researched zones of the lake.

But the profundal zone has presented its own obstacles. For decades, scientists relied on crude scoopers, known as grab samplers, to provide information about life in Superior's depths. Lowered onto the lake bottom to take indiscriminate bites out of its upper layers, this exploratory tool is about as precise as performing brain surgery with a butcher knife.

Scientists once relied on crude scoopers, like this grab sampler, for obtaining information about life at the bottom of the lake, one small, largely random bite at a time. Remote-sensing technologies have provided far more detail on the complexities of the profundal zone. NOAA Great Lakes Environmental Research Laboratory.

Although grab samplers allowed researchers to identify the animals that inhabit the profundal reaches, they offered few clues about daily life at the lake bottom. For example, scientists long assumed that the lake's profundal zones were homogenous, quiescent places that stood in stark contrast to the noisy, restless world of nearshore waters. But the advent of remote-sensing technologies such as sonar revealed an underwater topography that was far more rugged and varied than previously thought. Fine-grained pictures of deep-lake sediments show that tiny particles of settling clay and organic debris do not evenly blanket the lake floor from end to end. Some basins trap sediments, while other benthic environments are routinely scoured down to bedrock by powerful currents. (See also "Searching High and Low: The Science of Lake Superior Exploration.")

Remote sensing has yielded startling insights into the phenomena of deepwater biology as well. Take, for example, the migrations of opossum shrimp *(Mysis relicta)*. Mysids regulate their buoyancy by incorporating air bubbles into their bodies. Perhaps more aptly named *Mysis vampira,* these tiny crustaceans rise from the lake bottom each day around sunset to feed on zooplankton that are attracted to algae growing near the surface. Mysids often ascend in such huge numbers that the shadows of their passing show up on ship sonar screens. After spending the night feeding near the surface, these denizens of the deep slowly sink to the bottom again about an hour before sunrise.

But such glimpses into the lives of deepwater animals have been rare. Exploration by scuba diving, for example, has been limited to about 250 feet, allowing only superficial reconnaissance of a lake where 77 percent of waters are deeper than 250 feet. The intense pressure, numbing temperatures, and darkness of profundal zones (only 1 percent of light reaches a depth of 325 feet in even the clearest lake) have been as effective in barring humans as locked gates.

That changed in summer 1985 when scientists on board the submersible *Johnson-Sea-Link II* traveled to the bottom of Lake Superior and got an up-close look at this mysterious world for the first time. (See also "Searching High and Low: The Science of Lake Superior Exploration.") Included on the itinerary was a trip to the deepest, most inhospitable reaches of the lake located forty miles off the coast of Munising, Michigan. Here, 1,265 feet below the surface of the water, the researchers plumbed the perpetual darkness where temperatures hovered just above freezing and pressures of forty atmospheres were so intense that a Styrofoam cup tethered to the sub's exterior was compressed (with all its proportions intact) to the size of a thimble.

Instead of finding a barren, muck-covered moonscape, researchers discovered a world of creatures flourishing in their spartan circumstances. During its descent, the submersible drifted through planktonic clouds of mysids, each no bigger than a child's fingernail. In the sub's headlamps, these tiny, nearly transparent crustaceans glowed like milky white orbs. Scanning the bottom, scientists for the first time observed small pug-like fish called deepwater sculpins. The fish are known to possess sensitive lateral lines, a series of nerve endings on each side of their bodies that register the movement of prey. In the deep water's dark realm, sculpins are thought to use this sensory capability to detect the squirming of prey that are flushed by the fish's repeated plunges into the soft muck of the lake floor.

Other netherworld fish seem to have developed more sophisticated stratagems. Scientists observed the top predator of the deep—a freshwater cod known as a burbot—use its tail to dig trenches measuring about one-half foot wide and several yards long.

Benthic surveys using submersibles made it possible to view the inhabitants of lake-bottom environments for the first time. A top predator of the profundal zone, the burbot *(right)* was observed building elaborate runways in the mud of the lake floor, which are thought to funnel prey such as deep-water sculpins *(left)* into their hungry mouths. *(left)* Photograph by Justin G. Mychek-Londer. *(right)* Copyright Minnesota Department of Natural Resources–Steve Geving.

Then they watched the fish quietly take up residence in the trench. The burbot may use this construction in a deadly game with its primary prey, the sculpin. Researchers theorize that as they skim along the surface of the lake bottom, unsuspecting sculpins drop into the troughs, which lead like runways straight into the hangars of hungry burbots' mouths.

The unexpected drama at the bottom of the lake was enough to astonish even veteran scientists such as zoologist Jim Bowers, one of the sub's passengers. Quoted in an article that was written about the voyage, he exclaimed, "The abundance of life at the bottom was the most surprising discovery of the trip."

Grazing the Lake's Floating Pastures: Superior's Plankton

In addition to the communities of bottom-dwelling organisms, scientists have identified a second realm—species that live in the open water, also known as the pelagic zone or water column. New technologies have revealed startling insights into these open waters as well. Scientists once believed that Superior's waters contained few species and low numbers of phytoplankton. Then in 1973, two Canadian researchers published an in-depth study of phytoplankton that revolutionized scientific thinking about this staple of Superior's food web.

Prior to this study, scientists had sampled Superior's waters with plankton nets whose mesh size measured sixty-four microns (less than one-thousandth of an inch). Aided by new sampling techniques that could measure on a much smaller scale, the Canadian researchers presented a far more complicated picture. Their studies revealed that on average 65 percent of Superior's phytoplankton can be classified as nanoplankton, that is, two to twenty micrometers in size. (Up to three thousand of the smallest individuals could fit on the tip of a number 2 pencil lead.) Previous studies had missed more than half the phytoplankton that lived in the lake, largely because the plants simply slipped through the mesh of traditional nets.

Smaller may be better in Superior's waters. Nanoplankton seem to have a competitive advantage over their larger counterparts for a variety of reasons. They can take up nutrients and remove wastes more efficiently. Small cells have higher rates of productivity since they do not need to produce as much protoplasm as large cells in the process of cell division. Small cells also do not sink as quickly as large cells. This added buoyancy gives them a longer residence time in the upper layers of the lake where photosynthesis takes place.

Because they depend on sunlight for photosynthesis, living phytoplankton are largely limited to the upper strata of the lake. Grazing in these aquatic pastures—and sometimes preying on each other—are tiny zooplankton that have mastered life in surface waters.

These interactions are some of the most important in Lake Superior since they form the basis of the pelagic food web. The mechanisms that govern them, however, are largely unexplored, especially when compared to the amount of study given to grazing in terrestrial ecosystems, says aquatic biologist Karen Glaus Porter. "The effects of cattle grazing on pastures are part of the farmer's everyday knowledge," she points out. "The awareness that cattle prefer some plant species to others undoubtedly formed the basis for the earliest animal-husbandry practices, and it is not surprising that our knowledge of selective grazing and plant defenses in terrestrial ecosystems is highly developed. Our understanding of grazing in the open water of lakes, however, is far less complete."

As a result, future research could revolutionize today's assumptions about these fundamental workings of life within the lake. For example, scientists once thought that because of their limited mobility, both plant and animal plankton drifted more or less randomly in water. (The word *plankton* is derived from the Latin word for "wanderer.") Nothing could be further from the truth, say biologists B. J. Malone and D. J. McQueen, who studied zooplankton in two Ontario lakes. They point out that

Large phytoplankton were once thought to dominate Superior's floating plant communities. With the advent of finer-scale sampling devices, shown here, researchers have since discovered that tiny plankton, known as nanoplankton, make up 65 percent of Superior's total phytoplankton. NOAA Great Lakes Environmental Research Laboratory.

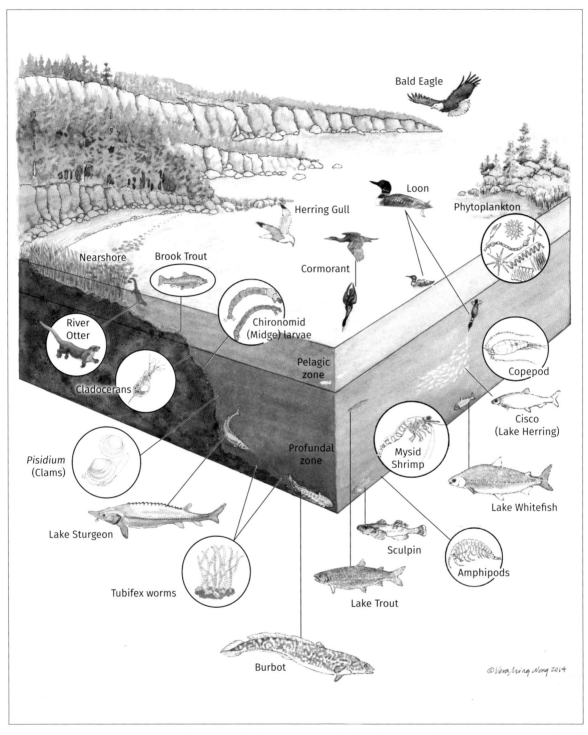

Lake Superior's pelagic food web has a complex biological and spatial structure. Populations of both plant and animal plankton are patchy and pursued by aggregations of zooplankton and fish predators. Illustration by Vera Ming Wong.

"pelagic environments are spatially complex with at least as much biological structure as terrestrial environments." Populations of both plant and animal plankton, they say, "are patchy and are in turn pursued by aggregations and schools of zooplankton and fish predators. The result is an organized and dynamic system as opposed to one comprising populations of randomly distributed individuals each acting as an independent unit."

This spatial structure occurs both horizontally and vertically in the lake. Copepods and cladocerans, two common pelagic crustaceans, coexist in Superior's waters by developing ecological adaptations to their respective zones within the lake. Using the bristles on their legs as food filters, the lake's thirty-five native species of cladocerans, also called water fleas, spend the better part of their lives beating their five pairs of legs to create tiny currents that draw plankton-laden waters to their mouths. Because they must work hard to procure sustenance, these crustaceans tend to be found in warmer, shallower zones where algae production is more abundant. This strategy allows them to gain the maximum return on their great investment of energy.

Not just any phytoplankton, however, will do. Many algae have developed defenses against grazers, including hardened cells walls, spines, or large chain formations. Nearshore areas favor the growth of smaller, more palatable algae, a boon to cladocerans since more unwieldy varieties can clog their filters. (They are equipped with several mechanisms, including a claw that can clear their filters of algal chunks. These large morsels go uneaten by cladocerans, however, since they cannot tear them into bite-sized pieces.)

But food in the lake's nearshore reaches is only seasonally plentiful, so cladocerans have developed mechanisms for coping with these boom-and-bust cycles. Most are parthenogenetic; that is, they do not rely exclusively on sexual reproduction. During spring and summer, the cladoceran population is dominated by females that are able to produce unfertilized eggs that hatch as other females. This strategy allows the animals to quickly build their numbers at a moment's notice and take advantage of sudden blooms of algae without wasting valuable time in mating. In the fall, however, as water temperatures drop and food becomes scarce, their populations decline. Females respond with a reproductive flip of the switch, producing unfertilized ova that hatch as males. The females then mate with these males. Encased in a hard shell, the resulting fertilized egg is able to overwinter even in a frozen or dried state. When favorable times return, these eggs can hatch quickly into females and start the process all over again.

Copepods, on the other hand, are the dominant zooplankton in the offshore waters of the lake. Here, cold waters dampen phytoplankton productivity. As a result, the pinhead-sized copepods grow much more slowly than cladocerans. They make up for the greater scarcity of food, however, by evolving body parts that enable them to utilize multiple and more energy-efficient strategies for procuring their dinner. Like water fleas, the lake's fifty-some species of copepods have the ability to filter small algae from the water. Shaped like the overturned hull of a boat with its rowers captive inside (hence the nickname "oarsmen"), copepods also beat their legs to draw food. But rather than simply discard the large algal chains that are abundant in open water, the oarsmen use an appendage to grasp food chunks. Biting mouth parts enable them to tear apart and chew these rafts of floating food, particularly long chains of diatoms, which aquatic biologist Porter says, "are held and eaten like candy canes." Some species of zooplankton even use these anatomical adaptations to cannibalize their neighbors.

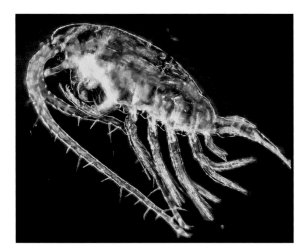 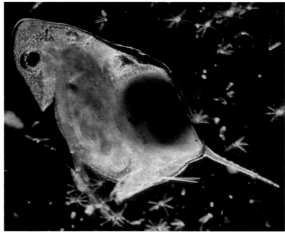

Copepods *(left)* and cladocerans *(right)* are two common zooplankton in Superior's pelagic zone. These crustaceans coexist in Superior's waters by utilizing ecological adaptations that allow them to occupy separate nearshore and offshore habitats within the lake. Photographs by Robert Megard, University of Minnesota.

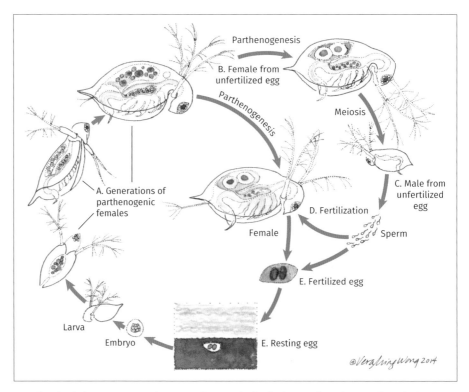

Most female cladocerans are parthenogenetic, reproducing both sexually and asexually. This characteristic is part of a successful strategy for surviving the seasonal nutrient cycles of nearshore waters. In spring and summer when food is abundant, females quickly build their numbers by producing unfertilized eggs that become females *(A and B)*. In fall as food becomes scarce, their unfertilized eggs produce males *(C)* to mate with *(D)* to create fertilized eggs *(E)*. These eggs can overwinter and hatch into females when food is again abundant *(A)*. Illustration by Vera Ming Wong.

Unlike copepods and cladocerans, mysids divide their habitats vertically, maintaining dual citizenship in both deepwater and pelagic realms. Mysids are believed to have evolved from *Mysis oculata,* a brackish-water cousin that adapted to life in freshwater after becoming isolated from the sea by the glaciers. Mysids are known as glacial relicts, species that arrived during the last glaciation and became trapped in basins at the southern boundary of the ice front. These remnant populations persisted, and today several glacial relicts—populations that include both plants and animals—live in the Superior basin, separated from the main range for their species or other closely related species by hundreds of miles. (See also "Hay Pickers and Grass Gatherers: Botanical Exploration along the Lakeshore.")

These tiny crustaceans were dispersed by meltwaters throughout the Great Lakes. In fact, their distribution boundaries coincide with the geographic limits of the great ice sheets. (Lake Michigan forms the natural southern limit of their range in North America.) Favoring temperatures of 34 to 43 degrees Fahrenheit, mysids continue to occupy the deepest and most well-oxygenated reaches of the Great Lakes. They are so dependent on cold conditions that when subjected to temperatures of about 55 degrees Fahrenheit in the lab, they rapidly died.

Perhaps no other invertebrate is so exquisitely calibrated to its particularities of place. Mysids have evolved ingenious solutions to a particularly tricky existential dilemma. A critical portion of their diet—phytoplankton and the tiny organisms that feed on them—is located in the upper strata of the lake, where algae receive the sunlight they depend on for photosynthesis. Yet, if mysids graze in these layers during the day, they are vulnerable to their sight-feeding predators, even though they have the advantage of the ultimate camouflage in Lake Superior—near transparency. Measuring about an inch long, mysids are among the largest zooplankton in Lake Superior. They are also some of the most abundant. Jack Kelly of the Environmental Protection Agency's Mid-Continent Ecology Division observes that the total weight of the lake's mysids would outstrip that of the basin's 650,000 humans by five or six times, a tip-off to their importance as a dietary staple for nearly all species of fish, most notably lake trout. (So important are they as fish food that fisheries biologists have introduced mysids to Lake Tahoe and other oligotrophic lakes in Canada and Scandinavia in the hopes of stimulating fish production.)

To evade predators, mysids practice what animal behaviorists refer to as "risk-sensitive foraging." Mysids spend their days in the darkness of deep water. During his submersible dives, biologist Bowers discovered mysids in three distinct haunts. The smallest, least visible animals congregate in the water column at depths greater than 650 feet. The largest and most vulnerable individuals hover just above the lake bottom or bury themselves in the sediments of the lake floor.

No hideout is risk-free. Among the mysid's deepwater predators is the sculpin, which locates its prey by sensing disturbances in the interface between the water and bottom sediment. So closely bound are sculpins to the sediment substrate that they rarely rise more than two to four inches above the lake bottom to feed. This may explain why Bowers found the mysids floating anywhere from less than an inch to ten feet above the lake floor during the day. By stationing themselves just out of reach, mysids may avoid hungry sculpins on dinner patrol.

Opportunistic omnivores, mysids are not above ambushing a few morsels for themselves while they wait for mealtime to roll around. Among the contents of mysid guts, researchers have found the remains of *Diporeia*. Researchers now believe that mysids form an important link in the nutrient cycle between benthic and pelagic systems. By switching to detritus and bottom-dwelling zooplankton such as *Diporeia* when phytoplankton populations are low, mysids even out the peaks and troughs of food cycles for fish. Unfortunately, its supplemental bottom feeding may also make mysids one of the elevators by means of which pollutants are shipped from the lake's basement and redistributed to its upper floors.

Yet life is no picnic for mysids either. Within an hour after sunset, most of the population has floated up into the water column to feed on algae and other zooplankton. (The movement of aquatic organisms in response to such ecological triggers as light and food is known as diel vertical migration.) It is a metabolically costly strategy for avoiding predators. Hiding out in cold bottom waters slows growth and reproduction. So mysids float to shallower waters. But in making this migration, mysids not only strike a costly trade-off between eating and being eaten, they also must dip into their precious energy stores to subsidize these daily rounds. The distance traveled, zoologist Stanley Dodson observes, is proportional to the clarity of the lake's waters. "The deeper light penetrates into a lake," he says, "the deeper the zooplankton must sink during the day to avoid fish predation." In a 1972 study of mysids in Green Lake, Wisconsin, researchers found that to undertake a vertical migration of 165 feet, mysids expend about 8 percent of their total daily caloric content. In the great expanse of Lake Superior, migrations of up to 330 feet are the norm, more than the distance of three football fields or the "equivalent to a person commuting about 34 miles to work and back each day," observes Minnesota Sea Grant writer Sharon Moen.

When mysids do finally land a good meal, they provide leftovers for many smaller

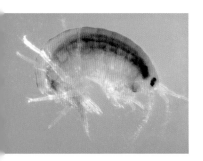

Mysis relicta, also known as opossum shrimp, are tiny crustaceans the size of a child's fingernail. The ability to regulate their buoyancy by incorporating air bubbles into their bodies enables the shrimp to make a daily migration from the bottom of the lake to its upper waters in order to feed in relative safety from predators. Courtesy of Minnesota Sea Grant.

Depth (meters)

copepod equivalents per cubic meter

50,000
10,000
5,000
1,000
500

In this sonar echogram, the bottom blue layer represents mysids (opossum shrimp) floating close to the lake floor. Around sunset, clouds of mysids rise from this deep level to feed on zooplankton attracted to algae growing near the surface. Photograph by Robert Megard, University of Minnesota.

animals, much like terrestrial predators such as wolves. In the lab, they have been observed lunging at their food in a series of quick, jerky jabs. When offered large phytoplankton, they set upon them with great gusto, shredding the plants into many smaller fragments. Researchers suspect that mysid feeding may increase the supply of sustenance for zooplankton that are capable of filtering only the smallest particles. This appears to be especially important during the spring diatom bloom, when food is mostly available in large chains that clog the filters of small zooplankton.

Way Down Under: Life in Superior's Depths

While crustaceans such as mysids, cladocerans, and copepods divvy up the open waters, another suite of organisms partitions the real estate of the profundal zone. Here, the force of gravity works with currents and wave action to sweep fine silts and particles of clay into the deep recesses of the lake floor, where they accumulate at a rate of up to only 0.012 inch per year—a tiny fraction of the thickness of your fingernail. In a sampling conducted by zoologist Ann Heuschele, 96 percent of benthic animals were found in the top 1.5 inches of sediment. Only the top 1 inch of these sediments contains oxygen, a reminder of the precarious knife edge on which the lake's bottom dwellers survive. This uppermost layer of the lake floor is so heavily worked by animals burrowing into it for food or shelter that it becomes extremely porous. Samples dredged from the lake floor are as soft and fluffy as Christmas tree floc, so lightweight that they can be blown away by the slightest breeze.

This slim margin of life lies so far beyond the warming influence of the sun that the temperature of the lake bottom varies only from 36 to 40 degrees Fahrenheit. Living beyond the sun's reach also prevents Superior's benthic organisms from partaking of the living soup of photosynthesizing phytoplankton, whose need for sunlight consigns them to regions near the lake's surface. Filter feeders and detritivores (consumers of detritus), these scavengers of the deep largely depend instead on the "rain" of material that slowly falls to the bottom of the lake, including the remains of dead plants and animals, sediments, and zooplankton feces.

Benthic organisms have evolved remarkable strategies to make the most of the thin veneer of oxygenated sediments at the bottom of the lake. It is as if they have become citizens in a particular locale and have developed a proficiency in the biological customs needed to live there. For example, *Diporeia,* the majority of which occupy the upper three-quarter inch of sediments, probe fine bottom sediments or migrate into overlying waters. Studies of amphipods in Lake Michigan show that they literally eat the world around them—99 percent of their gut contents is silt and other sediments. From this matrix, the animals extract more-nutritious tidbits such as bacteria, diatom fragments, pollen, and fungal spores. So adapted are they to eating leftovers that when researchers fed live algal cells to laboratory *Diporeia,* the amphipods were unable to digest them. Not surprisingly, *Diporeia* put on most of their biomass in spring, when the snowmelt and spring floodwaters wash loads of nutrients into the lake, where they fuel population surges of single-celled algae known as diatoms—and increase the rain of algal die-offs that settles on the lake floor. (See also "Amphipods and Diatoms: The Big Lake's Bread and Butter.")

This seasonal feast stokes the furnace for the entire lake. When *Diporeia* eat, the lake is fed, since they are the most abundant invertebrate on the lake bottom and a food

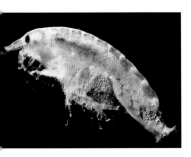

Bottom-dwellers like the amphipod *Diporeia* scavenge material that slowly falls to the bottom of the lake. They are the most abundant invertebrate on the lake bottom and a food staple for many species of fish at nearly all stages of their life cycles. NOAA Great Lakes Environmental Research Laboratory.

staple for many species of fish at nearly all stages of their life cycles. Predators such as burbot and sculpins patrol the lake bottom, alternately flushing and ambushing their prey. *Diporeia* mount a defensive campaign by using a single long antenna that can detect vibrations of their mud-rooting predators at a distance. This early warning device gives them a slight defensive edge.

While amphipods occupy the top layer of sediments, other animals such as the larvae of chironomids (commonly known as midges) construct burrows below the sediment surface. Closely related to the more well-known dipteran insect the mosquito, midge larvae spend up to two years in the lake sediments, where they obtain all of the energy they need to complete their life cycle. From their tube-shaped burrows, which are dug three to four inches into the bottom sediments, midges wave their tails, creating a current of water from which they filter food. They supplement their diet by scraping bacteria and other detritus from the fine-grained sediments around the entrances to their burrows or by swimming up into the water column as high as three feet above the lake bottom to retrieve food.

Chironomids share the lake sediments with another common resident, the pisidiids, or tiny clams. Like their neighbors, they also have evolved ingenious ways to feed while maintaining cover. Some species bury themselves in the upper 0.8 inches of the sediments, where they create a current that allows them to filter algae, microorganisms, and detritus. Others create a more elaborate infrastructure for feeding by digging long, horizontal canals into the sediments that funnel food-laden water in and sweep waste out.

The real earth movers of the profundal zone, however, are the oligochaetes, or freshwater worms. Comprising four families and at least thirteen species, they are the most taxonomically diverse invertebrates in Lake Superior. Of these groups, the species in the genus *Tubifex* "are easily the most important, if importance is measured by widespread distribution, population density, and biomass, and ability to alter sediment properties," say geologists Peter L. McCall and Michael J. S. Tevesz.

These slender worms, the size of a straight pin, evolved several adaptations that allow them to exploit an underutilized niche in the lake—the oxygen-poor bottom sediments. Tubificids occupy burrows in the lake bottom that can reach eight inches or deeper. With their "heads in the sand" and their tails swaying in the current, they feed primarily in the top four to five inches of the sediments, consuming silt and clay particles and stripping them of their attached microorganisms, particularly bacteria, as the material moves through their guts. Known as conveyor-belt feeders, tubificids then deposit fecal pellets loaded with undigested sediments onto the surface of the lake bottom.

Burrowing is one way that tubificids avoid predators. The downside is that the worms must contend with the reduced oxygen levels of subsurface sediments. Tubificids compensate by producing hemoglobin in their blood, thereby increasing their capacity to carry oxygen. These same adaptations enable the worms to survive the more oxygen-depleted conditions commonly found in polluted waters. Today, their excessive numbers on the lake floor are considered a tip-off of serious organic pollution.

Superior: Pristine or Imperiled?

The Great Lakes' minute biological workings became the focus of concerted scientific investigation beginning in the 1960s. New federal, provincial, and state initiatives, coupled with the expansion of university research programs, drew the public's attention to the priceless reservoir of freshwater that lies at the heart of North America. The flowering of knowledge lasted well into the 1970s, leading some researchers to call these decades the golden age of Great Lakes research.

Unfortunately, it was ecological catastrophe, not scientific curiosity alone, that sparked this great era of inquiry. In the history of scientific research in the Great Lakes, "every new era of investigation has been the offspring of the parents of adversity: Chaos and Crisis," writes aquatic ecologist Wayland Swain. "Always a child of desperate need—though often unloved, unwanted, and unnurtured by those in authority—in each case, the tenacious research infant has matured and risen to the challenge set before him."

And the Great Lakes have presented no simple challenges. Because they are large systems that are driven by complex physical forces, the Great Lakes often operate more like oceans than lakes. As such, they are difficult to study without similarly large, complex—and costly—research infrastructure.

But by the time scientists had acquired such capabilities in the Great Lakes, humans had contributed their own twist to nature's already complicated plot. In 1940, Swain observes, the eight Great Lakes states accounted for more than two-thirds of the nation's industrial output. During World War II, they had become what he calls the "free world's arsenal," producing everything from rubber tires and tanks to bullets and bombs. In the postwar boom many of these facilities were converted to peacetime manufacturing. Not surprisingly, by the mid-twentieth century the waters of the Great Lakes had become

Portions of the Great Lakes have experienced heavy industrial development. Resulting environmental impacts to water quality and human health led to binational initiatives to address ecological crises. NOAA Great Lakes Environmental Research Laboratory.

perilously polluted. As Eugene F. Stoermer, one of the pioneers in phytoplankton research, recalls, "Grossly visible signs of pollution such as oil slicks, massive algal blooms, and sediment and detritus plumes were common at most ports. Great Lakes fisheries were in a state of almost total collapse. Severe taste and odor problems were prevalent at many, if not most, municipal water supplies. Many beaches and waterfronts in the Great Lakes region were avoided because of bacterial contamination, massive *Cladophora* blooms, and generally unsanitary and unpleasant conditions."

It was clear that nothing short of a major binational initiative—both in terms of research and remedial action—would reverse the lakes' downhill course. Pressing ecological need arose, however, at a time of gross ecological ignorance. As a graduate student in the late 1950s, Stoermer decided to focus on diatoms, one of the most fundamental links in the Great Lakes' food web, only to discover that he was venturing into uncharted scientific waters. No university in the United States had a specialist in freshwater diatoms on its faculty. The field lacked even a standard reference volume on diatoms. "No one had more than the foggiest concept as to the extent and diversity of the diatom flora in the Great Lakes," he recalls. So like many of his colleagues in the 1960s, Stoermer found himself confronting unprecedented pollution problems while struggling to fill in major gaps in some of the most basic knowledge about the lake.

Yet this was not the first time that the United States and Canada had faced an ecological crisis and a vacuum of scientific knowledge in the waters along their borders. In 1909 they joined forces to formally tackle the first in a series of threats to their shared aquatic resources—bacterial pollution from untreated sewage that was causing deadly outbreaks of disease, including typhoid fever, in nearly every major population center from Duluth to Buffalo, New York. Clean water had become a matter of national interest since it was vital not only to human health but also to the well-being of a variety of economic enterprises from shipping, recreation, hydropower, and fishing to providing water for each nation's growing urban and industrial centers.

That year the United States and Canada signed the historic Boundary Waters Treaty. The agreement was both an affirmation and an exhortation: Each country was accorded the right to use the shared waters along its borders, but they were to refrain from polluting them "to the injury of health or property of the other." To help resolve issues arising from the treaty and to make recommendations regarding the management of the north country's aquatic commons, the treaty established the International Joint Commission (IJC), a binational group of appointed overseers.

It was a remarkable feat of international cooperation. By 1950 the United States and Canada had wiped out typhoid epidemics by investing in the construction of sewers and in research to develop effective treatment techniques for municipal drinking water.

But a host of unforeseen changes in the Great Lakes basin would eclipse such early successes. For example, between 1901 and 1980, the population in the Great Lakes quadrupled, growing from ten million to a staggering forty million people. Studies conducted by the IJC in 1946–48 showed that the massive sewer projects that were built earlier in the century to solve problems of bacterial contamination were being overtaxed by the growing demand. Wastewater flowed into the lakes with alarming efficiency. Topping the list of new pollution concerns was the enrichment of the Great Lakes' deep, cold glacial waters.

Cultural Eutrophication: Overfertilizing the Great Lakes' Aquatic Pastures

Before the settlement of the region by Euro-Americans, these inland seas were a great reservoir of oligotrophic water. As such, they shared a number of chemical and biological characteristics. Though poor in nutrients and low in suspended particles, their deep waters were rich in dissolved oxygen, a life requirement for such desirable—and commercially valuable—fish species as lake trout. With the exception of a few bays and some shallower portions of Lake Erie, which tended to be naturally mesotrophic, or moderately productive, the Great Lakes hosted a surprisingly uniform complement of aquatic plants and animals that reflected their common glacial heritage.

In 1966 biologist E. Bennette Henson observed that the "Great Lakes and other lakes within their drainage basins represent essentially the last natural supply of oligotrophic water in this part of the United States. These waters support a unique fauna that may symbolize pristine water. The continental glaciers of not long ago shaped their basins and brought with them certain species of fauna that characterize these waters." But, he cautioned, "Biological succession is now taking place, and changes are occurring with extreme rapidity."

Henson was referring to the much-publicized concerns at the time about a human-caused acceleration of the natural aging process of lakes known as eutrophication. Put simply, lakes are works in progress, ecosystems in transition. Depending on their geographic settings and the characteristics of their drainage basins, some lakes gradually become richer in nutrients, a condition that sets off a sequence of physical, chemical, and biological changes. The Great Lakes, Henson says, have been undergoing eutrophication since their creation, largely due to the natural enrichment contained in runoff and warmer climatic trends in postglacial times. The process is infinitesimally slow, however. On their own, the Great Lakes probably would not have become eutrophied for at least tens of thousands of years, if ever.

But "during the last two centuries," Henson points out, the aging of the Great Lakes "has been significantly accelerated by human civilization," a process known as cultural eutrophication. Human activities have dramatically altered these inland seas, particularly the lower lakes. Converting forests to farm fields increased the erosion of soils into streams and lakes, where they acted like a fertilizer to stimulate phytoplankton growth. In many parts of the Great Lakes, such as around Lake Ontario, Lake Erie, and the southern shores of Lake Michigan, agricultural settlement was followed by rapid urbanization.

The poster child for ecological degradation in the Great Lakes, however, was Lake Erie. Its once-clear waters, particularly those in its warm, shallow subbasins, grew murky with the repeated blooms of tiny free-floating blue-green algae. Fouling beaches and many of the open waters were colonies of *Cladophora,* a green alga whose long filaments sprout like luxuriant tresses from rocks, piers, and buoys or grow in tangled, free-floating mats. When these plants died and began to decompose, they sapped the lake's oxygen stores. Making matters worse were organic wastes discharged from municipal and industrial sources. As they settled and decayed, they too depleted Erie's oxygen reserves. Populations of fish, such as lake trout, whose incubating eggs require high levels of dissolved oxygen and clean gravel or rock substrates, perished in the mucky, anoxic lake bottom. Mass die-offs of adult fish were common. By the 1960s, the media dubbed Lake Erie as North America's Dead Sea.

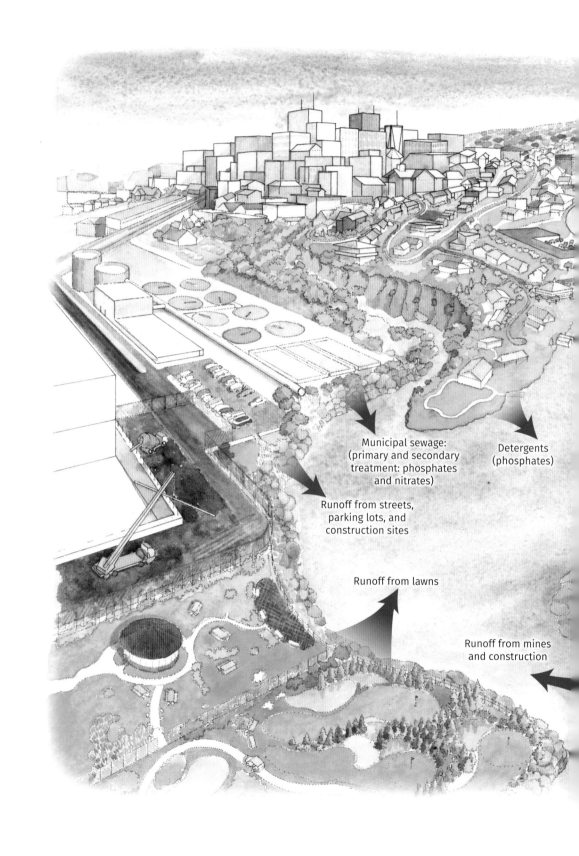

Municipal sewage:
(primary and secondary
treatment: phosphates
and nitrates)

Detergents
(phosphates)

Runoff from streets,
parking lots, and
construction sites

Runoff from lawns

Runoff from mines
and construction

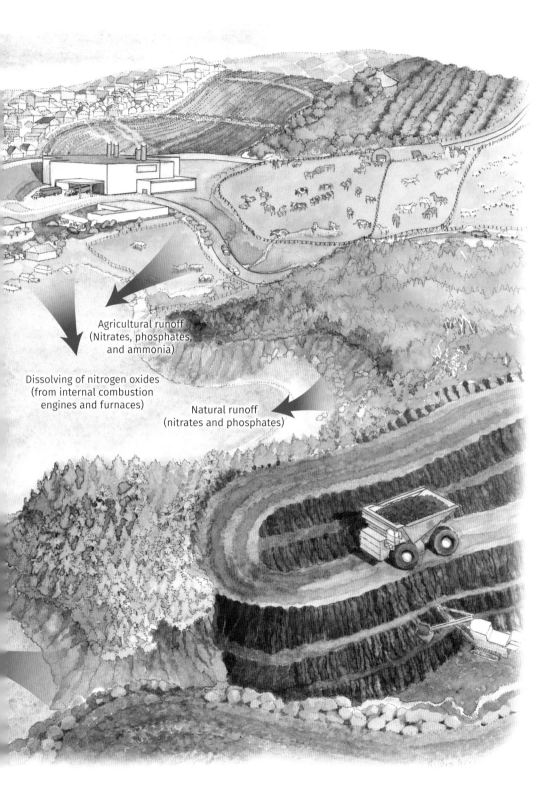

Agricultural runoff
(Nitrates, phosphates,
and ammonia)

Dissolving of nitrogen oxides
(from internal combustion
engines and furnaces)

Natural runoff
(nitrates and phosphates)

The eutrophication, or "aging," of the Great Lakes has been significantly accelerated by human activities. Enriched runoff leads to nutrient overload and a breakdown of biological and chemical cycling in lake ecosystems. Illustration by Vera Ming Wong.

Perhaps most alarming was the discovery that pollution was not simply posing a periodic nuisance for swimmers and boaters and causing intermittent fish kills, but it also was erasing the chemical and biological signature of entire ecosystems. In the mesotrophic community of benthic animals in western Lake Erie, for example, the larvae of the mayfly *Hexagenia limbata* had been a historic staple of the lake's food web (just as *Diporeia* dominate benthic communities in oligotrophic waters). By the mid-1950s, however, the oxygen-loving mayfly larvae had disappeared from that part of the lake. In their place came an explosion in the population of *Tubifex* worms, particularly the pollution-tolerant species *Tubifex tubifex*.

Faced with such widespread ecological degradation, in 1964 the U.S. and Canadian governments petitioned the IJC under the terms of the 1909 Boundary Waters Treaty to study water-quality problems in the Great Lakes. In a 1969 report that summarized nearly a decade of research, the IJC declared that of all the challenges facing the Great Lakes, the "most serious water pollution problem in the lower Great Lakes, having long

Overfertilizing of the lower Great Lakes from agricultural runoff and discharges of organic wastes has led to impacts, both nuisance and deadly. Shown here are *Cladophora*, filamentous algae, fouling a beach on Lake Erie, young loons dead of botulism poisoning *(opposite, above)*, and blue-green algae blooms *(opposite, below)*. NOAA Great Lakes Environmental Research Laboratory; *(opposite, above)* photograph by Damon McCormick; *(opposite, below)* NOAA Great Lakes Environmental Research Laboratory.

term international significance, is the increasing eutrophication of the lakes."

By then the United States and Canada had identified a common enemy—phosphorus, some 60 percent of which came from the phosphates that were used to boost the cleaning power of household detergents. In 1973 the two countries signed the Great Lakes Water Quality Agreement. In their goal to limit the release of phosphorus into Great Lakes waters, the United States and Canada passed legislation that reformulated detergents, reducing their phosphate content or substituting it with other chemicals. Together the two countries also invested $8 billion in infrastructure upgrades ($43.9 billion in 2014 dollars), including modifying sewage treatment plants to remove phosphorus from the effluent.

These efforts produced startling improvements in Great Lakes waters. Total phosphorus levels in the most severely polluted lakes—Erie and Ontario—dropped by more than two-thirds, from highs of 25 micrograms per liter in the mid-1970s to 10 micrograms per liter in the mid-1990s.

But instead of stabilizing or continuing to drop, phosphorus levels once again began to rise. By 2011, according to a *New York Times* report, toxic algae blooms slimed across one-sixth of Lake Erie's waters, disrupting the lake's $10 billion tourism industry, killing fish, and fouling the drinking water for 2.8 million of the basin's citizens with high levels of the liver toxin microcystin, which is produced by *Microcystis,* a type of blue-green algae. This time the phosphorus poured not from kitchen sinks and factory pipes but primarily from the fields of corn and soybeans around Toledo, Ohio. Ironically, new techniques aimed at delivering smaller, more efficient doses of fertilizer to farm fields are partly to blame for the polluted runoff. Since many farmers now practice no-till agriculture to conserve soil, fertilizer is applied to fields in pellets that sit on the surface. Rain and snowmelt wash a small portion of this fertilizer away, eventually draining nutrient-rich stormwater into the Maumee River and then into the western end of Lake Erie. In recent years, the problem has been exacerbated by a 13 percent increase in the intensity of spring storms, which are likely to become the norm according to several climate change scenarios. And the infestation of Erie with nonnative zebra mussels, which gorge on beneficial algae and excrete the algae's phosphorus, produces a double whammy: the zebra mussels both reduce the competitor algae for *Microcystis* and also supply the extra nutrients that this toxic alga needs for growth and reproduction. "We've seen this lake go from the poster child for pollution problems to the

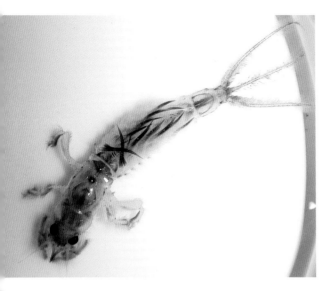

An aquatic nymph (*above*) and large hatch of adults (*below*) of the mayfly *Hexagenia limbata*. Once a historic staple of Lake Erie's food chain (similar to *Diporeia*'s dominance of benthic communities in Lake Superior), mayflies have largely disappeared from western Lake Erie due to pollution. Photographs by Fred L. Snyder.

best example in the world of ecosystem recovery," observes Jeffrey M. Reutter, director of the Sea Grant College Program at Ohio State University. "Now it's headed back again."

Phosphorus: Too Much of a Good Thing?

Why has excess phosphorus wreaked such havoc in the Great Lakes, particularly when nutrients are in short supply in these systems? After all, wouldn't stimulating the growth of phytoplankton—the core organisms of the food web—ultimately result in greater overall biological productivity? To use a terrestrial analogy, wouldn't growing more grass in the pasture increase the production of cattle?

Indeed, it is an argument that Lake Superior researcher Mary Balcer often hears from members of the general public, particularly anglers who would like to see an increase in fish yields. "People joke with me about fertilizing the lake to increase fish production," says Balcer, a professor of biology at the University of Wisconsin-Superior. "They even talk about the virtues of building nuclear power plants on the Superior shore so that the waters they discharge can warm up the lake and make it more productive. The problem is, they're only half-joking."

To understand how and why excess nutrients are so disruptive in oligotrophic waters, researchers have focused on the numbers of phytoplankton and their species makeup. Of all the organisms in the lake, phytoplankton stand on the front line of chemical change. As such, they are the lake's "first responders" to disturbances such as influxes of excess nutrients.

Among the best bioindicators of eutrophication and other environmental disturbances are diatoms, which leave behind evidence of their existence in the form of frustules, or hard shells. Made of the glass-like substance opaline silica, the frustules drift to the lake bottom and become preserved in its sediments. The frustules bear distinctive markings that can be used to accurately pinpoint the species of diatoms that produced them. Because of this durability, phytoplankton researchers can reconstruct ecological changes over long periods of time, whether it is by examining diatom samples from museum collections that date to the late nineteenth century or by using paleolimnological techniques to analyze fossilized specimens contained in sediment cores that are thousands of years old. Diatoms also have other virtues: there are lots of them. A single sediment core, Stoermer says, may contain the identifiable remains of individuals representing hundreds of species.

As of 2009, scientists have described more than 1,800 species of diatoms and several hundred additional varieties in the Great Lakes. In one teaspoon of sand collected from a harbor off Michipicoten Island in Lake Superior, Stoermer counted more than 470 diatom taxa. Great Lakes diatom communities are among the "most diverse known from any place in the world," Stoermer observes. And this number represents only the tip of the iceberg.

With the evolution of such great diversity comes a high degree of habitat selectivity. Some diatoms are planktonic, while others prefer particular substrates, ranging from rocks, sand, and mud to larger aquatic plants. They also have specific chemical tolerances, many of which have been well documented. Researchers therefore can use diatoms' extreme sensitivity to their surroundings as a way to help determine the causes of environmental change.

The question remains, if limited supplies of nutrients cap the growth potential of diatoms, then why is diatom production impaired—rather than enhanced—when more food in the form of fertilizing phosphorus is added to the lakes? In the 1970s Stoermer and his colleague Claire L. Schelske at the University of Michigan, Ann Arbor, studied that question and developed what has become known as the silica-depletion theory, which links elevated levels of phosphorus to diminished diatom growth.

Of all the nutrients needed by phytoplankton for growth, phosphorus is the most quickly depleted. Scientists refer to this as the primary limiting nutrient. Once aquatic plants have used up all the available phosphorus, there is still a generous supply of other nutrients left over, such as oxygen, hydrogen, carbon, silica, and nitrogen. It is like having only enough flour to bake one cake even though you have enough eggs, butter, and milk to bake ten cakes. Additional inputs of phosphorus into aquatic systems, however, are like having a nearly unlimited supply of flour. You can keep baking cakes until you run out of another ingredient, known as the secondary limiting nutrient.

For Great Lakes' diatoms, that second ingredient is silica, a component of soil and terrestrial plant remains that enters the lake by way of runoff. Diatoms depend on silica for growth and reproduction. Under presettlement conditions, diatoms ran out of phosphorus long before they ran of out of silica. Consequently, the lakes contained a large volume of untapped soluble silica, known as the silica reservoir. With the exception of periodic natural disturbances such as forest fires within the watershed, which temporarily disrupted nutrient levels in the lakes, the silica concentrations in this reservoir remained steady year after year.

Excess phosphorus, however, sets off a chain of adverse reactions. First, diatoms bloom in greater-than-normal numbers. As they grow, diatoms draw soluble silica from the water column to construct their frustules. Large blooms, therefore, can quickly exhaust a lake's natural reservoir of silica. Some of the silica contained in the frustules, known as biogenic silica, is released back into the water column and used by a new crop of diatoms when the shells of dead diatoms dissolve. But this can take a long time. In studies conducted in Lake Michigan, for example, researchers found that it takes about one year for biogenic silica to break down into the soluble form that living diatoms can use. During this long interim, many diatoms sink to the bottom of the lake, where they become buried in sediments. This process can lock up biogenic silica for thousands of years, if not sequester it permanently.

Over time, the stepped-up diatom growth cycle removes more silica from the water than is replenished from terrestrial sources or recycled from the lake sediments. The problem becomes especially acute in the summer when temperature-stratified waters largely block the chemical and biological exchange between the lake's upper and lower layers. During these periods, silica depletion develops, and diatom production in the epilimnion (the top one hundred feet of water) is severely diminished.

In artificially eutrophied waters, however, there is still plenty of phosphorus left over for species that do not require silica for growth and reproduction. In addition to ongoing inputs from human sources, the phosphorus in aquatic plant remains is efficiently taken up by new plants and recycled several times during the annual production cycle, a process that helps to maintain high levels of this nutrient once it builds up in lake waters.

In this new phosphorus-rich environment, members of the phytoplankton community that once were minor players, such as the green and blue-green algae, step up to fill the vacuum left by the silica-limited diatoms and grow to nuisance levels. In severely disturbed systems such as Lake Erie, even foreign algae species that have been introduced into the lake in the bilge water of sea-going freighters now have displaced some indigenous species and emerged as dominant members of the reshuffled phytoplankton assemblage. Unfortunately, some of these exotic species are too large or too distasteful for consumption by native zooplankton. Moreover, even when edible, many introduced algal species are not particularly nutritious since they are richer in carbohydrates than fats, the form of energy to which organisms in the Great Lakes food web are adapted. Reverberations are felt throughout. Indeed, Stoermer points out that wholesale reorderings in fish communities have predictably followed radical changes in phytoplankton assemblages by thirty to fifty years.

In subsequent research, Stoermer and Schelske discovered some surprising twists to their original silica-depletion theory. The scientists found that inputs of even relatively modest amounts of phosphorus were enough to disrupt the lake's silica dynamics, in part because the effects of phosphorus were magnified by other substances present in wastewater and industrial discharges such as vitamins, trace metals, and compounds known as chelating agents, which bind with silica and prevent it from being in a form that diatoms can use.

Most scientists have also long assumed that wholesale ecological change in the Great Lakes has been a relatively recent phenomenon. Not so, says Stoermer. Such waters as the Bay of Quinte, located on the western end of Lake Ontario, went from oligotrophic to eutrophic in the mid-nineteenth century when the cutting of forests and clearing of land within its watershed sent soils coursing into its waters. Perhaps most surprising of all was that the bay had undergone these changes in less than a decade.

Lake Ontario was not alone for long. Lake Michigan's extraordinary assimilative capacity helped to buffer it against radical changes until 1955. Between 1955 and 1970, however, the scale was tipped. The overly enriched waters of Lake Michigan lost one-half—fourteen million tons—of the silica from its silica reservoir. At the same time, the biogenic silica captured in the lake's sediments increased tenfold. For oligotrophic organisms that had adapted over thousands of years to seasonal food cycles largely determined by the limited availability of phosphorus, the change in nutrient regime took place in the ecological equivalent of a blink of an eye.

Superior: The Exception to the Rule?

Researchers have long held that Superior's waters most closely resemble those that first streamed from the glaciers more than ten thousand years ago. Intact, too, are the communities of plants and animals that can trace their lineage to glacial times. True, they acknowledge, Superior has its environmental hot spots. The bay that lies at the mouth of the Kaministiquia River in Thunder Bay, Ontario, for example, has long been polluted by discharges from pulp-paper plants and wood-preservative factories. In a 1967 study, researchers tallied numbers of the pollution-tolerant worm *T. tubifex* of up to 226,000 individuals per square foot in the bay's grossly contaminated bottom. By contrast, in samples taken from the bottom in the clean, oligotrophic waters of the open lake, scientists have found that *T. tubifex* is represented by only a few individuals. In a native assemblage of worms, *T. tubifex* are far outnumbered by the more dominant species of the genera *Limnodrilus* and *Stylodrilus*.

Still, with the exception of Thunder Bay and polluted harbors around Duluth in the United States and Jackfish Bay, Peninsula Harbour, and Nipigon Bay in Ontario, Superior has avoided the afflictions that have plagued its sister lakes. Concentrations of major ions—calcium, magnesium, sodium, sulfate, chloride, bicarbonate, and potassium—remain constant. Silica concentrations in the water are high. For the most part, the members of the phytoplankton community are representative of those found in oligotrophic conditions. Thus, many scientists regard Lake Superior as the gold standard by which the success of ecological remediation efforts in the lower lakes should be measured.

But studies have shown that Lake Superior—despite its oligotrophy and particularly cold temperatures—is just as vulnerable to disruptions in its phytoplankton community as the other Great Lakes. Researchers have found that although some species of filamentous algae such as *Cladophora* prefer warmer waters, others, such as *Ulothrix zonata,* thrive in Superior's cold temperatures when subjected to phosphorus-enriched conditions. Growing *U. zonata* in constructed tanks in a 1982 study along the Minnesota North Shore, researchers found that the algae responded readily to inputs of phosphorus under normal lake conditions. Their experimental observations were borne out by a reconnaissance of the shore in which researchers discovered *U. zonata* carpeting a bedrock ledge near the outfall of the Grand Marais wastewater treatment plant. Not surprisingly, correspondingly high levels of phosphorus had been recorded in the water at the site. So far, the sparse settlement of the Lake Superior basin has helped to preserve the high quality of its waters. But according to the researchers, that could easily change. "Increased activity by man in Lake Superior would invariably result in increases in the level of phosphate," and "if the wave-washed, nearshore areas of Lake Superior were to receive large amounts of phosphorus-rich effluents," they warned, "a luxuriant growth of *U. zonata* and other filamentous green algae could develop."

Low human density and a greater volume of water may have helped Superior to avoid the wholesale ecological transformation of the other Great Lakes, but research suggests that the lake's biotic community too bears an irreversible imprint of change. In 1993 Stoermer and Schelske published the results of a study in which they analyzed sediment cores taken from the offshore basins in all five Great Lakes. Based on the kinds of siliceous microfossils, primarily diatoms, found in the sediments, they were able to construct a history of ecological change in each lake.

In sediments representing the period 1945 to 1979 in Lake Superior, the researchers found some disturbing irregularities. Notable was an increase in the total biomass of diatoms. And while the species recorded were still representative of those found in oligotrophic conditions, the species makeup of the lake's indigenous assemblage nonetheless had changed: species of the genus *Cyclotella* had become more abundant. The scientists viewed this increase as a troubling development since, among the oligotrophic diatoms, *Cyclotella* are known to respond most readily to environmental change. In reviewing the algal record in the other Great Lakes, the researchers noted that this new *Cyclotella*-dominated assemblage in Lake Superior corresponded to the one that occurred in Lake Ontario, now the most altered of the Great Lakes, shortly after European settlement. A closer look at the algal record showed that Superior was following in the footsteps of its sister lakes and was, in fact, in a kind of first stage of eutrophication.

The increased abundance of one species of *Cyclotella*—*C. comensis*—in particular caught the researchers' attention. The exact cause of this increase is unknown, but Schelske and Stoermer suggest that the diatoms may be responding to recent changes in global carbon dioxide levels or other related climate-warming phenomena. Another plausible explanation is that populations of *C. comensis* may indicate growing nitrate pollution in Lake Superior since increases in this species in Lake Huron have been linked to elevated levels of nitrate, a product created in the combining of nitrogen and oxygen. In the past century, the burning of fossil fuels and the manufacture and use of nitrogen fertilizers have doubled the amount of nitrogen in the atmosphere. Superior may have largely escaped the pollution that followed from the intense development of the shorelines of the lower lakes, but of all the Great Lakes, it is the most vulnerable to contamination from the atmosphere. In the future, the biggest threats to the lake may come not by land or by water but in the fall of gentle rain or the silence of a snowy night.

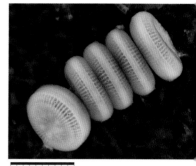

L   x6,0k   10 um

An increase in the abundance of diatoms in the genus *Cyclotella*, especially *Cyclotella comensis*, between 1945 and 1979 marked a change in Lake Superior's ecosystem. The exact cause is unknown, but diatoms may be responding to global climate change or to growing nitrate pollution in Lake Superior. Photograph from External Quality Assessment Trials: Phytoplankton, State Reservoir Administration of Saxony (CC BY-NC-SA 3.0).

Out of Thin Air: Pollution from the Sky

In 1774 the Swedish scientist Carl Wilhelm Scheele blended a cocktail of hydrochloric acid and manganese dioxide to create a heavy gas with a strange greenish hue. Some thirty-six years later, the English chemist Sir Humphry Davy isolated the same gas using a different process. He named it *chlorine,* after the Greek word for "greenish yellow."

Little did either scientist know that chlorine would change the course of history, becoming one of the most versatile and commonly used gases of modern times. Chlorine's meteoric rise began on the outskirts of places such as London, where laundries once kept vast open fields as part of their operations. Here at the city's edge, clothes were spread on the ground and soaked in buttermilk to hasten the natural bleaching action of sunlight. But all that changed when the British textile industry mixed chlorine with potash to create a liquid bleach.

With the advent of electricity, manufacturing chlorine became cheap and easy. By running an electrical current through a solution of sodium chloride, that is, saltwater, scientists could convert the stable chloride ions into highly reactive molecules of chlorine gas.

The availability of chlorine stimulated laboratory experimentation. As one building block in a chemical kit of parts, chlorine could be readily combined with other substances to create whole new compounds. By hooking one or more chlorine atoms onto either chains or rings of carbon atoms, for example, scientists constructed a large and versatile family of chemicals known as organochlorines (*organo* referring to the inclusion of carbon).

Organochlorines would transform nearly all sectors of modern society, from agriculture to heavy industry. Take DDT. First synthesized by a German chemistry student in 1874, DDT ramped up production after 1939 when Swiss scientist Paul Müller discovered its potential as a deadly insecticide. Its efficacy was borne out during World War II, when various militaries used the chemical to successfully combat typhus-bearing lice among their troops. DDT was considered such a miracle substance that Müller was awarded the 1948 Nobel Prize for Physiology or Medicine.

Private companies and U.S. government labs were eager to put such chemical know-how to civilian use following the war. DDT, they discovered, could be mixed into spray form and applied to forests, farm fields, wetlands, and city streets to control mosquitoes and other insects. By 1962, manufacture of the pesticide hit an all-time high with U.S. factories pumping out about 176 to 187 million pounds per year. DDT became one of the most freely dispensed organochlorines in the postwar environment.

New compounds continued to be added to the suite of chlorinated chemicals. They soon became ubiquitous in products ranging from plastics, industrial solvents, and pharmaceuticals to refrigerants. Today, writes Joe Thornton in *Pandora's Poison: Chlorine, Health, and a New Environmental Strategy,* more than eleven thousand organochlorines currently are in use. Thousands more have been released into the environment as accidental by-products of their manufacture and application.

But in the honeymoon euphoria over chemicals, little to no thought was given to their long-term effects on the health of humans or the biosphere. Then in 1962 came the publication of Rachel Carson's *Silent Spring.* Carson's jeremiad against the postwar love affair with toxic chemicals, particularly DDT, shocked the nation. Passages that described birds dying violent, convulsive deaths in a pesticide fog warned that poisons were creating ever-widening dead zones that threatened to unravel the very fabric of life.

By then an even more unsettling picture began to emerge: populations of top avian predators such as bald eagles and peregrine falcons were plummeting. Especially alarming was the discovery of DDT in the tissues of birds that occupied some of the most pristine and remote reaches of the nation. In most cases, the pesticide exposure was not killing adult birds outright. Instead, as scientists later discovered, the chemical hampered the metabolism of calcium in female birds, resulting in eggshells that were so thin they cracked under the weight of incubating adults. Scientists came to a terrifying realization: in subtle and far-reaching ways, synthetic chemicals were disrupting the biological processes on which life depends. They had tripped a cascade of multigenerational effects over which humans had lost control.

Carson's book prompted the nation to reexamine its reckless embrace of the mantra "Better living through chemistry." Nowhere was the reckoning perhaps more intense—or pressing—than in the Great Lakes, the heart of the U.S. industrial belt.

One of the earliest warnings in the region came from an unexpected source: the fur industry. In 1965 G. R. Hartsough published an article in the trade journal *American Fur Breeder* in which he noted a growing number of reports of massive reproductive failure on mink ranches in the north-central states and Ontario. At the time Hartsough, an official with the Great Lakes Mink Association, cautioned that the cause of the problem could not be pinpointed with any certainty. But preliminary evidence, he wrote, "pointed the finger at Great Lakes fish as being the culprit." Tests of female mink that were fed a steady diet of Great Lakes fish revealed high levels of pesticides. As a result, Hartsough and his colleagues suggested that pesticides were to blame for poor mink reproduction, but he conceded, "The offending material may . . . possibly be some entirely different chemical contaminant."

In 1968 Richard J. Aulerich and Robert K. Ringer, animal scientists at Michigan State University, East Lansing, began a series of experiments to investigate the industry's complaints. The initial results showed that adult animals had no problems with breeding and whelping when fed a diet comprising up to 15 percent Lake Michigan coho salmon. Mortality rates among their offspring, however, reached an alarming 80 percent. Subsequent tests, which compared health outcomes of animals fed a diet of Pacific Ocean fish with those fed Great Lakes fish, determined that a toxic agent was present in the Great Lakes fish.

Like Hartsough, scientists initially suspected that excessive body burdens of DDT or dieldrin caused the reproductive anomalies. But previous laboratory experiments showed that mink with even higher loads of these pesticides were able to reproduce normally. Additional sleuthing revealed significant concentrations of several other chlorinated hydrocarbons, among them polychlorinated biphenyls, or PCBs, a family of highly toxic chemicals in which biphenyl, a crystalline aromatic hydrocarbon, is combined with chlorine. Aulerich and Ringer refocused their attention, carrying out additional tests in which animals were fed fish that were deliberately laced with these chemicals. The test subjects exhibited similar health and reproductive problems as those that ate coho salmon from Lake Michigan. Furthermore, the animals developed serious conditions at doses of PCBs that were far below the lethal threshold. Even minute quantities of PCBs were enough to nearly shut down reproduction among ranch mink altogether. To make matters worse, high rates of reproductive failure also occurred when the animals were fed other species of Great Lakes fish, indicating that PCB contamination was widespread in the Great Lakes. According to Canadian fisheries scientist Michael Gilbertson, the mink-ranch research "represents one of the very few cases in which low-level exposure to organic chemicals has been shown to be the cause of a subtle epidemic in a mammal population, other than humans."

American mink, *Mustela vison*, were raised as part of the fur industry in the Great Lakes region. Massive reproductive failure of farmed mink fed fish from the Great Lakes was one of the early warnings of the toxic effects of polychlorinated biphenyl (PCB) pollution in the Great Lakes environment. Photograph by Jesper Clausen / Fur Commission USA.

Mink, which are particularly sensitive to PCB contamination, were canaries in the coal mine, the tip of a looming iceberg. Reports began flooding in from the field indicating anomalies in other animals as well. Census takers in Lakes Michigan, Ontario, and Huron noted unusual numbers of birds such as cormorants and gulls with congenital deformities ranging from crossed beaks to duplicate limbs. They also witnessed behavioral abnormalities, including higher levels of aggression among adult birds and large numbers of untended nests. Mortality among mature animals appeared to be on the increase, particularly during times of stress such as migration and molting.

By 1978, researchers had pieced together a puzzle showing a pattern of damaging, long-term side effects from the use of organochlorine chemicals. The pollution of the Great Lakes was no trifling matter. Twenty percent of the entire U.S. population depended on the Great Lakes for such essential resources as drinking water. Pollution also threatened to destroy lucrative resource-based enterprises such as recreational and commercial fisheries.

The United States and Canada took action. Six years after officials from both countries signed the historic Great Lakes Water Quality Agreement of 1972, which successfully reduced the flow of excess nutrients in the lakes, they returned to the negotiating table. This time they addressed a new threat. As part of their renewed pledge "to restore and maintain the chemical, physical and biological integrity of the waters of the Great Lakes Basin Ecosystem," the United States and Canada vowed to meet the growing dangers of toxic chemicals, which had eclipsed excess nutrients and thermal pollution from power plants as the top water-quality issue. According to aquatic ecologist Swain, the amended 1978 agreement "officially heralded the period of scientific concern in the Great Lakes for toxic xenobiotic chemicals of anthropogenic origin."

By 1983, the IJC's Great Lakes Water Quality Board had identified nine hundred chemicals and heavy metals that posed potential dangers to human health and Great Lakes biota. In its 1985 report, the board singled out eleven "critical pollutants" of special concern: PCBs, DDT, dieldrin, toxaphene, mirex, methylmercury, benzo[a]pyrene (a member of a class of chemicals known as PAHs), hexachlorobenzene (HCB), furans, dioxins, and alkylated lead. Eight of the targeted problem contaminants were chlorinated chemicals.

That same year, four states bordering Lake Michigan issued consumption warnings for fish taken from the lake. In May 1986 the governors of the eight Great Lakes states signed the Great Lakes Toxic Substances Control Agreement, which identified the problem of persistent toxic substances as "the foremost environmental issue facing the Great Lakes." In 2012 the United States and Canada renewed their commitment to the health of the Great Lakes by signing an updated amendment to the historic Great Lakes Water Quality Agreement.

Piecing Together the Puzzle of Lake Superior Pollution

To determine levels of persistent pollutants, federal agencies established a series of monitoring programs. These efforts target bioaccumulative organic compounds that are widespread in the environment, such as PCBs, and measure them in a range of natural media, from lake sediments and animal tissues to the air and precipitation.

One of the longest running and most useful data sets of pollution trends ever collected is the Great Lakes Fish Monitoring Program (GLFMP), which was created in the early 1970s by the U.S. Fish and Wildlife Service. This annual survey analyzes tissue samples taken from top predator fish throughout the Great Lakes. The highly successful project is administered by the U.S. Environmental Protection Agency (EPA) and the U.S. Geological Survey (USGS).

The findings of GLFMP and other programs have been anything but straightforward. Time and again, signals gathered from the field have repeatedly stumped Great Lakes scientists, presenting them with a series of challenging research puzzles.

Take PCBs. Researchers were perplexed by the fact that samples of water from the Great Lakes often contained levels of PCBs that were within margins considered safe for human consumption. In some cases, they could barely be detected. Nonetheless, high concentrations of the very same problem chemicals—twenty-five million times higher than concentrations in the surrounding water—were found in the tissues of commercially valuable fish such as coho salmon and lake trout. So great was the disparity that Swain and his colleague William C. Sonzogni estimated that "in terms of exposure potential, it is possible to breathe the air in the Lake Michigan basin and drink its water for a period of more than five years before achieving the same effective exposure as from eating a single pound of Lake Michigan lake trout or coho salmon."

Just what was going on? Clues to the behavior of these new toxins in the environment came from studying their fundamental chemical structure. Nature has evolved myriad ways of cycling compounds through the environment, be it degradation by ultraviolet light, microbes, precipitation, or chemical reactions. Designed to beat Mother Nature at her own game, many organochlorines were purposely concocted to resist breakdown by natural forces. These characteristics made them extremely desirable for industrial and agricultural purposes. For example, PCBs, whose hardiness is legendary, are largely immune to breakdown by acid-base reactions, hydrolysis, chemical oxidation, photodegradation, thermal changes, and the vast majority of chemical agents.

Organochlorine pesticides have earned a similar reputation as the Arnold Schwarzeneggers of the chemical world. Many of the poisons used to kill noxious weeds and insects in agriculture fields, for example, share key traits that contribute to their effectiveness and persistence. At the atomic level their properties make them hydrophobic, or insoluble in water. This feature extends the working lifetime of pesticides in environments that are soaked by rain or irrigation. A second critical factor is the inclusion of hydrocarbons (the biphenyl in PCB, for example), which form very strong bonds within the molecule as well as with any organic material they come across.

The advantages of this formulation for pesticides are clear: the ability to both cling to organic particles and resist dissolution and transport by rain or snowmelt allows poisons to stay put longer in the field to work on agricultural pests. The addition of chlorine further boosts the toxicity of these compounds.

Conversely, other organochlorines were designed to be highly reactive, such as the common industrial chemical known as trichlorophenol, which degrades rapidly when exposed to light. Unfortunately, many of these unstable compounds break down into other substances that are both extremely persistent and highly toxic.

That high levels of organochlorine contaminants would be discovered in the environment was not surprising. Engineered to persist, their buildup was a simple case of math: more chemicals were being released than were being broken down and neutralized. Some chemicals, it was found, could resist degradation for decades, even centuries—others virtually forever.

But why were ambient concentrations in the air and in the water relatively low while levels in living organisms soared off the charts? Again, environmental scientists discovered clues in the chemistry lab. By their nature, organochlorines are lipophilic, that is, they have an affinity for fat molecules. PCBs, for example, are extremely soluble and durable in oils. As a result, they were widely used in coolants for capacitors and transformers in the utility industry. Others found their way into products and materials as varied as pesticides, plasticizers in paints, copying paper, adhesives, sealants, printing ink, rubber manufacturing, waterproofings, fungicidal insulations, and nail coatings.

Organochlorines continue to be commonly used in degreasing solvents for industrial machinery and in dry-cleaning chemicals. Indeed, Thornton writes, the affinity of the dry-cleaning chemical perchloroethylene for oily, fatty substances is so great that butter sold in grocery stores located near dry cleaners has been found to contain levels of perchloroethylene that are hundreds of times higher than those in food outlets located farther away from these establishments.

Because organochlorines dissolve so readily in oily substances—in general, the

Through biomagnification, contaminant loads increase dramatically as small organisms pass their contaminants to larger ones throughout the food web. Top fish predators, such as lake trout, concentrate high levels of persistent pollutants like PCBs and DDT in their tissue, prompting public health officials to issue fish-consumption advisories. Illustration by Vera Ming Wong.

greater the number of chlorine atoms, the greater the solubility in fat—they become incorporated into the fatty tissues of living organisms. Thus, because of their affinity for fat molecules, organochlorine stowaways slip past some of the body's protective structures, such as the blood barrier around the brain and the placenta, which shields the developing fetus. But because the body is slow to metabolize these compounds—or, in some cases, has not yet evolved mechanisms for metabolizing them—they remain sequestered in fatty tissue, building up over time in a process known as bioaccumulation.

Top predators—eagles, dolphins, polar bears, and humans—carry the heaviest pollutant loads, largely due to the greater contamination of the foods they eat. Scientists call this phenomenon biomagnification. Put simply, contaminant loads increase dramatically as small organisms pass their contaminant loads to larger ones throughout the food web . Herring, for example, carry greater loads than the zooplankton on which they feed. Lake trout, in turn, are far more contaminated than their herring prey. The biggest doses are ingested by animals that consume large predator fish, such as lake trout. Great Lakes sports anglers who eat their catch of lake trout or coho salmon have been shown to harbor levels of such dangerous chemicals as dioxins, furans, and PCBs that are more than double that of other people in the basin who consume little to no fish or fish at a lower rung on the biomagnification ladder.

Once they enter the body, organochlorines behave like terrorists who have developed passkeys to infiltrate the body's most sensitive command-and-control centers. Some contaminants take over the directing of traffic on the body's electrical grid, jamming neurological signals. Still others relay false messages or block transmissions in the hormonal system. Others aid and abet the proliferation of cancer cells. Because they resist being metabolized, these chemicals remain lodged for decades, wreaking havoc in the fundamental processes of life. As in the environment, some organochlorines that do undergo biodegradation in the body produce breakdown products that are even more toxic than the parent compounds. "This may seem a paradoxical effect," Thornton writes. "Chlorination makes some compounds more persistent, increasing their toxicity, but it makes others less persistent, also increasing their toxicity. This 'damned-either-way' phenomenon makes sense if we view it from an evolutionary perspective. The body's detoxification enzymes evolved to break down potentially hazardous organic compounds in plants and other organisms we might eat, but not to handle organochlorines, which have never been naturally abundant in our food. Enzymes optimized by evolution to degrade one kind of molecule may produce harmful metabolites when they encounter novel substances. As a result, chlorine chemistry creates a biological catch-22: reactive organochlorines will probably be converted into toxic metabolites, while the stable ones are often already very toxic and will remain so for a very long time."

Exposure to chlorinated chemicals need not be great to produce adverse health effects. Even low doses of organochlorines, Thornton observes, "can reduce sperm counts, disrupt female reproductive cycles, cause endometriosis, induce spontaneous abortion, alter sexual behavior, cause birth defects, impair the development and function of the brain, reduce cognitive ability, interfere with the controlled development and growth of body tissues, cause cancer, and compromise immunity."

Only one segment of the animal world is capable of quickly reducing its body load of poisons: nursing females. Indeed, writes Florence Williams in a 2005 issue of the *New York Times Magazine,* "Nursing a baby, it turns out, is the ultimate detox diet."

Congenital deformities in fish-eating birds caused by PCBs in the aquatic food chain range from cross beaks, like the one on this nestling bald eagle chick from Lake Erie, to duplicate limbs. Photograph by James G. Sikarskie.

Here Today, Here Tomorrow: Pollution's Long Life in Lake Superior

Recognizing the threat to wildlife and human health, the U.S. government banned or restricted the manufacture of some organochlorines, including DDT in 1972 and PCBS in 1976. As expected, the data show a stark reduction of contaminants that corresponds to legal bans on their production. To date, concentrations of PCBS and DDT in Great Lakes fish, for example, have plummeted by 85 to 95 percent from their all-time highs in the 1970s.

But even with these dramatic reductions, contaminant loads in Great Lakes fish are still high enough to warrant fish-consumption advisories. Even more troubling, research indicates that the rate of decontamination for some outlawed compounds has slowed or even reached a plateau in the Great Lakes. In some cases, pollution trends are on the increase.

The discrepancies sent Great Lakes scientists back to the drawing board. Why have so many synthetic pollutants remained stubbornly lodged in the environment, they asked, some lingering decades after their phaseout?

Research shows that a large reservoir of contaminants, including PCBS, still exists in the environment. From 1929, when the production of PCBS ramped up on a large scale, to 1976, when they were discontinued, scientists estimate that some six million tons of PCBS were produced in the United States. Today, the runoff and discharges from sewage treatment facilities, sanitary landfills, transformer storage yards, and abandoned industrial sites continue to leak PCBS into the environment.

These residual sources have helped to explain some of the ongoing contamination of the lower Great Lakes, whose shores were lined with factories and sprawling cities. But scientists have been baffled by the higher-than-expected levels of toxic pollutants in Lake Superior, a region that is relatively free of heavy industry.

Some initial clues to solving this mystery came in 1970 when atmospheric scientist Joseph Prospero and his colleagues published research showing that dust from Saharan Africa traveled on trade winds across the Atlantic and was deposited on islands in the Caribbean. Hitchhiking on some of these dust particles were chemical contaminants that could be traced to Africa and Europe.

Prospero's work prompted researchers to explore the potential for long-range atmospheric transport of pollutants to the Great Lakes. Pieces of the puzzle began to emerge in the early 1970s when a group of scientists demonstrated that tributary streams could not fully account for elevated levels of some trace metals in Lake Michigan. The researchers suggested that the atmosphere was a significant source of these materials to the lake.

Then, in 1978 Swain published the shocking results of work conducted on Lake Superior. His data provided the first clear evidence showing that the atmosphere could transport not only inorganic substances, such as metals, but also organic compounds, including PCBS.

From 1974 through 1976, Swain and fellow researchers collected sport fish and commercially targeted species from the nearshore waters off the mainland and analyzed their tissues for residues of persistent organic chemicals. To establish a baseline for comparison, they also sampled water and fish from a remote, pristine lake that was free from obvious human influence. The researchers chose Siskiwit Lake. This self-contained body of water is perched fifty-six feet above Lake Superior in the inner reaches of Isle Royale, which lies twenty miles from the nearest mainland shore and far from major industrial sites. The area surrounding Siskiwit Lake had never been mined or logged. Indeed, the only machinery ever operated near or on the lake was an outboard motor used to power the scientists' research boat.

The researchers were unprepared for what they found. Siskiwit Lake contained concentrations of PCBs that were thirty times greater than those in the open waters of Lake Superior. Correspondingly, fish in pristine Siskiwit Lake had body burdens of the pollutant that were greater than those in fish taken from the open waters surrounding

Research conducted in the 1970s showed that remote Siskiwit Lake, a self-contained body of water perched fifty-six feet above Lake Superior in the inner reaches of Isle Royale, had concentrations of toxic chemicals such as PCBs that are thirty times greater than those in the open waters of Lake Superior. Photograph by Emily Lord.

Isle Royale. And PCBs were not the only toxic chemicals on the list. Siskiwit fish were also contaminated with other persistent pollutants, including the insecticides DDT and dieldrin.

In a follow-up study, Swain analyzed snowfall samples from the Duluth-Superior metropolitan area and Siskiwit Lake. Surprisingly, PCB concentrations in snow from pristine, remote Isle Royale were five times greater than those from the Duluth-Superior area. Swain's work proved that pollution in the Great Lakes was not just pouring out of factory pipes but was also falling from the sky.

Further research showed that for some chemicals, long-range transport on atmospheric pathways was the primary route into the lake. In a paper published in 1981, environmental chemist Steven Eisenreich and his colleagues determined that more than 85 percent of the total annual input of PCBs into Lake Superior—about 13,860 to 17,600 pounds—came from atmospheric deposition. Tributaries into the lake supplied the remainder.

Atmospheric pollution enters the lake in two forms. Dry deposition refers to air pollutants that reach Earth's surface as gas or particles. Wet deposition includes pollutants in gas or particulate form that are captured in the air and deposited on land and water bodies by rain, snow, or fog.

The Great Lakes, it turns out, are extremely sensitive to inputs from the atmosphere. For one thing, their large surface areas act like big collecting dishes. Increasing their vulnerability is the fact that a greater proportion of the water that replenishes the Great Lakes comes in the form of precipitation, which effectively captures pollutants from the sky and delivers them to the lakes. Adding to the problem is the long water-retention time of the Great Lakes. (Compared to its sister lakes, Superior has by far the slowest rate of natural outflow.) Factor in the persistence of pollutants such as PCBs, which linger in the environment or in the tissues of living organisms for years, even decades, and the Great Lakes "yield the potential for bioaccumulation and bioconcentration virtually unparalleled elsewhere in North America," write Paul Rodgers and Swain.

But one of the biggest risk factors for pollution turns out to be a simple accident of geography. Not only are the Great Lakes located near major sources of pollution—sprawling industrial centers such as Duluth, Chicago, Cleveland, and Detroit—but they also lie downwind of continental sources. As scientists have recently discovered, colder regions, such as the Great Lakes, serve as a kind of continental dumping ground for a wide variety of pollutants. That is because the warmer temperatures in more southerly locales tend to evaporate many chemicals from the land and water. Once they enter the atmosphere, they can ride on air currents that circulate between the equator and the poles. As contaminated air masses hit cool temperatures, the chemicals are more likely to associate with particulates in the air and fall to earth with them, both as dry material and within precipitation. These pollutants can rise and fall many times in and out of the atmosphere as they slowly work their way north. Once deposited in the north, they are much less likely to leave. Cooler temperatures at these latitudes inhibit their departure both by favoring attachment to particles and limiting transformation of contaminants into the gas phase through evaporation. Consequently, the northern latitudes and polar regions of the globe become sinks for many of the world's toxins. Scientists variously refer to this phenomenon as the global distillation effect, condenser effect, grasshopper scenario, or Cold Finger Theory.

Among the Great Lakes, Lake Superior may be the most sensitive to this kind of atmospheric deposition. Take, for example, Superior's response to loadings of the pesticide toxaphene. After DDT was outlawed, toxaphene became the pesticide of choice for cotton and other row crops in farm fields of the southern United States. Before it too was outlawed in 1990, toxaphene was the most heavily used pesticide in the United States. At peak production in 1974, output topped some 1.2 trillion pounds per year.

Despite its widespread use in the southern tier of the United States, toxaphene was little used in the Great Lakes basin. Yet when Swain sampled fish from Siskiwit Lake in the 1970s, he discovered residues of the pesticide in their tissues. In follow-up tests, Lake Superior lake trout and smelt have been found to contain the highest concentrations of toxaphene in Great Lakes fish. Furthermore, while research documented declines of toxaphene in fish from 1982 to 1992 in the other Great Lakes, levels in fish from Lake Superior remained constant.

Lake Superior appears to be a rest stop for toxaphene from two primary sources: Sunbelt soils that were dusted decades ago, and deposits in the lower Great Lakes that continue to offload their chemical burden to nearby Lake Superior. In a 2001 issue of *Environmental Science & Technology,* a group of scientists concluded that "the colder temperatures, larger surface area, and larger volume [of Lake Superior] led to greater concentrations [of toxaphene] historically and a slower rate of loss by volatilization."

But the question remains: If the atmosphere is the major source of pollutants to the Great Lakes, why doesn't their vast quantity of water simply dilute them? After all, together the lakes contain some 5,500 cubic miles of water, nearly 20 percent of all the freshwater on earth. Superior alone holds three quadrillion gallons of water.

It is a question that has puzzled scientists for decades starting with researchers such as H. P. Nicholson, who pondered the widespread occurrence of DDT in oceans in a 1967 issue of the journal *Science.* "We cannot at this time explain how DDT became so universally distributed," he wrote. "It has been speculated that DDT entered the oceans through runoff from the land by way of rivers, and from the air, but surely the tremendous dilution which would have occurred through mixing with waters of the sea would prevent direct exposure of most of these fish and other animals to measurable quantities."

It turns out that the waters of the Great Lakes, like those of the oceans, treat pollutants much as they do other natural inputs—they sequester them in discrete locations where organisms at the base of the food web derive much of their sustenance. One emerging theory being tested suggests that for many materials, the port of entry is through a distinct aquatic habitat known as the surface microlayer (SML). No thicker than a film of oil, the microlayer is a complex biotic soup whose biology and chemistry are markedly different—and more enriched—from the bulk waters beneath it. Inhabiting this aquatic veneer is an entire collection of microscopic life forms—ranging from zooplankton and bacteria to viruses, fungi, and algae. Together they are known as neuston, from the Greek word *neustos,* meaning "floating."

Although the term was coined in 1917 by the Swedish researcher Ernst Naumann, who isolated protozoans from the ocean's surface, little was known about the SML when scientists began to investigate Great Lakes pollution in the 1960s and 1970s. Most existing knowledge came from marine environments, where a surge in freighter traffic coincided with a growing number of accidental oil spills. Mitigating damaging oil slicks became a research priority. Yet scientific inquiries into the ecology of SMLs in freshwater environments lagged such that in 1982 microbiologist G. Wolfgang Fuhs observed that "hardly any work on the microbiology of the surface film has ever been performed in the Great Lakes."

Scientists have begun to take a closer look at this microscopic world, and the results have been nothing short of astounding. For starters, this diaphanous layer is remarkably tough and very elastic. Biologist John T. Hardy, who has studied some two hundred microlayers from freshwater and ocean habitats, points out that they can remain intact under eighteen-mile-per-hour winds and four-foot waves.

And the organisms that live in this surficial film have evolved highly specific adaptations that enable them to thrive in its fertile, but sometimes extreme, conditions. Some species of bacteria, for example, contain special pigments to help them cope with the intense solar radiation that bombards the water's surface. The microflora exhibit unique survival strategies as well. Planktonic algae, for example, optimize the use of the SML's tight quarters. Some algae, known as epiphytoneuston, live on the upper surface of the microlayer, while others, called hypophytoneuston, live on its underside.

Despite its meager proportions, the SML is no bit player in the ecological drama of oceans and lakes. Indeed, this microcompartment has been shown to play a fundamental role in the cycling of nutrients. The water's waxy, oily surface traps and concentrates chemicals and organic debris from the sky. Rising bubbles from the water column below transfer inorganic ions (such as ammonium, nitrate, and phosphate) and various metals. Also rafted to the surface are fine-grained sediments, hitchhiking bacteria and algae, and the particle remains of dead plants and animals. The lipids and proteins that are excreted by living organisms in the SML feed legions of bacteria, some of which are attacked and killed by opportunistic neighbors such as viruses. Their decay, in turn, fuels the growth of new generations of microorganisms.

Materials contained in SMLs can be further concentrated under certain wind conditions when alternating areas of upwelling and downwelling water converge in what are known as Langmuir cells. These phenomena are visible as telltale windrows of foam and debris on the surface of the water that parallel the direction of prevailing winds. Studies on Lake Mendota in Madison, Wisconsin, have shown that Langmuir cells cluster populations of *Daphnia* zooplankton. Schools of white bass cruise the foam tracks to exploit these dense clouds of prey.

Compared to subsurface waters, the SML so teems with microorganisms that their numbers may be up to ten thousand times more abundant. But the very qualities that make the microlayer so biologically productive can turn this ecological niche into a toxic dumping ground. Dissolved organic matter and lipids capture and concentrate hydrophobic pollutants from the atmosphere, collecting them, writes biologist John T. Hardy, "like nutmeg powder sprinkled on an eggnog." The surface microlayer, he concludes, "is becoming a soup of toxic metals, organic pollutants, bacteria, pesticide residues, and the byproducts of combustion-derived hydrocarbons from cars, trucks, airplanes, refuse incinerators, and power plants."

Although the magnitude of the SML's role in the cycling of contaminants is still being examined, it is clear that these problem chemicals do not remain sequestered in the SML. This microcompartment serves as a kind of warehouse of contaminated goods that are distributed by an extremely efficient network of physical, chemical, and biological pathways throughout the lake. For example, breaking waves or rain or snow pelting the water create bubbles that break at the surface, scattering droplets several inches into the air. Coated with material from the SML, these aerosols can reseed the atmosphere with contaminants, such as PCBs.

The plants and animals that occupy this surficial realm also are exposed to dangerous chemicals. Analyses of phytoplankton in the SML revealed concentrations of contaminants that were one thousand times that of lake water. Zooplankton feeding in this enriched zone carried body loads of pollutants that were one million times that of the water.

Although aquatic animals can pick up pollutants in a variety of ways, including absorbing them from the water that passes over their respiratory organs, the primary source of their contamination is the foods they eat. Once organochlorines, for example, enter the lake, they ordinarily do not float freely for long. Their molecules are engineered to both resist dissolving in water and to be strongly attracted to particles high in organic carbon and lipids. Among the most attractive are the surfaces of algae, a fundamental building block in the lake's energy structure.

The algae of Lake Superior possess several features that make them especially vulnerable to these dangerous materials. In Lake Superior, as in oligotrophic waters in general, the phytoplankton are overwhelmingly dominated by microalgae (less than five microns in size) and ultraplankton (five to twenty microns in size). In the nutrient-poor conditions of Lake Superior, these tiny organisms have evolved a large surface-to-volume ratio. This adaptation increases their exposure to their environment, thereby maximizing their ability to capture sunlight for photosynthesis and to absorb scarce nutrients.

But such physiological advantages also have severe downsides when processing damaging inputs from human activities. The plants' large surface area compounds the accumulation of toxic metals and chemicals. The potential for loading up on pollutants is further increased by Superior's cold temperatures. In warm waters, where algae grow quickly, chemicals, such as PCBs, have little time to pass the barriers of cell walls and become incorporated into the cell matrix. But when algal cells are dormant, as they are for much of the year in the cold waters of Superior, PCBs have ample opportunity to insinuate themselves into the matrices of algal cells. Those with a high fat content, such as diatoms, are especially susceptible to being invaded.

To make matters worse, plant-eating zooplankton in Superior prefer these smaller algal morsels. As a result, these "size fractions may well be the most important pathways for the entrainment of metals and other contaminants into the aquatic food chain," conclude aquatic scientist Mohiuddin Munawar and his colleagues in a 1987 article in the *Archives of Hydrobiology.*

Zooplankton also graze on detritus, which, like algae, attract persistent pollutants. Food crumbs that have escaped the mouths of hungry animals do not linger in the lake's surface waters. They can settle out from the euphotic zone, where light is sufficient to support photosynthesis, and drift into deeper reaches, carrying toxic chemicals that will wreak havoc on resident biota for decades, even millennia.

This rain of material is known as particulate flux. Analyses of sediment traps in Lake Michigan show that living phytoplankton made up about one-third of these settling particles. The remainder was dominated by the fecal waste of zooplankton but also included eroded clays, shredded algae, and other phytoplankton along with the carcasses and molted exoskeletons of zooplankton.

As a result, PCBs from the water column efficiently bond with settling particles. According to Brian Eadie and his colleagues Joel Baker and Steven Eisenreich, on an annual basis 11 percent of the PCBs in Lake Superior drift to the bottom of the lake attached to these settling particles. "In the long term," they point out, "burial of contaminants in bottom sediments serves as an important self-purification mechanism in lakes, estuaries, and oceans." But given the characteristically slow rate of sediment accumulation in Lake Superior, it takes decades or more for contaminated sediments to be buried under newer layers.

In the meantime, the journey of particulate matter to a grave in lake bottom sediments is rarely a direct route. Instead, it is typified by sharp bends and numerous detours along the major thoroughfares that are the recycling routes for the lake's nutrients. As it sinks, particulate matter is actively recycled by organisms ranging from coprophagous copepods (zooplankton that feed on fecal pellets) to colonizing bacteria. In time, the floating detritus is broken down into smaller and smaller entities, slowing their settling rates even further and making them available to aquatic organisms for relatively long periods of time. Few tidbits go unsampled. Hungry filter-feeding zooplankton, according to aquatic ecologist Karen Glaus Porter, "do not reject food of even the poorest nutritional value, probably because particulate matter collected from lake water contains bacteria and detritus as well as algae and therefore always has some nutritive value."

Those particles that have run the gauntlet of hungry mouths in the water column face whole new populations of organisms on the lake floor, most notably *Diporeia,* which process large amounts of settling particles. In the summer, according to some calculations, amphipods that live in deep offshore waters may assimilate up to 30 percent of settling organic matter. Their buffet includes bits of diatom fragments, insect remains, dead zooplankton, bacteria—and their associated contaminants.

They also ingest loads of fine-grained sediments, which make up the vast majority of sediment types that settle in the deepest basins of the Great Lakes, in the process acquiring even more contaminants. That is because these sediments often have higher levels of organic carbon than larger particles and therefore are magnets for persistent pollutants. And like tiny phytoplankton, they have large surface-to-mass ratios, which allow them to accumulate significant amounts of contaminants relative to their size.

Research has shown that animals occupying the benthic regions of the lake—even part-timers such as mysids, which feed in the water column at night and prey on their neighbors in the sediments during the day—face a risk of accumulating far higher pollution loads than animals that occupy only the water column. Studies of PCB accumulation in Lake Michigan invertebrates, published in 1998 by Leland J. Jackson and his colleagues in the *Journal of Great Lakes Research,* show that between 1980 and 1991 total concentrations of PCBs in the water column had declined by 61 percent. However, the decrease was not reflected in the tissues of mysids. "Relative to copepods and cladocerans," the authors write, "Mysis may now be more contaminated than in the early 1980s."

Organisms that spend most of their time in and around bottom sediments got an even more sobering report card. For example, in *Diporeia,* which live where contaminants are most concentrated, PCB levels were 20 percent higher than in mysids and eight times higher than in the copepods and cladocerans that occupy the water column.

But benthic residents do more than simply sequester contaminants in their tissues. By going about the daily business of life, they stir up sediments and recirculate contaminants. Researcher John A. Robbins describes the bottoms of well-oxygenated oceans and freshwater lakes as a "pageantry of life" with "organisms of many forms and sizes diving, ploughing, channeling, establishing burrows, feeding, irrigating, metabolizing, respiring, defecating, reproducing, preying on their neighbors, dying and disintegrating." This disturbance of lake-bottom sediments is known as bioturbation. *Diporeia,* for example, occupy the upper three-quarter inch of sediment, which in Lake Superior can represent decades of sediment accumulation. As they plunge into the fluffy top layers to hunt or escape predators, they send puffs of silt up into the water.

Far more methodical—and influential in the amount of earth that is moved—are oligochaete worms. Oligochaetes favor fine materials that are rich in organic matter, exactly the kind of material to which many persistent contaminants adhere. In laboratory studies, these "conveyor-belt" feeders have been found to be responsible for the movement of about 90 percent of hydrophobic contaminants in the sediments—with seemingly few ill effects for some species. Tubificid oligochaetes, for example, are so highly tolerant of pollution that they can thrive in sediments that are too contaminated to support any other benthic species.

The burrowing of animals such as tubificids also increases the water content of the sediments, making them more porous, less compact, and more easily eroded and resuspended by natural physical forces. During summer, when the lake waters become stratified, many of these fine materials, those measuring two to eight microns, float in what is known as the benthic boundary layer (BBL). A zone of visibly turbid water, the BBL extends some sixteen to ninety-eight feet above the lake bottom. It is composed of two parts. At the very bottom of this layer, about 0.2 inches above the sediment surface, lies a distinct compartment called the sediment boundary layer. Fluffy in texture, it consists of resuspended sediments and organic particles that have drifted down from the lake's surface waters. A second layer, containing less dense concentrations of these materials, known as the benthic nepheloid layer (BNL), hovers above it.

Studies in Lake Superior have shown that more organic matter is degraded in the BNL than in any other portion of the water column even though planktonic bacteria in the upper strata of the lake are eight times more productive than those that live on the frigid, murky bottom. According to Eadie, Baker, and Eisenreich, 80 to 90 percent of the organic matter that reaches the lake bottom is degraded near the sediment-water interface.

Bottom-dwelling bacteria make up for the sluggish productivity of the BNL zone by having prolonged access to food—seven to fifty times longer than planktonic bacteria in the upper reaches of the lake where particles sink more quickly out of sight. Because organic matter is exposed for greater lengths of time to bacterial degradation in bottom waters, there are more opportunities for the release of contaminants, such as PCBs, that cling to these particles. This metabolic activity, scientists say, may account for the increased concentrations of PCBs that have been measured in the BNL.

The thermal stratification that forms in summer and winter isolates most of this material along with essential ecosystem nutrients in the lower reaches of the lake. But in the spring and fall turnovers, the lake's waters are mixed from top to bottom. These life-giving cycles sweep biological and chemical nutrients from the lake bottom into the epilimnion. At the same time, they loft contaminants back into the water column for up to nine years, where they can reenter the food web.

The New Generation of Great Lakes Contaminants

Scientists refer to the older generation of chlorinated compounds as legacy pollutants. The reckless use of everyday chemicals such as PCBS, DDT, and dieldrin offered a dramatic cautionary tale: that once persistent, toxic pollutants find their way out of the laboratory and into the world, they often evade our capacity for control. Insinuating themselves into even the tightest compartments of the Great Lakes environment, these contaminants effectively slip out of reach. Stealthy saboteurs of the most fundamental processes of life, they wreak silent havoc on the health of people and the planet long after their release.

Much of the damage is subtle and quietly compounded over generations, as Joseph and Sandra Jacobson have shown. From 1980 to 1991, this husband-and-wife team of psychologists from Wayne State University, Detroit, carried out a longitudinal study of the children of mothers who ate fish from Lake Michigan. The Jacobsons examined the 242 infants born to women who ate three meals of contaminated fish from Lake Michigan each month for six years prior to their pregnancy and also throughout their pregnancy. They compared the results to 71 infants whose mothers had not eaten Lake Michigan fish. The children were tested at designated intervals until they reached age eleven.

The researchers reported that newborns of fish-eating mothers had marked physical differences ranging from shorter gestations and lower birth weights to smaller heads. They also exhibited behavioral and cognitive deficits including weaker reflexes, memory impairments, and lags in motor control.

By age eleven, the IQs of those children who had the highest exposure to PCBS (an exposure that was only slightly higher than that of the general population) were lower by an average 6.2 points. Their verbal and reading performance lagged behind the children of mothers who had not eaten fish, as did their short-term memory and planning ability. The children also had greater trouble with concentration, which led researchers to question whether the growing epidemic of Attention Deficit Hyperactivity Disorder (ADHD) could be triggered by exposure in the womb to toxic chemicals such as PCBS.

The results stunned even the study's authors. "I thought that once they reached a structured school environment, whatever minor handicaps they had would be overcome," said coauthor Joseph Jacobson. "So I was quite surprised to find that, if anything, the effects were stronger and clearer at age 11 than they had been at age 4."

The health risks posed by legacy pollutants are not confined to the more industrialized lower lakes. In 2011, the Minnesota Department of Health (MDH) released the results of a study of 1,465 newborns in the Lake Superior Basin, more than 1,200 of whom were born in Minnesota. On average, 8 percent of the infants tested registered blood mercury levels above the safety limit set by the EPA. In Minnesota alone, that number rose to a troubling 10 percent. In 2012, the EPA awarded a $1.4 million grant to the MDH to step up health screenings of mothers and infants along the Minnesota North Shore and to mount a public campaign to better educate citizens about fish-consumption advisories. The grant was part of a $320 million program known as the Great Lakes Restoration Initiative (GLRI), a 2009 appropriation from the Obama administration to clean up toxic contamination in the Great Lakes. According to the EPA, the GLRI "is the largest investment in the Great Lakes in more than two decades."

Unfortunately, these and other lessons have not tamed our chemical profligacy. As the lakes labor to shed their load of first-wave pollutants, new chemical troublemakers enter their waters. In the United States alone, writes Nena Baker in her 2008 book *The Body Toxic,* "more than eighty thousand industrial substances [are] registered for commercial purposes with the EPA. About ten thousand of these chemicals are widely used in everything from clothing, carpeting, household cleaners, and computers to furniture, food, food containers, paint, cookware, and cosmetics. But *the vast majority of them have not been tested for potential toxic effects.*"

The growing catalog of new pollutants in the waters of the Great Lakes does not surprise Deborah Swackhamer, a professor in the School of Public Health and codirector of the Water Resources Center at the University of Minnesota. "The list of contaminants in the Great Lakes is a dynamic one, and not simply the familiar legacy of chlorinated chemicals of the past," she warns. "As commerce changes, so do the chemicals in our environment, and it is anticipated that we will continue to see the appearance of new chemicals in the Great Lakes."

Many of them are exhibiting ominous patterns that are strikingly similar to those of DDT and PCBs. In recent years, the residues of whole new classes of chemicals have been building up in the environment and in our "fat, bones, blood, and organs, or [passing] through us in breast milk, urine, feces, sweat, semen, hair, and nails," Baker writes. They include brominated compounds, such as the flame retardants used in textiles and carpeting; fluorinated compounds that make fabrics resistant to stains; chlorinated paraffins, which are blended into industrial cutting oils, commercial paints, adhesives, sealants, and caulks; and phthalates, which, among other uses, are added to plastics to increase their flexibility.

As the list of persistent chemicals grows, so too does the discovery of an array of new health threats. In the *Silent Spring* era, Carson and other scientists drew attention to the acute toxicity of pesticide fogs that sent birds into agonies of fatal nerve spasms, and to the more chronic damage that led to maladies such as cancer and birth defects. In recent years, however, researchers have begun to focus on a novel syndrome of disorders that arises from our exposure to everyday chemicals. It is called endocrine disruption.

Concerns that human-made compounds could be interfering with the body's chemical-messaging system arose in 1996 with the publication of *Our Stolen Future.* Authors Theo Colborn, Dianne Dumanoski, and John Peterson Myers raised the alarm about the potential for synthetic chemicals to wreak havoc with the delicate hormonal traffic that controls growth and the healthy functioning of a range of processes that govern reproduction, metabolism, behavior, and immunity from disease.

Triggered by a signal from the brain, glands—such as the testes, ovaries, thyroid, pancreatic islets, and adrenals—release hormones into the blood stream. These hormones circulate until they locate cells with receptors that are precisely fitted to their particular chemical design, much like a lock-and-key system. When hormones bind with the correct receptors, the cell either activates or shuts down the expression of certain genes. Some synthetic compounds can act like hormonal impersonators and bypass signals from the brain, tripping their own cascade of unintended cellular changes. Others interfere with this careful choreography of stimulus and response by preventing natural hormones from docking with their receptors. These chemical saboteurs have been linked to a wide range of health problems including not only cancer but also lowered fertility, impaired immune function, and neurological and behavioral problems.

Avoiding endocrine-disrupting compounds (EDCs), Swackhamer says, may be impossible. They have become the chemical backbone of the products we use everyday. Take, for example, atrazine, a weed killer that is liberally sprayed on everything from corn crops and golf courses to suburban lawns. In a 2009 *New York Times* article, writer Charles Duhigg points out that atrazine is "among the most common contaminants in American reservoirs and other sources of drinking water." Even at current legal levels, he writes, the research shows that atrazine "may be associated with birth defects, low birth weights and menstrual problems." Laboratory animals briefly exposed to the chemical at critical junctures in fetal development appear to have a greater vulnerability to cancer later in life.

Phthalates, a family of compounds used as plasticizers, leach from a wide range of everyday objects, including cosmetics, food containers, teething rings, and the intravenous bags used in hospitals. In 2008 Bisphenol A (BPA) made international headlines when news reports revealed that this known endocrine disruptor was a common component of plastic bottles, including baby bottles. Even ordinary household cleaning agents contain hormonally active ingredients, such as the phenols in detergents, and the antimicrobial chemicals in hand and dish soaps. Scientists from the University of Minnesota and Canada added the effluent from pulp mills to the list. The processing of tree fiber for paper, they discovered, releases natural plant estrogens.

Many of these compounds are present in the environment in low concentrations, such as parts per trillion. (For example, one-twentieth of a drop of water in an Olympic-sized pool is equivalent to one part per trillion.) Only recently have scientists even developed technologies that are capable of detecting such trace amounts. The presence of hormonally active chemicals at such low concentrations, however, does little to assuage the concerns of scientists and public health policy makers. That is because hormonally active chemicals have been shown to interfere with endocrine function at low doses; indeed, some chemicals wield the greatest effects in minute quantities. In a laboratory study published in the June 2005 issue of *Environmental Health Perspectives,* for example, a team of international scientists showed that male fathead minnows, a common freshwater fish species, produce a protein known as vitellogenin (VTG) when exposed to a synthetic hormone in birth control pills at doses of less than one part per trillion. (This reproductive protein, which is a component of egg yolk, occurs naturally in female fish. When male fish produce VTG, however, it can signal exposure to feminizing estrogens in their environment.) Levels of five parts per trillion in surface waters are enough to cause dramatic changes in these fish populations. Even more worrisome is the fact that the effects of hormonally active chemicals can be amplified when combined with others in the environment.

Equipped with new detection technologies, in 1999 and 2000 seven scientists from the USGS mounted one of the largest reconnaissances of hormonally active chemicals in U.S. waters. The team sampled 139 streams in thirty states in an effort to measure the presence of organic waste contaminants, many of them hormonally active. Targeted were ninety-five commonly used compounds. Since wastewater treatment plants currently lack the capacity for removing these contaminants, the study focused largely on waterways that were downstream of urban centers or large feedlot operations.

The USGS study showed that 80 percent of the streams contained one or more organic waste contaminants. Among the most frequently detected were steroids, insect repellants, fire retardants, caffeine, detergent-related phenols, and triclosan, an antimicrobial disinfectant used in personal-care products such as liquid soaps, cologne, and cosmetics. In most cases, the concentrations of these compounds were present in low, background levels and not high enough to kill aquatic organisms outright.

Concerned about such reports, in 2007 the Minnesota State Legislature commissioned a study of endocrine-disrupting chemicals in state waters. A team of scientists from the Minnesota Pollution Control Agency and USGS sampled 110 compounds in a dozen lakes and four rivers. They included major bodies of water such as the Mississippi River and Lake Superior, which receive effluent from municipal wastewater treatment plants, and northern lakes that catch runoff from the septic systems of vacation homes. Remote, undeveloped lakes in northern Minnesota, such as Northern Light Lake in Cook County, were used as study controls. In addition to sampling surface waters, researchers also analyzed sediments and tested male fish for a variety of abnormalities, including the presence of VTG.

The team of scientists released the results of their study in November 2009. As in waters nationwide, contamination by endocrine-disrupting chemicals was commonplace throughout Minnesota's lakes and rivers. The compounds that occurred most frequently in the state's waters included hormones (of both natural and human origin). Researchers detected BPA in 82 percent of the sampled lakes. Residues of BPA were also common in sediments, along with triclosan and various phenol ingredients in detergents. Pharmaceuticals also were routinely detected, including the pain medication acetaminophen in half of the sediments from both lakes and rivers. In more than one-third of these samples, researchers also found carbamazepine, a drug that is used to treat ADHD.

The list is not altogether surprising given the fact that most of the samples came from waters and sediments that were contaminated by wastewater from sewage treatment facilities and septic systems. Researchers were baffled, however, by the presence of some of these contaminants in the study's reference lakes, which they defined as "'pristine' lakes without any surrounding development." In these lakes, researchers detected the insecticide DEET, detergent phenols, BPA, and estrogens. The sediments of these reference lakes also included several contaminants found in more urbanized lakes and rivers, including the drugs acetaminophen and carbamazepine.

The fish in these so-called pristine lakes exhibited health effects similar to those in more contaminated environments. The presence of VTG in fish collected from these reference lakes indicates a likely exposure to these estrogenic chemicals. New generations of fish in these pristine lakes could continue to be exposed as well. "Concentrations of contaminants in sediments appear to be much higher than the lake water at the same locations," the authors write. "This suggests that these chemicals are accumulating over time in the lake sediment. More study is needed to determine how persistent these chemicals are in sediment and to understand the impact of this accumulation to aquatic ecosystems."

Also puzzling to the researchers was this: the effluent flowing into Lake Superior from the wastewater treatment plant at Two Harbors consistently ranks at or near the top for the highest levels of many hormonally active compounds. Unlike PCBs and other organochlorines, many of these compounds do not associate with particles, enabling them to flow unchecked through wastewater treatment systems. As a result, the study's authors write, "Lake Superior is receiving a continuous stream of EDCs and wastewater-associated contaminants from WWTPS [wastewater treatment plants] along its shores with unknown consequences for the lake."

As commerce changes, these wastewater treatment plants continue to seed the lake with new forms of pollution. Raising red flags among environmental agencies and public health officials is contamination by microplastics. The problem is this: when released into the environment, petroleum-based plastics do not biodegrade but physically break down into ever-smaller units, even down to the molecular level. In aquatic ecosystems, these particles have been shown to attract and concentrate persistent organic pollutants. Scientists have largely focused their research on the cycling of microplastics in marine food webs, from plankton to albatross. In December 2013, however, an article in the journal *Marine Pollution Bulletin* showed that concentrations in the Great Lakes may be far higher than in the oceans; researchers tallied 1.1 million bits of microplastics per square mile in parts of the most contaminated Great Lakes—Lakes Erie and Ontario. Sixty percent of the microplastics floating in the waters of the Great Lakes comes in the form of tiny plastic beads used as abrasives in a wide range of toiletries, from facial scrubs to toothpaste. These beads are small enough to bypass wastewater-treatment safeguards and enter lakes and streams in treated effluent. They also spill into water bodies when the plants discharge raw sewage during stormwater overflows.

Superior's Sediments: A Record of Human History

The sediments of the Great Lakes provide a report card on human activities. The widespread appearance of ragweed pollen, which accompanied the clear-cutting of forests and shift to agriculture around 1850, is so abrupt in sediment horizons that researchers use it as a reliable marker of time in soil profiles. The disappearance of chestnut pollen around 1930 to 1935, following the widespread destruction of these trees by chestnut blight, also is clearly demarcated. Legible too in the sediments is the history of nuclear weapons testing. The radionuclide cesium-137, for example, debuted in the stratigraphic record in the early 1950s. Following the test ban treaty between the United States and Soviet Union in 1963, this pollutant begins to disappear. The appearance of dioxins in the sediments after 1940 can be traced to the production and combustion of chlorinated compounds, including those from backyard burn barrels as well as from municipal and chemical incinerators. Cores taken from the bottom of Lake Superior date peak concentrations of PCBs to the early 1970s.

With the recent discovery that some EDCs accumulate in aquatic sediments, scientists may also be able to use cores from the Great Lakes to precisely date the exponential rise of antimicrobial agents in our hand soaps, the growing use of Tylenol to relieve our headaches, our reflexive turn to prescription drugs to help calm our children, and our love affair with throwaway plastic objects.

Throughout our evolution as a species, we have employed our collective ingenuity and creative genius to make our lives more secure and comfortable—with great success. Too often, however, our uncritical pursuit of single-minded technological fixes has led to the destruction of the very support systems that make life possible. The sedimentary record of the Great Lakes bears witness to our folly.

Can this record now compel us to proceed with greater deliberation and restraint about the chemicals we introduce such that we live in sync with our life-support systems rather than deconstruct them with tools of our own making?

As the legacy of synthetic chemicals in the Great Lakes so vividly illustrates, we must err on the side of caution before unleashing the contents of our chemical arsenals on the world. The Great Lakes serve as a kind of keystone ecosystem. If we take good care of the health of the world's largest source of freshwater and the people who live within its watershed, chances are good that we will be safeguarding that of the rest of the world as well. ❧

SUGGESTIONS FOR FURTHER READING

Arimoto, Richard. "Atmospheric Deposition of Chemical Contaminants to the Great Lakes." *Journal of Great Lakes Research* 15 (1989): 339–56.

Aulerich, Richard J., Robert K. Ringer, Harry L. Seagran, and William G. Youatt. "Effect of Feeding Coho Salmon and Other Great Lakes Fish on Mink Reproduction." *Canadian Journal of Zoology* 49 (1971): 611–16.

Baker, Joel E., and Steven J. Eisenreich. "PCBs and PAHs as Tracers of Particulate Dynamics in Large Lakes." *Journal of Great Lakes Research* 15 (1989): 84–103.

Baker, Joel E., Steven J. Eisenreich, and Brian J. Eadie. "Sediment Trap Fluxes and Benthic Recycling of Organic Carbon, Polycyclic Aromatic Hydrocarbons, and Polychlorobiphenyl Congeners in Lake Superior." *Environmental Science & Technology* 25 (1991): 500–508.

Bierman, Victor J., Jr., and Wayland R. Swain. "Mass Balance Modeling of DDT Dynamics in Lakes Michigan and Superior." *Environmental Science & Technology* 16 (1982): 572–79.

Brian, Jayne V., Catherine A. Harris, Martin Scholze, Thomas Backhaus, Petra Booy, Marja Lamoree, Giulio Pojana, Niels Jonkers, Tamsin Runnalls, Angela Bonfa, Antonio Marcomini, and John P. Sumpter. "Accurate Prediction of the Response of Freshwater Fish to a Mixture of Estrogenic Chemicals." *Environmental Health Perspectives* 113, no. 6 (2005): 721–28.

Carlson, Daniel L., and Deborah L. Swackhamer. "Results from the U.S. Great Lakes Fish Monitoring Program and Effects of Lake Processes on Bioaccumulative Concentrations." *Journal of Great Lakes Research* 32 (2006): 370–85.

Charles, M. Judith, and Ronald A. Hites. "Sediments as Archives of Environmental Pollution Trends." In *Sources and Fates of Aquatic Pollutants,* ed. Ronald A. Hites and Steven J. Eisenreich. Washington, D.C.: American Chemical Society, 1987.

Clarke, Arthur H., Jr. "Some Aspects of Adaptive Radiation in Recent Freshwater Mollusks." *Malacologia* 9, no. 1 (1969): 163.

Czuczwa, Jean M., Bruce D. McVeety, and Ronald A. Hites. "Polychlorinated Dibenzo-P-Dioxins and Dibenzofurans in Sediments from Siskiwit Lake, Isle Royale." *Science* 226 (1984): 568–69.

Davis, Charles C. 1966. *Plankton Studies in the Largest Great Lakes of the World.* University of Michigan, Great Lakes Research Division, Publication No. 14. Ann Arbor: University of Michigan, 1966.

Delfino, Joseph J. "Toxic Substances in the Great Lakes." *Environmental Science & Technology* 13 (1979): 1462–68.

Dodson, Stanley. "Predicting Diel Vertical Migration of Zooplankton." *Limnology and Oceanography* 35 (1990): 1195–1200.

Duhigg, Charles. "Debating How Much Weed Killer Is Safe in Your Water Glass." *New York Times,* 23 August 2009.

Eadie, Brian J., Thomas F. Nalepa, and Peter F. Landrum. "Toxic Contaminants and Benthic Organisms in the Great Lakes: Cycling, Fate and Effects." In *Chronic Effects of Toxic Contaminants in Large Lakes,* vol. 1 of *Toxic Contamination in Large Lakes,* ed. Norbert W. Schmidtke. Chelsea, Mich.: Lewis Publishers, 1988.

Eadie, Brian J., and John A. Robbins. "The Role of Particulate Matter in the Movement of Contaminants in the Great Lakes." In *Sources and Fates of Aquatic Pollutants,* ed. Ronald A. Hites and Steven J. Eisenreich. Washington, D.C.: American Chemical Society, 1987.

Eisenreich, Steven J. "The Chemical Limnology of Nonpolar Organic Contaminants: Polychlorinated Biphenyls in Lake Superior." In *Sources and Fates of Aquatic Pollutants,* ed. Ronald A. Hites and Steven J. Eisenreich, 393–497. Washington, D.C.: American Chemical Society, 1987.

———. "Overview of Atmospheric Inputs and Losses from Films." *Journal of Great Lakes Research* 8, no. 2 (1982): 241–42.

Eisenreich, Steven J., Paul D. Capel, Joel E. Baker, and Brian B. Looney. "Chemical Limnology of PCBs in Lake Superior—A Case Study." In *Sources, Fate, and Controls of Toxic Contaminants,* vol. 3 of *Toxic Contamination in Large Lakes,* ed. Norbert W. Schmidtke. Chelsea, Mich.: Lewis Publishers, 1988.

Eisenreich, Steven J., Paul D. Capel, and Brian B. Looney. "PCB Dynamics in Lake Superior Water." In *Physical Behavior of PCBs in the Great Lakes,* ed. Donald Mackay, Sally Peterson, Steven J. Eisenreich, and Milagres S. Simmons. Ann Arbor, Mich.: Ann Arbor Science Publishers, 1983.

Eisenreich, Steven J., Gregory J. Hollod, and Thomas C. Johnson. "Atmospheric Concentrations and Deposition of Polychlorinated Biphenyls to Lake Superior." In *Atmospheric Pollutants in Natural Waters,* ed. Steven J. Eisenreich. Ann Arbor, Mich.: Ann Arbor Science Publishers, 1981.

Eisenreich, Steven J., Brian B. Looney, and Gregory J. Hollod. "PCBs in the Lake Superior Atmosphere 1978–1980." In *Physical Behavior of PCBs in the Great Lakes,* ed. Donald Mackay, Sally Peterson, Steven J. Eisenreich, and Milagres S. Simmons. Ann Arbor, Mich.: Ann Arbor Science Publishers, 1983.

Eisenreich, Steven J., Brian B. Looney, and J. David Thornton. "Airborne Organic Contaminants in the Great Lakes Ecosystem." *Environmental Science & Technology* 15 (1981): 30–38.

Evans, Marlene S., Ralph W. Bathelt, and Clifford P. Rice. "PCBs and Other Toxicants in *Mysis relicta.*" *Hydrobiologia* 93 (1982): 205–15.

Evans, Marlene S., Brian J. Eadie, and Rebecca M. Glover. "Sediment Trap Studies in Southeastern Lake Michigan: Fecal Pellet Express or the More Traveled Route?" *Journal of Great Lakes Research* 24 (1998): 555–68.

Evans, Marlene S., Michael A. Quigley, and James A. Wojcik. "Comparative Ecology of *Pontoporeia hoyi* Populations in Southern Lake Michigan: The Profundal Region versus the Slope and Shelf Regions." *Journal of Great Lakes Research* 16 (1990): 27–40.

Fahnenstiel, Gary L., Linda Sicko-Goad, Donald Scavia, and Eugene F. Stoermer. "Importance of Picoplankton in Lake Superior." *Canadian Journal of Fisheries and Aquatic Sciences* 43 (1986): 235–40.

Farrand, William R. "The Quaternary History of Lake Superior." In *Proceedings of the 12th Conference on Great Lakes Research.* International Association for Great Lakes Research, 1969: 181–97.

Ferrey, Mark, Angela Preimesberger, Heiko Schoenfuss, Richard Kiesling, Larry Barber, and Jeffery Writer. *Statewide Endocrine Disrupting Compound Monitoring Study, 2007–2008.* St. Paul: Minnesota Pollution Control Agency, 2010.

Fields, Scott. "Great Lakes: Resource at Risk." *Environmental Health Perspectives* 113, no. 3 (2005): 165–73.

Garcie-Reyero, Natalia, Ira Adelman, Li Liu, and Nancy Denslow. "Gene Expression Profiles of Fathead Minnows Exposed to Surface Waters above and below a Sewage Treatment Plant in Minnesota." *Marine Environmental Research* 66 (2008): 134–36.

Gilbertson, Michael. "Epidemics in Birds and Mammals Caused by Chemicals in the Great Lakes." In *Toxic Contaminants and Ecosystem Health: A Great Lakes Focus,* ed. Marlene S. Evans. New York: John Wiley and Sons, 1988.

Glassmeyer, Susan T., David S. De Vault, Tanya R. Myers, and Ronald A. Hites. "Toxaphene in Great Lakes Fish: A Temporal, Spatial and Trophic Study." *Environmental Science & Technology* 31 (1997): 84–88.

Great Lakes Sea Grant Network, and Michigan Sea Grant Program. *Zero Discharge and Virtual Elimination in the Great Lakes: A Collection of Viewpoints from Prominent Great Lakes Specialists.* Vol. 93. Michigan Sea Grant College Program, MICHU-SG-93-702, 1993.

Habermann, Russell. "Readers Want to Know: Should We Worry about Microplastics in Lake Superior?" *Seiche,* September 2013: 4.

Halden, Rolf U., and Daniel H. Paull. "Co-occurrence of Triclocarban and Triclosan in U.S. Water Resources." *Environmental Science & Technology* 39 (2005): 1420–26.

Halfman, Barbara M., and Thomas C. Johnson. "Surface and Benthic Nepheloid Layers in the Western Arm of Lake Superior, 1983." *Journal of Great Lakes Research* 15 (1989): 15–25.

Hardy, John T. "Where the Sea Meets the Sky." *Natural History* 5 (May 1991): 59–65.

Hartsough, G. R. "Great Lakes Fish Now Suspect as Mink Food." *American Fur Breeder* 38 (1965): 25–27.

Hatcher, Robert F., and Bruce C. Parker. "Microbiological and Chemical Enrichment of Freshwater-Surface Microlayers Relative to the Bulk-Subsurface Water." *Canadian Journal of Microbiology* 20 (1974): 1051–57.

Heidler, Jochen, Amir Sapkota, and Rolf U. Halden. "Partitioning, Persistence and Accumulation in Digested Sludge of the Topical Antiseptic Triclocarban during Wastewater Treatment." *Environmental Science & Technology* 40 (2006): 3634–39.

Henson, E. Benette. *A Review of Great Lakes Benthos Research.* University of Michigan, Great Lakes Research Division, Publication No. 14, 1966.

Heuschele, Ann S. "Vertical Distribution of Profundal Benthos in Lake Superior Sediments." *Journal of Great Lakes Research* 8 (1982): 603–13.

Hicks, Randall E., Peter Aas, and Christine Jankovich. "Annual and Offshore Changes in Bacterioplankton Communities in the Western Arm of Lake Superior during 1989 and 1990." *Journal of Great Lakes Research* 30 (2004): 196–213.

Hicks, Randall E., and Christopher J. Owen. "Bacterioplankton Density and Activity in Benthic Nepheloid Layers of Lake Michigan and Lake Superior." *Canadian Journal of Fisheries and Aquatic Sciences* 48 (1991): 923–32.

Hudson, Matthew J., Deborah L. Swackhamer, and James B. Cotner. "Effects of Microbes on Contaminant Transfer in the Lake Superior Food Web." *Environmental Science & Technology* 39 (2005): 9500–9508.

Jackson, Leland J., Stephen R. Carpenter, Jon Manchester, and Craig A. Stow. "Current Concentrations of PCBs in Lake Michigan Invertebrates, a Prediction Test, and Corroboration of Hindcast Concentrations." *Journal of Great Lakes Research* 24 (1998): 808–21.

James, Ryan R., Jeffrey G. McDonald, Daniel M. Symonik, Deborah L. Swackhamer, and Ronald A. Hites. "Volatilization of Toxaphene from Lakes Michigan and Superior." *Environmental Science & Technology* 35 (2001): 3653–60.

Jeremiason, Jeff D., Steven J. Eisenreich, Joel E. Baker, and Brian J. Eadie. "PCB Decline in Settling Particles and Benthic Recycling of PCBs and PAHs in Lake Superior." *Environmental Science & Technology* 32 (1998): 3249–56.

Jeremiason, Jeff D., Keri C. Hornbuckle, and Steven J. Eisenreich. "PCBs in Lake Superior, 1978–1992: Decreases in Water Concentrations Reflect Loss by Volatilization." *Environmental Science & Technology* 28 (1994): 903–14.

Kerfoot, W. Charles, and Jerome O. Nriagu. "Copper Mining, Copper Cycling and Mercury in the Lake Superior Ecosystem: An Introduction." *Journal of Great Lakes Research* 25 (1999): 594–98.

Kolpin, Dana W., Edward T. Furlong, Michael T. Meyer, E. Michael Thurman, Steven D. Zaugg, Larry B. Barber, and Herbert T. Buxton. "Pharmaceuticals, Hormones, and Other Organic Wastewater Contaminants in U.S. Streams, 1999–2000: A National Reconnaissance." *Environmental Science and Technology* 36 (2002): 1202–11.

Lee, Jennifer. "E.P.A. Orders Companies to Examine Effects of Chemicals." *New York Times,* 15 April 2003.

———. "Second Thoughts on a Chemical: In Water, How Much Is Too Much? *New York Times,* 2 March 2004.

Malone, B. J., and D. J. McQueen. "Horizontal Patchiness in Zooplankton Populations in Two Ontario Kettle Lakes." *Hydrobiologia* 99 (1983): 101–24.

Marcus Eriksen, Sherri Mason, Stiv Wilson, Carolyn Box, Ann Zellers, William Edwards, Hannah Farley, and Stephen Amato. "Microplastic Pollution in the Surface Waters of the Laurentian Great Lakes." *Marine Pollution Bulletin* 77 (2013): 177–82.

Martinovic, Dalma, Jeffrey S. Denny, Patricia K. Schmieder, Gerald T. Ankley, and Peter W. Sorensen. "Temporal Variation in the Estrogenicity of a Sewage Treatment Plant Effluent and Its Biological Significance." *Environmental Science and Technology* 42 (2008): 3421–27.

McCall, Peter L., and Michael J. S. Tevesz, eds. *Animal-Sediment Relations: The Biogenic Alteration of Sediments.* New York: Plenum Press, 1982.

McNaught, Donald C. "Overview of Contaminant Interactions with Surface Films, Zooplankton, and Fish." *Journal of Great Lakes Research* 8 (1982): 358–59.

———. "Short Cycling of Contaminants by Zooplankton and Their Impact on Great Lakes Ecosystems." *Journal of Great Lakes Research* 8 (1982): 360–66.

Meyers, Philip A., Clifford P. Rice, and Robert M. Owen. "Input and Removal of Natural and Pollutant Materials in the Surface Microlayer on Lake Michigan." *Ecology Bulletin* 35 (1983): 519–32.

Minnesota Sea Grant. "How Much Pollution Falls on the Great Lakes?" *Seiche,* Spring 1991: 1–10.

Moen, Sharon. "All Plankton Great and Small." *Seiche,* October 2005: 4–6.

———. "Drugs, Sex and the Seamier Side of 'Fresh' Water." In *Superior Science: Stories of Lake Superior Research.* Duluth, Minn.: Minnesota Sea Grant, 2004.

———. "Our Lake Has Fleas." *Seiche,* June 2000: 1–5.

———. "Trout, Ciscoes, and Shrimp: News from the Serrated Edge of Science." *Seiche,* September 2009: 2–3.

Munawar, Mohiuddin, Iftekhar F. Munawar, P. E. Ross, and Colin I. Mayfield. "Differential Sensitivity of Natural Phytoplankton Size Assemblages to Metal Mixture Toxicity." *Archives of Hydrobiology* 25 (1987): 123–39.

Mychek-Londer, Justin G., and David B. Bunnell. "Gastric Evacuation Rate, Index of Fullness, and Daily Ration of Lake Michigan Slimy *(Cottus cognatus)* and Deepwater Sculpin *(Myoxocephalus thompsonii)." Journal of Great Lakes Research* 39 (2013): 327–35.

Nalepa, Thomas F., and Peter F. Landrum. "Benthic Invertebrates and Contaminant Levels in the Great Lakes: Effects, Fates, and Role in Cycling." In *Toxic Contaminants and Ecosystem Health: A Great Lakes Focus,* ed. Marlene S. Evans. New York: John Wiley and Sons, 1988.

Nicholson, H. Page. "Pesticide Pollution Control." *Science* 158 (1967): 871–76.

Parker, Bruce, and George Barsom. "Biological and Chemical Significance of Surface Microlayers in Aquatic Ecosystems." *Bioscience* 20 (1970): 87–93.

Parker, Robert D. R., and David B. Drown. "Effects of Phosphorus Enrichment and Wave Simulation on Populations of *Ulothrix zonata* from Northern Lake Superior." *Journal of Great Lakes Research* 8, no. 1 (1982): 16–26.

Pascoe, David A., and Randall E. Hicks. "Genetic Structure and Community DNA Similarity of Picoplankton Communities from the Laurentian Great Lakes." *Journal of Great Lakes Research* 30 (2004): 185–95.

Pearson, Roger F., Deborah L. Swackhamer, Steven J. Eisenreich, and David T. Long. "Concentrations, Accumulations, and Inventories of Toxaphene in Sediments of the Great Lakes." *Environmental Science & Technology* 31 (1997): 3523–29.

Perlinger, Judith A., Matt F. Simcik, and Deborah L. Swackhamer. "Synthetic Organic Toxicants in Lake Superior." *Aquatic Ecosystem Health & Management* 7, no. 4 (2004): 491–505.

Porter, Karen Glaus. "The Plant-Animal Interface in Freshwater Ecosystems." *American Scientist* 65 (1977): 159–70.

Rainey, Robert H. "Natural Displacement of Pollution from the Great Lakes." *Science* 155 (1967): 1242–43.

Raloff, Janet. "Redefining Dioxins." *Science News* 155 (1999): 156–58.

Rios Mendoza, Lorena M., and Chi Yeon Evans. "Plastics Are Invading Not Only the Ocean but Also the Great Lakes." American Chemical Society Meeting, New Orleans, 2013.

Robertson, Andrew. "The Present Status of Research on the Zooplankton and Zoobenthos of the Great Lakes." *Journal of Great Lakes Research* 10 (1984): 156–63.

Rodgers, Paul W., and Wayland R. Swain. "Analysis of Polychlorinated Biphenyl (PCB) Loading Trends in Lake Michigan." *Journal of Great Lakes Research* 9 (1983): 548–58.

Rossman, Ronald. "Horizontal and Vertical Distributions of Mercury in 1983 Lake Superior Sediments with Estimates of Storage and Mass Flux." *Journal of Great Lakes Research* 25 (1999): 683–96.

Schroeder, William H., and Douglas A. Lane. "The Fate of Toxic Airborne Pollutants." *Environmental Science & Technology* 22 (1988): 240–46.

Sicko-Good, Linda, and Eugene F. Stoermer. "Effects of Toxicants on Phytoplankton with Special Reference to the Laurentian Great Lakes." In *Toxic Contaminants and Ecosystem Health: A Great Lakes Focus,* ed. Marlene S. Evans. New York: John Wiley and Sons, 1988.

Smith, Daniel W. "Analysis of Rates of Decline of PCBs in Different Lake Superior Media." *Journal of Great Lakes Research* 26 (2000): 152–63.

Sonzogni, William C., and Wayland R. Swain. "Perspectives on U.S. Great Lakes Chemical Toxic Substances Research." *Journal of Great Lakes Research* 6 (1980): 265–74.

Stoermer, Eugene F. "Thirty Years of Diatom Studies on the Great Lakes at the University of Michigan." *Journal of Great Lakes Research* 24, no. 3 (1998): 518–30.

Stoermer, Eugene F., John P. Kociolek, Claire L. Schelske, and Daniel J. Conley. "Siliceous Microfossil Succession in the Recent History of Lake Superior." *Proceedings of the Academy of Natural Sciences of Philadelphia* 137, no. 2 (1985): 106–18.

Stoermer, Eugene F., Julie A. Wolin, and Claire L. Schelske. "Paleolimnological Comparison of the Laurentian Great Lakes Based on Diatoms." *Limnology and Oceanography* 38, no. 6 (1993): 1311–16.

Swackhamer, Deborah L. "The Past, Present, and Future of the North American Great Lakes: What Lessons Do They Offer?" *Journal of Environmental Monitoring* 7 (2005): 540–44.

———. *Trends in Great Lakes Fish Contaminants.* Chicago: U.S. Environmental Protection Agency, Great Lakes National Program Office, 2004.

Swackhamer, Deborah L., Roger F. Pearson, and Shawn Schottler. "Toxaphene in the Great Lakes." *Chemosphere* 37 (1998): 2545–61.

Swackhamer, Deborah L., Shawn Schottler, and Roger F. Pearson. "Air-Water Exchange and Mass Balance of Toxaphene in the Great Lakes." *Environmental Science & Technology* 33 (1999): 3864–72.

Swackhamer, Deborah L., and Robert S. Skoglund. "The Role of Phytoplankton in the Partitioning of Hydrophobic Organic Contaminants in Water." In *Organic Substances and Sediments in Water,* ed. Robert Baker. Chelsea, Mich.: Lewis Publishers, 1991.

Swain, Wayland R. "Chlorinated Organic Residues in Fish, Water and Precipitation from the Vicinity of Isle Royale, Lake Superior." *Journal of Great Lakes Research* 4 (1978): 398–407.

———. "The Great Lakes: An Example of International Cooperation to Control Lake Pollution." *GeoJournal* 5 (1981): 447–56.

———. "Great Lakes Research: Past, Present and Future." *Journal of Great Lakes Research* 10 (1984): 99–105.

Thornton, Joe. *Pandora's Poison: Chlorine, Health, and a New Environmental Strategy.* Cambridge, Mass.: MIT Press, 2000.

Trowbridge, Gail. "Lake Superior Recycles Its PCBs." *Seiche,* April 1999: 3.

Turner, Jefferson T., and John G. Ferrante. "Zooplankton Fecal Pellets in Aquatic Ecosystems." *BioScience* 29 (1979): 670–77.

White, David S. "Persistent Toxic Substances and Zoobenthos in the Great Lakes." In *Toxic Contaminants and Ecosystem Health: A Great Lakes Focus,* ed. Marlene S. Evans. New York: John Wiley and Sons, 1988.

Williams, Florence. "Toxic Breast Milk?" *New York Times Magazine,* 9 January 2005. 21–24.

Wines, Michael. "Spring Rain, Then Foul Algae in Ailing Lake Erie." *New York Times,* 14 March 2013.

Zhuikov, Marie. "Pulp Mill Effluent and Fish Don't Mix (Well)." *Seiche,* August 2008: 5–8.

Zimmer, Carl. "Scientists Find a Microbe Haven at Ocean's Surface." *New York Times,* 28 July 2009.

INTERNET RESOURCES

International Joint Commission, Great Lakes Water Quality Initiative 2012, http://www.ijc.org/
    en_/Great_Lakes_Water_Quality
U.S. Environmental Protection Agency, Great Lakes Restoration Initiative, www.glri.us

MAPPING
LAKE SUPERIOR

The Early Years

WHEN THE FIRST EUROPEAN EXPLORERS PUSHED THEIR CANOES INTO THE waters of Lake Superior, they had little more than a prayer and a paddle to guide them. But it would not be long before other navigational aids became available. Remarkably, as early as 1671, a reasonably accurate map of the lake had been published in France.

The credit for developing such useful tools fell not to fur traders, however, but to the Jesuit missionaries who plied the waterways in search of souls. The 1671 map of Lake Superior, for example, is largely attributed to the French Jesuit Claude Dablon, who was trained as a geographer and likely gathered topographical information as he crisscrossed the lake in 1669–70 with his countryman and fellow cleric Father Claude Jean Allouez.

That early travelers were able to draft a credible approximation of the shores of the world's largest lake is nothing short of remarkable, considering that they possessed only the most rudimentary surveying tools, if any at all. To this day, writes geographer Conrad Heidenreich, "surprisingly little is known about the methods and instruments used by explorers to gather the data for their maps." In some cases, they obtained verbal accounts, even drawings, of such physical features as inland waterways from the native people who were familiar with them. They recorded observations and sketches in their diaries. Their notes suggest that some may have carried a compass or, more rarely, instruments such as an astrolabe, a medieval tool that calibrates the position of the sun and other celestial bodies for navigation. Most readings were likely supplemented by the popular technique of using travel time to estimate distances.

Other travelers simply invented their own survey methods—with varying degrees of success. In 1766, before setting out on a twenty-two-month journey to the upper Great Lakes and beyond, the Englishman Jonathan Carver decided that he would rely solely on his judgment to measure distance. To hone his perceptual skills before setting out on his travels, Carver noted that he "practised [sic] several times the pacing out of lengths of a mile by land and then would go the same distance in a battou or canoe by water." His self-described goal was to "judge prety [sic] exact as to the distance of a mile." Carver's method—not surprisingly—proved ineffective in calculating distances, not to mention pinpointing topographical features with accuracy. Instead of measuring distance to the nearest half mile, as was his intention, he overshot the reach of Lake Erie from its eastern to western ends by fifty-five miles. In Lake Superior, he inexplicably added a bogus island, known as Mauropas Isle, twenty miles off the coast of Canada's northern shore.

By the mid-1800s, commerce on the lake had changed and so did the need for greater navigational precision. During the fur trade era, when the canoe was the principal mode of transportation, travelers sought out only the location of rivers and settlements along the lake edge. Bathymetric, or water depth, surveys were largely unnecessary. In time, however, steamers loaded with timber, fish, minerals, and tourists began to crisscross the lake. Vessels with deeper drafts ran aground on offshore shoals, exacting a precious toll—in lucre and human lives.

The need for greater navigational safety led to the first attempt to comprehensively study the lake. Under a charge from the British Royal Navy, Lieutenant Henry Wolsey Bayfield helped to survey the entire coastline of Lake Superior in 1824 and 1825. His map, which for decades served as the definitive navigational aid to Lake Superior, was a remarkable achievement for its time. It included not only water depths in selected harbors and the locations of shipwrecks, trading posts, and settlements, but it also incorporated geological data and information about the kinds of trees found growing along the lake.

Although Bayfield's map was a cartographic milestone, it rarely made its way into the pilothouses of cruising ships. The size and complexity of Lake Superior rendered even state-of-the-art maps like Bayfield's useless in all but a few well-trafficked places. In 1835 Ramsey Crooks, president of the American Fur Company, wrote that although Lake Superior had been "fully & scientifically explored by the British Government, and a chart [had] been published," the chart's "circulation has hitherto been confined pretty much to the public Bureaus and the Officers of Government."

As a result, many captains continued to rely on on-site scouting rather than paper to guide them through unknown waters. In 1864, Alexander McDougall, a second mate aboard the ship *Ironsides,* described the navigational methods commonly used in his day. "Exploration was rampant," he wrote. "Everyone was excited by wonderful prospects or great discoveries, and people were spending money as they do in flush times. . . . We had to land passengers at all sorts of out-of-the-way places. It would be my duty as second mate to go ahead in a small boat when we were making a landing in strange waters and take the soundings [the practice of measuring water depths by throwing lines weighted with lead overboard]. I learned the pilotage of lakes and rivers and became expert very early in heaving the sounding lead. . . . The only reliable map was one made by Lieut. Bayfield, which had very few soundings, and even so it was hard to get copies of it. All the time I sailed the Lakes I never had a chart in my hands."

For decades, further exploration of Lake Superior languished as politically powerful East Coast interests diverted federal dollars to fund navigational improvements along the Atlantic Seaboard. But in 1841, faced with mounting losses on the Great Lakes, the U.S. Congress turned its attention to the upper Midwest, authorizing the Corps of Engineers (later renamed the Army Corps of Engineers) to begin a major exploration of the nation's inland seas. The help from Washington could not have come at a better time. "There were few light-houses and beacons to indicate the positions of danger to navigation, and, in the absence of charts, pilots were obliged to rely upon their own knowledge, which was frequently only acquired by the vessel's grounding on a shoal or striking a hidden rock," writes General Cyrus B. Comstock of the Corps of Engineers in his 1882 history of the project. On such uncharted waters with few improved harbors and gales the size of ocean storms, he observes, there was, not surprisingly, "much loss of life and property each year."

The task of mapping the lakes, known as the U.S. Lake Survey, was itself fraught with inconvenience, not to mention the same kinds of danger that dogged commercial traffic. With cities few and far between, surveyors hauled their own supplies, transporting them by rowboat—and later by steamer—from campsite to campsite as the workers inched their way up coastlines.

Progress was painstakingly slow. During the field season, which took place between May and October, one group of surveyors mapped the shoreline and nearby topography while another headed into the lake. Neither assignment was simple. General William F. Raynolds, who supervised the Lake Survey from 1864 to 1879, described some of the conditions he encountered during the 1868 reconnaissance of Lake Superior: "The character of the country in which the surveys are being prosecuted forbids that attention to the details of topography which would otherwise be desirable. It is the exception to find anything but a dense forest, in which it is impossible to make an accurate survey without opening every foot of the lines of sight. No sketching can be done that is reliable. Parties within easy hearing distance cannot see each other. And last, though by no means least,

FIRST "BAYFIELD" SOUNDING CREW
1884

Sounding crew of the survey steamer *Bayfield*. In the mid-nineteenth century, surveyors traveled the coastlines of the Great Lakes, dividing their tasks between mapping the shoreline and nearby topography and taking depth measurements of the lake. Photograph by the Canadian Hydrographic Survey. Courtesy of Fisheries and Oceans Canada.

To minimize the mounting losses of life and property, in 1841 the U.S. Congress authorized the Lake Survey, a decades-long project to improve navigation on the Great Lakes. Included in the documents is this 1873 survey map of the western basin of Lake Superior. Courtesy of the American Geographical Society Library / University of Wisconsin Sea Grant.

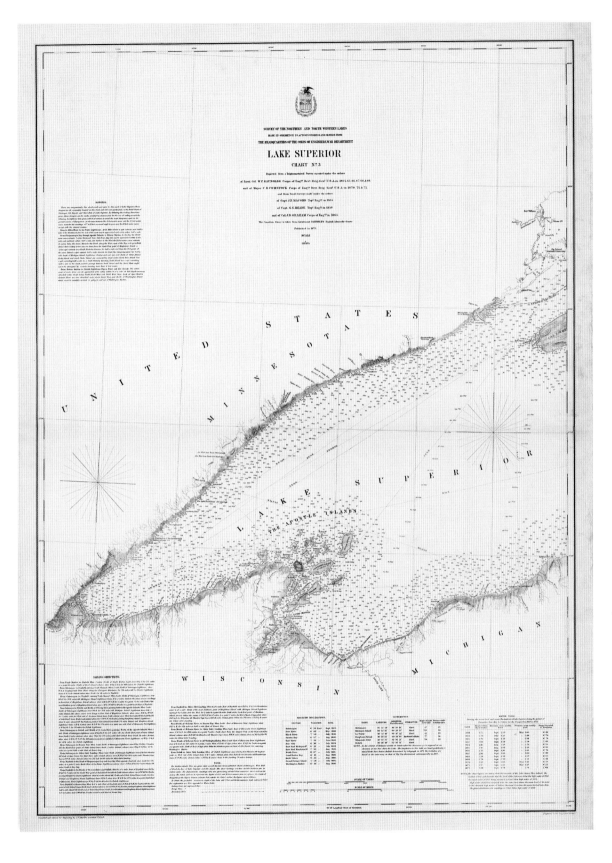

during the summer season, which is the only one in which work can be done at all, the forests are so full of venomous insects that it is next to impossible for an instrument to be used."

Those who worked offshore in rowboats may have escaped noxious insects and impassable forests, but they faced other challenges. The work of taking soundings was laborious. Wind and waves often compromised their accuracy. And most of them, no doubt, kept one eye on the task at hand and another on the sky, especially when traversing rocky shorelines that offered few places to safely land a boat in a sudden storm. Anxiety was their constant companion.

For some workers on the project, the end to their labors could not have come soon enough. "The whole survey of this portion of the lake, extending from Duluth to opposite the western end of Isle Royale," Raynolds writes, "was attended with privations and difficulties not met with in more favored localities; the almost entire want of communication, the absence of harbors, and to a great extent of even boat landings, the rough mountainous country covered by a dense forest, all combined to render the operations of the survey not only difficult, but dangerous, and I cannot but rejoice that the work has been finished successfully without an accident." ∾

SEARCHING
HIGH AND LOW

The Science of
Lake Superior
Exploration

By the mid-1600s, at a time when European settlements on the New England coast were little more than faint stars in a dark galaxy of forest and sea, French mapmakers had already delineated Lake Superior's shoreline with a startling degree of accuracy. Barely more than a decade after traders Médard Chouart des Groseilliers and his brother-in-law Pierre Esprit Radisson unloaded the first packs of Lake Superior furs onto Montreal's docks, Superior's wolfish profile had taken shape on paper—the snout pointing westward, the south shore tucked under the north's great spinal curve like an animal curled in sleep. Even the entrances to many rivers had been noted with tiny lines that sprouted from the boundaries of the lake like fine hairs.

Since this nascent attempt to scribe the circumference of Superior, scientists have refined survey techniques on both land and water. Massive ore carriers are now able to safely thread their way through treacherous shoals. Using maps that are available in any sporting-goods store, kayakers can locate the nearest sheltered cove in an oncoming storm.

The early European explorers would have marveled at such achievements. Yet they seem almost crude when compared to the even more recent milestones in our ability to both measure and imagine Lake Superior. Consider the revolutionary changes brought about by the advent of space flight alone. Only since the 1960s have humans been able to view images of the Great Lakes in their entirety. Since then, the technological tools for mapping the lakes have increased exponentially in number and sophistication, as well as in their temporal and spatial scope. Today, the infrared sensors of satellites keep track of Superior's temperatures and ice cover in real time. Other satellite-mounted instruments read the spectral reflectance of light from lake waters and tabulate concentrations of phytoplankton, particulate matter, and dissolved substances, all key components in determining water quality. With the aid of geographic information system (GIS) databases, vast gigabytes of information can be mixed and matched with information on the ground, such as land-use surveys of forest clearing and shoreline development. GIS allows users to compile these disparate data in easy-to-read, color-coded representations that can be called up at the click of a computer key by anyone with access to the Internet.

Earth-based instruments are no less spectacular in their ability to gather information about the lake. Penetrating more than one thousand feet of water, sonar instruments detect lake-floor sands rippled by deepwater currents. Some ten thousand years ago, the keels of glacial icebergs etched grooves into the lake bottom. Today, their outlines can be laser printed on paper in three-dimensional black-and-white images as fine grained as graphite drawings. Scientists travel to the bottom of the lake in bubbles of acrylic and steel, drifting through clouds of tiny shrimp to witness firsthand the handiwork of fish known as burbot. These deepwater architects construct elaborate runways and pueblo-like structures in total darkness.

Bringing state-of-the-art technology to bear on their explorations, scientists now are discovering what Lake Superior lovers have long intuited: the lake is a living community whose networks of getting and spending are as intricate and compelling as those of the most advanced human civilization. Here are some of the mysteries they have uncovered.

What Lies Beneath: Mapping the Nearshore Zone

At about one o'clock on the afternoon of July 28, 1993, the driver of a semitrailer truck was conducting a routine haul past Taconite Harbor near the town of Schroeder when he looked up to see an avalanche of muck coursing toward him. Only minutes before, a dike that held back twenty-five years' worth of waste from a coal-fired power plant had been breached. Torrential overnight rains had so saturated the ash pile that it finally "just slid down the hill" from its perch one-half mile above Highway 61, said a spokesperson from the Minnesota Pollution Control Agency (MPCA). An estimated four hundred thousand cubic yards of coal ash—the equivalent of forty thousand dump-truck loads of material—surged toward the lake. The driver was able to leap free of the truck before it was swept from the road and buried in muck, some of which covered the highway to a depth of twenty-seven feet. The sludge continued its rampage on the lake side of the highway, tearing through an electrical transformer and spilling eight thousand gallons of mineral oil. Before it was all over, some two thousand cubic feet of ash laced with heavy metals, such as mercury, had poured into Lake Superior. Spot checks by divers revealed that some parts of the lake floor had been covered to a depth of two and a half feet.

Within hours trucks began clearing the highway and hauling the ash waste back to the inland landfill. But when it came to Lake Superior, cleanup options were far more limited. MPCA officials could do little more than set up booms on the lake to contain the slick of mineral oil that stretched some four miles along the coast. Lacking a practical means to effectively remove the sludge from the lake bottom, they decided to leave it in place.

With the exception of a few isolated reports from divers, no one knew the full extent of the damage. Even more troubling was the uncertainty of the pollutants' fate in the lake. Would the runaway ash be picked up by coastal currents and flushed westward, adding to the contaminant load of the Duluth-Superior harbor? Did it destroy prime spawning habitat for recovering fish species such as lake trout? No one had a clue, in part, because the details of Superior's physical features and functions—from the composition of the lake bed to the lake's water-circulation patterns—had not been thoroughly studied or mapped with any precision.

Obtaining a more detailed understanding of Superior's coastal waters is no mere exercise in scientific curiosity. Nearshore zones are the biological powerhouses of the lake. Here, in these relatively shallow waters, sunlight can penetrate to the bottom of the lake and promote the growth of algae, which feeds a wide array of aquatic organisms. Rivers and streams also contribute to the productivity of nearshore zones by flushing warmer water and life-sustaining supplies of chemical nutrients and plant and animal food into the lake. As a result, "much of the lake's biological activity occurs in a very narrow zone around the edges. Many of the fishes have some part of their life history that is tied to that zone," says Carl Richards, an aquatic ecologist and former director of Minnesota Sea Grant at the University of Minnesota Duluth.

Among the species of greatest concern to fisheries managers are lean lake trout. Once the top predator of the Great Lakes aquatic community, lake trout populations suffered steep declines in the 1960s due to overfishing and invasion of the parasitic sea lamprey. Biologists are especially keen to conserve lake trout stocks in Superior since it is the only lake in the Great Lakes system in which the fish has dramatically rebounded.

Key to the trout's survival is the protection of its nearshore spawning grounds. Each fall these deepwater fishes move into shallow water of one hundred feet or less to reproduce. The fish seek out lake bottoms of mixed cobble—a catchall category that includes materials ranging from fist-sized rocks to giant boulders—as nurseries for their young. These habitats not only support aquatic organisms that supply food for emerging offspring, but they also provide more protective crevices for developing young than open areas of sand or bedrock. And nearshore zones also offer an added benefit—turbulence. Storm waves trigger bottom currents that routinely sweep fine sediments such as silts and clays into the deeper basins of the lake and away from shallower spawning areas, where they can clog fish gills and choke oxygen-loving eggs. Lake trout are faithful to these natal grounds. Through countless generations, sexually mature adults have returned to their birthplace to spawn.

Not surprisingly, those keenest to investigate harm from the ash spill were fisheries biologists. Along the Minnesota North Shore, even the loss of small stretches of reproductive habitat can have big repercussions since the shallow-water habitats favored by lake trout are at a premium. Bathymetry (water-depth) surveys depict the nearshore area along Minnesota's coast as a narrow shoal of relatively shallow water with precipitous drops into deep zones. One-half mile off the coast of Silver Bay, for example, water depths plummet to nearly one thousand feet, unlike the eastern shore, say, whose bottom slopes more gently into the lake, offering spawning trout a far more extensive complex of nearshore shoals.

But in the aftermath of the spill, bathymetric maps offered scientists only the crudest information about the lake bottom that lay just offshore. Because bathymetric charting is time-consuming and expensive, only the subsurface contours of bays and harbors used for commercial shipping have been mapped with any precision. Information between ports often is very generalized. Between Duluth and Grand Marais, for example, coastal waters have been surveyed at intervals of only 2.6 data points per square mile. These charts give captains enough information to safely pilot a fishing craft or an ore carrier. But they offer only a very limited picture of the underwater topography.

Information about the composition of the lake floor was even more nebulous. On topographic maps, for example, the lake bottom that had been covered by the ash spill was simply marked with such vague descriptors as "sand bottom" or "rock bottom." Using them to locate trout spawning grounds is the aquatic equivalent of trying to find the highly specialized jack-pine breeding habitat of the Kirtland's warbler on a vegetation map generically marked "coniferous forest."

So when $240,000 in settlement money became available from LTV Steel, the owner of the breached waste dump, the MPCA decided to fund a project to identify lake-trout spawning habitat along the Minnesota shoreline, a fundamental, long-standing need outlined in the Minnesota Department of Natural Resources' (DNR) *Fisheries Management Plan for the Minnesota Waters of Lake Superior*. The first step was compiling a detailed map of the lake floor spanning some eighty miles between Duluth and Grand Marais.

The execution of this daunting assignment involved the cross-disciplinary expertise of scientists from the University of Minnesota's Natural Resources Research Institute (NRRI), the Minnesota DNR, and the U.S. Geological Survey. Together, they assembled an innovative combination of new and existing technologies. The team cruised offshore waters in depths of one hundred feet or less using an echo-sounding device. Much like a high-tech fish finder, the instrument sent acoustic pulses to the lake bottom and measured the time it took for them to bounce back to the equipment on board the boat. These signals—collected at a rate of two per second—were filtered through a new marine software program that could translate the sound-wave information into both water depth and substrate type. (Soft materials such as muds, for example, tend to send less energy back than hard materials such as rocks or sands. To verify the accuracy of these results, the researchers compared the acoustical data to actual images of the lake bottom that were taken by a video camera fixed to the underside of the boat.) As they traveled along the shoreline, a satellite-based global positioning system (GPS) simultaneously kept track of their location in the lake. The combination of these methods resulted in a high degree of specificity and accuracy, allowing scientists to pinpoint the lake's substrates to within several square feet.

Later, in the NRRI's computer labs, technicians painstakingly entered the data into a comprehensive GIS mapping program. Using color codes to represent eight different categories of substrates, the scientists created easy-to-read maps of the underwater topography of most of Minnesota's coastline and posted them on the Internet. At the click of a mouse, North Shore residents now can call up images of the lake floor that lies just outside their windows.

Does knowing what lies offshore dispel some of the mystery of Lake Superior? Not at all, says Richards, the lead scientist of the project. The researchers were astounded by what they found. Instead of unmodulated expanses of nondescript sand or rock bottom, the scientists discovered an astonishing variety in the configuration of lake-floor materials that reflected the dazzling geological diversity of the terrestrial environment. Off the shore of Grand Marais, for example, the bottom gently slopes into the lake before dropping into deep water, providing the Minnesota North Shore with some of its most extensive shallow-water nearshore habitat. Indeed, the nearshore zone around Grand Marais contains twice as much shallow-water habitat (depths less than one hundred feet) than other comparable sections in Minnesota waters. On the color-coded GIS maps, the underwater landscape looks like an intricate patchwork quilt. Microhabitats of sand (a categorical term in the study that includes sand, silt, and clay) and bedrock outcrops are fitted into large swaths of cobble and conglomerate.

A survey of the lake bottom along the Minnesota North Shore in the 1990s revealed an extraordinary variability in substrate types and topography, as seen in this mapping of the lake bottom at the Silver Bay Safe Harbor.

Even in some of the most limited stretches of shallow water, such as the area just off the shore of Split Rock Lighthouse, the researchers uncovered a startling degree of diversity. In this narrow ribbon of shallow water, plains of cobbles and conglomerate give way to fields of boulders the size of jumbo refrigerators and great underwater beaches of sand. "We thought the lake floor would be either all rocks or all sand. No one thought there was this level of heterogeneity out there," Richards says.

Although originally intended to identify lake-trout spawning grounds, the maps will make life easier for a wide array of professionals working in the region. Fisheries biologists now have a valuable tool for increasing the success of fish-stocking efforts. Research has demonstrated that when hatchery-raised lake trout are released into the lake in their earliest life stages—as fertilized eggs or fry—they become imprinted on the sites in which they were planted and later return to them to spawn. Releasing developing fish in optimal spawning habitat increases the likelihood of their future reproductive success.

Having a more up-close view of the bottom also will enable researchers to better understand the less glamorous, though no less important, animals that make fish life possible in the lake. Invertebrates such as crustaceans, snails, clams, worms, and insect larvae, for example, feed on microscopic plants and animals, forming an important intermediate link between small and large organisms in the lake's food web. Like trout, many of these tiny animals have specific habitat preferences. "Whether the bottom is sand, clay, bedrock, or cobble has a tremendous influence on the biological activity that takes place there," Richards says. "There are many people who would love to tie the different parts of Lake Superior's ecosystem together, but it's not an easy thing to do." Until now, scientists have not been able to consistently survey populations of invertebrates or thoroughly research their relationships to their lake habitats because the composition of the nearshore lake floor was largely unknown.

The new maps can help to preserve this biologically active zone by giving people in the region the information they need to make better decisions about what happens not just in the water but also on the land. For example, researchers found that the best trout-spawning habitat lies in waters that are less than sixty feet deep. These are also the waters most heavily affected by upland land uses in the watershed, of particular concern to the Minnesota North Shore, which has become one of the most heavily developed coastlines on Lake Superior.

Scientists have had few tools, however, for assessing and monitoring the impact of human activities on the lake. As a result, many land-use decisions have taken place in an ecological vacuum. One of the special virtues of the GIS system is that it allows researchers to merge this offshore information with other sets of different geographic data and to present

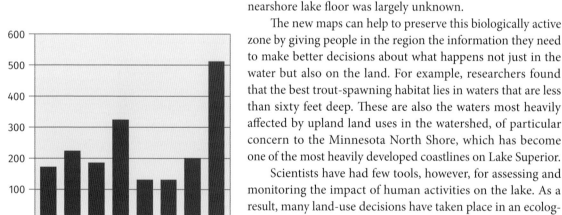
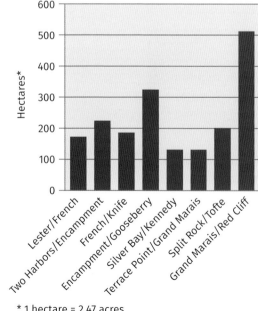

* 1 hectare = 2.47 acres

Detailed substrate mapping is a useful tool for fisheries management. One application has been identifying prime lake-trout spawning habitat in the nearshore reaches along the North Shore. Among the best is a stretch of coast between Grand Marais and Red Cliff.

reams of numbers in comprehensible graphic formats. The nearshore maps, for example, can be paired with maps of the terrestrial coastline to better visualize the potential effects of not just another toxic spill but also a wide array of other human activities, such as road building, housing construction, and forest management. Already the nearshore maps have been used to help engineers protect lake-trout spawning grounds in their siting of a new marina in Silver Bay. The maps have also identified spawning habitat at risk from siltation from shoreline erosion, enabling the MPCA to make decisions about where to most effectively allocate scarce dollars for shoreline-stabilization projects. "It's given us a whole new way of thinking about what happens in the nearshore zone," Richards says.

Master Carvers: Glaciers and the Topography of the Lake Bottom

In summer 1848 scientist Louis Agassiz led an expedition to the Canadian North Shore of Lake Superior. A decade before, Agassiz had made a name for himself championing the theory that ice sheets had once blanketed the northerly reaches of the globe, including huge portions of Europe. On a tour of the Rhone Valley in 1837, Agassiz noted signs of what he believed were ancient glacial activity: great mounds of rocky gravel, isolated boulders stranded in open fields, and long, linear scratches etched into exposed rocks. His North American travels would lend only further credence to his radical new idea. No sooner had his party pushed off into Lake Superior from Sault Ste. Marie than a delighted Agassiz discovered a large flat rock along the shoreline that bore the telltale striations of an ice sheet's scraping.

Agassiz's enthusiasm for the north country was captured in his 1859 book *Lake Superior*. Although immensely popular, his account did not incite a stampede of geologists eager to test his provocative assumptions. Remote and rugged, the area discouraged casual forays. Even expeditions that enjoyed logistical and financial support, such as scouting parties in search of valuable ores, faced daunting challenges. John W. Foster, member of a geological exploration team that explored Isle Royale in the late 1840s, wrote in a diary of his journeys: "The shores [of Isle Royale] are lined with dense but dwarfed forests of cedar and spruce, with their branches interlocking and wreathed with long and dropping festoons of moss. While the tops of the trees flourish luxuriantly, the lower branches die off and stand out as so many spikes, to oppose the progress of the explorer. So dense is the interwoven mass of foliage that the noonday sunlight hardly penetrates it. The air is stiffled [sic]; and at every step the explorer starts up swarms of musquitoes [sic], which, the very instant he pauses, assail him."

Indeed, nearly half a century would pass before geologists systematically explored the legacy of the glaciers in the Lake Superior basin. Even then, they confined their investigations to the lake edge, mapping coastal moraines, for example, and the beaches and wave-cut cliffs of the various glacial lakes that preceded modern-day Lake Superior. Well into the twentieth century, the basin's population remained sparse and isolated, making inland explorations arduous and technically challenging.

But no barrier to geological investigation proved more intractable than the deep waters of Lake Superior. In 1870, scientists identified minuscule crustaceans (known also as opossum shrimp or mysids) in the stomachs of Great Lakes whitefish. At the time, mysids were thought to be an exclusively marine species. The discovery fueled curiosity about the ancient geologic events that enabled them to enter the freshwater interior of North America. Many a geologist in the early years of exploration no doubt gazed across the lake's blue expanse and wondered what chapters in the great story of the ice sheets lay unread in its depths.

Scientists would not get their first real clues about the glacial history of Superior's lake floor until 1961, when the Universities of Michigan and Minnesota teamed up to

Glacial ice sheets shaped the Lake Superior basin and lake bed, leaving a variety of tracks, among them striations in the bedrock. Photograph by Gary Alan Nelson.

probe the bottom of Superior using new underwater technologies, many of which were first developed to study the deep-sea floor. These studies extracted cores from bottom sediments, some of them measuring up to nearly seven hundred feet long. At the same time, researchers obtained seismic-reflection readings of the lake bottom. This sonar-based technology could penetrate the great drifts of sediments that had been deposited by the glaciers and reveal the contours of the subbottom, or bedrock topography, that lay beneath them. By the end of the 1960s, as other universities and government agencies joined in with research projects of their own, a detailed picture of the lake floor began to emerge.

These initial investigations were motivated purely by academic interest, although the data would later be analyzed by scientists exploring the oil and mineral potential of the subsurface and those studying the fate of pollutants in the lake. Geologists hoped to find areas on the lake floor that managed to escape the grinding and plowing of the glaciers and erosion by their meltwaters. Unfortunately, they discovered that the glaciers had reworked the lake bed as thoroughly as the surrounding landscape.

Still, the lake floor did not disappoint, either in its topographical majesty or geological complexity. It turns out that the character of the lake floor, from the contours and composition of Superior's bedrock to the distribution of sediments around the lake, largely falls into a kind of geological yin and yang. When scientists examined sediment cores extracted from the lake bottom, they noted a curious pattern. The clays at the bottom of the cores, namely, the first sediments to be deposited by the receding glacier, were red or reddish brown in color. As time went on, however, the red clays gradually gave way to gray clays. What was the reason for the transition? According to geologist William R. Farrand, the glacier crossed a geological line from one bedrock province to another. The boundary, he speculates, is located somewhere on the lake bed in the northern part of Superior. Sediments to the south and west of a line extending from Sault Ste. Marie, Michigan, to the city of Thunder Bay on the Canadian North Shore were derived from the grinding of reddish sedimentary bedrock that dates to Protero-zoic times some 1 to 1.1 billion years ago. As it continued retreating, the ice sheet shed its excavations from a different bedrock province—the 3-billion-year-old gray granitic bedrock of the Canadian Shield. This granite underlies the portion of the Lake Superior basin to the north and east of this line.

Like the geological nature of the sediments, the topography of the lake-floor bedrock also exhibits a dual personality. Separating the eastern from the western basin is a subsurface ridge trending north-northeast from the tip of the Keweenaw Peninsula to the Slate Islands. To the east of this line lies a topography that can be described as Superior's Rocky Mountains. Here, a series of rugged ridges and valleys trending in a north–south orientation dominates the lake floor. It is as if a set of giant claws raked across the bedrock, leaving deep lacerations on the lake floor.

The topographical relief of this submerged mountain range is dramatic. The valleys, which measure three to six miles wide and more than sixty-two miles long, are flanked by towering ridges. At one site, scientists have measured as much as an 820-foot drop from a ridge crest to the bottom of an adjoining trough.

These complex features create unique ecological niches and add to the habitat diversity of the deepwater reaches of the lake. Here, mountain peaks, which resemble the ridgetops of the terrestrial highlands that surround the lake, reach for the surface.

Forming shallow, isolated reefs and shoals in the midlake zone, they are sought out as spawning grounds by a race of lean lake trout known as a humper. These reef waters, ranging from depths of 164 to 492 feet, can suddenly plunge to nearly 1,000 feet. Indeed, in a valley just twenty-three miles north of Grand Island, Michigan, lies Superior's deepest spot. First recorded in 1948 as 1,333 feet, this measurement was taken by new, more precise instruments and amended to 1,265 feet. Due to its rugged lake-bed topography, says geologist Farrand, the eastern basin likely contains the most complex bathymetry in the entire Great Lakes.

What caused this dramatic complex of submerged peaks and valleys? The answer remains tantalizingly unclear. The north–south orientation, which corresponds to the directional movement of the ice sheet, suggests to some geologists that the glaciers cut this corrugated pattern into the bedrock. Others scientists argue that the valleys were eroded by streams that flowed far beneath the surface of the ice sheet. Still others contend that the dendritic pattern of this peak-and-trough complex is the remnant of an ancient river system that predated the glaciers. A major drawback to this argument, however, is that only two of the valleys in the eastern basin extend onto the south shore of the lake; the others abruptly halt when they hit a steep rise in the lake floor about nine miles from the South Shore.

The lake's western basin exhibits a far gentler topography. Here, deep bedrock valleys also score the lake bed, but scientists point out that they are neither as abundant

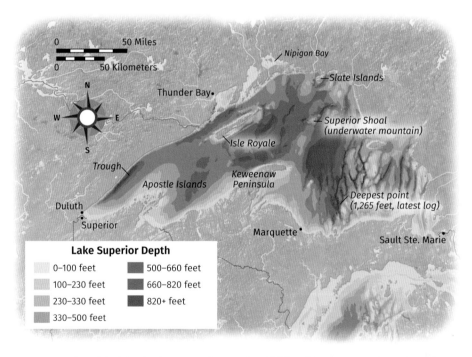

Contour mapping of the lake floor's topography reveals a dividing ridge that extends north-northeast from the tip of the Keweenaw Peninsula to the Slate Islands. The basin along the eastern side of this ridge is characterized by rugged ridges and valleys. The western basin has a far gentler topography, as deeper deposits of glacial drift soften the extremes of the underlying bedrock.

434

nor is their configuration as complex as the submerged system in the eastern half of the lake. Furthermore, the lake floor in the western basin is mantled by much deeper deposits of glacial drift, which tend to fill in topographic lows and soften the extremes of the underlying bedrock. As a result, the western floor of the lake is characterized by relatively broad, undulating valleys that lie in waters with an average depth of nearly 500 feet. Forming the walls of some of these valleys is a series of moraines that lie in a roughly perpendicular orientation to the long axis of the lake. Hundreds of feet high and wide, these ridges of gravel can snake across the lake floor for many miles. Deposited when the glacier paused in its retreat, these moraines have allowed geologists to track the progress of the ice sheet as it made its way across the Lake Superior basin.

At several locations, however, the gentle relief changes abruptly, and the lake floor suddenly opens up to abyssal canyons that can plunge to depths of nearly 1,000 feet. The most dramatic of these are two deep and nearly continuous linear troughs that parallel the northern shoreline between Duluth and Thunder Bay. Several others can be found along Isle Royale. They measure between three and eighteen miles across and twenty-five to eighty-one miles long. Thick blankets of sediments mask the true depths of the bedrock foundation that underlies them. The deepest valley in the western basin lies about one-half mile off the coast of Silver Bay on the Minnesota North Shore, where seismic-reflection readings detected bedrock buried under more than 1,640 feet of sediments. Add to this another 985 feet of water, and this bedrock valley plunges nearly one-half mile beneath the water's surface.

The eastern and western basin can also be differentiated by yet another duality: their vastly unequal endowments in glacial sediments. True, thick drifts of glacial sediments, some as deep as 984 feet, have amassed in the abyssal valleys of the eastern basin. In general, however, sediment accumulations are far scantier, averaging less than 100 feet, unlike those in the western basin, where accumulations routinely reach almost 200 feet. Especially thinly mantled are the submerged ridges where sediments form little more than a veneer between 23 and 49 feet thick.

What accounts for the discrepancy? Many of the towering ridges in the eastern basin lie close enough to the surface to allow the turbulence caused by large waves to flush sediments. And in some of the deep troughs of the eastern basin, mystery currents regularly sweep the lake bottom.

Much of the disparity in sediment distribution, however, is due to events that occurred thousands of years ago. The glacier melted back in a northeasterly direction. As a result, the western end gave way to open water long before the ice released its hold on the eastern basin. Sediments from the mountain of glacial ice poured into the fledgling lake, the first precursor to modern-day Lake Superior, known as Glacial Lake Duluth.

By today's standard, the sediment loads were immense. The average sedimentation rate for late-glacial times was about 0.2 inches per year, with deposits of up to 4 inches during peak years. Once the glacier finally retreated from the basin, allowing plants to take root on the scoured earth and stem erosion, these rates declined substantially. Today, the high water-to-land ratio in the Lake Superior basin has resulted in a meager sedimentation rate of 0.004 to 0.012 inches per year, the lowest in all the Great Lakes.

But even though the ice had retreated from the modern-day lake bed by 9,500 years ago, the glacier remained perched on Superior's northern shoreline. Indeed, glacial debris continued to enter the lake until about 8,500 years ago. The fact that huge

In the northern hemisphere, the Coriolis force causes winds and water currents to flow in a counterclockwise fashion. This phenomenon is visible here, as rivers carrying sediments form muddy plumes that are deflected to the right as they enter the lake. Copyright John and Ann Mahan.

sediment loads poured into the lake from the drainage area to the north rather than from the south played a crucial role in their distribution.

Forces of nature, along with accidents of geology, worked together to ensure an uneven distribution of these sediments. Meltwater streaming off the ice sheet and into the lake, much of it flush with fine sediments known as rock flour, would have been deflected in a counterclockwise circulation pattern due to the Coriolis force. (This phenomenon is produced by Earth's rotation, which causes water currents and winds such as hurricanes to move in a counterclockwise fashion in the Northern Hemisphere and in a clockwise direction in the southern half of the globe. In Lake Superior, this law of physics is especially visible to the unaided eye in the spring, when river waters flush with snowmelt and sediments form muddy plumes that are deflected to the right as they enter the lake.)

Entrained between the Ontario mainland on one end and the rocky headlands of Isle Royale's northern shore on the other, these waters flowed into a natural sediment trap. Not surprisingly, researchers discovered that major drifts of sediment settled out in the deep trough between Thunder Bay and Isle Royale and farther west into the troughs along the present-day Minnesota shoreline. After millennia, many bedrock features in the western basin became buried in drift.

Because these sediments were deposited underwater, the lake floor contains some glacial features that are very different from those found on land. In 1973 scientists studying the bottom using side-scan sonar discovered a network of linear grooves crosshatched into the sediments just northwest of the Keweenaw Peninsula. The intersecting grooves ranged from 16 to 246 feet wide and 6.5 to 16 feet deep. Some stretched more than 1.25 miles across the lake floor. Researchers hypothesize that during low water levels the keels of icebergs, calved from the nearby ice sheet, cleaved the thick deposits of sediments like plowshares as they floated around the lake. Similar linear gouges have been found near Isle Royale and the Michigan mainland between Grand Marais and Whitefish Point. Superior's low sedimentation rate in the aftermath of the glaciers helped to preserve the contours of these features.

That same year researchers documented even more peculiar markings on the floor of Lake Superior. On the lake bottom off Grand Marais, Minnesota, they discovered a series of what scientists simply call ring structures. It is as if someone pressed a giant doughnut over and over again into a batter of soft sediments on the lake bed, says Nigel Wattrus, a geologist from University of Minnesota's Large Lakes Observatory

(LLO). Using a new multibeam sonar technology, Wattrus, along with scientists from the University of Cardiff in Wales, have revisited work conducted on the ring structures in the early 1970s. Developed for studying ridge systems in the deep waters of the midocean, this mapping device covers wide swaths of the lake bottom with a fan of acoustic energy that relays information about both water depth and the shape of the lake floor.

A wider reconnaissance of the lake bottom has revealed that the structures, spanning 330 to 1,640 feet in diameter, are common features in the fine sediments throughout most of the western basin and in waters deeper than 500 feet. Separating their outlines from the surrounding sediments on the lake floor are large cracks, which measure up to 16 feet deep. Over time, sediments have buried portions of many ring structures, so that in some parts of the lake the bottom is scored with crescent shapes, as if a huge hand scribed a series of commas.

The fact that ring structures show evidence of being filled in slowly over time suggests to scientists that they were formed in the distant past. The cause, however, remains puzzling. One explanation posits that the rings reveal the uneven contours of Earth's underlying deep crust. Another interpretation is that the rings were created during the retreat of the last ice sheet. The torrential volume of glacial debris that poured into Lake Superior could have rapidly compacted bottom sediments. As water was expelled from these sediments, particularly those with a high clay content, they shrank, causing cracks to form.

Based on data collected from their multibeam sonar mappings, however, Wattrus and his colleagues may provide fresh evidence to verify an old hunch put forward by the scientists who first explored the structures in the 1970s. They believe the ring impressions were formed via a process known as syneresis. Wattrus points out that the molecules of a fine-grained clay found on the bottom of Lake Superior, known as smectite, have a strong affinity for one another, causing them to bind tightly together. In the process of contraction, they expel water trapped between the sediment particles. As the sediments shrink, they form the cracked circular patterns on the lake floor.

Results of the new imaging technology could provide a new twist on this hypothesis, however. Data collected by the scientists suggest that the ring structures may be the surface expression of what is known in geology circles as polygonal fault structures (PFS). Beneath these formations, which occur on ocean floors throughout the world, is a spiderweb of cracks that extends deep into the oceans' subbottom. Research reveals a similar pattern of faulting beneath the lake bed in Superior, suggesting that the ring structures may mark the beginning stages of PFS development in Lake Superior. If so, they could provide valuable new information about the formation of these geological features, which are commonly associated with marine petroleum reserves.

Multibeam sonar technology is also helping scientists to better understand several brief, but catastrophic, chapters in the geological history of Lake Superior. In the 1980s, Canadian geologist James Teller and his colleagues uncovered five groups of channel complexes in the forests northwest of Lake Nipigon. Scientists believe that during the waning days of the last ice sheet, these causeways provided temporary connections between Glacial Lake Agassiz and Lake Superior. In Nipigon Bay off Canada's north coast, Wattrus and his colleagues have located what may be a great fan of sand and gravel that was dumped as the deluge of Glacial Lake Agassiz's sediment-laden waters poured into Lake Superior.

What intrigues scientists is not simply the sheer magnitude of these events but also their effects on the global climate. Throughout most of its lifetime, Glacial Lake Agassiz, a monolithic body of water that once covered portions of present-day Minnesota, North Dakota, and much of southern Canada, drained south through what is now the Mississippi River. But as the Laurentide ice sheet began to recede, it opened other outlets to the north and east. Beginning about 10,900 years ago, the breaking of a series of ice dams allowed water from Agassiz to pour into Lake Nipigon, which, in turn, overflowed through a series of channels into Lake Superior. The size of these deluges is nearly unimaginable. One by one over time, the ice dams that lay across each of the five channel complexes were breached. Scientists estimate that initial flood bursts, which lasted a year or two, measured some 3.5 million cubic feet per second, the equivalent of twenty-four times the volume of Niagara Falls. As they coursed to Lake Superior, floodwaters cut spillways measuring an average of one-half mile wide and tens of feet deep, filling them from bank to bank. In one period alone, torrential waters, with a volume equivalent to that of present-day Lake Michigan, poured into Lake Superior. The deluge, which lasted nearly two years, may have temporarily raised Superior's lake levels by 164 feet. By 9,900 years ago, these and other floodwater surges had drained Lake Agassiz, lowering levels by nearly 500 feet.

The force of these massive flows is suspected to have eroded down to bedrock a massive barrier of glacial drift that lay across Whitefish Bay on the southeastern end of Superior, thereby opening what is today the modern outlet for the lake through the rapids of Sault Ste. Marie. Floodwaters eventually made their way into the Atlantic Ocean via the St. Lawrence River. So great was the inundation of freshwater that it halted traffic on the great highway of ocean currents that moved heat from equatorial regions to higher latitudes. The disruption sent Earth plummeting into a cold spell known as the Younger Dryas. The temporary deep freeze allowed the glacier to reinvade portions of its former territory, including the Lake Superior basin. Around 9,900 years ago, the ice returned, once again blocking the flow of water between Agassiz and Superior.

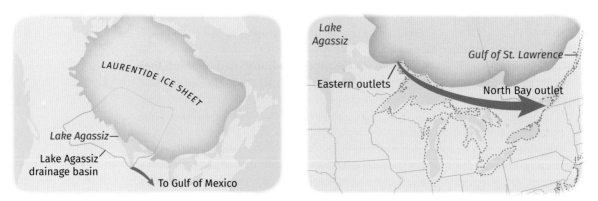

Throughout most of its lifetime, Glacial Lake Agassiz drained south through what is now the Mississippi River *(left)*. But as the Laurentide ice sheet began to recede, it opened other outlets to the north and east. Beginning about 10,900 years ago, the breaking of a series of ice dams allowed water from Agassiz to pour into Lake Nipigon, which in turn overflowed through a series of channels into Lake Superior *(right)*. Torrential flows eventually made their way into the Atlantic Ocean via the St. Lawrence River, resulting in a planetary cold spell known as the Younger Dryas.

But the ice age was quickly coming to an end. By 9,500 years ago, the glacier had melted back again, uncovering the former channels that connected Agassiz with Superior. For a brief time, floodwaters once again surged into Lake Superior. But the waters of Lake Agassiz soon found other spillways as the retreating glacier exposed outlets farther to the north. Freed from the weight of the ice, the land began to rise in a phenomenon known as isostatic rebound. As the Nipigon channels rose higher, they formed a barrier between the world's two great inland seas, permanently shutting off any exchange of water between Lakes Superior and Agassiz around 8,500 years ago.

Filling in the gaps in our knowledge of the lake bottom—whether it is solving the mystery of ring structures or deciphering the calling card of Glacial Lake Agassiz on the floor of modern Lake Superior—is the stuff of intrigue to scientists such as Tom Johnson, who has conducted geological investigations of the lake bottom since the early 1970s. "It's not going to affect the Dow Jones averages at all to know what went on here," he observes. "But it's neat to wonder about the past and about what's now going on at the bottom of the lake. It's the excitement of big unknowns, of discovering what makes this system work the way that it does."

Stirring Things Up: Probing the Lake's Muck and Mire

In late summer 1871, biologist Sidney I. Smith accompanied U.S. Lake Survey crews on a mission to map water depths in Lake Superior's eastern basin. While surveyors dropped sounding lines overboard, Smith lowered dredges that took bites out of the lake floor. His mission: to sift through the bottom sediments and identify the animals that lived in them in an attempt to compare the "food of the fishes with the fauna of the waters which they inhabit." Smith was especially keen to learn more about the tiny insect larvae, worms, clams, and crustaceans that inhabited the lake bottom, since he hoped that they might help to explain why many species of commercially valuable stocks were declining, including the desirable whitefish.

Despite its crude and episodic nature, Smith's sampling yielded some surprisingly accurate insights into life at the lake bottom. In the nearshore zone, the lake floor was a patchwork of cobbles, sand, gravel, and mud. Faunal assemblages differed considerably depending on the substrate they inhabited.

The deepwater regions of the lake, on the other hand, surprised him with their uniformity. Dredges in waters deeper than 180 feet routinely contained clays that varied from location to location only in their ruddy brown or gray colors and proportion of sand mixture. Water temperatures deviated only slightly from 39 degrees Fahrenheit. And not only were there far fewer species living in the deepwater zones as compared to nearshore reaches, but Smith also found the same basic complement of animals from site to site.

Such constancy gave rise to the notion that the lake's profundal zone was a stable, immutable environment whose depths isolated it from the forces of change. For nearly a century, researchers possessed few technological tools to challenge this perception.

Then in 1960, they filed their first eyewitness reports as scuba divers descended to the floor of the Great Lakes to study its underwater geology and to collect biological samples. But the lakes' challenging conditions made such deepwater reconnaissance an expensive, time-consuming, and dangerous job. In the frigid waters of the Great Lakes, divers lose body heat so rapidly that they must increase their daily food intake to about five thousand calories. Protective clothing is vital but cumbersome. For example, without gloves, divers quickly lose dexterity in their fingers. But their bulk hinders even the most basic research functions, such as note taking and manipulating equipment. At extreme depths, even the most adaptive hand gear may be irrelevant since "the effects of nitrogen narcosis considerably reduce man's ability to write legibly at depths in excess of 150 feet," writes geologist Lee H. Somers in a 1967 description of the use of diving techniques for Great Lakes investigation.

In addition to brain fog, poor visibility is another frustrating (and potentially dangerous) limitation for divers. As a result, only features that can be captured by the camera lens at close range are legible in photographs. "It is like doing ordinary dry-land field work, on a cold January night, without a moon, during a dust storm, by the light of a flashlight of variable power," Somers observes. "[The diver] can only see those materials within the range of his flashlight beam. This might be as much as 60 or 70 feet, or as little as six or seven inches. In the latter instance, he would have to work with his face to the ground; fortunately that is a convenient position for the diver."

Perhaps the greatest limitation of scuba-assisted exploration in the Great Lakes is that most of the profundal zone is too deep to access. Lake Superior's average depths are 483 feet, far out of the reach of divers, whose recommended maximum depth is about 250 feet.

With the advent of remote-sensing technologies came a more sweeping and detailed picture of the deep lake bottom. Beginning in the 1950s, scientists began to deploy a whole new array of technologies, many of them adapted from those used to study the world's oceans. Scientists mapped the bathymetry of the lake using sonar technology and used drills to extract long sediment cores from various lake-bottom locations. Their research results presented evidence that the profundal zone was a far more dynamic environment than previously suspected.

During the summers of 1977 and 1978, geologist Tom Johnson of the University of Minnesota and his colleagues obtained seismic-reflection profiles of nearly two thousand miles of Lake Superior's bottom, including large areas off the coast of the Keweenaw Peninsula. The uneven distribution of sediments in the profundal zone provided telling clues that conditions were anything but serene on the lake bottom. Sediments "do not rain down uniformly onto a placid lake-floor environment," the researchers concluded, adding that "lake beds may be more dynamic sedimentary environments than generally believed."

Learning more about the forces at work on the lake floor was no mere academic exercise. By the late 1970s, enough information had amassed to suspect that the deep-water reaches of the Great Lakes had become repositories for chronic and dangerous pollutants. Sediments were identified as one of the means by which problem contaminants such as PCBs and dioxins traveled from the atmosphere and the lake's polluted tributaries into the profundal zones of the lakes.

When PCBs enter lakes, for example, they separate out from the water column by developing strong chemical bonds with organic-rich particles or minute sediments, such as silts and clays. By and large, these materials do not settle for long in shallow water. That is because they are easily caught up in surface waves, which slowly sweep them toward the center of the lake basin. Indeed, in Lake Superior deposits of fine-grained sediments are patchy and thin in waters less than four hundred feet, if they exist at all. But in some of the profundal basins of southeastern Lake Superior, these particles have accumulated to depths of thirty-three feet. Scientists estimate that about 75 percent of the fine sediments that enter Lake Superior are swept into such deepwater zones.

When they drift into the profundal zone, the particles carry with them toxic hitch-hikers, most notably PCBs. With the phaseout of PCB use in the industrial sector, scientists speculated that deepwater areas would become tombs for contaminated particles as they become naturally buried under new layers of sediments. But throughout large portions of the deep lake bottom, as they discovered, sediments not only accumulate but also are routinely eroded. Under these conditions, the profundal zone serves as both a repository and an ongoing source of old pollutants.

The cause of some of these disturbances? Deep and powerful currents that rake across the lake bed. "If bottom currents are strong enough," writes geologist Roger D. Flood, "then lake-bed sediments can be eroded and, along with any pollutant load, be reintroduced into the lake." Given Lake Superior's slow sedimentation rate and the longevity of many pollutants, some of which can persist in the environment for decades, if not centuries, subaqueous erosion can play a major role in the recycling of contaminants.

This discovery revolutionized the way researchers viewed the dynamics of the profundal zone. Scientists have known since the late nineteenth century that currents course through Lake Superior. But they assumed that such currents never penetrated into the deepwater reaches of the lake. Indeed, early studies of the Keweenaw Current, a powerful coastal jet that parallels the northern coastline of the Keweenaw Peninsula, concluded that its flow was confined to the upper one hundred feet of water. In their reconnaissance off the Keweenaw coast, Johnson and his team identified what appeared to be current-modified features on the lake floor at depths well below the maximum recorded for the Keweenaw Current, suggesting that previous measurements had underestimated the current's reach.

Among the features targeted by the team were the ring structures in the lake bed. In some places, most notably in the vicinity of Grand Marais, Minnesota, the deeply indented outlines of many of the rings have not been filled in with sediments. As a result, whole rings are still legible. On the lake bottom some twelve miles off the Keweenaw Peninsula, however, sediments have accumulated in portions of most ring structures such that only half circles remain. Superimposed on these ring structures are a series of parallel lineations scribed into the surface sediments. Similar clusters of lines have been found on the ocean floor where currents are known to be active. Scientists have interpreted these lineations as telltale signs that even in 655 feet of water, currents course along with sufficient strength and regularity to rework the lake bottom and cause significant infill.

Investigations at the base of a slope off the Keweenaw Peninsula provided further evidence that bottom currents, like great underwater plows, were sculpting the lake floor. Seismic-reflection profiles revealed numerous well-developed furrows excavated into the silty lake bed similar to those found on the current-swept portions of the deep-sea floor. Measuring 10 to 16 feet wide and 1.6 feet deep, some furrows were strewn with debris, including twigs, leaves, and the odd paint can or large log, that had been deposited by powerful flows of water.

Researchers got their first in-person look at these and other features in 1985 from the windows of the *Johnson Sea-Link II* submersible. During the summers of 1985 and 1986, the *Sea-Link II* made seventy-eight dives at thirty-one different locations throughout the central and eastern regions of Lake Superior. Capable of carrying four people in dives of up to four hours, the submersible explored the lake floor at depths ranging from 50 to 1,265 feet.

Previous sonar reconnaissance of the rugged underwater topography in the southeastern quadrant of the lake revealed extremely variable patterns of sediment deposition. In some areas, sediments accumulated at a rate of 0.04 to 0.08 inches per year, while other areas were swept clean. Using video cameras mounted on the submersible, researchers documented evidence that such variability existed even at some of the deepest locations in the lake. For example, in a north–south-trending trough in the southeastern quadrant of the lake, whose deepest point lies more than a thousand feet under the lake's surface, scientists found no measurable deposition of recent sediments. When they explored the southern end of the trough, however, they detected ripples on the lake bed indicating that currents coursed through the trough and likely prevented sediment buildup along its length. Curiously, however, meters placed on the lake floor picked up no evidence of active water flows at the time. The question is, what causes the flushing of the trough and when does it occur? "Here's this great big lake," Johnson says,

The *Johnson Sea-Link II* submersible gave researchers their first in-person opportunity to explore the lake floor at depths up to 1,265 feet. Photograph by Joel Baker, Minnesota Sea Grant.

"and it's inhaling and exhaling. Does it make one big puff every ten years or does it clean the trough out every winter?"

Researchers obtained clues about the timing of some deepwater currents from experiments conducted on the lake floor off the Keweenaw coast. They hypothesized that many bottom currents occur seasonally. During the summer, Superior's waters are stratified; that is, cold, deep layers become isolated from warmer surface waters by a sharp density gradient. Most storm waves are not powerful enough to overcome the inertia of stratified waters to reach deep to the lake bottom. Stratification is one reason why the Keweenaw Current remains close to the surface during the summer. Most studies on the lake are undertaken during the summer months, when the lake is relatively calm and safe for research excursions in small boats. But measuring the current's depth only during the summer gives a false impression of its full reach.

To find out what happens in the lake during the rest of the year, scientists on board the submersible placed a current meter about thirty-three feet above lake bottom near the furrows off the Keweenaw coast. Their suspicions—that currents were more active in the fall when stratification breaks down and fierce storms are able to generate strong flows far below the surface—were borne out by the meter readings. Results tallied between August 1986 and August 1987 showed that current speeds slowed considerably during summer stratification to a maximum hourly average of one-half foot per second. During the peak storm season in November, when the lake's waters were isothermal (that is, a uniform temperature from top to bottom and therefore more easily churned up throughout the entire water column), the maximum hourly average of current speeds nearly doubled.

It turns out that currents are not the only source of disturbance on the lake bed. On a sonar reconnaissance off the Keweenaw Peninsula, Johnson's team picked up shallow, dish-shaped structures some 3 to 4 inches in diameter pockmarking the lake floor. Also legible in the soft sediments were numerous V-shaped impressions approximately 9 inches long.

The team suspected that they were biogenic in origin, that is, created by living organisms. But the tiny invertebrates that occupy the lake-bottom sediments—clams, crustaceans, worms, and insect larvae—normally measure no more than 0.39 inches in length. Besides, these creatures largely rework only the top 0.39 to 0.8 inches of sediment, where nearly all the organic nutrients are concentrated.

Using cameras lowered by cable to the lake bottom, Johnson and his team captured images of the occasional fish, probably the bottom-dwelling sculpin, lying in the spherical hollows. The scientists guessed that the V-shaped gouges in the soft sediments also were created by sculpins. In their peculiar pattern of locomotion, the fish hop, rather than swim, along the bottom. In a resting position, they use their pectoral fins to anchor themselves in the silty lake bed, leaving behind chevron-shaped impressions as their calling cards.

But scientists realized the full extent to which living organisms disturbed deepwater sediments (a process known as bioturbation) only after traveling to the lake bottom in the *Johnson Sea-Link II*. For the first time, researchers observed the burbot, a freshwater cod that occupies the deepest recesses of the lake, in its native habitat. Long hailed as the top predator of the profundal zone, the fish earned a new—and much deserved—reputation as the master architect of the deep. During their dive, scientists videotaped

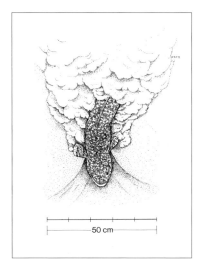

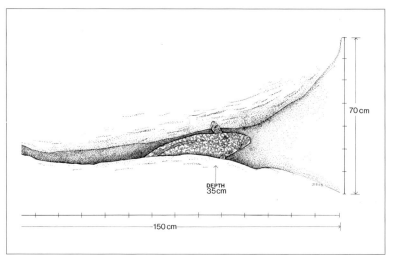

Burbot, the top predators of the lake bottom, fan their pectoral fins in the soft sediments to excavate trenches. Illustrations by Richard A. Cooper, Undersea Research Center, University of Connecticut.

three different structures created by burbot. Among the most common were trenches dug into the lake bottom. To create them, burbot use their pectoral fins as grappling hooks in the soft sediments. Once anchored, they can excavate a depression by fanning sediments with their bodies. The stylized trenches they create routinely include a flared front opening with numerous rear exits.

Researchers also observed burbot swimming back and forth under bank overhangs to create crescent-like undercuts or grooves. Perhaps the most spectacular, however, were burrow structures honeycombed in the banks of clay cliffs. Their structural complexity led researchers to compare them to the clustered architecture of swallow nests. Since the walls of the burrows are excavated into soft clay, which is prone to slumping, scientists speculate that burbot must invest considerable time and energy in their maintenance. Such diligence is likely to pay important dividends. Since adult burbot are not known to have any predators, the fish probably do not use their biogenic infrastructure for cover but as funnels for prey, which include mysid shrimp and the bottom-hugging sculpin.

Voyagers aboard the *Johnson Sea-Link II* returned to the surface with both stunning and disconcerting news. The far reaches of Lake Superior amazed researchers with their abundance of aquatic life, whose richness far exceeded anyone's expectations. Yet stored here, even in the most remote depths of the lake, is the chemical evidence of many decades of human folly, which will continue to swirl in the darkness for generations to come.

Where East Meets West in the Bottom of Lake Superior

Between 1892 and 1894, Mark Harrington of the Weather Bureau, U.S. Department of Agriculture, released 5,000 blown-glass bottles of "unusual color" into the Great Lakes. Measuring 7 by 2.5 inches, each bottle was weighted with sand and corked so that it floated just below the surface of the water and out of the reach of the wind. Each bottle also contained a message. Inside was a small map of the lake noting the date, latitude, and longitude of the bottle's release and a printed card that asked finders to record the pickup location. Enclosed for their convenience was a stamped envelope. In 1895, based on cards mailed in by obliging strangers, supplemented by interviews with sailors and commercial fishermen, the Weather Bureau published the first comprehensive survey of currents in the Great Lakes.

To draft these maps, Harrington utilized the best available technology of his time. (Indeed, scientists studying currents in Lake Superior used variations of this drifter technique—albeit a more technologically souped-up version—as recently as the late 1990s.) Nonetheless, as scientific experiments go, his study was seriously flawed. Of the 5,000 bottles that were released in the lakes, only 672 were recovered, 101 of them in Lake Superior. Some sank upon release, while others were ground to pieces on rocky beaches and cliffs. Countless bottles no doubt simply languished unopened on sparsely populated or inaccessible shores. In some cases, nature simply refused to cooperate. None of the 250 bottles floated between Thunder Bay and Duluth in Lake Superior, for example, were retrieved. The prevailing offshore winds and frequent upwellings along this stretch of coastline likely kept bottles from stranding. And given the long stretches of rocky headlands, many drifters that did make it to shore probably did not survive intact.

From the bottles that were recovered, Harrington made several calculations,

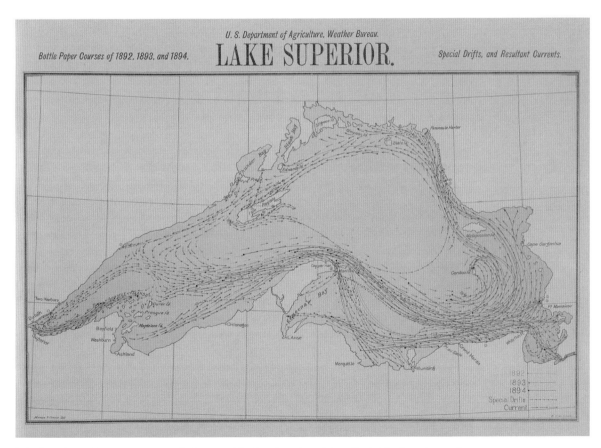

Lake Superior surface currents were first mapped in 1895, based on data collected by the release of 5,000 blown-glass bottles between 1892 and 1894, 672 of which were returned.

including the pathways and minimum velocities of currents. It was, to say the least, an exercise in fuzzy math.

But while Harrington may have miscalculated many of the details, he did correctly identify some general principles about the movement of currents in the Great Lakes. For example, Harrington showed that currents primarily move in a counterclockwise direction around the lake.

Harrington also made plain the fact that the lake's waters were anything but a featureless, uniform system. He noted a current hugging the southern shore of the lake, for example, a powerful coastal jet that later became known as the Keweenaw Current. According to physical oceanographers, it is the strongest current in the Great Lakes and, quite possibly, the strongest of any lake in the world. Remarkably, Harrington's crude investigations not only detected this pronounced flow but also revealed far more subtle features of the lake's circulation pattern. He provided graphic illustration that lake currents, like rivers in the terrestrial realm, spawn eddies and gyres and merge with other currents, creating a series of microhabitats in what on the surface can look like an undifferentiated body of water.

For three-quarters of a century, Harrington's map served as the only comprehensive investigation of currents in Lake Superior. Such neglect of even the lake's most fundamental processes was not unusual. Many aspects of physical limnology and geology, including bottom sediments, their underlying bedrock topography, and the lake's water circulation patterns, would not be studied in greater detail until the 1950s. Within the next two decades, however, the number of published studies on these topics suddenly spiked.

Some research was undertaken for purely commercial purposes. The discovery in 1967 of ferromanganese (iron-manganese) nodules on the floor of Lake Michigan's Green Bay, for example, sparked an interest in exploring the underwater mining potential of these minerals throughout the Great Lakes. In Lake Superior, the exhaustion of copper reserves in the Keweenaw copper-producing district prompted researchers to investigate the waters off the Keweenaw Peninsula for lode copper in the lake bottom's bedrock.

What galvanized the scientific community more than any other issue, however, was the growing eutrophication of the Great Lakes and the collapse of their once-thriving fisheries. According to the U.S. Public Health Service, by 1960 twenty-three million people—14 percent of the population of the United States—lived in the Great Lakes basin. With the exception of the city of Chicago, communities discharged their nutrient-rich sewage into the lakes or their connecting waterways. Industries followed suit, fouling waters with persistent chemicals that became lodged in the food web. It quickly became apparent that both the ecological health of the lakes and the well-being of the people who depended on their waters were imperiled.

Committed to reversing the degradation of the waters along their borders, in 1972 the United States and Canada signed the Great Lakes Water Quality Agreement. "It was a historical event for Great Lakes studies," writes biologist Mohiuddin Munawar, particularly for the much-neglected Lake Superior. One of the practical outcomes of the treaty was the initiation of a binational effort to investigate the effects of economic and land-use activities on the upper lakes, which included Lakes Superior and Huron.

In 1973 a consortium of U.S. and Canadian researchers undertook the first lake-wide, cross-disciplinary study of Superior in an effort to obtain baseline readings of the most biologically and chemically intact lake before it too was altered beyond recognition. In 1978 the *Journal of Great Lakes Research* devoted an unprecedented two issues to presenting their findings. Prior to this point, observes Munawar, who served as the publication's editor, knowledge of Lake Superior had been "purely descriptive and [had] lacked scientific interpretation. Moreover, lakewide surveys of Lake Superior, for various disciplines of limnology, [had] been sadly lacking."

Researchers planned to repeat their measurements of Superior's chemistry, biology, geology, and physics every ten years, but the funds for additional comprehensive surveys never materialized. Instead, scientists would add to the knowledge about the lake's circulatory and metabolic systems in more piecemeal research. Glaring gaps confronted them at every turn. For example, scientists had little idea how contaminants once they entered the lake were dispersed or how the lake processed them via physical action or biological activity.

Still, they had their suspicions. More than a century ago, residents along the shores of the Great Lakes learned—the hard way—that currents play a critical role in the movement of pollution. In the 1890s, the city of Duluth, like most other burgeoning towns on the Great Lakes, took the "dilution is the solution" approach to the problem of sewage disposal—they piped wastewater offshore in the hopes that contaminants would be reduced to negligible levels after mixing with huge volumes of clean water in the open lake. To ensure that pathogen-laced sewage would not contaminate intake pipes for drinking water, sanitarians located siphons 2.5 miles north of the nearest sewage outlet. Despite such precautions, the city suffered from mass outbreaks of waterborne diseases, including one typhoid episode that in 1896 killed a record 2,020 citizens. What went wrong?

Officials discovered that wind action and currents could periodically change flows such that bacteria-rich waters backwashed into drinking-water pipes. A 1929 report from the Minnesota State Board of Health observed: "When the water is piled up in the [western] end of the lake by wind action, seiches, etc., it will tend to rush out through the channel along the Minnesota shore line. This will carry pollution from the harbor and all Duluth sewer outlets past the [Lakewood Water Station] intake." Waterborne epidemics abated when the city instituted chlorine treatment in 1912. But while chemical treatments could combat microbial pollution, they could not cleanse water of the growing load of industrial waste that poured into the Duluth-Superior harbor from upstream paper mills on the St. Louis River and industrial plants along the harbor's shores.

Yet despite the need to more clearly understand the behavior of currents, few further inquiries were made into the fundamentals of water movement and temperature in western Lake Superior. When Theodore Olson of the University of Minnesota's School of Public Health reopened limnological investigations in the mid-1950s, he decried the "limited character of the available information and the great void of knowledge," declaring that "the surface had merely been scratched." At stake, Olson pointed out, was not simply human health but also Minnesota's economic well-being. Lake Superior is a water resource with "an extremely important bearing on the long range economy of Minnesota," he wrote in 1958. "At a time when all the world is plagued with problems relating to pollution of water and the adequacy of water needed for irrigation projects, domestic use, industrial purposes, and recreation, a large body of clean fresh water such as Superior is a natural asset which cannot be squandered. To all this we can add the value of potential commercial fisheries and the beneficial effects which such a large mass of water may have on the climate."

In summer 1956, Olson conducted research in the lake's west end on behalf of the Minnesota Water Pollution Control Commission. Part of his goal was to amass enough water-circulation data so that officials could craft what he called "an intelligent approach to solving the problems of waste disposal." Olson's report concurred with studies conducted decades before, warning that "periods may occur during the year when weather and currents, combined in a different sequence, may establish a set of circumstances under which materials discharged from the Duluth-Superior jetties will remain concentrated in the western end of the lake." At the same time, he discovered evidence of physical forces that kept these discharges from stagnating. He sampled bacterial populations as well as conducted water-visibility tests to determine levels of suspended or dissolved materials. The results showed that harbor effluents were rapidly diluted within one mile of shore. Olson hypothesized that the "great volumes of water moved in and out of the area by lake currents exert a continual flushing effect."

But scientists soon realized that relying on currents to whisk contaminants out into the open lake was not the solution to pollution. The waters of the Great Lakes, they discovered, were not giant sponges that could safely absorb pollutants indefinitely; rather, the lakes were more akin to bathtubs whose contents were drained and replenished one tiny drop at a time. When persistent contaminants, that is, those that resist breakdown into harmless components, enter the lake and are incorporated into the tissues of the lake's biota or into its sediments, they can be cycled and recycled throughout the system for decades, centuries, and, quite possibly in some cases, the lifetime of the lake. The problem is especially troubling in Lake Superior, whose waters turn over only once every 190 years. Because the narrow outflow at Sault Ste. Marie restricts the volume of water that can drain into the lower lakes, Lake Superior has the longest retention time of all the Great Lakes.

The problems resulting from this closed-loop circuitry became clear in the case of taconite tailings that were discharged by Reserve Mining into the lake from the company's ore-processing facility in Silver Bay. Between 1955 and 1980, until halted by court order, Reserve Mining dumped some five hundred million tons of tailings (the finely crushed matrix of rock from which iron ore has been magnetically removed). Experts say one-half of the tailings—the coarse, heavier particles—settled out in the nearshore zone to form an underwater delta whose surface area covers more than 0.5 square miles. The remainder, composed primarily of fine-grained materials, settled over a 386-square-mile area at the bottom of a deep trench that lies just off the coast.

Officials anticipated that the majority of tailings would not travel far from their point of deposition. The fine fragments, they believed, would quickly drift into the thousand-foot-deep trough off the Minnesota shore, followed in time by the heavier, more slowly eroded grains. In the abyssal reaches of the lake, researchers contended that the tailings would undergo a gradual burial by new tailings or the lake's natural process of sedimentation, far from the churning of more turbulent shallow waters.

But tests of Duluth's drinking-water supply taken in 1969 showed that currents had carried fine tailings particles 50 miles southwest into the western end of Superior, where significant amounts slipped through municipal water filters and into household taps. More surprising still was the detection in April 1975 of similar asbestiform fibers in the municipal drinking-water supply of Thunder Bay, Canada, which is located 132 miles northeast of Silver Bay.

Such findings prompted scientists in the 1960s and 1970s to revisit Harrington's preliminary work on the pattern of surface currents within the entire lake. One of the most sustained investigations was carried out by Robert A. Ragotzkie of the University of Wisconsin–Madison's Department of Meteorology and his colleagues. Realizing that a key to a holistic understanding of Lake Superior's currents lay in learning more about the Keweenaw Current, Ragotzkie published the first in-depth profile of the current in 1966. (Scientists speculate that the asbestiform fibers found in Thunder Bay's drinking water were carried by currents along the Minnesota North Shore into the western end of the lake near the Apostle Islands from which they were slowly funneled into the Keweenaw Current and transported via other currents northwest to Thunder Bay.)

But the studies, while they yielded startling insights into behavior of the current, offered only a snapshot view of its workings. Fieldwork was largely restricted to the more clement summer months, when small research vessels could be operated safely on the lake. The limited technological tools at their disposal also made research expensive and time-consuming. Because the lake's nearshore surface waters are considerably warmer than those of the open lake (a difference of up to 14 degrees Fahrenheit), the scientists were able to study portions of the current's flow using temperature recorders and current meters fixed to buoys or by using shipboard instruments. But to get a more comprehensive view of its workings, researchers took to the air, flying laborious missions to obtain temperature readings from aircraft-mounted radiation thermometers.

It was not until 1998, when the National Science Foundation awarded a $5.3 million research grant to a multidisciplinary team of fifteen scientists that the Keweenaw Current would receive sustained, year-round scrutiny. Known as the Keweenaw Interdisciplinary Transport Experiment (KITES), the project not only benefited from a substantial federal grant but also from a whole array of remote-sensing technologies that would have seemed luxurious, if not a bit like science fiction, to the scientists who pioneered studies of the Keweenaw Current in the 1960s. For example, each hour buoys record the temperature of the air and water surface, sea-level pressure, wave heights, and the direction and speed of the wind. At fifteen-minute intervals, underwater moorings register the speed and direction of the current as well as its temperatures from the lake floor to the water's surface with an accuracy of 0.0018 degrees Fahrenheit. Nets set up in the water column collect a range of organisms transported by the current, from microscopic bacteria and phytoplankton to larval-stage fish. Sediment traps capture suspended particles for later chemical analysis.

Surface-water temperatures that were once measured in limited intervals by plane-mounted thermometers are now remotely monitored twice daily by the Advanced Very High Resolution Radiometer (AVHRR) instrument, a thermal-infrared sensor mounted on a polar-orbiting government weather satellite. Another satellite-mounted sensing device collects data on the turbidity of the current's waters by measuring such characteristics as sediment density.

With the wide array of technologies has come an equally extensive roster of scientific specialties, with researchers from microbiologists and fisheries scientists to physical oceanographers and limnologists all contributing their respective expertise to the puzzle. Why is the current of such concern?

Like blood in the circulatory system of the human body, currents determine how nutrients—and the tiny organisms such as plankton and larval fish that seek them out— are distributed around the lake. "Knowing how the blood flows helps in the care of the body," says Elise Ralph, a physical oceanographer at the University of Minnesota Duluth, and one of the scientists on the KITES team.

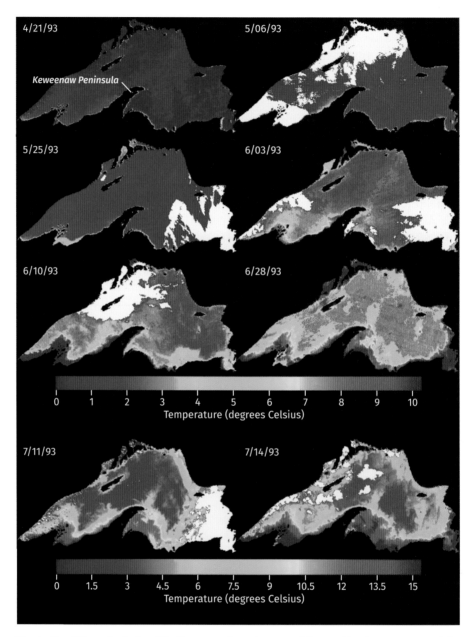

The depth, width, and velocity of the Keweenaw Current, a powerful coastal jet that wraps around the Keweenaw Peninsula on Superior's southern shore, change with the seasons. In spring and summer, as warming waters from the shallow western basin drift to the east, the current develops a pronounced flow, visible in the June and July satellite images as a bright orange band. Courtesy of Judith Wells Budd.

The Keweenaw Current's power and magnitude suggest that it plays a major role in the ecological workings of the lake. Measuring an average of one-half mile wide, at its peak flow the Keweenaw Current carries as much volume of water as the outflow of the Mississippi River into the Gulf of Mexico—some seventy billion gallons per day. It travels at an average speed of one foot per second, although researchers have clocked maximum velocities of three feet per second, which match average speeds of the Gulf Stream in the Atlantic Ocean.

Such powerful flows broker biological and chemical exchanges across the borders of distant—and sometimes highly segregated—aquatic realms. The current "acts like a funnel," observes KITES director Sarah Green, a professor of chemistry at Michigan Technological University, Houghton. "It can broadcast larval fish, zooplankton eggs and other organisms. It has a scouring and depositing action."

During the summer, warm waters from shallow areas within the western basin slowly drift to the east (the warmest waters originate in the Apostle Islands). At the base of the Keweenaw Peninsula, these waters form a pronounced flow of water whose outer boundary parallels the Keweenaw coastline within 2 to 6 miles of shore. As it courses up the northeastern shore of the Keweenaw Peninsula, the current generally wraps around the tip and takes a southeasterly path into the lake's eastern half. This great aquatic highway links the eastern and western basins of the lake, which span a distance of some 350 miles. According to Ralph, the Keweenaw Current is the "connective artery that communicates what happens in western Lake Superior with eastern Lake Superior."

At the same time, the current's flow sheds eddies or occasionally is deflected into the central lake, where it also makes a vital connection between nearshore and offshore zones. Such conduits are especially important in Lake Superior. More so than the other Great Lakes, Superior is characterized by a relatively narrow nearshore area that drops off precipitously into deep water close to shore. A variety of factors—from differential heating of nearshore and offshore waters and lake-bottom topography to prevailing wind patterns—work to keep the lake's nearshore reaches and its profundal zone highly segregated throughout much of the year. "Of all the Laurentian Great Lakes," say Lake Superior researchers W. Charles Kerfoot and John A. Robbins, "Lake Superior contains the strongest development of a separate coastal regime, chemically and biologically distinct from cooler offshore waters."

Indeed, the sharp demarcation between nearshore and offshore zones has a lot to do with why the current forms along the Keweenaw Peninsula in the first place. Near the shoreline, a coastal shelf with shallower waters suddenly drops off into deep water, resulting in an abrupt bathymetric shift. The Keweenaw Current shows up as a coastal flow of warmer, and therefore more buoyant and faster-moving, water flanked on its lake side by colder, more sluggish offshore waters.

Bathymetry also dictates many of the current's behavioral characteristics. The current quickens and slackens depending on the width of the coastal track in which it flows. In the stretch from Eagle Harbor to Copper Harbor, for example, the lake floor disappears into depths of 655 feet within only one mile of shore, forming one of the steepest coastal drop-offs in the Superior. Here, constrained between the severe bottom slope and land, the current narrows and gathers strength. But where the lake bottom slopes more gently off the Keweenaw Peninsula, the current flattens, and its vigor slackens.

But like props that compose a stage set, bathymetry and lake-bottom topography serve as the background fixtures for a living drama that unfolds in seasonal acts. In early summer, a phenomenon known as a thermal bar develops in Lake Superior. Often visible on the surface of the water as a line that collects foam and debris, the thermal bar marks the place where the water temperature is 39 degrees Fahrenheit. At this temperature, water reaches its maximum density and begins to sink, causing the surrounding water to flow toward the bar to replace the sinking waters. This action forms a kind of vertical curtain that parallels the shore. In early summer, the thermal bar takes shape near shore, gradually moving offshore into deeper water as the season progresses. The warming of the whole lake surface eventually causes it to dissipate.

During this time, the current whipsaws between the coast and the thermal bar depending on the direction of the wind. Southerly or westerly winds, for example, can push warmer surface waters of the current away from the coast. When that happens, cold water rises to replace it, a phenomenon known as upwelling. Winds blowing from the north or east can jam the current up against the coast, where waters are forced downwards, causing what is known as a downwelling event. Such occurrences facilitate abrupt chemical and biological exchanges throughout the water column. Downwelling events, for example, can plunge into the deep water organisms such as phytoplankton, whose reliance on light for photosynthesis normally relegates them to the upper zones that are penetrated by sunlight.

In studies of the current conducted during July 1973, Ragotzkie and his colleagues discovered an uncanny rhythm to these hydrologic events that corresponded to the passage of high and low pressure systems through the region. These systems, which alternated every four to six days, caused a corresponding alternation in upwelling and downwelling episodes, such that the current became what the researchers called a "pulsating coastal jet."

As the season progresses, another thermal phenomenon develops: a horizontal separation known as stratification sets in between the warmer surface waters and the deeper and colder reaches of the water column. Winds are able to easily push this buoyant upper layer, which in summer is largely confined to the upper one hundred feet of water. Not surprisingly, the greatest average speeds for the Keweenaw Current have been clocked during this season.

In fall, however, this temperature stratification breaks down and lake waters become isothermal. In the 1980s, scientists discovered that the current expands its vertical reach during this time. Encountering less inertia in deep waters during isothermal conditions, the energy of strong winds, particularly during intense fall storms, is able to reach deep into the water column and accelerate current velocities there. The vigor of these bottom flows spikes during late October through early February, when prevailing westerly winds build up steam over long fetches of open water. During this season, researchers have documented disturbances of lake-bottom sediment as far down as two hundred feet. Stirred up and caught in the current's bottom flows, which have been clocked at speeds of nearly one foot per second, organic and inorganic materials can be transported far into the eastern reaches of the lake. "Water in the lake is constantly moving," Ralph says. "It's not random and, most important, it's not one-dimensional."

Such processes, while they facilitate vital exchanges of food and chemical nutrients, also perform a darker function: They help to spread toxic pollutants and invasive species. The Duluth-Superior harbor, for example, is a major port of entry for cargo from around the world—including plant and animal hitchhikers that are introduced to the lake through discharges of ballast water.

The Keweenaw Current serves as an aquatic freeway on which contaminants, as well as living organisms, can travel to penetrate some of Superior's most pristine reaches. Some contaminants move in ingenious ways. The International Joint Commission estimates that 95 percent of persistent, toxic pollutants enter Lake Superior in the falling of dust, snow, and rain. The heavily eroded cliffs along the Wisconsin shoreline provide a major source of fine clay particles to which some pollutants, such as PCBs, bind and then hitchhike around the lake via currents. Once caught up in Superior's water flows, such contaminants can be carried vast distances.

But currents also can remobilize pollutants that were deposited in the lake decades—even a century—ago, such that Superior becomes the agent of its own poisoning. Take copper, for example. Researchers have detected elevated levels of copper in many places around the lake. Despite the fact that copper emissions into the atmosphere have increased five hundred-fold worldwide between 1900 and the mid-1980s, scientists have determined that less than 10 percent of the excess copper found in the lake comes from the sky. The majority is derived from local sources, including present-day, as well as historic, mining operations. According to researcher Kerfoot, tailings piles in and around the Keweenaw Peninsula can be blamed for a large portion of the ongoing copper pollution of the lake.

These spoils are the legacy of North America's first great mining boom. Between 1850 and 1929, the peninsula was second largest producer of copper in the world. By the time the mines closed in 1968, the area had exported 5.3 million tons of copper.

Environmental havoc resulted not from the amount of ore removed but from the heaps of copper-rich rock left behind. Mining operations no sooner sank their first shafts than the easily accessible high-grade veins were exhausted. As a result, companies began to exploit poorer-grade deposits, which contained copper at concentrations that ranged from 0.5 and 6.1 percent of the total mass. To separate the ore from its matrix of basalt and conglomerate, steam-driven presses pulverized great quantities of rock. Left behind were huge piles of waste rock known as stamp sands, whose fractions ranged from the size of sand grains to extremely fine claylike particles called slime clay. From 1850 to 1968, some five hundred million tons of stamp sands were either funneled into the waters of Lake Superior and interior waterways and lakes or piled along their shores. Near Freda and Redridge on the western shores of the peninsula, fifty million tons of waste rock were unloaded into Superior; near Gay on the peninsula's eastern shores some twenty-five million tons were discharged into the lake, becoming one of the biggest dumping sites for stamp-mill wastes from 1902 to 1932.

The stamp sands were sluiced into coastal waters less than 164 feet deep, an area known as the scour zone. Physical characteristics—their distinctive color and high concentrations of metals, among other attributes—have allowed researchers to distinguish the mine tailings from natural lake sediments and to trace their pathways through the lake. What was the fate of this material?

The high-energy nearshore zone with its scouring currents, pounding storm waves, and powerful upwellings eroded particles that had settled in shallow waters and dispersed them at great distances from their point of origin. Kerfoot points out that nearly all the coarse-grained waste rock from the Freda-Redridge operation was swept fourteen miles northeastward by the Keweenaw Current and redeposited on the sandy lake floor in shallow waters. Likewise, a major portion of the Gay stamp-sand pile was carried five miles south, where it washed up as a black-sand beach. Sediment cores taken from the deepwater reaches of the Caribou Basin indicate that the finer slime clays were carried much farther into the lake, settling out in the lake's deep basins.

In the meantime, Superior's handiwork is far from complete. The waste rock in underwater sand bars continues to be further dislodged and redeposited or washed up on beaches. New copper-laden particles are routinely introduced into nearshore waters as rain and snowmelt, along with groundwater leaching, erode the stamp-sand piles that mining operations heaped onto shorelands.

But flowing water is not the only agent of change at work. In winter, nearshore ice may freeze throughout the water column, becoming anchored in bottom sediments. When the solid pack ice breaks up during storms or floes are carried away by currents during the spring melt, sediments embedded in the ice can be transported long distances. Along the way, their keels also rake bottom sediments, stirring up particles that can be caught up in currents. Scuba divers have recorded ice keels that have penetrated depths of nearly one hundred feet to score the lake bottom.

The surface of lake ice also serves as rafts for materials that are deposited by storm waves, eroded from bluffs, blown by the wind, or carried by tributary rivers and streams. Scientists hypothesize that this is one means by which copper has moved into the deep basins of the lake. Satellite monitoring of Superior's ice cover has shown that the Keweenaw Current picks up steam in the spring under the influence of strong westerly winds, causing ice floes to drift past the tip of the Keweenaw Peninsula.

Little is known about the long-term effects of copper pollution in the lake. In fact, the first detailed study of the nearshore sediments and benthic organisms around Freda and Redridge, the site of a major tailings dump, was not published until 1979. Why has copper in the Superior environment received so little attention, particularly after a complex of water bodies on the Keweenaw Peninsula that were used as tailings dumps were declared a federal Superfund site in 1984?

One reason is that high concentrations of copper pose few direct threats to humans. In mammals, excess copper is only moderately toxic. And unlike mercury and lead, copper does not accumulate to dangerous levels in fish.

But copper does threaten organisms lower on the food chain, such as plankton, invertebrates, and larval fish, when chemical and biological weathering transforms the inert metal-bearing minerals to more soluble forms. When exposed to air and water, copper first becomes oxidized to cuprite. Cuprite is then further oxidized to copper and malachite, which are then released into the water. In studies of sediments and stamp sands around the Keweenaw Peninsula, researchers found concentrations of copper that were one hundred to one thousand times greater than those found in soils uncontaminated by mining activities.

Copper is of concern in aquatic ecosystems because of its high toxicity to phytoplankton. In fact, it has long been used as an algicide. Metals such as copper can be so toxic to plants that mine tailings piles often remain unvegetated for decades or longer.

Evidence points to an impairment of invertebrate species as well. In the inland streams and lakes and the nearshore areas of Lake Superior where elevated levels of copper have been detected, populations of benthic animals including insect larvae, zooplankton, and bacteria are reduced and in some cases eliminated. Even in places where copper concentrations registered far lower than those found in stamp-sand piles, organisms are severely affected. In a 1999 study that examined a large area off the west coast of the Keweenaw Peninsula, scientists found that amphipods were entirely absent from a sixteen-square-mile area near the Freda-Redridge dump site, and their numbers were severely reduced in the surrounding lake bottom. Such population impacts are of great concern to scientists since amphipods, which live on the lake bottom and ingest copper-laced sediments, are the dominant benthic macroinvertebrate in the Great Lakes and a mainstay of the entire food chain.

Once copper reaches the deep midlake zones, it can be effectively sequestered in sediments and its future effects minimized. But it may take decades, even longer, for copper to be safely buried on the lake bottom. In the meantime, it continues to be recycled. Plankton, particularly phytoplankton, are quick to scavenge copper from the water column. When these plants and animals die, their remains settle to the lake bottom, where they become incorporated into the sediments. As microbes degrade the organic matter of dead biota, some of the copper that has been bound up with this organic matter is released. But scientists estimate that less than 10 percent of this is remobilized. If left undisturbed, the majority of copper is buried over time as layers of sediment gradually accrue on the lake bottom and remove it from reach of benthic organisms that occupy the top layers of sediment on the lake bottom.

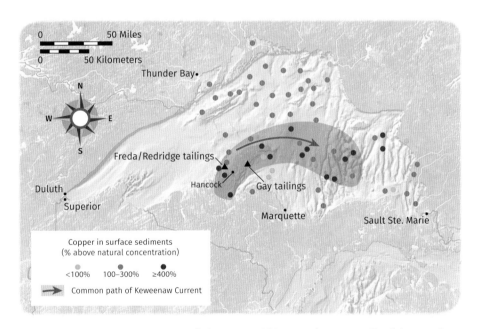

The Keweenaw Current serves as an aquatic freeway on which contaminants, as well as living organisms, can travel widely. Tailings piles in and around the Keweenaw Peninsula are the source for a large portion of the ongoing copper pollution of the whole lake. A 1979 study of surface sediments in the Caribou Basin demonstrated that high levels of copper enrichment closely corresponded with a common track of the Keweenaw Current.

Lake ice can serve as rafts for materials that are deposited by storm waves, eroded from bluffs, blown by the wind, or carried by tributary rivers and streams. Scientists hypothesize that pancake ice like this can also carry associated contaminants into the far reaches of the lake. NOAA Great Lakes Environmental Research Laboratory.

But for decades, even centuries, to come, ice floes and waves will continue to erode old tailings piles along the shores of Lake Superior, and currents will carry them around the lake, where, Kerfoot says, mining spoils "potentially threaten some of the most important fish rearing grounds and productive nearshore benthic habitats."

The studies on Superior's currents have given us essential insights into how to change our land-use practices around the lake. They demonstrate that harmful pollutants are not safely diluted in the great expanse of Superior's waters but funneled on vast arterial highways into its very heart—the nurseries and pastures that nurture life in the lake. What we know—and may learn in the future—about the movements of Superior's currents argue that we exercise extreme caution and care. ❧

SUGGESTIONS FOR FURTHER READING

Adams, C. E., and R. D. Kregear. "Sedimentary and Faunal Environments of Eastern Lake Superior." In *Proceedings of the 12th Conference on Great Lakes Research,* 1–20. Ann Arbor: International Association of Great Lakes Research, 1969.

Beeton, Alfred M., James H. Johnson, and Stanford H. Smith. *Lake Superior Limnological Data 1951–1957.* Special Scientific Report, Fisheries No. 297. Washington, D.C.: U.S. Fish and Wildlife Service, 1959.

Beeton, Alfred M., and R. Stephen Schneider. "A Century of Great Lakes Research at the University of Michigan." *Journal of Great Lakes Research* 24, no. 3 (1998): 495–517.

Berkson, Jonathan M. "Possible Syneresis Origin of Valleys on the Floor of Lake Superior." *Nature* 245 (1973): 89–91.

Berkson, Jonathan M., and C. S. Clay. "Microphysiography and Possible Iceberg Grooves on the Floor of Western Lake Superior." *Geological Society of America Bulletin* 84 (1973): 1315–28.

Boyer, Larry F., Richard A. Cooper, David T. Long, and Timothy M. Askew. "Burbot *(Lota lota)* Biogenic Sedimentary Structures in Lake Superior." *Journal of Great Lakes Research* 15, no. 1 (1989): 174–85.

Boyer, Larry F., and Robert B. Whitlatch. "*In Situ* Studies of Organism-Sediment Relationships in the Caribou Island Basin, Lake Superior." *Journal of Great Lakes Research* 15, no. 1 (1989): 147–55.

Broecker, Wallace S., Michael Andree, W. Wolfli, Hans Oeschger, Georges Bonani, James Kennett, and D. Peteet. "The Chronology of the Last Deglaciation: Implications to the Cause of the Younger Dryas Event." *Paleoceanography* 3 (1988): 1–19.

Callender, Edward. "The Economic Potential of Ferromanganese Nodules in the Great Lakes." In *Proceedings of the Sixth Forum on Geology of Industrial Materials*. Ann Arbor, Mich., 1970.

Cartwright, Joseph, Nigel Wattrus, Deborah Rausch, and Alastair Bolton. "Recognition of an Early Holocene Polygonal Fault System in Lake Superior: Implications for the Compaction of Fine-Grained Sediments." *Geology* 32 (2004): 253–56.

Clayton, Lee. "Chronology of Lake Agassiz Drainage to Lake Superior." In *Glacial Lake Agassiz,* ed. James T. Teller and Lee Clayton, 291–307. Geological Association of Canada Special Paper 26, 1983.

Comstock, Cyrus B. *Report upon the Primary Triangulation of the U.S. Lake Survey.* Professional Papers of the Corps of Engineers, U.S. Army, no. 24, 1882.

Dell, Carol I. "Sediment Distribution and Bottom Topography of Southeastern Lake Superior." *Journal of Great Lakes Research* 2, no. 1 (1976): 164–76.

Evans, James E., Thomas C. Johnson, E. C. Alexander Jr., Richard S. Lively, and Steven J. Eisenreich. "Sedimentation Rates and Depositional Processes in Lake Superior from 210Pb Geochronology." *Journal of Great Lakes Research* 7, no. 3 (1981): 299–310.

Farrand, William R. "The Quaternary History of Lake Superior." In *Proceedings of the 12th Conference on Great Lakes Research,* 181–97. Ann Arbor: International Association of Great Lakes Research, 1969.

Farrand, William R., and Christopher W. Drexler. "Late Wisconsinan and Holocene History of the Lake Superior Basin." In *Quaternary Evolution of the Great Lakes,* ed. Paul F. Karrow and Parker E. Calkin, 17–32. Geological Association of Canada, Special Paper 30. St. John's, Newfoundland: Geological Association of Canada, 1985.

Flood, Roger D. "Submersible Studies of Current-Modified Bottom Topography in Lake Superior." *Journal of Great Lakes Research* 15, no. 1 (1989): 3–14.

Flood, Roger D., and Thomas C. Johnson. "Side-Scan Targets in Lake Superior—Evidence for Bedforms and Sediment Transport." *Sedimentology* 31 (1984): 311–33.

Green, Sarah A., and Elise A. Ralph. "Current beneath the Waves: Studying the Keweenaw's Water Movement." *Lake Superior Magazine,* February/March 2000: 35–38.

Harrington, Mark W. "*Surface Currents of the Great Lakes, as Deduced from the Movements of Bottle Papers during the Seasons of 1892, 1893 and 1894.* Washington, D.C.: Weather Bureau, U.S. Department of Agriculture, 1895.

Heidenreich, Conrad E. "Mapping the Great Lakes/The Period of Exploration, 1603–1700." *Cartographica* 17, no. 3 (1980): 32–64.

Henson, E. Bennette. "A Review of Great Lakes Benthos Research." Great Lakes Research Division, publication no. 14. Ann Arbor: University of Michigan, 1966.

Hughes, John D., John P. Farrell, and Edward C. Monahan. "Drift-Bottle Study of the Surface Currents of Lake Superior." *Michigan Academician* 3 (1971): 25–31.

Johnson, Thomas C. "Late-Glacial and Postglacial Sedimentation in Lake Superior Based on Seismic-Reflection Profiles." *Quaternary Research* 13 (1980): 380–91.

Johnson, Thomas C., Thomas W. Carlson, and James E. Evans. "Contourites in Lake Superior." *Geology* 8 (1980): 437–41.

Johnson, Thomas C., John D. Halfman, William H. Busch, and Roger D. Flood. "Effects of Bottom Currents and Fish on Sedimentation in a Deep-Water, Lacustrine Environment." *Geological Society of America Bulletin* 95 (1984): 1425–36.

Kerfoot, W. Charles, Sandra Harting, Ronald Rossmann, and John A. Robbins. "Anthropogenic Copper Inventories and Mercury Profiles from Lake Superior: Evidence for Mining Impacts." *Journal of Great Lakes Research* 25, no. 4 (1999): 663–82.

Kerfoot, W. Charles, and George Lauster. "Paleolimnological Study of Copper Mining around Lake Superior: Artificial Varves from Portage Lake Provide a High Resolution Record." *Limnology and Oceanography* 39, no. 3 (1994): 649–69.

457

Klump, J. Val, Robert Paddock, Charles C. Remsen, Sharon Fitzgerald, Martin Boraas, and Patrick Anderson. "Variations in Sediment Accumulation Rates and the Flux of Labile Organic Matter in Eastern Lake Superior Basins." *Journal of Great Lakes Research* 15, no. 1 (1989): 104–22.

Kolak, Jonathan J., David T. Long, W. Charles Kerfoot, Tina M. Beals, and Steven J. Eisenreich. "Nearshore versus Offshore Copper Loading in Lake Superior Sediments: Implications for Transport and Cycling." *Journal of Great Lakes Research* 25, no. 4 (1999): 611–24.

Kraft, Kenneth J. "*Pontoporeia* Distribution along the Keweenaw Shore of Lake Superior Affected by Copper Tailings." *Journal of Great Lakes Research* 5, no. 1 (1979): 28–35.

Landmesser, Charles W., Thomas C. Johnson, and Richard J. Wold. "Seismic Reflection Study of Recessional Moraines beneath Lake Superior and Their Relationship to Regional Deglaciation." *Quaternary Research* 17 (1982): 173–90.

Maher, Louis J., Jr. "Palynological Studies in the Western Arm of Lake Superior." *Quaternary Research* 7 (1977): 14–44.

Minnesota State Board of Health, Minnesota Commissioner of Game and Fish, and the Wisconsin State Board of Health (1928–1929). "Investigation of the Pollution of the St. Louis River below the Junction of the Little Swan, of St. Louis Bay, and Superior Bay, and of Lake Superior Adjacent to the Cities of Duluth and Superior," 1929.

Moen, Sharon. "Lake Superior's Deep-Water Donut Mystery." *Seiche,* December 2002: 1–3.

Moore, J. R., Robert P. Meyer, and Richard J. Wold. "Underwater Copper Exploration in Lake Superior—Prospects Mapped in 1971." In *Proceedings of the Fourth Annual Offshore Technology Conference,* Houston, Tex., 1972.

Perkins, Sid. "Once upon a Lake." *Science News* 162 (2002): 283–84.

Ragotzkie, Robert A. "*The Keweenaw Current: A Regular Feature of the Summer Circulation of Lake Superior*. Technical Report 29, Department of Meteorology, University of Wisconsin, Madison, 1966.

Richards, Carl. "Mapping the 'Secret' Spawning Ground of Lake Trout." *Seiche,* September 2000: 1–5.

Smith, Sydney I. *Sketch of the Invertebrate Fauna of Lake Superior*. Report of the U.S. Commission of Fish and Fisheries (1872–1873), Part 2 (1874): 690–707.

Somers, Lee H. "Diving Techniques as Applied to Geological Investigations of the Great Lakes." In *Proceedings of the 10th Conference on Great Lakes Research*, 149–156. Great Lakes Research Institute, 1967.

Sydor, Michael, Richard T. Clapper, Gordon J. Oman, and Kirby R. Stortz. *Red Clay Turbidity and Its Transport in Lake Superior*. EPA-905/9-79-004. Chicago: Great Lakes National Program Office, U.S. Environmental Protection Agency, 1979.

Teller, James T. "Glacial Lake Agassiz and Its Influence on the Great Lakes." In *Quaternary Evolution of the Great Lakes*, ed. Paul F. Karrow and Parker E. Calkin, 1–16. Geological Association of Canada, Special Paper 30. St. John's, Newfoundland: Geological Association of Canada, 1985.

Teller, James T., and L. Harvey Thorleifson. "The Lake Agassiz-Lake Superior Connection." In *Glacial Lake Agassiz,* ed. James T. Teller and Lee Clayton. Geological Association of Canada Special Paper 26, 1983.

Wold, Richard J., Deborah R. Hutchinson, and Thomas C. Johnson. "Topography and Surficial Structure of Lake Superior Bedrock as Based on Seismic Reflection Profiles." *Geological Society of America Memoir* 156 (1982): 257–72.

Zumberge, James H., and Paul Gast. "Geological Investigations in Lake Superior." *Geotimes* 6 (1961): 10–13.

INTERNET RESOURCES

"Mapping Lake Trout Habitat along Minnesota's North Shore," http://d-commons.d.umn.edu/jspui/handle/10792/1021

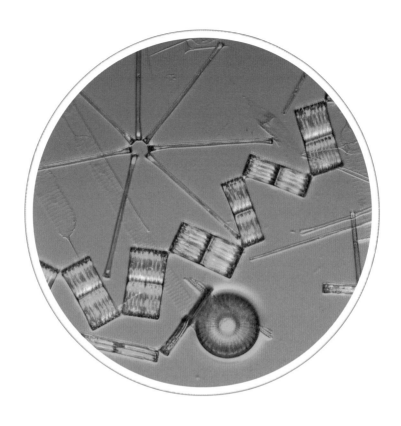

AMPHIPODS
AND DIATOMS

The Big Lake's
Bread and Butter

Sometimes when the sun hits the lake at just the right angle and a breeze creates a tight chop on the surface of the water, Lake Superior will appear as a sea awash in fist-sized nuggets of cut glass. From the shore to the horizon, the lake swarms with the glinting of tiny spearpoints of light.

Of course, the scene is nothing more than an optical illusion. Still, the impression is

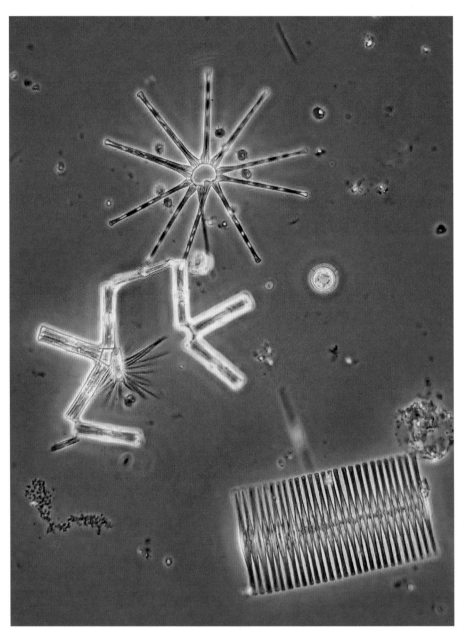

Diatoms are single-celled algae. Unlike other forms of algae, they are characterized by rigid cell walls made of a glass-like substance called opaline silica. Photograph by Robert Megard, University of Minnesota.

not entirely misleading, since Lake Superior does indeed shimmer with bits of glass. Put a drop of lake water under a microscope, and chances are you will find triangles, rods, and pinwheels of crystalline filigree.

They are known as diatoms, single-celled algae that contribute to the soup of tiny drifting plants in Lake Superior called phytoplankton. Unlike other forms of algae, diatoms are characterized by rigid cell walls made of a glass-like substance called opaline silica. As the natural history writer Richard Headstrom rhapsodizes, diatoms are "some of the most beautiful and delicate things found in the microscopic world."

Diatoms have attracted such enthusiastic admirers largely because of their elaborate structure and decoration. They are composed of two symmetrical but overlapping parts that fit together like a lid on a box. This beautiful architecture is formed when the algal cell divides, causing the glass housing to split in half. Using silicic acid (a combination of silicon, oxygen, and hydrogen) from the surrounding water, each new cell fashions a replacement for the missing shell piece.

When it comes to intricacy, the patterns that embellish their surface rival those of snowflakes and rock crystals. It is as if a master jeweler had worked the cell walls with a fine chisel and plotted their divots and ribs, spines and bristles with mathematical precision. These patterns of ornamentation are so dependably exact that scientists study them, along with an individual's size and shape, to identify the diatom species.

Their transparent walls, known as frustules, have led many a writer to wryly observe that diatoms live in glass houses. But, as Headstrom more accurately points out, diatoms inhabit a "majestic sarcophagus, for the beautiful shells persist long after the living part has died."

The siliceous remains of diatoms drifted to the floor of ancient seas and lakes and accumulated to great depths. Over time—more than eighty million years in marine environments and some fifty million years in freshwater systems—the piles became compressed into a porous and lightweight whitish rock known as diatomite. Today the United States leads the world in the commercial mining of diatomite, with most of the activity centered in California. When ground into a fine powder, diatomite has found hundreds of industrial applications, most commonly as filters in the production of beverages and pharmaceuticals and as fillers in products ranging from paints and insulation to agricultural chemicals.

Arguably, the most famous use of diatomite came in the mid-nineteenth century when Alfred Nobel mixed the crushed rock with nitroglycerin to create dynamite. The addition of diatomite made possible the safe transport of what was otherwise a dangerously unstable explosive. "It is clear that, without this discovery of dynamite by Nobel," writes geologist David M. Harwood, "the advancement of railroads across the U.S. and other countries, the construction of canals for shipping, tunnels, dams, highways, and the extraction of the coal and raw materials that fueled the industrial revolution would not have occurred at the same pace."

These tiny jewels of nature may spark other revolutionary changes. Take, for starters, the invention of tough new materials. In a 2003 issue of *Nature,* researchers reported results of experiments showing that the frustules of some diatom species could withstand forces equivalent to the weight of 150 to 1,000 pounds per square inch.

And these same frustules could someday become the darlings of places like Silicon Valley. The manufacture of semiconductor chips, for example, involves carving various nanoscale features into blocks of silicon. The process is laborious, expensive, and energy intensive. It also uses, as well as produces, harmful chemicals and other wastes. Diatoms, on the other hand, self-assemble their own intricate silicon structures far more efficiently and safely. In the future, bioengineers say that diatoms may be used to help build inorganic structures such as sensors and devices for the biomedical, telecommunications, and energy industries. In fact, their potential importance in high-tech industries led a group of international scientists to publish the first genome sequence of a diatom in 2004.

Still, researchers possess only the most rudimentary understanding about the organisms that are responsible for more than 20 percent of Earth's oxygen that is produced by photosynthesis and may comprise 25 percent of the world's plant biomass! Scientists began to study diatoms only about two hundred years ago. Most of our knowledge about these tiny plants has been collected in the past century, however, thanks in large part to the invention of technologically sophisticated microscopes.

Even estimating the number of the world's diatom species is like shooting at a moving target. Since around 1970, about four hundred new species of diatoms have been described each year—and the pace of discovery shows no sign of slowing. The most current inventory puts the number of diatom species at one hundred thousand. With further discoveries and a little taxonomic reordering, researchers say the number could easily jump to one million.

In the Great Lakes, scientists have studied diatoms with systematic rigor only in recent times. The first comprehensive survey of phytoplankton in Lake Superior, for example, was not conducted until 1973. Phycologists documented eighty-eight species of diatoms, most of which had never been reported in Lake Superior or any of the other Great Lakes before. Several species were new to science.

In 1999 scientists published an update of a 1978 checklist of diatoms from the entire Great Lakes. According to the current inventory, the Great Lakes are home to 2,188 diatom species and varieties (groups of diatoms that exhibit significant differences from others in their species but whose distinctive characteristics are not sufficiently evolved to warrant their reclassification as a separate species). And the list keeps growing as phycologists painstakingly peer through their microscopes to analyze the many thousands of organisms that are collected during each lake survey. In 2005, for example, Grand Marais made science headlines when a boomerang-shaped diatom species that was collected from the city's harbor in 1992 was identified as a new species—*Hannaea superiorensis*.

The amended tally, however, still only covers an estimated 70 percent or less of the existing species in the Great Lakes. "Although a great deal remains to be done before we have more than a beginning inventory of the flora present," says the pioneering Great Lakes phycologist Eugene F. Stoermer, "what we do have indicates that Great Lakes diatom communities are amongst the most diverse known from any place in the world."

Why have the Great Lakes served as the incubators for such great evolutionary flowering? Diatoms were helped, in no small measure, by the glaciers that created vast lakes with lots of vacant real estate—and enormously varied ecological opportunities. Different groups of diatoms began to select specific haunts from among the lakes' multifarious habitat niches. Some adapted to life in the open water, while others took up residence on the granite reefs in nearshore waters and on the offshore's mucky bottom. Over time they evolved physical and behavioral characteristics that were geared to their particular circumstances of place.

Top Down and Bottom Up: Diatoms and Amphipods

Scientists are only just beginning to understand the role of this complex and rapidly evolving community of diatoms in the Great Lakes food web. Each new investigation seems to reveal another surprising twist. For example, in terms of biomass, diatoms were once thought to dominate Superior's phytoplankton community largely because researchers used sampling procedures that favored their measurement over other kinds of algae. After applying more inclusive and finer-grained techniques in 1973, however, scientists found that on a lake-wide average phytoflagellates (algae with whiplike appendages used for locomotion) comprised 53 percent of the phytoplankton biomass, followed by diatoms at 38 percent.

Nonetheless, this demotion in the ranks does not diminish the role that diatoms play in the biological workings of the lake. Quite the contrary. When it comes to feeding the animals that form the base of the Great Lakes food chain, few organisms can match the versatility and availability—not to mention the metabolic wallop—of diatoms.

Most of these virtues have evolved in response to the conditions of their habitats. Particularly important in deep lakes and lakes that are ice covered for many months of the year is the ability to survive long periods with little or no light. To tide them over, diatoms store lots of high-energy food in the form of lipids, or fats. Diatoms are botanical switch-hitters, photosynthesizing when they are in the light and then drawing on their phenomenal metabolic reserves when they drift into the darkness of deepwater zones.

They also cross spatial boundaries in the lake, feeding not only organisms that live in the planktonic region but also those in the benthic region of the lake. Protected by their rigid frustules, many diatoms are able to survive a descent to the lake bottom intact. Here they can persist for long periods of time—up to centuries in some cases!— buried in lake sediments.

One of the most important links in the Great Lakes food chain is the relationship between diatoms and the bottom-dwelling amphipod of the genus *Diporeia*. The most abundant and widespread invertebrates on the lake floor, *Diporeia* are a dietary staple of every fish species in the Great Lakes. (Researchers have tallied some 650 amphipods per square foot in the benthic regions of Lake Michigan. So numerous are they that amphipods account for up to 65 percent of the total benthic biomass in some parts of the Great Lakes.)

No bigger than a coffee bean, these furtive, shrimplike creatures provide an irreplaceable link in the lake's food chain. Feeding on microscopic bacteria and algae, they, in turn, become forage for larger organisms such as fish. Any change in amphipod populations, biologists say, would result in a restructuring of the entire Great Lakes food web.

How *Diporeia* achieved such a prominent status in the food chain remained something of a mystery to scientists. For most of the year, amphipods are scavengers, grazing intermittently on meager quantities of low-quality foods such as the bacteria on the decomposing remains of algae and zooplankton that settle to the bottom of the lake. How did the amphipods then obtain enough nutrition to build such significant and wide-ranging populations?

Clues to this conundrum came from research conducted in the Baltic Sea, where scientists discovered that an annual population spike in diatoms during the spring allowed the Baltic's *Diporeia* species to thrive, not merely survive, in its nutrient-poor waters. Taking their cues from this research, in the late 1980s scientists from the Great Lakes Environmental Research Laboratory, Ann Arbor, demonstrated that a similar

463

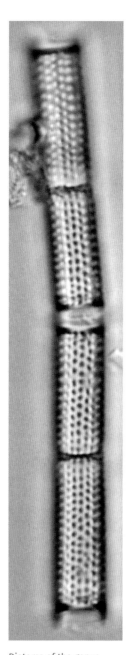

Diatoms of the genus *Aulacoseira* are the dominant species during the spring diatom "bloom" in Lake Superior. Photograph by Eugene Stoermer.

mechanism exists in the North American Great Lakes. They found that *Diporeia* bulk up on the highly nutritious remains of diatoms during the spring diatom bloom and then draw on fat reserves to tide them over the lean times throughout the remainder of the year. The dominant diatoms during this time are species of the genus *Aulacoseira,* barrel-shaped diatoms that are attached to one another at each end to form long colonies. In some of the more eutrophied parts of the Great Lakes, such as harbors, *Aulacoseira* are so abundant that the water turns golden brown and takes on an oily appearance and fishy smell.

Still, another question remained: How did so many highly nutritious diatoms, which bloom in the top one hundred feet of water, the area of maximum sunlight penetration, manage to survive the gauntlet of hungry zooplankton and bacteria that live in the photic zone and drift in such great numbers down into the deep recesses of the lake?

To answer that question, you need to first understand the thermal regime of Lake Superior. In a normal year, the lake undergoes stratification in winter and summer; that is, warm and cold waters separate and form two distinct layers in the lake. Sandwiched between these two zones is a narrow, third layer of water known as the thermocline. (When the lake has achieved maximum stratification, usually in August, the thermocline occurs at a depth of about thirty to sixty feet.)

In the waters above the thermocline (known as the epilimnion) and in those below it (known as the hypolimnion), temperatures change gradually in relatively wide intervals. Within the thermocline, however, temperature changes occur far more rapidly, and these intervals are compressed. This temperature compression forms a density gradient—a kind of invisible thermal barrier—that slows the settling of food particles down to organisms that live in perpetual darkness on the lake bottom.

In spring and fall, however, as the lake warms up or cools down, the water reaches a uniform temperature from top to bottom. The thermal layering temporarily disappears. Lacking the heavy inertia of stratification, the lake's waters are more easily pushed by winds and currents. The result is a deep mixing of waters known as turnover. During this time, the lake experiences an increase in nutrient levels, especially in the levels of phosphorus and silica that diatoms need for growth, as materials are dredged from the deep and recirculated into the water column, and the surge of snowmelt in rivers and streams flushes inputs from the surrounding land. (In contrast to the lower lakes, where stratification may set in by mid-May, the spring turnover in Lake Superior, particularly in the open water, may stretch into mid-July. In general, compared to its sister lakes, Superior's waters undergo longer intervals of mixing in the spring and fall with much shorter periods of stratification in winter and summer. As a result, peaks in the lake's diatom blooms are not as pronounced.)

Diatoms take advantage of this springtime bounty of nutrients and undergo a reproductive spurt. On first glance, however, it seems that nature has created a lavish banquet but has forgotten to mail out invitations to the guests, since most diatoms in the upper waters of the lake go uneaten. That is because it is too early in the season for algae-grazing zooplankton to have built up significant populations. Without a thermocline to impede their descent, the remains of the large, heavy diatoms drift to the bottom, many of them surviving the journey to the lake bottom relatively intact. Not surprisingly, in a springtime sampling of settled algae in lake-bottom sediments, researchers found that diatoms comprised 95 percent of all algal remains.

Benthic creatures such as *Diporeia* devour the rain of manna from above. Such nutritious bounty is brief, however, in the life of amphipods, which for much of the remainder of the year must contend with far less satisfying fare. Like other animals, such as black bears, which also depend on a seasonal feast to build up the levels of body fat that enable them to survive a five-month fast during their winter torpor and to successfully reproduce, amphipods opt for an ingenious metabolic option. In an extremely efficient transfer of energy, *Diporeia* take up the lipids found in diatoms and store them in their bodies with little modification—for good reason. These lipid reserves provide up to two times the metabolic energy per unit of weight as proteins or carbohydrates. *Diporeia* are so rich in fats—lipids constitute up to one-half of their total dry weight—that researchers have observed whole droplets within their body cavities.

But *Diporeia* do not stock up on just any kind of fat. Of the six different kinds of lipids, up to 85 percent of the amphipods' body fat is composed of triglycerides, a form that animals use for the express purpose of storing energy. A slow metabolism allows the amphipods to draw sparingly—and efficiently—on this surplus of concentrated energy. As a result, *Diporeia* are able to feed intermittently or even go for months without eating at all. This ability to fast is critical in an environment in which food is scarce or of low nutritional quality for much of the year.

Just how much do amphipods rely on this brief season of bounty? Scientists estimate that *Diporeia* derive up to one-third of their annual energy budget from it. In samples of *Diporeia* collected in early spring from Lake Michigan, for example, lipids comprised 20 to 25 percent of the animals' biomass. Within a few weeks, however, a period of time that coincided with the spring diatom bloom, their lipid content had jumped to 40 to 50 percent. By December, these lipid stores had dwindled, indicating that the amphipods drew upon them for sustenance.

But there is an especially ingenious twist to the conclusion of the amphipod-diatom story. Research has shown that *Diporeia* may play a critical role in sowing the seeds for their own future diatom harvests. With the onset of summer, the diatom bloom subsides as the algae deplete the supply of the soluble silica they need to build their frustules. Phytoflagellates supplant diatoms as the dominant phytoplankton. They are easier prey for the burgeoning zooplankton populations, and fewer crumbs of food settle into the bottom reaches of the lake. Even fewer tidbits reach the bottom once the lake undergoes summer stratification and the thermocline impedes particle settling. The brief window of opportunity for *Diporeia* closes until the fall turnover. Researchers have demonstrated that as they feed on diatoms, amphipods fracture the frustules, helping to speed up the process of chemical dissolution. Once it is broken down, the biogenic silica (the silica bound up in diatom frustules) is released into the hypolimnion. During the fall turnover, bottom waters once again are mixed into the water column and the silica that was sequestered in bottom waters during stratification is stirred up to fuel a second—albeit far less robust—diatom bloom.

Will the Circle Be Unbroken?

Throughout many regions of the Great Lakes, the cycle of give-and-take is being disrupted by a newcomer on the lakes' ecological scene—zebra mussels. The problem is this: *Diporeia* burrow in the top three-quarters of an inch of bottom sediments. Their movements function like a plow, tilling the freshly settled algae into the fluffy sediments, where they are able to feed undercover from predators. The mussels, on the other hand, tender their filtering siphons about one-third inch above the lake bottom, where they intercept particles such as diatoms before they settle out of the water column. This feeding strategy, coupled with the fact that zebra mussels grow in spectacular densities, enables them to efficiently harvest the rain of food. The result: mass famine for *Diporeia*. In some parts of Lake Michigan, researchers have documented an 88 percent decline in the number of amphipods between 1979 and 1993, the interval in which zebra mussels were introduced and became established in the lake.

Researchers postulate that these mussels will never reach such a nuisance level in

Accidentally introduced into the Great Lakes in 1979, zebra mussels have disrupted the nutrient cycle, contributing to an 88 percent decline of the native amphipod *Diporeia* in some parts of Lake Michigan. NOAA Great Lakes Environmental Research Laboratory.

Lake Superior since the calcium needed by the invaders to build their shells is in short supply and cold water temperatures inhibit breeding. But this should not encourage people in the region to let down their guard. In fall 1998, following a year of mild weather, researchers discovered a sudden increase in the number of zebra mussels in the Duluth-Superior harbor. Some areas reported mussel densities of 465 animals per square foot. Predictions that global warming may bring milder conditions and warmer-than-normal water temperatures to the region make it difficult to predict with any certainty the course of zebra mussel infestation in Superior.

Like *Diporeia,* diatoms too have been harmed by some human activities. As early as 1830, for example, logging, farming, and urbanization by white settlers altered the composition of diatom communities in parts of the lower lakes. Time has not only intensified the pace and scope of these stresses to the native diatom flora, but it has also introduced new, unexpected ones. Since the mid-twentieth century, microscopic hitchhikers, particularly from the Baltic Sea region, have escaped into the lakes from the ballast water of ocean-going freighters. As the Great Lakes freshwater seas have grown saltier from the wastes from human activities, such as the longtime use of road salts and factory effluent, the algae that once were native to marine or brackish-water environments have been able to adapt to life in the lakes. In some regions, these newcomers have supplanted native diatom species.

Over the past 150 years, human activities have reshuffled the original deck of the lakes' ecology. Predicting the outcome of the game is anyone's guess since wild cards—from sea lampreys and zebra mussels to alien species of diatoms—are regularly introduced. With the addition of each wild card, however, the original deck grows smaller. "Indeed," Stoermer says, "we are only beginning to appreciate the extent of local extinctions and exotic introductions which have taken place in the Great Lakes diatom flora." ❧

SUGGESTIONS FOR FURTHER READING

Frey, Luanne C., and Eugene F. Stoermer. "Dinoflagellate Phagotrophy in the Upper Great Lakes." *Transactions of the American Microscopical Society* 99, no. 4 (1980): 439–44.

Gardner, Wayne S., Michael A. Quigley, Gary L. Fahnenstiel, Donald Scavia, and William A. Frez. "*Pontoporeia hoyi*—A Direct Trophic Link between Spring Diatoms and Fish in Lake Michigan." In *Large Lakes: Ecological Structure and Function*, ed. Max M. Tilzer and Colette Serruya, 632–44. New York: Springer-Verlag, 1990.

Goho, Alexandra. "Diatom Menagerie Engineering Microscopic Algae to Produce Designer Materials." *Science News* 166, no. 3 (2004): 42–44.

Hamm, Christine E., Rufolf Merkel, Olaf Springer, Piotr Jurkojc, Christian Maier, Kathrin Prechtel, and Victor Smetacek. "Architecture and Material Properties of Diatom Shells Provide Effective Mechanical Protection." *Nature* 421, no. 6925 (2003): 841–43.

Mills, Edward L., Joseph H. Leach, James T. Carlton, and Carol L. Secor. "Exotic Species in the Great Lakes: A History of Biotic Crises and Anthropogenic Introductions." *Journal of Great Lakes Research* 19, no. 1 (1993): 1–54.

Munawar, Mohiuddin, and Iftekhar F. Munawar. "Phytoplankton of Lake Superior 1973." *Journal of Great Lakes Research* 4, no. 3–4 (1978): 415–42.

Nalepa, Thomas F., David J. Hartson, David L. Fanslow, Gregory A. Lang, and Stephen J. Lozano. "Declines in Benthic Macroinvertebrate Populations in Southern Lake Michigan, 1980–1993." *Canadian Journal of Fisheries and Aquatic Sciences* 55 (1998): 2402–13.

Scavia, Donald, and Gary L. Fahnenstiel. "Dynamics of Lake Michigan Phytoplankton: Mechanisms Controlling Epilimnetic Communities." *Journal of Great Lakes Research* 13 (1987): 103–20.

Stoermer, Eugene F. "Thirty Years of Diatom Studies on the Great Lakes at the University of Michigan." *Journal of Great Lakes Research* 24, no. 3 (1998): 518–30.

Stoermer, Eugene F., Russell G. Kreis Jr., and Norman A. Andresen. "Checklist of Diatoms from the Laurentian Great Lakes." *Journal of Great Lakes Research* 25 (1999): 515–66.

The Plumbing
of Lake Superior

I N THE 1980S LAKE MICHIGAN'S RECORD-HIGH LEVELS MADE FOR RIVETING TV news footage. Who can forget the surf that flooded Chicago's Lakeshore Drive, battered the foundations of shoreline high-rises, and took such big bites out of clay bluffs fronting the city's tony northern suburbs that many of their great rambling houses seemed poised for a tumble into the lake? Chicagoans clamored for a lowering of Michigan's levels by stopping the inflow of water into Lake Superior through the Long Lac and Ogoki diversions and increasing outflows through the Chicago diversion. After all, they argued, wouldn't the water be better used feeding the parched fields of the American heartland than chewing away at the city's famed shoreline?

But drawing down the Great Lakes is not as easy as pulling a plug or shutting a tap. Bodies of water that share boundaries with the United States and Canada are regulated by international compacts. Except during periods of exceptionally high or low water, Superior's water levels are maintained at 602 feet above sea level, a legal agreement that was struck in 1914 to protect navigation and shoreline interests after the International Joint Commission (IJC) permitted hydropower facilities in Canada and the United States to divert water from the St. Marys River.

But just because the Great Lakes are regulated does not mean that they can be controlled. The vast infrastructure of locks and dams and canals can fool the public into believing that engineers call the shots when it comes to Great Lakes water levels. "The term *regulation* is misleading in that it implies that lake levels can be completely controlled by humans," explains a 1986 report by the Great Lakes Commission. "For the most part, the Great Lakes act as a natural system and water will flow through the system only as quickly as nature will allow."

Indeed, nature—in the form of both water volume and basin morphology (that is, its physical contours)—has the upper hand. Together, the Great Lakes' immense basins form a vast natural reservoir that holds some sixty-five trillion gallons of water. Their connecting channels and outlets are relatively narrow, however, even after being widened

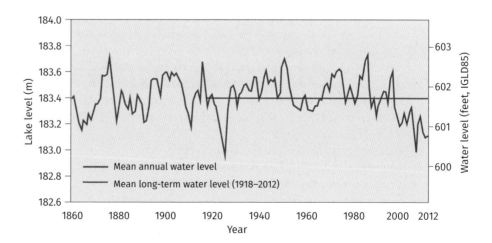

To protect navigation and shoreline interests, the United States and Canada signed a legal agreement in 1914 to set Superior's lake level at 602 feet. However, as demonstrated by this graph of wildly fluctuating water levels between 1860 and 2012, engineers have a limited ability to control water levels.

over time to accommodate shipping and hydropower. It is akin, say, to an Olympic-sized swimming pool whose waters are drained by an outlet the size of a dime.

As a result, there is a tremendous amount of inertia in the Great Lakes. The enormous lag time between action and reaction makes an overnight lowering of the lakes as impossible as an ore carrier taking a sudden 180-degree turn. "Because the Great lakes are so large and the discharge capacities of their individual outlets are limited," the commission's report points out, "extremely high or low lake levels persist for a considerable length of time after the factors which caused the extremes have changed." The replumbing that would be required to control such a massive amount of water boggles the mind. Even if another outlet the size of the Illinois and Michigan Canal (also known as the Chicago diversion) could be built on Lake Michigan, for example, the outflow in the short term would only lower lake levels by a few fractions of an inch, not nearly enough to stem damage to shoreline property when the lake reaches the high-water mark.

Furthermore, even if it were possible to open a plug and quickly drain off high waters in the Great Lakes, the law would quickly close it. Officials are prohibited from authorizing increases in interbasin discharges, even temporary ones, from the only artificial outlet that currently exists on the Great Lakes—the Chicago diversion, which sends Great Lakes water into the Mississippi River via the Illinois River for the purposes of obtaining municipal water supplies, sanitation, hydropower, and navigation. In 1967, the U.S. Supreme Court set a withdrawal limit of 2.1 billion gallons of water per day (enough to fill the Sears Tower, the world's second-tallest building, once every five hours) from Lake Michigan into the canal. Since its completion in 1900, the diversion has lowered the mean levels of all the Great Lakes, most dramatically the levels of Lakes Michigan and Huron, which have dropped an average of 2.5 inches. By U.S. law, the construction of any new diversion would require the unanimous approval of the governors of the eight Great Lakes states. In the unlikely event that a consensus could be reached, a project of such magnitude undoubtedly would languish in the ensuing economic, political, and regulatory wrangling.

But even if the stepped-up pumping of water were legally feasible, many would consider it undesirable since it creates dependencies on a flow of water that might be difficult to maintain during average years or prolonged periods of low water. As anyone familiar with the historic record knows, the Great Lakes have their ups and downs. Since lake levels were first recorded in the 1860s, they have fluctuated by as much as 6.5 feet from their baseline average. Furthermore, dramatic changes in water levels defy prediction and can occur quickly. Rapid drops from extreme highs to extreme lows, for example, took place from 1985 to 1988 and again from 1997 to 1998.

Complicating matters is the host of stakeholders on the lakes with their own competing interests in maintaining certain water levels. The high water that caused flooding and tied up traffic for Chicago commuters allowed hydroelectric facilities to generate greater amounts of power. Water levels that damaged shoreline property enabled cargo ships to carry heavier loads. Indeed, high water is a blessing to transportation companies. With every drop of an inch during years of low to average lake levels, large lake-going vessels must reduce their cargo loads by 270 tons.

Several measures are in place to ensure that existing uses are protected before any new or increased diversions of Great Lakes water are approved. Among them is the 1985

Great Lakes Charter, which prohibits such diversions if they "individually or cumulatively . . . have any significant adverse impacts on lake levels, in-basin uses and the Great Lakes Ecosystem."

These are laudable goals, but it is extremely difficult to definitively link any one diversion to lowered water levels, much less assign blame for any economic and ecological harm that might result. According to a 2000 report by the IJC, "The dynamic nature of the Great Lakes–St. Lawrence system and the multiplicity of physical, chemical, and biological processes affecting ecosystem status challenge science's ability to establish and characterize causal relationships between a given water use and its impact on levels, flows, and fluctuations, on any observed changes in the ecosystem, and on economic uses of the system."

Because of the unknowns and ambiguities, including the possibility that global warming will cause lake levels to bottom out at record lows, the IJC has joined a host of scientists to urge extreme caution in approving any new diversions or consumptive uses of Great Lakes water. Contrary to popular perception, these waters, they point out, are not a renewable resource. In reality, only 1 percent of the water in this great sweetwater reservoir is replenished through snowfall, rain, and the flow of rivers, streams, and groundwater. The remainder is historic water, runoff from the glaciers that receded from the Great Lakes region twelve thousand years ago. In Lake Superior, for example, only 2.5 feet of water falls on the lake each year in the form of rain and snow. Another 2 feet enters the lake via tributaries and groundwater. The lake has its own accounting system for these annual inputs. Approximately 1.6 feet is lost to evaporation. The remainder leaves Lake Superior through the St. Marys River to recharge the lower lakes. Siphoning more than 3 feet of water each year would permanently lower the lake's levels.

What ecological damage would result? Good question. The areas of the lake subject to the greatest changes from ongoing disturbances, such as fluctuating water levels, are extremely difficult to study. The result, the IJC observes, is glaring knowledge gaps, particularly with regard to fish, coastal wetlands, biodiversity, and the nearshore zone's water quality and habitats.

The research that does exist suggests that coastal wetlands, semi-enclosed bays, and nearshore zones could be among the ecosystems in the Great Lakes that would be most heavily impacted by stabilized or permanently lowered water levels. Studies of the Mink River estuary along the northwestern shoreline of Lake Michigan have shown, for example, that a wide array of wetland plants and plant communities is dependent on alternating cycles of flooding and water recession. During prolonged periods of low water, mudflats are exposed, allowing the seeds of many wetland species to germinate. In time, trees and shrubs from the surrounding uplands gradually begin to invade the lowlands and crowd out many of these species. The flooding that occurs during years of high water, however, kills the woody plants and clears the stage for a different mix of wetland species to emerge when parched conditions return. According to Mink River researcher Janet Keough, preliminary studies "suggest that disturbance from occasional high water periods tends to increase both community and species richness in coastal marshes."

In addition to these long-term fluctuations of water, there are shorter-term hydrologic cycles that also produce beneficial variations in water levels, such as seiche-driven pulses that occur throughout the day and the more intermediate ups and downs that accompany a change in seasons. Together they choreograph a complex but rhythmic dance of water and plants. In the process, they provide food and shelter for the young of many species of Great Lakes fish, including those that are valuable to commercial and recreational fishing. A host of organisms, from phytoplankton to juvenile fish, depend on lake waters to inundate coastal wetlands and not only deliver seasonal moisture but also distribute and replenish nutrients. (See also "The Rise and Fall of Seiches.")

Water quality, too, is at risk by permanently lowered water levels. Like backwater areas, bays rely on the circulation of water from the larger lake system for a supply of oxygen and nutrients and a flushing of wastes. Without such chemical and biological exchanges, bay waters can stagnate, and pollutants become concentrated.

And a drop in water levels can further shrink the already limited habitat in the nearshore zone. Throughout most of the Great Lakes, the water depth increases rapidly within a relatively short distance of the shore. As a result, unlike the deep offshore reaches of the lakes, the "shallow habitats of the nearshore and coast are disproportionately more influenced by lake levels," says a 2001 report by the IJC. Even small changes "can alter the extent, structure, and functions of coastal habitats, and alter the extent of interaction between coastal and nearshore habitats."

Reducing or degrading nearshore reaches could have disastrous consequences for the entire lake. That is because the coastal zone is far more productive—and hosts greater biological diversity—than offshore waters. There is some evidence to suggest that nutrients flushed into the lakes' nearshore zones from adjoining wetlands play a major role in stimulating the growth of phytoplankton, the bread and butter of the lakes' entire food chain. Permanently lowered waters could sever this connection and reduce food supplies in this critical zone.

Also included among the organisms most heavily affected by lowered lake levels, says fishery biologist Bruce A. Manny, would be fish that have very specific spawning requirements. Some of them, including lake trout, "build nests or use a particular substrate, such as submersed vegetation or rock rubble, in combination with a certain water depth." If waters become shallower, many of these habitats could become too warm for reproduction or damaged by repeated exposure to storm waves.

The construction of new mechanisms for transferring water in and out of the Great Lakes—either by pumping or the digging of new canals—also adds to the potential for

Mudflats occur in coastal wetlands during low-water periods. The drier conditions allow the seeds of many wetland plants to germinate, enhancing biodiversity in Superior's coastal reaches. Photograph by Douglas Wilcox.

Flooding inundates coastal wetlands during high-water years. By killing trees and shrubs, these floods create openings for a different mix of wetland species to emerge when parched conditions return. Photograph by Douglas Wilcox.

ecological harm. Some of the lakes' most destructive—and costly—pests, such as the sea lamprey, gained entry into the lakes following the construction of shipping canals. And depending on the size and location of intakes, the pumping of water also can remove large numbers of larval and juvenile fish, as thermal-electric power-generating plants already do.

Even in times of record highs, when waves are lapping at the knees of high-rise apartment buildings, the Great Lakes do not have water to spare. "Seemingly 'wasted,'" the IJC points out, "the infrequent very high waters do, in fact, serve a purpose by inundating less frequently wetted areas and renewing habitat for their biotic occupants. Major outflows from the Great Lakes provide needed freshwater input to fish populations as far away as the Gulf of Maine." As the IJC emphasizes, once the needs of all users, from ore carriers to wild rice and larval fish, are fairly accounted for, "there is never a 'surplus' of water in the Great Lakes system."

In the future, state and federal governments increasingly will be called upon to defend the line between necessary and surplus water in the Great Lakes. As water supplies in many densely populated parts of the world, including Asia, Africa, and the Middle East, not to mention the American High Plains and Desert Southwest, have become perilously depleted or too polluted for human consumption, the Great Lakes, which contain 20 percent of the world's freshwater resources, can appear to contain a vast surplus of a precious commodity. The issue has already been tested by a controversial 1998 proposal by the Nova Group, a Canadian business-consulting firm that obtained a permit from Ontario's Ministry of the Environment to pump a maximum of 158 million gallons of Superior's water each year for five years. The cargo, free for the taking, would have been transported to Asia in the holds of big ships and sold as drinking water. Fortunately, before the Nova Group even had a chance to line up customers, a tidal wave of binational protest quickly swamped officials at the highest levels of government, including then-U.S. Secretary of State Madeleine Albright and Canada's then-Foreign Affairs Minister Lloyd Axworthy. In the wake of the controversy, the plan was scuttled.

To date, the United States and Canada have demonstrated considerable political will to resist schemes to drain the Great Lakes through pipelines and cargo holds. But citizen groups in the basin continue to press for greater political and legal protections. As of this writing, the most comprehensive legislation to safeguard North America's inland freshwater seas includes the 2005 Great Lakes–St. Lawrence River Basin Sustainable Water Resources Agreement, a nonbinding agreement between the United States and Canada, and the Great Lakes–St. Lawrence River Basin Water Resources Compact, a binding agreement among the U.S. Great Lakes states that was signed into law by President George Bush in 2008.

Though far-reaching, the new laws are not watertight. The compact, for example, largely bans the wholesale diversion of water to areas outside the Great Lakes basin. But an exemption, known as the "bottled water loophole," allows the sale of Great Lakes water so long as it is packaged in containers of 5.7 gallons or less. As Jeffrey Dornbos writes in a 2010 article in *Case Western Reserve Law Review*, "Las Vegas would be prohibited from building a pipeline to pump water out of the Great Lakes to meet its growing water demand. Because of the bottled water exemption, however, private companies would not be prohibited from selling the same Great Lakes water to Las Vegas, as long as the water was incorporated into bottles rather than pumped directly."

It may be politically and ethically easy to derail megaprojects whose purpose is to deliver Great Lakes water for the greening of golf courses in Arizona or the production of microchips in the Nevada desert. "Unfortunately," writes aquatic ecologist Wayland R. Swain, "I believe that we are not going to be faced with a simple decision of the availability of industrial water in Wyoming, Montana, Texas, or Arizona. Instead, the question will be of even greater import, certainly to the point of economic consequence, perhaps even affecting national security, and potentially to the point of national or even world-wide hunger."

The authors of the 1997 report *The Fate of the Great Lakes: Sustaining or Draining the Sweetwater Seas?* concur: "The most difficult dilemmas that the residents of the Great Lakes will be faced with are ethical ones. In a world of increasingly scarce water supplies, how can we deny access to the waters in this region to those who are in desperate need for water? How do we balance the needs of human beings with those of the fish, birds and animals for whom the Great Lakes Basin is also their home?" ∾

SUGGESTIONS FOR FURTHER READING

Annin, Peter. *Great Lakes Water Wars*. Washington, D.C.: Island Press, 2006.

Barringer, Felicity. "Growth Stirs a Battle to Draw More Water from the Great Lakes." *New York Times*, 13 August 2005.

Bixby, Alicia A. *The Law and the Lakes: Toward a Legal Framework for Safeguarding the Great Lakes Water Supply*. Chicago: The Center for the Great Lakes, 1986.

The Center for the Great Lakes. *Effects of Global Warming on the Great Lakes: The Implications for Policies and Institutions*. Chicago: The Center for the Great Lakes, 1988.

Deutsch, Claudia H. "There's Money in Thirst." *New York Times*, 10 August 2006.

Dornbos, Jeffrey S. "Capping the Bottle on Uncertainty: Closing the Information Loophole in the Great Lakes–St. Lawrence River Basin Water Resources Compact." *Case Western Reserve Law Review* 60 (2010): 1211–40.

Egan, Timothy. "Near Vast Bodies of Water, Land Lies Parched." *New York Times*, 12 August 2001.

Farid, Claire, John Jackson, and Karen Clark. *The Fate of the Great Lakes: Sustaining or Draining the Sweetwater Seas?* Toronto: Canadian Environmental Law Association and Great Lakes United, 1997.

Frerichs, Stephen, and K. William Easter. "Regulation of Interbasin Transfers and Consumptive Uses from the Great Lakes." *Natural Resources Journal* 30 (1990): 561–79.

Gamble, Donald J. "Commentary: The GRAND Canal Scheme." *Journal of Great Lakes Research* 15, no. 3 (1989): 531–33.

Great Lakes Commission. *Water Level Changes: Factors Influencing the Great Lakes*. Boyne City, Mich.: Harbor House Publishers, 1986.

Hartmann, Holly C. "Climate Change Impacts on Laurentian Great Lakes Levels." *Climatic Change* 17 (1990): 49–68.

International Joint Commission. *Draft Plan of Study for Review of the Regulation of Outflows from Lake Superior*. Windsor, Ont.: Great Lakes Regional Office, 2001.

———. *Great Lakes Diversions and Consumptive Uses*. Windsor, Ont.: Great Lakes Regional Office, 1985.

———. *Protection of the Waters of the Great Lakes: Final Report to the Governments of Canada and the United States*. Windsor, Ont.: Great Lakes Regional Office, 2000.

———. *Protection of the Waters of the Great Lakes: Review of the Recommendations of the February 2000 Report*. Windsor, Ont.: Great Lakes Regional Office, 2004.

Keough, Janet R. "The Mink River—A Freshwater Estuary." *Wisconsin Academy of Science, Arts and Letters* 74 (1986): 1–11.

————. "The Range of Water Level Changes in a Lake Michigan Estuary and Effects on Wetland Communities." In *Wetlands of the Great Lakes: Protection and Restoration Policies; Status of the Science*, ed. Jon Kusler and Richard Smardon, 97–110. Proceedings of an International Symposium, Niagara Falls, N.Y., May 16–18, 1990.

Leslie, Jacques. "Running Dry." *Harper's Magazine* 300 (July 2000): 37–52.

Manny, Bruce A. "Potential Impacts of Water Diversions on Fishery Resources in the Great Lakes." *Fisheries* 9, no. 5 (1984): 19–23.

Moen, Sharon. "IJC Suggests the Great Lakes Are Not Exportable." *Seiche,* June 2000: 7.

————. "Never-Ending Motion: Lake Superior's Wetlands." *Seiche,* November 2003: 6–7.

Perkins, Sid. "Crisis on Tap?" *Science News* 162 (2002): 42–43.

Prince, J. David. "State Control of Great Lakes Water Diversion." *William Mitchell Law Review* 16, no. 1 (1990): 107–70.

Sachs, Jeffrey D. "The Challenge of Sustainable Water." *Scientific American*, December 2006.

Santos, Fernanda. "Inch by Inch, Great Lakes Shrink, and Cargo Carriers Face Losses." *New York Times,* 22 October 2007.

Sengupta, Somini. "In Teeming India, Water Crisis Means Dry Pipes and Foul Sludge." *New York Times,* 29 September 2006.

Smith, Craig S. "Saudis Worry as They Waste Their Scarce Resource." *New York Times,* 26 January 2003.

Swain, Wayland R. "Great Lakes Research: Past, Present, and Future." *Journal of Great Lakes Research* 10, no. 2 (1984): 99–105.

Tyler, Patrick E. "Libya's Vast Pipe Dream Taps into Desert's Ice Age Water." *New York Times,* 2 March 2004.

Wilder, Julia R. "Questions of Ownership and Control," in *Perspectives on Ecosystem Management for the Great Lakes: A Reader,* ed. Lynton K. Caldwell. Albany: State University of New York Press, 1988.

Yardley, Jim. "Beneath Booming Cities, China's Future Is Drying Up." *New York Times,* 28 September 2007.

INTERNET RESOURCES

"Coping with Water Scarcity: A Strategic Issue and Priority for System-Wide Action," UN-Water Thematic Initiatives, 2006, http://www.unwater.org/downloads/waterscarcity.pdf

Great Lakes Commission, http://glc.org/about/

Great Lakes–St. Lawrence River Basin Sustainable Water Resources Agreement, http://www.cglg.org/projects/water/docs/12-13-05/great_lakes-st_lawrence_river_basin_sustainable_water_resources_agreement.pdf

Great Lakes–St. Lawrence River Basin Water Resources Compact, http://www.cglg.org/projects/water/docs/12-13-05/Great_Lakes-St_Lawrence_River_Basin_Water_Resources_Compact.pdf

The Lake Superior and Mississippi River Canal

I N A 1995 ESSAY FOR *TIME* MAGAZINE, HUMORIST GARRISON KEILLOR DESCRIBED a get-rich-quick scheme for the state of Minnesota known as Excelsior. The idea was simple: build a concrete channel the size of the Suez Canal to divert Lake Superior water into the Mississippi River via the St. Croix River. At Keokuk, Iowa, Keillor writes, direct the surplus west to the Colorado River to fill such natural waterways as the Grand Canyon with "enough water to supply the parched Southwest from Los Angeles to Santa Fe for more than fifty years."

Minnesotans would benefit handsomely from the sale of the water. (Real-estate mogul Donald Trump, Keillor quips, "will discover that he is owned—lock, stock and roulette wheel—by Lutheran Brotherhood and must renegotiate his debt load with a committee of silent Norwegians who don't understand why anyone would pay more than $120 for a suit.") And once the lake was drained, the people of Minnesota also could cash in on a new, more lucrative tourist attraction—the Superior Canyon, complete with wooded buttes and exposed shipwrecks.

OK, Keillor may have exaggerated the connection between the Lutheran Brotherhood and Donald Trump. And it is probably a bit of a stretch to say that the whole of Lake Superior could be emptied within a decade. As for the part about pumping Lake Superior into the Mississippi River, well, Keillor lifted that one straight out of the history books.

The Minnesota Territorial Legislature heard rumblings in support of just such a canal as early as the 1850s. Business interests argued that without an inland waterway from Superior to the Mississippi, the economy of the northern heartland would languish. Minnesota politicians bought into the idea. In 1857 the Mississippi River and Lake Superior Ship Canal Company was formed and given the power to condemn lands for the building of a canal and to collect tolls from users.

Procuring public funds for the surveying of the waterway, much less its construction and maintenance, would prove to be a challenge, however. Over the next three decades pro-canal measures would regularly surface on the legislative agenda, but the project never gained momentum.

Then in 1894 proponents finally got the attention of the U.S. Congress, which appropriated $10,000 (approximately $1.3 million in 2011 dollars, a whopping sum for 1894) for a reconnaissance of the region by the U.S. Engineers. The agency determined that the most feasible course for the canal was through Wisconsin's Brule River on the southwestern end of Lake Superior. From there water could be funneled into Upper St. Croix Lake and then into the main channel of the St. Croix River, where it would flow to its final destination—the Mississippi River. The cost for the 150-mile-long project was estimated at $7,815,000. In addition, citizens would be asked to shoulder an annual expenditure of $420,000 for the canal's operation and maintenance.

The payoff to taxpayers was the breakup of the railroad's monopoly on transportation. According to canal proponents, the new water route would create competition. Lower freight rates would result in cheaper goods. Twin Cities' businesses would also become more profitable, paying less to ship items to East Coast markets, such as fertilizers collected from the South St. Paul stockyards, binder twine and farm machinery manufactured by inmates at the Minnesota State Prison at Stillwater, and linseed oil and cakes manufactured in Minneapolis, the world's leader in linseed production.

The U.S. Engineers disagreed. After reexamining the results of its cost-benefit analysis in 1899, 1909, and 1912, the agency reaffirmed its position that the project was technically, but not economically, feasible.

Canal boosters, however, were unwilling to take no for an answer. In 1913 they organized the Lake Superior and Mississippi River Canal Commission. The following year the commissioners launched yet another plea to the Minnesota legislature, this time in the form of a lengthy refutation to the engineers' conclusions. Lending urgency to their cause was the scheduled opening of the Panama Canal in 1914. Midwestern businesses now had an alternate route to the U.S. West Coast (albeit a roundabout one). To reach eastern U.S. markets via the Great Lakes, however, commercial traffic in the upper Midwest continued to rely on overland rail lines to Lake Superior. If the northern heartland were to take full advantage of the new grand round of national and international waterways, canal proponents argued, this last stranglehold on transportation by the railroad would have to be broken.

The government remained—thankfully—unpersuaded. Had it succeeded, the plan would have paved over some of the most scenic riparian areas in northern Wisconsin and Minnesota. And since it was intended not only to facilitate the flow of goods from established urban centers but also to stimulate economic development along its length, the waterway would have devastated areas adjacent to the river corridors themselves. In its 1914 report the commission pointed out, for example, that the canal would have put the limestone ledges along the riverbanks of the upper Mississippi and St. Croix Rivers within handy reach for quarrying. Not only would the construction of a canal have opened up the far reaches of the St. Croix watershed to logging, but it would also have stepped up the pace of cutting by making it easier—and more profitable—to ship pulpwood and hardwood to eastern U.S. markets. According to the commission's report, "Development of the waterway will in itself develop these new industries."

In the rush of American industry to embrace new transportation technologies (the railroads soon saw themselves eclipsed by trucks and cargo planes), it seemed as if all further motions to construct the Superior-Mississippi canal were finally dead in the water. Perhaps that is why it was so surprising when in 1967 Duluth businessman Jeno F. Paulucci unveiled an economic development scheme for the north country whose centerpiece was, you guessed it, a canal linking Lake Superior with the Mississippi River. This time the project had a snazzier title—the Missing Link—but its rationale remained the same. According to a 1968 article in the *Duluth News Tribune,* Paulucci "predicted the low-cost freight generated by water-borne traffic would result in the new diversified industry along the canal route and throughout Northeastern Minnesota and Northwestern Wisconsin."

But things had changed since 1914. The Missing Link plan was reintroduced at a time when Americans were beginning to address not just the technical and economic feasibility of massive engineering projects but also their recreational, ecological, and aesthetic dimensions. In this changed climate, the nation's rivers had come to be seen as more than dumping grounds for human and industrial waste and as liquid energy for generating power or moving goods. In 1968, the St. Croix River received a national designation as a Wild and Scenic River, a federal protection that restricts new development along the river corridor that would mar its beauty for people visiting the river on water or on foot. Paving or quarrying its banks is prohibited.

Nonetheless, proposals for infrastructure projects of heroic proportions continued

The 1968 Missing Link, or Northern Canal Link, plan to connect Lake Superior with the Mississippi River revived several prior canal schemes, the first of which was proposed in 1914.

to be floated from time to time. A 1982 University of Michigan study, for example, examined the feasibility of constructing a 611-mile paved canal from Lake Superior to the Missouri River basin. Researchers calculated the costs for such a supply line at $27 billion (the price tag included the construction of seven power plants, at $1 billion apiece, to pump the water) and millions more annually for pumping and system maintenance. The purpose of the canal was to augment the water supplies of the Missouri River so that they, in turn, could be diverted to the thirsty corn and wheat belt in the High Plains states, where heedless irrigation had dangerously depleted the vital groundwater reserves of the Ogallala Aquifer.

The canal scheme followed on the heels of the Powder River coal slurry pipeline project, an even more far-flung proposal that would have stretched a 1,300-mile pipeline of water from Lake Superior across the northern United States. Initially, project engineers envisioned diverting water from Lake Superior to the Powder River Basin in Gillette, Wyoming, for the purpose of floating western coal in a pipeline back to electric utilities and industrial users in the Midwest and Great Lakes basin. Revised designs eventually called for a one-way pipeline that transported the coal using water from western sources. In either scheme, the wastewater, following removal of the coal, would have been treated and dumped into Superior. The purpose of the pipe? Cheap transportation. Anticipating a growing demand for the West's low-sulfur coal, proponents argued that the slurry method presented an economical alternative to rising rail costs.

Enjoying yet another revival—and the backing of Quebec's then-Premier Robert Bourassa—was a $100 billion intracontinental replumbing project known as the Great Recycling and Northern Development Canal scheme, or GRAND Canal, for short. Conceived in the 1930s, revisited in the 1960s, and dusted off again in the early 1980s, the plan called for building a dike across James Bay so that waters from the rivers flowing into the bay could be captured and ultimately back-pumped into the Great Lakes watershed. The project would have used the Great Lakes as a massive reservoir from which water could be diverted to parched areas ranging from the Canadian prairies to the Arizona desert.

Like the Missing Link, these projects have been deemed technically, but not commercially, feasible. But some fear that this may change as parts of the southern United States grow more thirsty—and desperate. Will Lake Superior water someday irrigate wheat fields in Texas or flow in the toilets of Phoenix, as Keillor's hypothetical Excelsior suggests? If it were also wildly profitable, would citizens and governments in the Great Lakes be able to resist? ❧

SUGGESTIONS FOR FURTHER READING

International Joint Commission. *Great Lakes Diversions and Consumptive Uses.* Windsor, Ont.: Great Lakes Regional Office, 1985.

Keillor, Garrison. "Minnesota's Sensible Plan." *Time,* 11 September 1995, 84–85.

Lake Superior and Mississippi River Canal Commission. *Report of Lake Superior-Mississippi River Canal Commission to the Legislature of Minnesota.* St. Paul: Lake Superior and Mississippi River Canal Commission, 1914.

"Waterway Link to Mississippi Studied." *Duluth News Tribune,* 21 January 1968.

The Rise and Fall
of Seiches

The morning of Thursday, July 13, 1995, dawned quietly on Lake Superior. By all accounts, it looked like just another ordinary summer day in the north country. By midafternoon, however, strange goings-on began to be reported from points all around the lake. In Ashland, Wisconsin, rapidly rising water levels briefly submerged docks in Chequamegon Bay. Across the lake in Rossport, Ontario, stunned marina workers watched as the nearshore waters receded before their eyes. Hung by their moorings like pendants, boats clung to a dock that jutted out into fifty feet of newly exposed sandy lake bottom. Twenty minutes later the waters returned, flooding the temporary beach as quickly as they had ebbed away.

Readings taken by instruments off the shore of Michigan confirmed that the sightings were not just tall tales to be added to the already prolific body of exaggerated lore about Lake Superior. According to scientists from the National Oceanic and Atmospheric Administration, lake levels dramatically rose and fell, fluctuating as much as three feet in the course of a single afternoon. Just what was going on?

The freak occurrences had all the classic signs of a phenomenon known as a seiche (pronounced "saysh"). The term was coined from the French in the nineteenth century by the father of limnology, Swiss scientist François-Alphonse Forel. It means "to sway back and forth"—which is exactly what Superior's waters did the afternoon of July 13.

Seiches are triggered by natural forces. In the oceans, they can be set off by any number of disturbances, including earthquakes, the eruptions of volcanoes, and landslides. In the Great Lakes seiches are most commonly caused by persistent winds or variations in barometric pressure. High-pressure systems, which play off low-pressure fronts moving through the lake basins, can act like a group of people sitting on the end of a water bed: They depress waters on one side of the lake and cause a related rise in those on the other side. Stiff, prolonged winds can actually push lake waters to the point where they pile up and cause a corresponding drop in water levels on the lee shore. When the air pressure plummets or the speed or direction of the wind changes, the forces of gravity and friction take over. The pent-up water suddenly relaxes and sloshes between shorelines until the wave energy is spent.

To illustrate this action, scientists commonly use the tilted-bathtub analogy. Tip up one end and the water rushes to the opposite side. Set it down and the water races back, rocking within the confines of the tub until its kinetic energy dissipates. A similar thing happens on a lake. Scientists suspect that the July 13 seiche was set off by a severe storm system that passed through the Lake Superior region. The lake is so vast and the effects of intense weather systems so far-reaching that observers sometimes can witness storm-induced phenomena—such as water disappearing from under a dock—even though directly overhead the sun shines through blue skies.

Seiches of such magnitude are relatively infrequent occurrences on Lake Superior, the deepest of the Great Lakes. When they do take place, the damage makes headlines. A seiche that surged through Two Harbors, Minnesota, in 1998 resulted in several hundred thousand dollars worth of damage to ore boats docked in the harbor. When caught in sudden seiche-driven currents out on the lake, such vessels have been thrown off course or run aground.

On some of the lower lakes, however, major seiches are fairly routine, particularly on Lake Erie. The combination of atmospheric pressure and stiff southwesterly winds have resulted in water levels up to twelve feet higher near Buffalo, New York, on the

The dramatic ebb and flow of water from a storm-induced seiche happened in the span of twenty minutes on Lake Superior at Munising, Michigan, in 1921.

lake's eastern end as compared to Toledo, Ohio, on Erie's western end. That is because Erie is much shallower and its bottom is relatively flat, enabling the wind to move a far greater volume of water. In fact, so common are large seiches on the lake that the National Weather Service issues regular forecasts for shoreline water levels for some of the most heavily impacted ports.

In 1960 the U.S. Weather Bureau (later renamed the National Weather Service) instituted such an early warning system for portions of the lower lakes based on forecasting techniques developed at the University of Chicago. The goal was to avert another catastrophe like the one that took place in the Chicago area on June 26, 1954, when a rapid seven-foot rise in the water level on Lake Michigan descended like a sudden swell on a group of unsuspecting anglers and swept them from a dock on the Chicago lakefront. Six people were drowned.

Modern seiche forecasts are just the latest in a series of efforts to chart the rise and fall of water levels in the Great Lakes. As early as the mid-seventeenth century, seiches attracted the attention of travelers, who noted extreme water fluctuations in their journals. In the mid-nineteenth century, seiches became the subject of considerable scientific curiosity, and there was lively debate in academic circles about whether seiches, like ocean tides, were influenced by lunar cycles or other natural forces.

The theory that seiches were lunar in origin was finally put to rest when scientists offered conclusive proof that seiches did not follow the twice-daily, ebb-and-flow pattern of ocean tides. As part of their duties, public officials and U.S. Army personnel in the frontier territories of the Great Lakes installed devices at the mouths of streams to monitor changes in water levels. In 1828, one of them, Lewis Cass, Michigan's territorial governor, made detailed measurements of water levels at the mouth of the Fox River in Lake Michigan's Green Bay. His charts revealed that lake levels fluctuated frequently and sporadically throughout the day, sometimes only by a few fractions of an inch. Indeed, as subsequent research has shown, small seiches are common on all the Great Lakes but are routinely masked by ordinary wind waves.

Today, seiches—both large and small—continue to command the attention of researchers. It turns out that they serve as something akin to the lungs of the lake, inhaling and exhaling with gentle regularity and stopping, on occasion, to yawn with great abandon. In the course of this breathing, the lake circulates fresh water and vital nutrients, particularly through the biologically fertile zones of the lake, such as coastal wetlands, bays, and nearshore waters.

Take, for example, the relationship between Lake Superior and the Kakagon and Bad River Sloughs, a pristine wetland complex located on the lake's southern shore. Approximately every two hours, the flow of water in the two main rivers that drain the sloughs changes direction. At "high tide," Lake Superior courses into the river channels and raises water levels by an average of four to five inches.

It may not seem like much, but researchers have good reason to suspect that these subtle fluctuations are vital to both the health of organisms in the sloughs and in the nearshore reaches of Lake Superior. During seiche-driven pulses of high water, inflowing waters from Lake Superior overcome the momentum of waters flowing into the lake, causing them to flood adjacent backwaters. In the temporary lull between pulses, sediments are able to settle to the bottom, where they fertilize plants, such as wild rice. At the same time, a biological factory of microorganisms is busy producing or transforming organic materials into beneficial substances. When the waters reverse, the vigorous ebbing of the seiche "tide" helps to sweep organic materials, nutrients such as carbon, nitrogen, and phosphorus—and, according to some fish ecologists, large numbers of larval fish and other organisms—from the sloughs into the coastal reaches of the lake. In a sense, seiches power the biological communication between fecund backwaters and Superior's littoral region, helping to refresh the narrow zone of nearshore waters that is critical to the growth and reproduction of many organisms, including lake trout.

Seiches also serve to replenish the life-sustaining conditions in another biologically active zone of Lake Superior—bays. Researchers have found that regular seiche pulses act in concert with other physical processes to set up currents that circulate nutrients and recharge the otherwise stagnant recesses of inlets and bays with fresh water. In a study of Chequamegon Bay in summer 1968, researchers recorded oscillations of the bay's water levels that ranged from two to thirty inches. These fluctuations occurred at intervals of 12, 21, 36, 62, and 140 minutes. Scientists estimate that each day these tiny pulses collectively may be responsible for moving an enormous volume of water—from 5 to 10 percent of the water contained in Chequamegon Bay—into the open lake.

Important biological and chemical exchanges take place in the vertical as well as horizontal reaches of bays. Located in the sunlit reaches of water, for example, is an aquatic pasture of photosynthesizing algae. Grazing on these algae—and each other—are tiny zooplankton that live, reproduce, and die in this upper zone. Bacteria break down the dead plants and animals into chemical components that serve as fertilizing agents for future crops of algae.

The problem is that these nutrients eventually sink and become trapped in bottom sediments. Seiches spark the formation of bottom currents that stir up fertilizing compounds and form a turbid, nutrient-rich zone of organic particles, known as the benthic nepheloid layer, suspended just above the lake floor. From time to time, large seiche events can enlarge the dimensions of the nepheloid layer, enabling settled nutrients to make their way back up into the water column, where they fuel the growth of algae, upon which all life in the lake depends. In the 1980s researchers discovered that currents in the Duluth harbor driven by big seiche events expanded the vertical dimension of the nepheloid layer from sixteen feet off the bottom of the lake to forty-eight feet, suffusing the water column with a great cloud of nutrients. "In the most seiche-prone bays of the Great Lakes," says Ben Korgen, a physical oceanographer at the Naval Oceanographic Office, "seiches may be the most important single physical process that determines how pollutants are distributed or how nutrients are lifted from below into the sunlit surface waters where living tissue is formed by photosynthesis."

For most of the year, the lake's rhythmic breathing supplies the plants and animals in its reach with a modest, but steady, supply of life-sustaining foodstuffs. Occasionally it draws a deep breath and lets loose a mighty exhalation. The next time you read about ore boats the size of Kmarts suddenly dry-docked in otherwise deep harbor waters, you can "rest assured that the living lake has been treated to a bonanza!" Korgen says. ∾

SUGGESTIONS FOR FURTHER READING

Dearborn, H. A. S. "On the Variations of Level in the Great North American Lakes, with Documents; Communicated for This Journal." *American Journal of Science and Arts* 16 (1829): 78–94.

Keough, Janet R. "The Range of Water Level Changes in a Lake Michigan Estuary and Effects on Wetland Communities." In *Wetlands of the Great Lakes: Protection and Restoration Policies; Status of the Science,* ed. Jon Kusler and Richard Smardon, 97–110. Proceedings of an International Symposium, Niagara Falls, N.Y., May 16–18, 1990.

Korgen, Ben. "Bonanza for Lake Superior: Seiches Do More Than Move Water." *Seiche* 3 (February 2000).

———. "Seiches." *American Scientist* 83, no. 4 (July–August 1995): 330–41.

Mather, William W. "Notes and Remarks Connected with Meteorology on Lake Superior, and on the Variations in Its Level by Barometric Causes, and Variations in the Season." *American Journal of Science and Arts* 51 (1848): 1–20.

Platzman, George W. "A Numerical Computation of the Surge of 26 June 1954 on Lake Michigan." *Geophysica* 6 (1958): 407–38.

Prince, Harold H., and Frank M. D'Itri. *Coastal Wetlands.* Chelsea, Mich.: Lewis Publishers, 1985.

Ragotzkie, Robert A., William F. Ahrnsbrak, and A. Synowiec. "Summer Thermal Structure and Circulation of Chequamegon Bay, Lake Superior—A Fluctuating System." In *Proceedings of the 12th Conference on Great Lakes Research.* International Association for Great Lakes Research, 1969: 686–704.

Whiting, Henry. "Remarks on the Supposed Tides, and Periodical Rise and Fall of the North American Great Lakes." *American Journal of Science* 20 (1831): 205–19.

ISLANDS

Of the North Shore's many natural features, perhaps none has been the subject of greater mystery—and misunderstanding—than its offshore islands. Given the inherent difficulties of exploring landfalls that lie in the midst of the world's largest freshwater expanse, such lapses are hardly surprising. In 1659 a group of Indians on the Keweenaw Peninsula told French explorer Pierre Esprit Radisson that a journey to Isle Royale in "faire and calme wether" required a day of continuous paddling "from sun rising to sun sett." Even under the best circumstances, such crossings were always a gamble. With a sudden shift in weather, they could become a life-and-death struggle. Indeed, today's paddler would recognize an unsettling accuracy in the Ojibwe description of one island (probably Michipicoten). To the Ojibwe, who labored to keep their featherweight canoes on course in stretches of open water, their destination appeared as a "floating island, which is sometimes far off, sometimes near, according to the winds that push it and drive it in all directions," wrote the French missionary Claude Dablon in the *Jesuit Relations* of 1669–70.

Such accounts were largely derided as superstition by European newcomers such as Dablon. But few were willing to test the veracity of the Ojibwe's observations by heading into the open lake themselves. As a result, much of what the early explorers knew about the lake's islands was based on hearsay, faulty translations of Indian accounts, or just plain wishful thinking. Blinded by greed or ambition, some newcomers simply invented their own extravagant notions about the lands that hovered like a blue smoke on the horizon. One of the most persistent rumors spoke of islands of pure copper that floated in the remote center of the lake. So solid was the ore that a man purportedly could hurl a rock on the ground and hear a metallic ring.

Like these mythical unanchored islands, many misperceptions simply came and went. But in some cases, ignorance would have historic consequences. In 1744, for

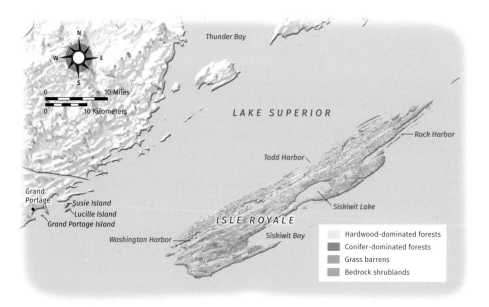

Located within ten miles of the mainland, the Susie Islands and Isle Royale archipelagos share many cultural, geological, and ecological connections with the Minnesota North Shore.

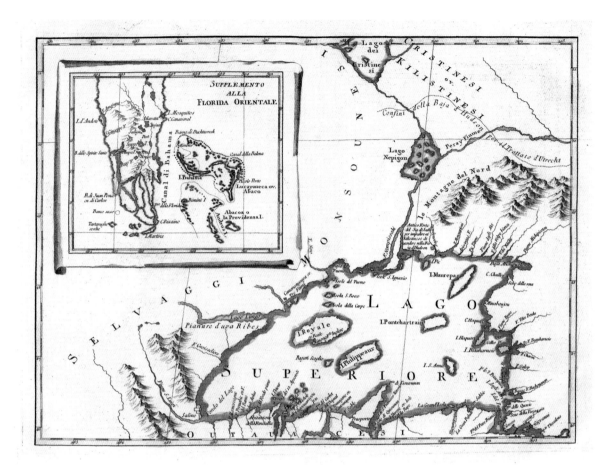

In his 1744 map of Lake Superior, French cartographer Jacques-Nicolas Bellin included a smattering of fictitious islands that persisted on official documents for more than a century. Courtesy of the David Rumsey Map Collection.

example, the French cartographer Jacques-Nicolas Bellin issued a map of Lake Superior in which he introduced a smattering of fictitious islands that persisted on official documents for more than a century. The largest, Isle Philippeaux, lay between Isle Royale and the Keweenaw Peninsula and measured some sixty miles long and twenty miles wide. In the 1783 Treaty of Paris, which delineated the boundary between the United States and British North America, the phantom Isle Philippeaux (spelled "Phelipeaux" in the document) officially fell to the United States, becoming one of the most celebrated errors in Great Lakes mapmaking.

Lake Superior's islands would not remain a mystery for long. With the development of a shipping infrastructure in the mid-nineteenth century, entrepreneurs would seriously begin to explore the natural-resource potential of the big lake's islands. In 1840 the Michigan geologist Douglass Houghton undertook a reconnaissance of the archipelago and within a few short years set off Isle Royale's first copper rush. By 1841, the American Fur Company employed some thirty fishermen to set their nets in the waters around Isle Royale.

But prospectors and commercial fishermen would not be the only treasure hunters. Following on their heels were seekers of other kinds of wealth—botanists, ornithologists, archaeologists, zoologists, and recreationists. As early as 1868 a young medical

Geologists began mounting scientific expeditions to Isle Royale in the early nineteenth century, along with natural historians who were interested in documenting its flora and fauna. This photograph was taken during an 1868 expedition to Michigan's Upper Peninsula and Lake Superior. Scientific Expedition to Lake Superior Photograph Collection, Bentley Historical Library, University of Michigan.

doctor and a handful of his students from the University of Michigan spent two months collecting samples of Isle Royale's flora and fauna. They would be among the first in a long series of distinguished scientists who found a biologically rich and beautiful world capable of sustaining the intensity of their curiosity.

So Close and Yet So Far Apart: Ecological Differences between the Islands and the Mainland

You can compare and contrast the biota of Superior's mainland and islands much as you would the facial features of fraternal twins: There are striking resemblances and conspicuous dissimilarities. From glacial times to the present, Lake Superior has been the primary force that determines the differences.

The simplest way to measure Superior's impact is through sheer numbers. As a general rule of thumb, the islands host fewer terrestrial species of vascular plants. Especially disadvantaged are plants that produce small seed crops or heavy seeds and those that rely on animals for transport and propagation. Not surprisingly, plant species that generate copious amounts of lightweight seeds that can be easily swept up by the prevailing winds are disproportionately represented in offshore locations.

The roster of mammal species is shorter too, forming a far more simplified complement than exists on a comparably sized territory on the mainland. Even on Isle Royale, where land takes up 210 square miles of an archipelago of 1,409 square miles, mammals number only eighteen species compared to the more than forty species found on the surrounding mainland. (The larger mammals are restricted to muskrat, beaver, snowshoe hare, short-tailed weasel, mink, river otter, red fox, marten, gray wolf, and moose.) The distance discourages all but the most hardy swimmers and flyers and those animals that manage to float to the islands' shores, hitchhike on the bodies of other species, walk across an ice bridge that occasionally forms during exceptionally cold winters, or are introduced by humans, either deliberately or accidentally. (Notably absent from Isle Royale are black bears, which hibernate during the cold season when ice bridges may become available.) Furthermore, to establish self-sustaining populations, animals must arrive in breeding pairs or be able to locate a suitable partner on the island, a task that may be next to impossible for species that have only a brief window of reproductive time. The odds against many creatures gaining a foothold under such conditions are formidable.

But physical distance is only one part of the islands' unique ecological story. Natural disturbance is another. Unlike the higher elevations of the Minnesota mainland, which have kept their heads above water for some eleven thousand years, the islands were buried under the ice sheet for a much longer period of time and then subjected to the dramatic rise and fall of the lake, which did not stabilize at the present level until about two thousand years ago. As a result, the islands are biologically younger than the Minnesota mainland since plant and animal communities have had less time to establish themselves and evolve in place.

In exposed reaches, the erosive forces of wind, waves, and ice push took up housekeeping duties where the ice sheet left off. Most heavily abraded were the smaller islands, many of which barely rise above the lake. Repeated disturbances have stripped the land down to a fairly homogenous terrain of erosion-resistant bedrock.

Only the higher elevations, sheltered recesses, and deep interiors of the larger landfalls are buffered from scouring action. Take, for example, the southwestern end of Isle Royale. The more extensive land mass and higher ridges have helped to protect the glacial deposits of the interior from being washed away by the lake. At the same time, they have moderated the local climate, allowing a deciduous forest of cool-temperate species such as sugar maple and yellow birch to take hold in the deep loamy soils.

But even in such hospitable pockets, soil development is slow. Cold, wet temperatures retard the cycles of plant growth and decay. "Late springs, cool summers, and early frosts have a marked effect on the growing season," write botanists Allison Slavick and Robert Janke about the general climatic conditions of Isle Royale. Adding to the obstacles for plants are fog blankets that frequently blot out the sun during summer's already short growing season.

Far more common on Superior's islands are boreal forests, like the one that has taken root on Isle Royale's northeastern end. Here, a more thorough scouring by the glaciers largely stripped away finer sediments such as clay, sand, and gravel. The trees best adapted to these nutrient-poor conditions are boreal conifers, in particular, black spruce and balsam fir. These hardy species also tolerate the local climate, which is chilled by the cold, dense air that slides down the isle's steep interior slopes and the frigid blasts that batter the land from the lake.

An estimated four hundred species of lichens inhabit Susie Island. Especially well represented are the rock-loving lichens, which form a nearly continuous carpet in places that are protected from ice and wave scour. Photograph by Beau Liddell, Images by Beaulin.

These trees do little, however, to remedy soil deficiencies. Unlike boreal hardwoods such as aspen, balsam poplar, and paper birch, which grow more quickly and produce a leaf litter that is easily decomposed and rich in nutrients, the slow-growing conifers shed needles that are low in nitrogen and highly resistant to decay. Elevated moisture levels dampen the ignition of wildfires that could break down the needle litter and release pulses of nutrients.

Disturbances such as large-scale windthrow also are less common in island forests. Without frequent visitations by wind and strafing by fire, the islands' forests are generally older than those found on the mainland and contain trees that are similar in age and species. (Ironically, although island forest trees are older, they typically are shorter and less robust.) Because of their homogeneity, island forests do not offer as many ecological opportunities for animals and other plants as do the more patchy and diverse forest communities of the mainland.

Although the islands may be biologically less diverse than the mainland, they nonetheless make an indispensable contribution to the biodiversity of the Lake Superior region as a whole. Due to the relative lack of disturbance compared to the mainland, the inland lakes of Isle Royale are home to an amazing diversity of invertebrates. Freshwater clams, sponges, bryozoans, snails, and insects occur in an abundance of forms and in huge sizes not seen on the mainland since the late 1800s. The cold, wet climate, bedrock shore, and extensive areas of thin soils mimic conditions that occurred in the immediate aftermath of glacial retreat. In other words, portions of the islands remain a kind of ecological frontier. As a result, they host a greater proportion of pioneer arctic and boreal plants than on the mainland, where many have become highly restricted in their ranges or have disappeared altogether. (See also "Hay Pickers and Grass Gatherers: Botanical Exploration along the Lakeshore.")

Indeed, the islands are rich in plants that thrive in the

Mosses attain an unparalleled diversity in island locations and include several rare species. High levels of year-round humidity allow mosses to accumulate to exceptional depths of three feet over the bedrock substrate. Copyright John and Ann Mahan.

contradictory conditions that resemble postglacial times. Living on shorelines that are alternately sun drenched and drought ridden are some of the best, if not the finest, assemblages of the 550 species of lichens found in the Lake Superior region. An estimated 400 species of lichen inhabit Susie Island alone. Especially well represented are the rock-loving lichens. A paddle along the islands' edges reveals whole rock faces that are encrusted with a thick impasto of gray-green starbursts, lemon-yellow dabs, and rusty-orange mottles. Adding texture to this abstract composition are foliose lichens, whose forms and color resemble curly-edge lettuce or blistered paint.

Also especially well represented in the islands' flora are the mosses, which attain an unparalleled diversity in these offshore locations. Not only are these terrestrial outposts rich in rare species—many of them arctic disjuncts—but they also harbor unusual community structures. On Susie Island, for example, the high levels of humidity that bathe the island year-round allow an exceptionally thick mat of sphagnum and feather mosses to form under the canopy of spruce trees. Engulfing boulders and deadfalls alike with an almost primordial lushness, the mosses form a soft, undulating carpet that can grow to depths of three feet. (Early accounts talk of similarly luxuriant moss growth on Isle Royale. Unfortunately, the repeated fires set by copper prospectors to expose ore bodies consumed these communities.)

The mossy profusion and padded silence of such forests give the impression that these places are set apart in space and time. Indeed, first-time visitors are likely to recall the description of Superior's lakeside forests given by J. Elliot Cabot, who accompanied the famed scientist Louis Agassiz on his tour of the Canadian north shore in summer 1848. "The woods are silent," Cabot wrote, "and as if deserted; one may walk for hours without hearing an animal sound, and when one does, it is of a wild and lonely character; the cry of a loon or the Canada jay, the startling rattle of the arctic woodpecker or the sweet, solemn note of the white-throated sparrow. . . . It is like being transported to the early ages of the earth, when the mosses and pines had just begun to cover the primeval rock, and the animals as yet ventured timidly forth into the new world."

No Island Is an Island: Human Interactions

But when it comes to human impacts on ecological systems, no island is really an island, even the twelve tiny landfalls that make up the Susie Islands archipelago. Measuring only eighty-eight acres, Susie Island is the largest landmass in the archipelago. In 1882, it attracted the attention of Major T. M. Newson, who burned over a portion of the island in search of copper. Newson burrowed into the island's volcanic rock, but his mine was a bust. In 1907 Ebenezer Falconer followed in Newson's footsteps, sinking a 210-foot shaft on Susie only to watch it flooded by a storm two years later. Falconer left behind an abandoned mine along with a more enduring piece of personal history: he named Susie Island, nearby Lucille Island, and neighboring Mount Josephine on the mainland after his daughters.

Such human interventions on the smaller islands, however, were relatively brief and sporadic for both native people and Euro-American visitors. Lacking sugar maples whose sap could be tapped in spring, beaver and caribou to hunt, and protected coves that might be used during fish harvests, the smaller islands did not offer the necessary sustenance that would make a visit worthwhile. As a result, says Tim Cochrane, an anthropologist and superintendent of the Grand Portage National Monument, native people likely utilized them only as emergency shelters during lake crossings.

Isle Royale, however, was different. Radiocarbon analysis of prehistoric mining sites dates the earliest known evidence of human activity to forty-five hundred years ago. (Sites that are some eight thousand years old have been documented around the Lake Superior basin, but older signs of human visitation have yet to be discovered on Isle Royale. Unfortunately, higher lake levels that persisted from five thousand to eight thousand years ago likely obliterated any potential evidence.)

The identity of these early visitors was a question that tantalized archaeologists ever since aboriginal mining pits were discovered on Isle Royale in the 1840s. Some gave into wild and extravagant speculation, even boasting that ancient cities once existed on the archipelago.

All too often outright racism trumped sound scientific judgment. Amateur archaeologist Henry Gillman was among the most outspoken proponents of the view that the prehistoric Isle Royale miners were part of a vanished, superior race whose refined civilization was evidenced in the elaborate burial-mound structures that extended across the United States, such as those concentrated in the river valleys of the Midwest. In an 1873 article in *Appleton's Journal,* he linked the mound builders with the lost ancient civilizations of Central and South America. Gillman considered the aboriginal miners to be so much more accomplished than the contemporary Indians of his day that he claimed the two groups were unrelated. According to Gillman, the "remarkable knowledge, enterprise, and patient endeavor" of Isle Royale's ancient miners was proof enough that they were "a people differing essentially from the North American Indian, as we have ever known him."

Minnesota state geologist Newton H. Winchell was among the first to counter what he called the "poor philosophy and poor science" of the subscribers to this theory. After examining a wealth of anthropological data and visiting Isle Royale in 1881, he concluded that the mound builders, and by extension the Isle Royale miners, were the ancestors of modern Indians. "It appears," Winchell wrote, "that every known trait of the moundbuilder was possessed also by the Indian at the time of the discovery of America."

The job of sorting out kernels of truth from the chaff of prejudice and fanciful thinking was made all the more difficult by the fact that native people appear to have stopped mining and trading copper from Isle Royale and other copper-rich areas in the Lake Superior region just prior to the arrival of Europeans in the mid-seventeenth century. Such an abrupt halt is especially confounding given the extensive history of copper mining and trade in which Indians once engaged. At least as early as 3000 BC, copper was widely distributed in North America. The distinctive high levels of silver in Lake Superior copper (so enriched that it was known as half-breed copper) have enabled archaeologists to trace the region's metal in objects that were traded as far west as the Great Plains and as far south as the Gulf of Mexico.

Indeed, the level of prehistoric mining activity on Isle Royale is remarkable. McCargoe Cove, located on the archipelago's northern shore, was one of the most actively mined areas, featuring 1,025 prehistoric mine pits alone. According to archaeologist Caven Clark, the great diversity of pottery styles in the shards recovered from sites near the mine and from numerous other locations around the island indicates that Isle Royale was a cultural crossroads. Indeed, the number of different Indian groups to visit the isle peaked during the Late Woodland Period (AD 1000 to 1400). In prehistoric times as well as during the historic period (post-1650), the archaeological record shows that these groups were drawn from Minnesota and Ontario populations that lived along the northern shore of the lake, including around the Pigeon River and Thunder Bay. (Despite long-held assertions to the contrary, Cochrane and Clark point out that there is no evidence to suggest that Indians along the southern shore of the lake routinely undertook the perilous forty-mile crossing to Isle Royale. They had no need to assume such risks, they say, since abundant copper reserves existed close at hand on the Keweenaw Peninsula.)

The early explorers and Jesuit missionaries offered various explanations for the absence of copper mining and trade among the native people they encountered. Influenced by the racist attitudes of their day, whites charged that Indians were either unaware of the copper resources in their midst or simply too superstitious to mine them. The Jesuit missionary Claude Allouez described what would become the prevailing stereotype about the relationship between native people and copper. In September 1666, while

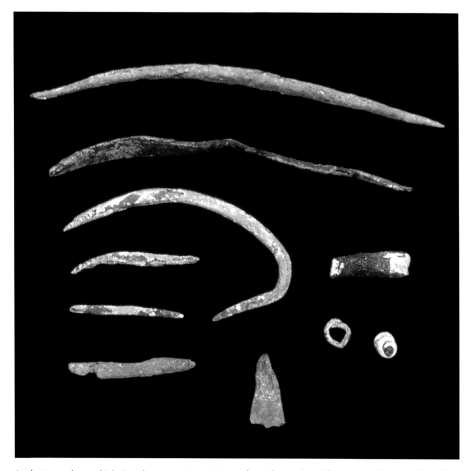

Ancient people used Isle Royale copper to create a variety of everyday objects. According to radiocarbon analysis of prehistoric copper mines on Isle Royale, human activity in the archipelago dates to as early as forty-five hundred years ago. National Park Service photograph by Caven Clark.

on his maiden voyage into Lake Superior, he observed that "the savages respect this lake as a divinity, and offer sacrifices to it because of its size . . . and also in consequence of its furnishing them with fish, upon which all the natives live when hunting is scarce in these quarters. It happens frequently that pieces of copper are found, weighing from ten to twenty pounds. I have seen several such pieces in the hands of savages; and since they are very superstitious, they esteem them as divinities, or as presents given to them to promote their happiness by the gods who dwell beneath the water. For this reason they preserve these pieces of copper wrapped up with their most precious articles. In some families they have been kept for more than fifty years; in others, they have descended from time out of mind—being cherished as domestic gods."

According to new scholarship, such simplistic explanations overlook the profound changes that European-introduced diseases wrought in the cultures and economies of Native Americans. Historians speculate that epidemics, like waves of great invisible armies, marched far in advance of the actual arrival of Europeans to decimate whole populations of native people, who lacked immunities to these diseases. Indeed, the populations of midcontinental Indian tribes may have succumbed to the first onslaught of disease as early as 1525, nearly a century before the first white explorer actually dipped his paddle into Superior. Not only would such a catastrophe have depleted the ranks of native miners, but it also would have dampened the demand for copper on the part of their most active customers. These included the agricultural groups in the valleys of the Ohio, Illinois, and upper Mississippi Rivers, whose dense settlement patterns would have rendered them especially vulnerable to epidemics.

Furthermore, white contact transformed the old copper economy by introducing a new currency: furs and European-made goods. Obtaining copper was a laborious and dangerous business. With the advent of the fur trade, Clark observes, Indians no doubt opted for less chancy and time-consuming means of obtaining metal implements. "Why make a copper awl," Clark asks, "when you could buy a whole box of them for a bale of furs?"

As for the charge that superstitious fears prevented Indians from visiting the islands in historic times, Cochrane says nothing could be further from the truth. Native people during both the prehistoric and historic periods experienced justifiable feelings of vulnerability as they steered their small canoes into the open lake. They embodied their apprehensions in the form of the often malevolent underworld figure known as Mishebeshu, which is literally translated as "Great Lynx." Although he assumed a rather cryptic form—Mishebeshu was pictured as both a huge aquatic feline and a hissing snake—the creature's effect on the lives of the Ojibwe, who were dependent on water for food and transportation, was real and unambiguous. One of the most terrifying aspects of Mishebeshu was his unpredictable power. Ethnographer Theresa S. Smith describes him as the "paradigmatic unbalancer of the world." Mishebeshu caused lakes to "grow suddenly rough and streams are transformed into a series of dangerous rapids and whirl-pools," she writes. "Mishebeshu is the uncanny element in this world, the hidden form beneath the ice, which may suddenly crack in winter. He is the one who pulls boaters and swimmers to their deaths and the one who makes the ground go soft beneath your feet." So feared was Mishebeshu that when a person dreamed of him, his tongue would be scraped with the bark of a cedar tree to loosen the monster's grip on the life of the unfortunate dreamer.

This image of Mishebeshu, literally translated as "Great Lynx," appears in a pictograph on the Canadian north shore of Lake Superior. A malevolent figure in Ojibwe mythology, Mishebeshu was blamed for sabotaging lake crossings with sudden storms. Photograph by Rolf Hicker, HickerPhoto.com.

Mishebeshu's domain was the big lake and its islands, including all the plants, animals, and minerals they contained. It was only natural, Cochrane says, for native people to engage in propitiation ceremonies designed to solicit safe passage. Lumps of copper, which were considered sacred objects, were routinely thrown overboard. Ritual sacrifices of dogs, especially white ones, often preceded journeys to and from the archipelago. In 1987 archaeologists discovered the burial remains of a dog on Isle Royale that they believe likely served as a sacrifice to Mishebeshu. As one Ojibwe man explained to the German ethnographer Johann Georg Kohl in 1860, dogs were selected as offerings since the "dog is our domestic companion, our dearest and most useful animal. . . . It is almost like sacrificing ourselves."

But Euro-Americans may have clung to the myth that Indians avoided the islands out of superstition, Cochrane says, to serve their own ends. Indeed, the portrayal of Isle Royale (not to mention the entire North American continent) as a virgin wilderness is part of what historians refer to as disinheritors' guilt. Denying the widespread occupation of the land by native people made it easier—psychologically and legally—for the new arrivals to take possession.

In prehistoric times, it is unclear whether copper mining was the main goal for undertaking the trip to Isle Royale or if it was simply considered an important part of an informal round of subsistence activities that included hunting, fishing, gathering berries and useful plants, and tapping maple trees for syrup. Cochrane insists that whatever their priorities, native people did not stop visiting Isle Royale simply because copper mining had come to an end. The archipelago held out other rewards that far outweighed the travel risks. The Indian name for Isle Royale, Minong, has been variously translated as the "place of berries," or a "good place to be." Indeed, Cochrane says, "it must have been the perfect refrigerator for people on the shore." Native people, he says, visited the archipelago in small family groups on a seasonal basis at least through the 1870s. At times, the isle's resources became a lifesaver for an entire people. For example, in September 1851, Cochrane points out, a Jesuit missionary traveled to Grand Portage and nearby Fort William and found the settlements virtually deserted. The Ojibwe inhabitants had paddled out to Isle Royale to fish the spawning runs of lake trout, a practice that continued in lean years until 1856.

According to Cochrane, the Ojibwe harvested more than just fish. The physical isolation of islands, what ecologists call the "fence effect," limits the easy dispersal of many species, concentrating the populations of some animals in densities far greater than those found on the mainland. Among the most abundant were woodland caribou. In the boreal forests of North America, caribou densities average 0.8 to 1.0 per square mile. Even with the added hunting pressures, which significantly thinned the archipelago's herd, the pre-1920 caribou population on Isle Royale was estimated at about 3.6 to 7.0 animals per square mile. (Like caribou, moose populations too have grown more concentrated on Isle Royale. At times, population densities on the archipelago have been more than five times the average densities on the mainland.)

While hunting, berry picking, and tapping maples for syrup undoubtedly were important subsistence activities, Cochrane believes that they took a back seat to the far greater resource appeal of Isle Royale: fish, especially during fall spawning runs, which can last from August to October. So central was the lake's fishery to the region's native inhabitants that Cochrane suggests they were maritime, rather than woodland, people. Indeed, the value placed on fish was encoded in the very specificity of the Ojibwe language. Ethnographer Kohl noted that the Indians in the Chequamegon area had "first, a general term for 'fishing,' and then a special word for every description of fishing." Kohl recorded multiple words used to denote variations on the act of fishing, including "I make fish," "I catch fish with nets," "I catch fish with a line on which there are many hooks," "I fish with a spear," "fishing with a spear in the [torch] light," and "I fish with a hook."

Then, as now, the vast complex of shallow reefs surrounding Isle Royale was considered among the most important, if not the best, preserve in all of Lake Superior for lake trout. Also coveted in the bountiful fish catch of Isle Royale was the lake sturgeon, which, according to Kohl, was known as the "king of fish" among the Ojibwe. This massive, prehistoric-looking fish was uncommon along much of the Minnesota North Shore largely because the steep gradient of its rivers and numerous rocky barriers offered meager spawning habitat. But sturgeons frequented the shallow mouths of Isle Royale rivers during spawning runs, where they were easy to detect and spear. Sturgeons were prized for their fatty flesh—and bulk. The size of a small whale, an adult fish can reach lengths of five feet and weigh up to eighty pounds.

Lake sturgeon frequented the shallow mouths of Isle Royale rivers during spawning runs. Easy to detect and spear, sturgeons were prized for their fatty flesh—and bulk. Adult fish in Lake Superior can reach lengths of five feet and weigh up to eighty pounds. Copyright Engbretson Underwater Photography.

Clark stresses that native people timed their expeditions to Isle Royale to exploit periods of seasonal resource abundance and then paddled back to the mainland. Relatively few people overwintered there, even in historic times. The reason, Clark says, is simple. Winter on Isle Royale could be just as harsh and unpredictable as on the mainland. And even though small family groups were thinly distributed through the forest in winter, appeals for help nonetheless could be made to relatives in times of stress. Those on the archipelago, however, could be perilously stranded, forced to rely on their own reserves until ice breakup in the spring.

The advent of large-scale commercial extraction of copper on Isle Royale, however, began to unravel this age-old subsistence pattern. Pressured by mining interests, the U.S. government negotiated a series of treaties to gain possession of Indian lands, particularly valuable ore-bearing territories. In the La Pointe Treaty of 1842, the Ojibwe relinquished thirty thousand square miles in Michigan's Upper Peninsula, parts of northern Wisconsin, and all of the U.S. islands in the lake. The Grand Portage Ojibwe, the band most entitled to lay claim to Isle Royale as ancestral lands, were not present, however, at the treaty negotiations. When they learned of the settlement, the band contested its terms, charging that the title to Isle Royale had been terminated without their knowledge or permission. Support for their cause came from the renowned Lake Superior missionary Father Frederic Baraga as well as from the Ojibwe negotiators along the southern and western end of the lake, who asserted that they spoke only for the islands under their jurisdiction, which did not include Isle Royale. Their protests fell on deaf ears. In the end, the Grand Portage band offered to sell Isle Royale for $60,000, but their bid was rejected out of hand. Instead, in what is known as the 1844 Isle Royale Compact, Indian agent Robert Stuart, under approval from the commissioner of Indian Affairs in Washington, D.C., pressured the band to cede Isle Royale for a mere $400 in gunpowder and $100 of fresh beef. To coerce the Grand Portage Ojibwe into signing the compact, Stuart withheld payments that were promised to the Ojibwe signatories of the La Pointe Treaty of 1842. "Isle Royale came under federal control for a very cheap price, but it was not a compact in which white Americans can take pride," write the authors of the 1988 *Narrative History of Isle Royale National Park*.

Unable to wait until the title to Isle Royale had been officially cleared, miners swept the isle in the first copper rush in 1843. It was followed by two additional mining booms, the last of which ended in 1892. The population influx peaked in the 1870s, so much so that Isle Royale was established as a county with its own county seat.

Native subsistence use of the archipelago dwindled in the wake of the ecological change that accompanied the miners. Forests were burned over to expose copper veins. Other trees were cut to build houses, shore up underground shafts, and fuel steam engines. Mainland-grown farm forage used to feed livestock and draft animals introduced nonnative seeds such as clover, thistle, plantain, dandelion, and grasses. Caribou and other game animals were hunted to stock the larders of mining camps, and their numbers plummeted.

Still, some Indian families continued to make seasonal treks to the archipelago. In her account of a year spent on Isle Royale in 1874–75 as the wife of a mining official, Sara Barr Christian recalls being startled by a group of Ojibwe who had silently crept up to her parlor window to listen to her piano playing. In the spring she and other mining families reciprocated by visiting the spring sugar bush of Indians camped on the north

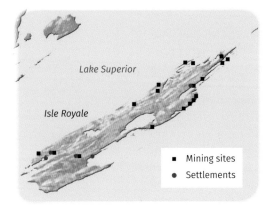

Isle Royale mining sites and settlements were established during three mineral booms that swept the island between 1843 and 1892.

Isle Royale copper has attracted humans for millennia. Excavated in 1874, this 5,720-pound mass of copper from the Minong Mine features the tool marks of prehistoric miners. Courtesy of the Burton Historical Collection, Detroit Public Library.

side of the island. "We were cordially welcomed," Barr writes, "and they showed us their process of making maple sugar."

White and Indian cultures would meet in other, jarring ways. For example, at most of Isle Royale's mines, including its most productive ones, the Minong and Island operations, workers simply blasted ever deeper into the veins of copper already exposed by the aboriginal pits. Some of the pits contained great boulders of pure copper that were last touched by native hands hundreds, perhaps thousands, of years ago. Having hammered off every protruding nodule with their tools of rounded lake cobbles, the prehistoric miners left behind huge, filed slabs, such as the one Barr witnessed during her stay. "It was almost as smooth as glazed pottery and perfectly beautiful," Barr writes. "It had been cleaned of all bits possible to remove, but the mass was beyond the power of these people to dislodge. . . . It was later removed by some interested party who brought a freighter into the cove, with hoisting machine and necessary equipment to move it into position to get it into onto the boat. . . . I suppose it went into the common melting pot, a smelter, along with other insignificant bars and ingots. But I do wish it might have gone to the Smithsonian Institute, it was so beautiful, and a silent monument to a lost tribe."

Compared to Michigan's Upper Peninsula, with its more accessible and easily transportable troves of copper, Isle Royale yielded little mineral wealth. As the *Ontonagon Herald* reported in 1879, "The ancients got the juice and left us moderns little but the acrid rind to nibble at." By the turn of the century, however, Isle Royale had developed a renown for other kinds of riches—clean, healthful air, rugged terrain to rejuvenate both body and mind, and abundant wildlife free for the taking. Its remote location gave it an aura of being removed in place and time. Not surprisingly, another myth— that of pristine wilderness—began to shape the archipelago's future.

The Making of an Isle Royale National Park: Wilderness or Working Landscape?

No sooner did the Wendigo Copper Company abandon its mining operations on Isle Royale in 1892 than the company announced plans to liquidate the timber on its remaining lands, sparing the most picturesque haunts for a luxury resort and game preserve. Officials boasted that Isle Royale would become "a paradise within a few years." And the builders of paradise made no small plans. According to a resort prospectus, "Not only will hotels be built in several places but bath houses, a summer theatre, pavilion, dance hall and all that goes to make a resort popular and attractive."

The company's grand scheme never materialized. But Isle Royale would not be without a luxury resort for long. In 1902 Chicago shipping magnate Walter H. Singer opened the Island House Resort, which offered guests twenty-two rooms, ten cabins, a barber shop, a pool room, and a bowling alley that could be converted into a dance floor. The steamships and railroads that hauled copper, timber, and lake trout out of the Lake Superior region now began to deliver riches as well: tourists. Indeed, by 1904 no fewer than five passenger boats serviced the resort.

Soon others would cash in on the growing tourism trend with their own vacation destinations: Tobin's Harbor Resort, Tourists' Home, Rock Harbor Lodge, Belle Isle Resort, and Johnson's Resort. Historian Lawrence Rakestraw observes that "a period came in the twentieth century when men found it more profitable to work tourists than ore bodies. The transition of Aspen, Colorado, from a silver mining center to a haven for philosophers and skiers took about forty years. On Isle Royale the transition was more rapid, with the end of the mining era coinciding with the beginning of tourism as a major economic activity."

Accommodations ranged from rustic to cushy, but they all shared a common promotional philosophy: restoration of body and soul in a remote and pristine natural environment. According to a 1914 prospectus from the steamer *America,* Isle Royale was geared to "the growing volume of near-to-nature resort lovers, who are tired of and are abandoning the worn-out and unsatisfying spas of the populous centers." The archipelago offered "wonderful places to explore, mountains to climb and rivers to trace to their source." Some resorts even offered guided moose-viewing excursions. An early brochure for the Rock Harbor Lodge perhaps best sums up the sales pitch that the burgeoning resorts made to their urban clientele: "Do you wish all the comforts of civilization while dwelling in the midst of nature's unspoiled wilderness—or do you prefer to rough it a bit?"

Lost in the tourism promoter's hype was the fact that Isle Royale was anything but isolated and untouched. Although Isle Royale was located far from polluted, noisy, and crowded urban centers, its fate had been—and would continue to be—integrally tied to them. Indeed, the very development of the archipelago as a wilderness getaway was as much shaped by urban needs and aspirations as was the sinking of a mine shaft. In the nineteenth and early twentieth century, patients suffering from a variety of chronic maladies were routinely prescribed an extended stay at a summer resort. Healers of all stripes, both quacks and reputable doctors, "advocated fresh air, temperate climate, sunlight, bathing, and exercise" for a variety of ailments ranging from consumption and asthma to gout and rheumatism, writes historian Cindy S. Aron. Physicians prescribed not only a change of air but also a regimen of quiet and isolation. By the turn of the nineteenth century, Aron points out, "medical opinion increasingly held that urban dwellers were suffering from nervous diseases brought on by too much civilization." As one editorial put it while extolling the virtues of a vacation in the Adirondacks, the "atmosphere of this whole region, and the rough out-door life and entire change from civilized habits, make it a perfect sanitarium for our worn-out city workers."

Early on, Isle Royale developed a reputation for a commodity it possessed in abundance—clean, healthful air. With good reason. Before the advent of antihistamines, the

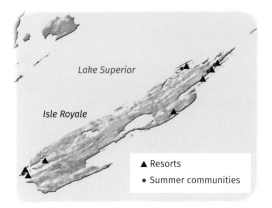

By the early twentieth century, Isle Royale had a reputation as a tourist destination that offered hay fever relief, rugged relaxation, and abundant wildlife.

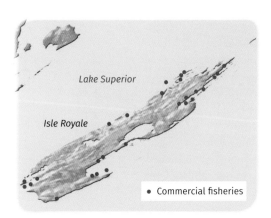

During the Roaring Twenties, Isle Royale supported seventy-five commercial fisheries.

only relief for asthma and hay fever sufferers was an excursion to a woodland retreat. Indeed, so relieved was he to see his asthma symptoms disappear during one visit to Isle Royale that Kneut Kneutson in 1922 abandoned his real-estate business on the Minnesota prairie to assume ownership of Park Place. (Renamed the Rock Harbor Lodge, it continues to house visitors to this day.) In the "fresh lake breeze," Kneutson noted, was "a fine brand of good medicine."

But the active, outdoor life was celebrated not just for its health benefits alone. President Theodore Roosevelt, an avid outdoorsman, would add a political and moral imperative. As the American frontier came to a close, Roosevelt feared that Americans were succumbing to "flabbiness" and "slothful ease," losing the "great fighting, masterful virtues" that created the nation's greatness and were essential to sustaining it. He appealed to Americans (a message primarily skewed toward men) to periodically indulge in a "life of strenuous endeavor" in the wilderness, believing that it promoted "that vigorous manliness for the lack of which in a nation, as in an individual, the possession of no other qualities can possibly atone." (See also "Nearshore.")

For midwesterners, the nearest "wilderness" was the Lake Superior region, particularly Isle Royale. In a 1926 article for *American Game* magazine, Frank M. Warren articulated a common attitude toward the archipelago: "Its remoteness and difficulty of approach have also protected it while the more available and profitable places have suffered," he writes. "Thus it is that Isle Royale, except for occasional efforts to find copper, has remained singularly free from change due to man's occupancy."

But Isle Royale was hardly the mythical Eden of Warren's rhapsody. For example, when archaeologist Fred Dustin visited the archipelago in the 1920s, he discovered a thriving fishing community. He tallied 75 seasonal commercial fishermen, who, along with the members of their families, accounted for some 250 people. Isle Royale catches not only showed up as food on the dinner table of islanders, but they also made up a significant portion of Lake Superior fish exports to wholesale centers in Chicago. So lucrative was this trade that, as one commercial fisherman recalls of the Roaring Twenties, "Everybody was making big money fishing."

Resorts and land developers also were turning a tidy profit. Dustin tabulated another 250 people in tourists and seasonal home owners, some of whom fell in love with the archipelago's rustic charm as resort goers and bought whole islands for their own private retreats. Among them was

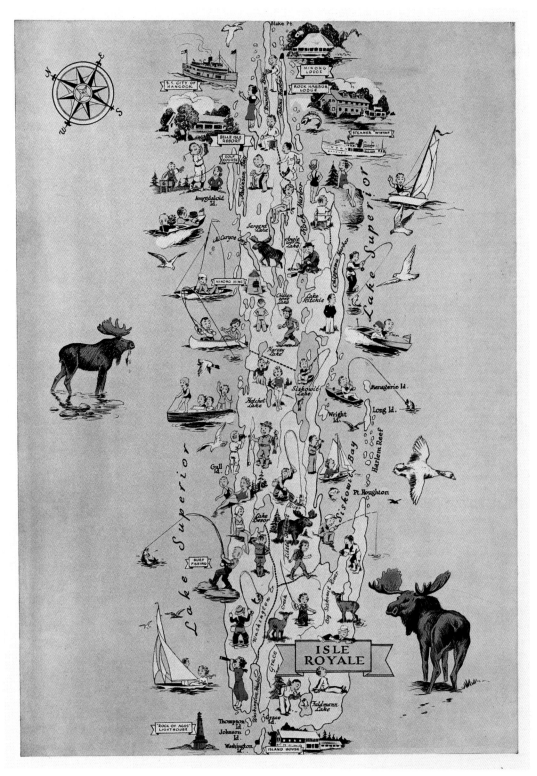

Despite its remote location, Isle Royale offered tourists amenities ranging from a bowling alley that could be turned into a floor for ballroom dancing to a groomed golf course. Courtesy of the David Rumsey Map Collection.

A historic fishing operation on Isle Royale. Photograph by Hibbard Studio, 1927. Courtesy of the Minnesota Historical Society.

Omaha coal executive George W. McGrath, who in the early 1920s built a family vacation compound on Amygdaloid Island for a reported $40,000 (nearly $500,000 in 2013 dollars).

Perhaps the biggest challenge to the argument that Isle Royale was a place set apart in place and time was the fact that in 1922 one-half of Isle Royale—some 66,500 acres—was owned by the Minnesota Forest Products Company. Periodic threats by the company to clear-cut its holdings galvanized efforts by seasonal residents, resort owners, and visitors to place the archipelago under federal protection. The outdoor editor of the *Detroit News,* Albert Stoll Jr., led the charge in calling for a wilderness park. In the first of many editorials, Stoll in December 1921 pled his case, describing Isle Royale as "practically the only bit of unspoiled nature east of the Mississippi."

The citizen-led effort to place the archipelago in federal hands was successful. In 1931 the U.S. Congress authorized the establishment of Isle Royale as a national park. But it was the bottom of the Great Depression, and monies to purchase private lands were scarce. Indeed, the government would not gain final title to all the land on Isle Royale until 1940, at which time the park was officially designated. In the interim, the Minnesota Forest Products Company cashed in on its holdings and logged its forest lands in 1935. Preservationists caught wind of the company's plans and scrambled to raise funds to purchase the uncut forest, but the money arrived a few months too late. Sheared of their valuable lumber, the cutover acres were sold to the government in 1936. During an ensuing drought, catastrophic blazes swept the archipelago, fueled in large part by the piles of slash left behind by logging operations. Some 20 percent of the island was consumed by fire.

All mining, logging, and private recreational development ceased once Isle Royale officially became a national park. In time, the island's cut and burned forests regenerated and matured providing habitat for surviving wildlife. In 1976 Congress set aside 98 percent of the park as wilderness. Today that number has grown to 99 percent.

According to the National Wilderness Preservation Act of 1964, such lands are "where the earth and its community of life remained untrammeled and where man himself is a visitor who does not remain."

But on Isle Royale, Cochrane says, wilderness is a relative term. "Isle Royale's 'natural' environment is much altered by human activity, despite its wilderness designation, which implies an 'untrammeled' environment," he observes. "For example, loggers and miners inadvertently introduced nonnative grasses when they brought draft animals to the Island. CCC [Civilian Conservation Corps] enrollees ate apples by the barrels and today a number of abandoned CCC camps are earmarked by moose-browsed apple trees. The Army Corps of Engineers has blasted out rock to make safe navigation into the fjord-like Chippewa Harbor. Copper miners repeatedly burned forests to expose rock for prospecting. The Belle Isle Resort blasted rock and hauled soil from McCargoe Cove to nurture their golf course grass. Timber was cut to shore up mine adits and shafts, to make barrel staves, to fuel steam ships and to build cabins. Trappers extirpated lynx, coyote and possibly beaver. Ojibwa caught the now-extinct passenger pigeon in aerial nets. Thus, it is best to speak of Isle Royale as an enduring untamed place, rather than an archipelago untouched by man."

Isle Royale was far from pristine when it was declared a national park in 1940. In the years leading up to its designation, at least half of the island was logged and portions were strafed by wildfires. Photograph by Hibbard Studio, 1927. Courtesy of the Minnesota Historical Society.

An Outdoor Laboratory for Studying Human Impacts

When the National Park Service was created in 1916, its stated goal was to "conserve the scenery and the natural and historic objects and the wildlife therein and to provide for the enjoyment of the same in such manner and by such means as will leave them unimpaired for future generations."

Over the decades, this mission has been debated and reinterpreted given changing pressures on the national parks as well as advances in scientific thinking about wildlife management and ecosystem processes. A sea change in park philosophy was inaugurated with a 1963 report commissioned by Stewart Udall, secretary of the interior under President John Kennedy. Known as the Leopold Report (so named for its chair, Dr. A. Starker Leopold), the document urged that national parks be managed as "vignettes of primitive America," that is, to preserve as much as possible the native complement of plants and animals that existed before Europeans set foot on the continent, either as they were or as they naturally would have evolved in the absence of white settlement.

The report marked a critical shift in the emphasis of national park management. Now instead of simply preserving their scenic qualities for the pleasure of tourists, managers were charged with serving as stewards for the many unheralded ecological interactions that create and support the parks' extraordinary beauty. Among the attitudinal shifts was the recognition that wildfires and predators play beneficial roles in many ecosystem processes. (Eliminated, for example, were hunting and trapping campaigns against such animals as wolves and grizzly bears.)

The job of preserving, managing, and restoring vignettes of primitive America to the national parks inevitably raised the bedeviling question: What are natural conditions, that is, those that exist outside the sphere of human influence? Indeed, this query has opened one of the most important scientific debates of our time. Paleobiologists have added complexity to the picture by discovering many natural processes that may have coevolved—and even become dependent on—the traditional land-management practices of native people that occurred over hundreds, sometimes thousands, of years. More troubling and difficult for managers to take into account are concerns about the transport of human-produced toxins to the far corners of the world, even to little-visited polar regions. The buildup of poisons in arctic mammals, for example, may be not only affecting the animals' health but also altering longtime behavior patterns, which may threaten the survival of some species. Even an activity as mundane as driving to the grocery store in Miami contributes to the surfeit of greenhouse gases that causes global warming, a phenomenon that is melting the ice pack on which polar bears depend.

National parks have occupied front-row seats in this often-contentious debate. Because the parks are commonly isolated or surrounded by sparsely developed lands, they are presumed to be insulated from the fallout of human activities. Many of them, including Isle Royale, have been selected as sites for scientific research and long-term field monitoring. They have been regarded as isolated outdoor laboratories in which scientists made every effort to minimize their own interventions in order to gather "pure" data on the "natural" unfolding of ecological processes around them.

Questioning the assumption that geographical remoteness equals pristine ecological conditions, however, some scientists have gone so far as to suggest national parks, including Isle Royale, are cultural artifacts, their biotas an artificial construct. The archipelago, they charge, is no more a wilderness, despite the relative absence of people, than the more heavily developed mainland. Indeed, far from being regarded as a laboratory of pristine nature, the isle should be used as a place in which to study the outcomes of human manipulation and management.

Look no further, they say, than the archipelago's moose and wolves, which have been the subject of one of the longest-running and most famous studies of animal interactions in the annals of science, not to mention in the popular press. In the minds of many, these animals are imbued with larger-than-life meaning. Indeed, the moose and wolf have become potent symbols of wilderness values.

The potential of the majestic moose to tap into the public's romantic perceptions of wilderness was not lost on early Isle Royale promoters. No sooner had the archipelago been repackaged as wilderness in the early twentieth century than the moose was proposed as its mascot. "What we are doing for the elk in Yellowstone National Park," Warren exhorted in 1926, "let us do for the moose on Isle Royale. Nature has made it a natural citadel for the moose. Surrounded as it is by Lake Superior, only man can get to and destroy them there."

The moose proved to be an effective marketing ploy for tourism promoters. As wildlife biologist R. Gerald Wright points out, "Before the wolves arrived, moose symbolized the island and were the primary concern of management. Isle Royale National Park was often described as America's greatest moose refuge, and visitors often came just to photograph them."

But revisionist biologists point out that the moose's presence on Isle Royale is due, indirectly or possibly even directly, to human intervention. Historically, woodland caribou were the dominant herbivore of the larger islands of Lake Superior, including Isle Royale. Indeed, there is no evidence in the archaeological record to suggest that moose ever inhabited the islands. When geologists John Foster and Josiah Whitney undertook a reconnaissance of the archipelago in 1850, they discovered a native complement of large mammals that was vastly different from the one found today. "The caribou, the lynx and the rabbit [snowshoe hare]," they noted, "are among the few animals that roam over its surface."

The bonds among this native triad, however, were soon broken. Moose are thought to have arrived on the archipelago in 1905. With abundant food and no predators, their numbers swelled. By the early 1930s their population was estimated between one thousand and three thousand animals. Wildlife biologists suspect that the numbers of snowshoe hares, whose populations fluctuate naturally, may have dropped even lower than normal after competing with moose for limited forage. In turn, lynx populations, which depended on the hares, plummeted. Hare numbers have since rebounded, mirroring cyclical patterns found on the mainland. But with few new individuals recruited from the mainland and their numbers already dwindling, lynx may have been finally eliminated through overtrapping.

By the late 1920s, caribou too had disappeared from Isle Royale. Within less than a century, the species that occupied Isle Royale for more than thirty-five hundred years had been extirpated.

The reasons for the disappearance of caribou remain speculative, says Jean Fitts Cochrane, a conservation biologist for Isle Royale National Park. But scientists have credible hunches. For starters, Isle Royale caribou likely were overhunted. "Given the great demand for meat and hides in the early 1800s," Cochrane says, "the availability of

Although moose are excellent swimmers, evidence suggests they did not swim to Isle Royale from the mainland but instead may have been transported by boat to the archipelago by a group of wealthy sportsmen. Courtesy of Superior National Forest.

firearms, the tradition of hunting on Isle Royale, and the close proximity of the island to the two largest trading posts in the region, Grand Portage and Fort William, I conclude that hunting on Isle Royale was likely heavy in the early to mid-1800s."

Caribou were under assault from other forces as well. Predation by lynx and coyotes took their toll on the caribou that managed to survive. All these factors likely contributed to the numbers of breeding adults dropping beneath a sustainable threshold. Compared to other ungulates, caribou are slower to reach sexual maturity and have relatively low reproductive rates, rarely giving birth to multiple offspring. It appears too that island populations relied on caribou migrating from the mainland to periodically replenish their numbers. But such recruits were not forthcoming, since caribou populations were subject to similar predation and overhunting there as well.

Furthermore, major ecological changes occurred that reshuffled the entire deck of species in the region. The logging of forests on the North Shore and subsequent scorching fires created huge swaths of early successional vegetation, such as aspen, balsam poplar, and paper birch. The abundance of their favored browse enabled white-tailed deer to expand their range into the region after 1900 and allowed moose populations to surge. The deer brought with them the meningeal brainworm, a parasite that is fatal to caribou. The greater numbers of deer and moose also increased the population of another animal—the wolf—whose predation may also have contributed to the mainland caribou's demise.

The swelling of moose numbers on the mainland due to human-caused ecological change may have increased the odds of some individuals dispersing to new territory on Isle Royale. Initially, wildlife biologists hypothesized that moose traversed an ice bridge that formed between Ontario and Isle Royale during the bitterly cold winter of 1912–13. This explanation, however, has not withstood scrutiny, says Bill Peterson, a retired wildlife manager for the Minnesota Department of Natural Resources. With their sharp hooves unable to gain purchase on the ice's slick surface, moose are naturally skittish

A winter staple for moose, mature balsam fir trees bear a telltale browse line *(above)* indicating the moose's farthest reach. Balsam fir in the forest's understory *(below)* has rebounded only in places were moose populations are low or have been intentionally excluded. Photographs by Rolf O. Peterson.

about heading out onto frozen lakes. Furthermore, moose rarely travel in large groups. It is impossible, wildlife biologists say, for a handful of animals to have produced a herd which by 1915 consisted of an estimated 250 to 300 animals.

One accepted explanation to date has been that several small groups swam to Isle Royale and took up residence. Such a feat is not altogether implausible since reports have documented moose swimming several miles offshore in Lake Superior. For moose numbers to have built to several hundred animals by 1915, such a marathon swim had to have occurred around 1905.

But new DNA evidence has muddied the already murky waters surrounding the moose's mysterious appearance on Isle Royale by lending greater credibility to a claim that moose were transported to the archipelago by a group of wealthy sportsmen. (Quite possibly they were members of the Washington Harbor Club, who had both the means and motive to undertake such a transplant. Composed of Duluth businessmen, many of them shipping and mining tycoons, the group established its own sportsman's club on Isle Royale in 1902.) According to the alleged story, moose were culled from herds around Baudette in northwestern Minnesota and transported to Two Harbors around 1905. From there they were shipped to Isle Royale, presumably to stock the archipelago with game. In 1998, as part of a larger genetics study of moose, officials from Alaska's Department of Fish and Game analyzed mitochondrial DNA (the genetic material passed from mothers to their offspring) from moose in Ontario, Minnesota, and Isle Royale. Test results showed that DNA from Isle Royale moose did not match any of the three types of DNA from Ontario moose. It did, however, match one of the three types of DNA found in Minnesota moose.

Proponents of the moose-transplant hypothesis say that such herculean efforts are not far-fetched. Some of the earliest pleas for Isle Royale's protection called for setting the archipelago aside as a game preserve, in part reflecting a growing constituency for woodland sports. At the turn of the nineteenth century, rod-and-gun clubs began to buy up acreage in the cutover lands of Minnesota, Michigan, and Wisconsin for the purpose of establishing private game reserves. The interest

in hunting prompted the Michigan State Game Commission to transport a small herd of white-tailed deer to the archipelago in 1910. The animals failed to flourish and disappeared from Isle Royale after 1936.

Regardless of their means of entry, moose began to exert profound changes in the composition of terrestrial and aquatic plant communities within a few years of their arrival. By 1930 the animals had so severely depleted their browse that their population was on the verge of collapse. A major postlogging fire, which swept over more than 20 percent of the archipelago in 1936, boosted herd numbers by providing the animals with their favored foods—aspen and paper birch—which often colonize landscapes following disturbances such as fire. But Isle Royale populations have not since surged to their 1930 highs, largely due to the arrival in the late 1940s of gray wolves, which have kept moose numbers in check.

Still, moose browsing has changed the look of the forest, the composition of its plant communities, and, according to research, the very fertility of the soil. The effects of moose on the forests of the main island have varied depending on such characteristics as soil depths, the type of forest and its disturbance regimen, and the density of moose. In general, however, shrubs and tree saplings have been cropped to within three to five feet above the ground. Shrubs such as red-osier dogwood, squashberry, and highbush cranberry have declined. With the shrub layer severely pruned or eliminated, more light penetrates to the forest floor to stimulate ground-cover plants. In heavily moose-browsed forests, studies have shown that this herbaceous layer covered twice as much area as in unbrowsed sites. Composed primarily of tall, older trees that had been established before moose set foot on the island, thick ground cover, and a dwarfed and thinned shrub layer, these forests take on a more open and parklike appearance than comparable forests on the mainland. Moreover, they bear little resemblance to the forests described by early Isle Royale visitors or to the woodlands on moose-free islands, which exist in the far reaches of the archipelago.

But in describing the archipelago's forests, botanists often refer more to what is missing than to what is present. American yew, for example, once the most abundant ground cover on the archipelago, was virtually extinct by the 1930s in all but a few remote islets that were unvisited by moose.

These hungry ungulates have had a dramatic impact on the future of balsam fir as well. Normally fir is abundant in northern forests. But on Isle Royale balsam fir is a winter staple for moose. On the western half of the island, only the fir trees that were well established before 1920 (the time when moose numbers reached a critical density) have survived into adulthood, their bottom branches bearing a telltale browse line indicating the moose's farthest reach. These tall, mature trees are nearing the end of their life cycles. It is unlikely that younger generations of balsam fir will tender their spires into the forest canopy any time soon since pruning by moose has generally kept them to less than five feet in height. In many cases, these trees have simply stopped growing in the face of sustained cropping. Even when given a reprieve from grazing during a severe moose decline in 1996, these stunted trees were so weakened that they were unable to rebound and died off in substantial numbers.

Also severely browsed in moose-occupied territories is American mountain ash and white birch. Between 1948 and 1950, University of Minnesota scientists erected four exclosures on the main island to monitor changes in the moose-occupied forests. In the control plots adjacent to the exclosures, researchers documented that more than 80

percent of aspen, birch, mountain maple, and mountain ash were consistently cropped by moose. Compared to the same species that grew within the exclosures and out of reach of moose, the browsed species were severely stunted.

In some of the most heavily grazed control plots, white spruce, a species that is nearly absent from nearby moose-free islands, are the only trees under eighty years of age that exhibit normal growth patterns. That is because moose largely overlook spruce. Like most conifers, spruce have adapted to the nitrogen-poor soils in which they live by growing slowly and living well within their nutritional budgets. They protect their meager nitrogen reserves by producing substances that discourage browsing. Spruce needles are especially high in resins and lignin and low in nitrogen, making them neither very palatable nor nutritious. Selective browsing by moose of other, more tasty and fortifying plant competitors has enabled spruce to attain higher-than-normal densities. Indeed, many heavily grazed forests have been turned into what some scientists have called "spruce savannas." Such monocultures reduce the overall biotic diversity of the archipelago's forests.

But the prodigious appetites of moose do more than simply limit the kinds of plants that are able to take root and mature. Forests are fed by the rain of litter that falls from above. By determining the species makeup of the forest—and the leafy debris that falls to the forest floor—moose browsing influences the very chemical constituents of the soil. Studies by University of Minnesota forest ecologist John Pastor and his colleagues have shown that spruce savannas may severely limit the amount of nitrogen available for forest regeneration and may have severe long-term effects on the forest's productivity. Spruce trees grow slowly, largely because they merely take small sips of nitrogen from the soil. This allows the unused portion of nitrogen to leach into the soil beyond the root layer of most plants. Spruce trees, Pastor points out, also retain their needles longer than

Moose largely ignore white spruce in favor of more nutritious vegetation such as balsam fir, aspen, birch, mountain maple, and mountain ash. Overbrowsing has resulted in simplified, parklike forests known as spruce savannas. National Park Service photograph by D. Paul Brown.

other trees, often up to seven years. The nitrogen they absorb is bound up for lengthy periods of time within the plants' tissues, thereby depriving other plants of a ready supply of this critical nutrient. When the trees do shed their needles, the litter contains less nitrogen than the leaves of northern hardwoods. And since the needles decompose far more slowly than those of deciduous trees, nitrogen is released at a far slower rate.

The chronically low levels of nitrogen may have long-term effects on the vigor of the forest, say Pastor and his colleagues. Low levels may slow down the rate at which plants are able to recover from severe browsing. Spruce may be able to take advantage of the reduced competition from fellow plants to increase their density. In time, however, browsed plants may evolve mechanisms for tolerating low nitrogen levels. But such coping mechanisms will only help to perpetuate the low-nitrogen cycle. As a result, the scientists say, moose may have changed not only the appearance but also the vitality of the forest for decades, perhaps even centuries, to come.

Managing the Future of a National Park

Lake Superior's islands teach a fundamental lesson about the region's ecology: that the boundaries between counties, countries, terrestrial and aquatic ecosystems, human and nonhuman realms are very porous, if they exist at all. Not surprisingly, the effects of our actions can have unintended consequences that are destructive, widespread, and long-lasting.

To this day, the islands continue to register the effects of human activities even though relatively few people set foot on them each year. Some are rarely visited at all. Even many miles of cold, unpredictable waters have not been able to insulate the islands from profound human effects. Take the seagull, whose populations, like those of the moose, have responded dramatically to human-caused ecological change. The ready access to year-round food in landfills as well as dependable nesting sites in the human-made structures of industrialized harbors have enabled gull populations to explode. As their numbers expand, biologists worry that these wide-ranging, highly adaptable birds will occupy ever more of the islands' sparsely vegetated areas for nesting sites and permanently change these unique island habitats. Studies have shown, for example, that nesting gulls not only destroy native vegetation, but they also change the soil chemistry of their nest sites. Native plant communities severely disturbed by nesting gulls are also open to nonnative species. Ferried there via regurgitated food pellets containing seeds of alien plants, they can take over vulnerable island habitats. (See also "Nature or Nuisance? Gulls in the Great Lakes.")

It turns out that the islands are vulnerable to another aerial assault. In 1974–76, Wayland R. Swain, a scientist with the Environmental Protection Agency, sampled fish from Siskiwit Lake on Isle Royale. His goal was to obtain baseline readings from the tissue of fishes reared in "pure" waters in order to compare contaminant loads in those taken from nearshore areas around the mainland that were polluted by human activities. Much to his shock and dismay, Swain found elevated levels of PCBs in Siskiwit Lake fish. And when he tested the waters, he discovered levels of PCBs in this isolated lake on the archipelago that were thirty times greater than those found in the open waters of Lake Superior.

Since then, PCBs have been joined by a long list of other dangerous chemicals. Indeed, scientists have determined that the largest sources of chemical contamination to the Lake Superior region comes via the atmosphere in the form of dust, rainfall, and snow. In places such as Isle Royale, the "fence effect," which concentrates animals in large numbers by preventing ready avenues of dispersal, ironically applies to pollutants as well.

Some scientists argue that the islands' best hope for surviving these and other assaults is to mitigate adverse human effects and restore the islands' native complement of plant and animal species wherever possible. But this hands-on approach raises especially difficult issues for managing designated wilderness areas such as Isle Royale National Park. For the most part, park managers have shied away from manipulating the biotic interactions on the archipelago, letting nature take its course, so to speak.

But dramatic changes, including vegetational shifts caused by moose overbrowsing, are testing this philosophy. The archipelago "is on a long, slow decline in terms of food for moose," says Rolf O. Peterson, a renowned scientist who has conducted landmark research on the interaction between Isle Royale's moose and wolves since 1958.

Since 2000, moose have had to cope with another stressor: a raging infestation of ticks. In an effort to rid themselves of these blood-sucking parasites, the animals literally tear out their hair, an action that distracts them from foraging. Weakened by weight, hair, and blood loss, moose have become more susceptible to disease, freezing conditions, and winter wolf kill. Between the winters of 2002–3 and 2004–5, moose numbers plummeted from an estimated 1,100 animals to less than 500. By 2009 the population made only modest gains to 530 individuals. The bonanza of frail and sickened prey caused a spike in the wolf population, which jumped from 19 in 2003 to 29 in 2004.

The feast for wolves, however, was short-lived. With declines in the moose population, wolf numbers dropped to 20 individuals by 2007. The ready availability of older, more easily culled moose (those over fourteen years of age) allowed wolf populations to rebound to 24 individuals by 2008. By 2013, the moose population once again approached its long-term average. Despite the increase in prey during this five-year period, however, wolf numbers declined by 66 percent.

What does the future hold for the isle's two faunal celebrities? As the past has shown, predicting their fortunes is a game of chance. "The last time it seemed like moose were at the end of their rope, parvovirus came in and wiped out the wolves," Peterson points out. He is referring to the time in the late 1970s when a mutant virus known as the canine parvovirus (CPV) swept domestic dog populations around the world, causing high mortality among young dogs due to dehydration brought on by violent bouts of diarrhea. By 1981 the case loads of veterinarians in Houghton, Michigan, were inundated with CPV-infected pups. That same year, humans brought the disease to Isle Royale, and the archipelago's wolf population collapsed. A record nine pups that were born in 1981 succumbed to the disease.

By 1988 wolf numbers had not yet recovered from the population crash of 1980–82 even though tests showed that CPV had run its course. Scientists discovered another potential reason for the decline of wolves on Isle Royale. Genetic analysis showed that all members of the archipelago's wolf packs are descended from a single female that arrived during the late 1940s. Over time, inbreeding has resulted in a loss of genetic variability such that today's wolves possess only half as much variability as their counterparts on the mainland. In 2009 Peterson and his colleagues reported their discovery of an increasingly high incidence of congenital bone deformities, which in domestic dogs can cause pain and paralysis as well as restrict mobility. According to the researchers, this birth defect is the "first strong evidence of negative physical impacts to wolves

Scientists speculate that the first gray wolf migrated to Isle Royale via an ice bridge that formed between the mainland and the island in the late 1940s. U.S. Fish and Wildlife Service Digital Library.

attributable to in-breeding." By 2012, the researchers tallied just nine wolves including only one confirmed female. They attributed the persistent decline to inbreeding, starvation as the number of old moose diminished, and a renewed outbreak of CPV.

According to Peterson, it is unlikely that wolves will be able to migrate from the mainland to rejuvenate the isle's genetic reserves. Their crossing, he points out, depends on the formation of a continuous ice pack that provides wolves with a bridge to the islands. Scientists predict that freeze-overs of the lake will become an increasingly rare occurrence in the future as global climate change progresses. "Given the long-term prospect of global warming," Peterson adds, "can we really assume that ice will form on Lake Superior with a 'natural' frequency? Throughout the warm decade of the 1980s, which followed a decades-long warming trend, there were almost no ice bridges from Isle Royale to the mainland." Without the migration of new individuals, either on their own or through the help of humans, Peterson believes that the Isle Royale wolves will face extinction.

Proponents of restoring caribou to the isle argue that park managers should take advantage of the moose decline to reintroduce these native ungulates. Peterson has argued for the importation of wolves from the mainland to shore up the genetic diversity of the isle's packs.

These and other opinions will sorely challenge the park's philosophy of letting "nature" take its course. In 2013, the National Park Service entertained three management options: to leave the wolves' fate to nature without any human intervention, to reintroduce wolves if the current population dies out, or to launch a "genetic rescue" by releasing new wolves on the island before the existing packs have disappeared. Arguing for the latter, Peterson says, "We must recognize that humans already play a role in every ecosystem on earth and that we have the capacity and insight to intervene softly on behalf of all life."

The nature of that intervention, however, remains to be determined on Isle Royale. During a June 2013 online public-comment forum convened by park managers, Peterson's was the lone voice in favor of intervention. For too long, say the proponents of ecological restoration, Isle Royale has been managed as a place that serves only the needs and aspirations of people, be that the burning of whole forests to expose ore bodies, the planting of animals to lure hunters or city-weary tourists, or, even, the perpetuation of moose—which some consider as exotic to the isle as dandelions—in the worthy pursuit of scientific research. In many cases, the result of these human-centered activities has been a degradation of the isle, that is, a simplifying of its ecological components. And in ecological terms, simplified systems often are dangerously unstable. Recent research suggests, for example, that overbrowsing of forests by moose is altering the course of lightning-caused fires across the archipelago. These blazes once created open habitats favored by sharp-tailed grouse, the only species of grouse that was capable of traversing the open water that separated Isle Royale from the mainland. Such wildfires also cleared out aging forests and created the young deciduous woodlands that supported beaver. Wildlife counts showed that by 2009 the number of active beaver sites had plummeted to a forty-year low, a downturn that scientists blame in part on the absence of fires. Fewer wolves—and therefore less pressure from predation—have allowed the beaver population to edge up slightly, but their long-term viability depends on access to the habitats they need to survive.

For some, however, downplaying the presence of moose—or even eliminating them altogether from the isle—is an attack on the very identity of the park, even though history shows that moose made an appearance on the archipelago only within the last one hundred years—and quite possibly through the agency of humans.

Yet our preoccupation with some species, such as the majestic moose, to the detriment of less charismatic ones, such as beetles and ferns, may have imperiled the future of them all. If we are to give places like Isle Royale the best long-term shot at surviving assaults such as global warming and atmospheric pollution, some restorationists say, then we must reintroduce and preserve as much as possible the original complement of plants and animals that formed a complex, sturdy matrix of relationships in presettlement times. After all, they observe, this intricately honed network allowed a multitude of species to survive for hundreds, if not thousands, of years and may offer the best hope for their perpetuation into the future.

Land managers face many perplexing questions. Is restoring the full array of native species a realistic goal given that many plants and animals in the Lake Superior watershed may be unable to survive in future conditions shaped by climate change? (See also "Epilogue: The Wild Card of Climate Change.") Is such a goal even desirable when our attempt to hold on to some of these species could pose long-term risks to the larger ecosystems they inhabit? Finally, regardless of the strategy we pursue, can we cultivate the necessary humility for admitting the limits of human knowledge and tempering our needs and aspirations in the service of maintaining the health of the whole watershed and the matrix of relationships on which we all depend? ⟳

SUGGESTIONS FOR FURTHER READING

Aron, Cindy S. *Working at Play: A History of Vacations in the United States.* New York: Oxford University Press, 1999.

Brodo, Irwin M., Sylvia Duran Sharnoff, and Stephen Sharnoff. *Lichens of North America.* New Haven, Conn.: Yale University Press, 2001.

Christian, Sara Barr. *Winter on Isle Royale: A Narrative of Life on Isle Royale during the Years of 1874 and 1875.* Okemos, Mich., 1932.

Clark, Caven P. *Archeological Survey and Testing at Isle Royale National Park, 1987–1990 Seasons.* Occasional Studies in Anthropology, no. 32. Midwest Archeological Center, National Park Service. Lincoln, Neb.: U.S. Department of the Interior, 1995.

Cochrane, Jean Fitts. *Woodland Caribou Restoration at Isle Royale National Park: A Feasibility Study.* Technical Report NPS/NRISRO/NRTR/96-03. Denver: U.S. Department of the Interior, National Park Service, 1996.

Cochrane, Tim. "Isle Royale: A Good Place to Live." *Michigan History* 74 (1990): 16–18.

Commission on Research and Resource Management in the National Park System. *National Parks: From Vignettes to a Global View.* Washington, D.C.: National Parks and Conservation Association, 1989.

Gillman, Henry. "The Ancient Men of the Great Lakes." *Proceedings of the American Association for the Advancement of Science* 24 (1878): 316–34.

———. "Ancient Works at Isle Royale, Michigan." *Appleton's Journal* 10, no. 9 (August 1873): 173–75.

Heidenreich, Conrad E. "The Fictitious Islands of Lake Superior." *Inland Seas* 43 (1987): 168–77.

Huber, N. King. *The Geologic Story of Isle Royale National Park.* Houghton, Mich.: Isle Royale Natural History Association, 1996.

Johnson, Nancy A. "Susie Island: Minnesota's Arctic Outpost." *Minnesota Monthly,* June 1984: 20–23.

Karamanski, Theodore J., Richard Zeitlin, and Joseph DeRose. *Narrative History of Isle Royale National Park.* Chicago: Mid-America Research Center, Loyola University, 1988.

Karrow, Robert W., Jr. "Lake Superior's Mythic Isles: A Cautionary Tale for Users of Old Maps." *Michigan History,* January/February 1985: 24–31.

Martin, Patrick E. "Mining on Minong." *Michigan History* 74 (1990): 19–25.

National Parks Conservation Association. *National Parks of the Great Lakes. A Resource Assessment.* 2007. http://www.npca.org/about-us/center-for-park-research/stateoftheparks/great_lakes/.

Pastor, John, Bradley Dewey, Robert J. Naiman, Pamela F. McInnes, and Yosef Cohen. "Moose Browsing and Soil Fertility in the Boreal Forests of Isle Royale National Park." *Ecology* 74, no. 2 (1993): 467–80.

Pastor, John, Robert J. Naiman, Bradley Dewey, and Pamela McInnes. "Moose, Microbes, and the Boreal Forest." *BioScience* 38 (1988): 770–77.

Peterson, Rolf O. *Ecological Studies of Wolves on Isle Royale.* Annual Report 1999–2000. Houghton: Michigan Technological University, 2000.

Rakestraw, Lawrence. *Historic Mining on Isle Royale.* Houghton, Mich.: Isle Royale Natural History Association, 1965.

Sellars, Richard West. *Preserving Nature in the National Parks.* New Haven, Conn.: Yale University Press, 1997.

Slavick, Allison D., and Robert A. Janke. *The Vascular Flora of Isle Royale National Park.* Houghton, Mich.: Isle Royale Natural History Association, 1993.

Smith, Theresa S. *The Island of the Anishnaabeg: Thunderers and Water Monsters in the Traditional Ojibwe Life-World.* Moscow: University of Idaho Press, 1995.

Snyder, John D., and Robert A. Janke. "Impact of Moose Browsing on Boreal-Type Forests of Isle Royale National Park." *American Midland Naturalist* 95 (1976): 79–92.

Strommen, Norton D. *Isle Royale National Park, Michigan.* Climatography of the U.S. No. 21-20-1. Climatic Summaries of Resort Areas. Silver Springs, Md.: U.S. Department of Commerce, Environmental Science Services Administration, 1969.

Swain, Wayland R. "Chlorinated Organic Residues in Fish, Water, and Precipitation from the Vicinity of Isle Royale, Lake Superior." *Journal of Great Lakes Research* 4, no. 3–4 (1978): 398–407.

Vucetich, John A., and Rolf O. Peterson. *Ecological Studies of Wolves on Isle Royale.* Annual Reports 1997–2013. Houghton: Michigan Technological University, 2013.

Vucetich, John A., Rolf O. Peterson, and Michael P. Nelson. *The Importance of Conserving the Wolves of Isle Royale National Park.* Houghton: Michigan Technological University, 2013.

Warren, Frank M. "The Wildlife of Isle Royale." *American Game* 15, no. 1 (1926): 15–17.

West, George A. *Copper: Its Mining and Use by the Aborigines of the Lake Superior Region, Report of the McDonald-Massee Isle Royale Expedition 1928. Bulletin of the Public Museum of the City of Milwaukee* 10 (1929): 1–182.

Whitman, Richard L., Meredith B. Nevers, Laurel L. Last, Thomas G. Horvath, Maria L. Goodrich, Stephanie M. Mahoney, and Julie A. Nefczyk. *Status and Trends of Selected Inland Lakes of the Great Lakes Cluster National Parks.* U.S. Geological Survey, 2000.

Wright, R. Gerald. "Wolf and Moose Populations in Isle Royale National Park." In *Science and Ecosystem Management in the National Parks,* ed. William L. Halvorson and Gary E. Davis. Tucson: University of Arizona Press, 1996.

INTERNET RESOURCES

Isle Royale National Park (U.S. National Park Service), http://www.nps.gov/isro/index.htm

UNESCO Biosphere Reserve Information, http://www.unesco.org/mabdb/br/brdir/directory/biores.asp?mode=all&code=USA+33

The Chorus Frogs
of Isle Royale

The sooner one leaves those maternal apron-strings—books—
and learns to identify himself with nature, and thus goes out of himself
to affiliate with the spirit of the scene or object before him—or, in other words,
cultivates habits of the closest observation and most patient reflection—
be he painter or poet, philosopher or insect-hunter of low degree,
he will gain an intellectual strength and power of interpreting nature,
that is the gift of true genius.

A. S. Packard Jr., *Our Common Insects* (1873)

Atypical workday on Edwards Island begins with stoking the cook's stove in the island's lean-to kitchen. On this sunny morning in June the job has fallen to Joan Edwards, a biology professor from Williams College in Williamstown, Massachusetts. Set under the eaves of an old storage shed, the tiny kitchen is open on three sides to the surrounding forest. It is as if a tornado had touched down and carefully peeled away the walls, leaving tins of food neatly stacked on the kitchen shelves and a chipped-enamel dipper hanging from a post over the rain barrel.

The domestic orderliness is a little jarring. We are, after all, surrounded by 571,796 acres of Isle Royale National Park, 99 percent of which are designated wilderness. Furthermore, Isle Royale lies smack dab in the world's largest freshwater sea, separated by rough and unpredictable waters from the nearest mainland fifteen miles away. Edwards Island lies another three miles from Rock Harbor, the main port of entry on the northeast end of Isle Royale. The distance discourages hauling even a few bags of groceries, much less an iron stove.

But Edwards makes life in the open air look easy, considering that there is no electricity or running water. She moves effortlessly from skillet to coffeepot to browning toast like flapjacks on the stove top. If Edwards seems perfectly at home in this makeshift kitchen under the trees, it is because she is. The island with its small compound of outbuildings has been in her family since her great-grandfather Maurice Edwards bought the island in the early twentieth century. (When Isle Royale became a national park, members of the Edwards family, which included her father's generation, signed a lease allowing continued family use of the island for the duration of the signatories' lifetimes.) As a child, Edwards spent her summers here. In the 1970s, while researching her doctoral dissertation, she returned to study the relationship between Isle Royale moose and their forage. During that time she introduced her husband and fellow doctoral student, biologist David Smith, to Edwards Island. On a foray to a neighboring North Government Island, he discovered a population of boreal chorus frogs, and for some two decades he has used its rocky shoreline as an outdoor laboratory for studying the ecology and evolutionary biology of these diminutive frogs.

Edwards too is studying some of Isle Royale's tiniest inhabitants. Between stirring a cauldron of oatmeal and loading firewood, she talks about her most recent project: monitoring populations of plants that have established a foothold on the very edges of the park's islands. They are known as arctic-alpine disjuncts because they are located hundreds of miles from their main ranges on the Arctic tundra and the upper reaches of the Rocky Mountains. Having persisted on Isle Royale since the retreat of the last glacier, their future is now imperiled. Forty-one percent of the archipelago's rare plants occupy rocky shoreline habitats, a location favored by most species of arctic disjuncts. (See also "Hay Pickers and Grass Gatherers: Botanical Exploration along the Lakeshore.")

Edwards is clearly eager to get on with the day's fieldwork and shows no signs of fatigue, even though she and two student assistants just arrived the night before in dense fog and rain. With them came the boxes of food and scientific equipment, including unwieldy cylinders of liquid nitrogen, all of it delivered intact after a twenty-four-hour drive from Williamstown to the northern tip of Michigan and another six hours across Lake Superior by ferry to the park.

No sooner do we down the morning's first cup of coffee than the students too emerge from the bunkhouse next door, accompanied by two others, who arrived with

Smith several weeks earlier to help with his frog research. Almost instinctively they gather around the stove even though the fire does little to take the edge off the morning chill. But the students do not seem to mind. They pick up bantering where they left off the night before. For the next six weeks, their verbal play will be their only source of social entertainment, water-bound as they are in the midst of a cold and fickle freshwater sea.

Smith and Edwards keep this in mind when interviewing prospective candidates. June is rainy season on Isle Royale, and the students often are out peering through fogs and persistent downpours to collect maddeningly detailed data. Accommodations are rustic—a shared bunkhouse and primitive outhouse. Dry clothes are a luxury, as evidenced by the damp shirts and jeans dangling from clotheslines that have been strung around the porch and through the trees. Hot showers are available only once a week when the group takes the forty-five-minute kayak trek into Rock Harbor. Self-reliance and good humor are a plus. The ill-tempered need not apply.

A sharp "Yo" from Smith a few hundred feet off in the forest cuts through our laughter like rifle shot. It is a signal that something unusual has wandered into camp. Everyone is suddenly hushed and alert. We rise from our chairs, creep single file through the brush until we reach him, his binoculars fixed on the top of a birch tree. A male rose-breasted grosbeak basks in the glow of the rising sun long enough for all of us to get a good look at his showy black, red, and white plumage. After several days of Lake Superior fogs that turned summer days into perpetual dusk and sent nighttime temperatures plummeting into the 40s Fahrenheit, the particolored bird appears shockingly out of place. It is as if Smith had suddenly pointed to a palm tree swaying in a grove of spruce and birch. For a moment, everyone is still.

Paying attention, as the students learn from observing Smith and Edwards, is the first order of business. The lesson is reinforced later during breakfast when one of the students takes the routine roll call of the previous day's sightings of plant and animals. The grosbeak is the first of its kind for the season. The following morning a luna moth will be added to the list after Edwards spots the pale-green tendrils of its hind wings amid the spines of devil's club that grows in thickets behind camp.

The surprise appearances of exquisite animals, workdays spent on the shores of one of the world's most beautiful lakes, dinner at day's end in the camp's screen house, where spirited conversation flickers as the sun drops behind the lake—these pleasures help to buoy the spirits of the students as they knuckle down to the rigors of fieldwork with its hours—even days and weeks—of drudgery. Waiting and enduring, they learn, is the stuff of scientific discovery. Noting the particulars, assigning them names and numbers, labeling vials, counting and measuring lengths and widths and heights and weights, and taking painstaking notes are part of the discipline required to make sense of the seemingly unruly, episodic world around them. Gaining insight is like practicing musical scales, fingertips following a choreographed routine until they recognize a pattern and in the pattern detect a faint melody. It is like staring at the monochrome of a northern forest until you think you can no longer bear its sameness and numbing familiarity and suddenly seeing the gleaming black head of a bird with a scarlet breast patch emerge from the leaves.

The Puzzle of the Frogs

At first glance, the southeastern shore of North Government Island looks identical to many other rockbound stretches of coastline on Lake Superior. A wide swath of basalt gently slopes toward the lake before dipping under its crystalline waters. It is the kind of shore edge prized by Lake Superior lovers—front-row and center seats on a 180-degree view of big sky and water.

Smith discovered this coastline back in the 1970s while exploring the neighborhood of islands around Edwards Island. But like most curious scientists, Smith was not content just to take in the scenery. Thousands of years ago the ice sheets had plucked out loose chunks of rock, leaving depressions the size of bathtubs. Fed by rain, lake waves, and seepage from the nearby forest, the water in these pools was so clear that much of the life within them was visible to the naked eye. Before long, Smith found himself down on his hands and knees, riveted by the goings on in these aquatic pockets.

Blue-green algae coated the bottoms like a plush velvet lining. (The color blue-green

Fed by rain, lake waves, and seepage from the nearby forest, bedrock pools along the shorelines of Lake Superior's islands provide chancy habitats that are subject to a wide range of disturbances. Photograph by Chel Anderson.

is something of a misnomer since the algae are in fact a dull greenish brown.) Grazing on the algal mat were tiny crustaceans scattered like specks of silvery glitter. And in some pools, black tadpoles, no bigger than lentils with tails, rasped on the algal feast with such gusto that their tails wriggled.

But not all the pools supported the same levels of biological activity. Some were crammed with tadpoles; others hosted only a handful. Pools near the water's edge were altogether devoid of visible signs of life, while those near the forest's edge contained dragonfly larvae, which sent up puffs of silt as they spurted into the bottom for cover.

Why did some pools have more tadpoles than others while many had none at all? Intrigued by the discrepancy, Smith began a simple investigative procedure: He mapped out a study area and started counting the tadpoles in each pool. Little did he know that the results of this rudimentary inventory would inspire him over the next twenty-some years to spend most of June and the better part of July around these pools, slowly piecing together an extraordinary narrative of life within their tiny watery realms.

The boreal chorus frog spends its adult life in the forest, leaving only in spring to mate and lay eggs in sites that include bedrock shore pools of Isle Royale. Photograph by Allen Blake Sheldon.

The Life History of Chorus Frogs

The tadpole phase is an amazing, but short, chapter in the drama of a chorus frog's life. Most of their life story begins and ends on land, where adult frogs overwinter as nuggets of ice in the dense fir and spruce forests that lie some 130 feet from the shore. Burrowed under a thin mantling of needles and leaf litter, the hibernating frogs survive temperatures as low as 22 degrees Fahrenheit.

A 1997 report by researchers at Miami University in Ohio showed that frogs are able to endure these frigid temperatures via a remarkable physiological adaptation: the conversion of glycogen, a carbohydrate, into blood glucose, a sugar. The glucose serves as a kind of antifreeze, lowering the freezing temperature within cells enough to prevent the formation of ice crystals that can rupture their delicate membranes. (The glucose does not, however, prevent the freezing of fluids around the cells. The frogs overwinter with an average of 65 percent of their bodies frozen as hard as marbles.) Comparable levels of blood glucose in mammals would be fatal, but frogs have developed an adaptive mechanism for dealing with the surfeit. When temperatures rise, the frogs are able to temporarily store glucose in their bladders. The glucose is then slowly absorbed into the blood, allowing the frog's liver to gradually convert the sugar back to harmless glycogen.

On Isle Royale chorus frogs are usually the first on the scene, the males emerging from the forest litter as early as April. By May mating is in full swing. Like other tree frogs in the family Hylidae, their vocal powers are legendary. The hormone-driven suitors rouse the females from their torpor with a shrill glissando that sounds like a finger sliding up an amplified comb. After analyzing recordings of these mating chirrups in the lab, researchers discovered that a single one-second call is composed of an astonishing nine different musical notes.

Although male chorus frogs are small enough to fit into a hollow the size of a tablespoon, their breeding calls can be heard nearly one-half mile away. The females have ears for their trills alone. Even in ponds with several different kinds of frogs (North Government Island hosts only chorus frogs and spring peepers), the females are indifferent to the grunts, peeps, croaks, and trills of males from other frog species.

Tadpoles have been described as the "universal snack food of aquatic life." Commonly found in the bedrock pools of Isle Royale, the young of chorus frogs (seen here) are preyed on by animals ranging from dragonfly larvae to flocks of migrating mergansers. Photograph by Allen Blake Sheldon.

Once enticed into the rock pools, female chorus frogs are mounted from behind by the males in a position known as amplexus (from the Latin *amplecti,* meaning, "embrace"). Using his front legs, the male clasps and gently squeezes the female's midsection to help her to extrude anywhere from two hundred to four hundred eggs. The eggs emerge in gelatinous clumps containing ten to seventy eggs each. As the eggs slip into the water, the male quickly deposits his sperm over them.

Hatched within six to fourteen days are tadpoles, the aquatic stage of a frog's life in which the animals, possessing tails and gills, are more fishlike than froglike. This dual citizenship in the aquatic and terrestrial realms has earned frogs their place in the family of amphibians (from the Greek *amphibios,* or "living a double life"). What follows seems like a process of rehearsed indecision, as if the features of the tadpoles are continually erased and redrawn by some restless, perfectionist force. In the life of a chorus frog, the tadpole phase lasts anywhere from five weeks to two months. Then it undergoes a metamorphosis into a froglet, which typically takes place in two to four days. During this time, the tail is resorbed; that is, the tissue dissolves and is assimilated by the body. Hind legs appear. Other, less visible changes take place as well. The animals develop lungs, while their gills close like healed wounds without scars. When the transformation is complete, the froglets silently head for the nearby forest. Although they started life as voracious vegetarians, they now will stalk the leaf litter for small invertebrates such as insects and worms.

Year after year the frogs make this miraculous trek into the forest as if schooled by some invisible mentor. They take up residence in a new country where they neither know the language nor the currency. It is as if a Kalahari Bushman had taken a red-eye to Manhattan only to wake up alone on Wall Street during Monday morning rush hour. Yet somehow the frogs quickly adapt to the customs of life on solid ground. The instinctive attachment to water, however, remains. Even the long months of cold, the darkness, the damp humus smell of the forest cannot erase it. Perhaps each spring the memory of water overwhelms them as the ice that has formed in the fissures of their bodies dissolves into running water, and they, like the generations before them, are carried by the powerful insistence of its flow out onto the rocks.

Uneasy Neighbors: Frogs and Dragonflies

Smith points out that the developmental biology of frogs has been well studied, largely because amphibians are small, inexpensive, and easy to care for within the confines of human-built laboratories. Few laboratory findings, however, have been tested in the field. Early on in his experiments, Smith discovered some of the reasons why.

By the time Smith and his students arrive, most of the chorus frogs have mated, laid their eggs, and returned to the forest, although for a few frogs the courtship season can extend into July. By mid-June the majority of eggs have hatched into tadpoles, providing the scientists with a dependable supply of research subjects. Right?

Wrong. Even remote North Government Island, with its isolation from human activities and stable rock shoreline, is subject to random, unforeseen events, the most minor of which can change the ground rules of Smith's research almost as fast as he drafts them. In the course of one summer, for example, he watched as a black duck and her brood of ducklings touched down on the shore and within hours picked the pools clean of tadpoles before moving on. Another year a chance landing of migratory mergansers did the same. In 1994 an unusually severe storm pounded the shore, flushing the subjects of his study into the big lake, where most of them either died of starvation in the nutrient-poor waters of Lake Superior or were swallowed by fish.

The seemingly immutable rocky shoreline too was subject to the forces of change. Thermometers that had been bolted to rock to record winter pool temperatures disappeared, ripped from their moorings by shifting and grinding ice. Smith noted that in the course of one winter, a block of basalt the size of a jumbo freezer had been hoisted from the water's edge far up onto the shore.

Smith was faced with a choice: Find a way to factor in the inevitable variables that alter even the best-laid research plans, or abandon his field studies in frustration. He chose to stick it out. In the process, he has assembled a rare profile of how a community of chorus frogs has evolved the capacity to respond to the challenges of a particular place.

Each year Smith and his students open the field season by taking a census of the tadpoles, a laborious task that is repeated two more times before they head back to Massachusetts. In each inventory, the team will capture, count, and release the tadpoles in the study area, which has grown to include nearly five hundred pools, some of which contain up to about one thousand tadpoles each.

When Smith reviewed the first population reports, he discovered some curious patterns in the data. Faced with a choice of where to lay their eggs, the female frogs seemed to favor pools located in the middle of the shore and tended to avoid pools on the lakeside or forest-edge margins.

That they avoided the pools closest to the lake did not seem too surprising. Here the tadpoles would be more likely to be carried off by storm waves before they had an opportunity to mature. What perplexed Smith, however, was the fact that fewer frogs laid eggs in the upper pools within twenty feet of the forest's edge, despite the numerous advantages. Not only did these pools afford tadpoles greater protection from storm waves, but their higher productivity also offered more abundant food supplies. Furthermore, the upper-zone pools were located within easy distance of the feeding and hibernating habitats of adult frogs.

Smith set out to discover what cues guided their selection process. Noting that the upper pools captured more undiluted runoff from the forest than the midzone and lower pools, he tested the water in each zone and found, not surprisingly, that water in the pools closest to the forest was more acidic. A lower pH alone did not seem to be a decisive factor, however, since some female frogs did lay eggs in the upper pools.

Then he noticed the dragonfly nymphs. Although North Government Island hosts several species of dragonflies, only the boreal species *Aeshna juncea* (more commonly known as the sedge darner) lays its eggs in the rock pools. In fact, the Isle Royale archipelago is the only locality in the lower forty-eight states where *A. juncea* is known to breed.

For reasons that are unclear, *A. juncea* favors pools near the forest's edge for its nurseries. In August the adults lay their eggs in the pools, where they overwinter before developing into nymphs, the equivalent of the intermediate aquatic stage of tadpoles. The nymphs spend another two and a half years underwater and, like tadpoles, undergo a series of dramatic morphological changes before they reach adulthood. When the insect is sexually mature, it climbs out of the water onto a rock or plant stem. Here it

Isle Royale is the only locality in the lower forty-eight states where the sedge darner *Aeshna juncea* is known to breed. Photograph by David Evans.

undergoes a final transformation so startling and bizarre that it can prompt even the most sober observer to break into raptures. It all begins when the back of the nymph's thorax splits down the middle as if being unzipped. Slowly, the compact, muscular thorax of the adult dragonfly emerges bearing the insect's legs and four minute nubs in which wings are as tightly folded as the most compact parachute. Then the head with its gleaming compound eyes clears the larval skin, a tough chitinous casing known as the exuvia, followed by the long needlelike abdomen. Within moments body fluids are pumped into the complex network of veins that crisscross the transparent wings, causing them to unfurl. During the hour or so it takes for the insect's body and wings to harden, the colors of the dragonfly's body intensify, and the insect practices for takeoff, flexing its thoracic muscles and whirring its wings. Should it survive this most vulnerable stage of life, the winged adult will gingerly step from its gunnysack–colored exuvia and take to the air.

Though lovely and diaphanous as any storybook princess, dragonflies are one of the most efficient killing machines ever designed by nature. The American entomological writer Edwin Way Teale points out that dragonflies have been known to cram more than one hundred mosquitoes in their mouths at one time. Indeed, the Australian dragonfly expert Robin J. Tillyard came across one specimen whose mouth was so stuffed with mosquitoes that it was unable to close its jaws. The nineteenth-century English naturalist Reverend John G. Wood noted that even when captured, dragonflies would continue to voraciously consume hand-fed insects.

Adolescent dragonflies—the water-dwelling nymphs—are no less efficient predators in satisfying their prodigious hunger. In his 1873 book *Our Common Insects,* Alpheus S. Packard observes that "from the moment of its birth until its death . . . [the dragonfly]

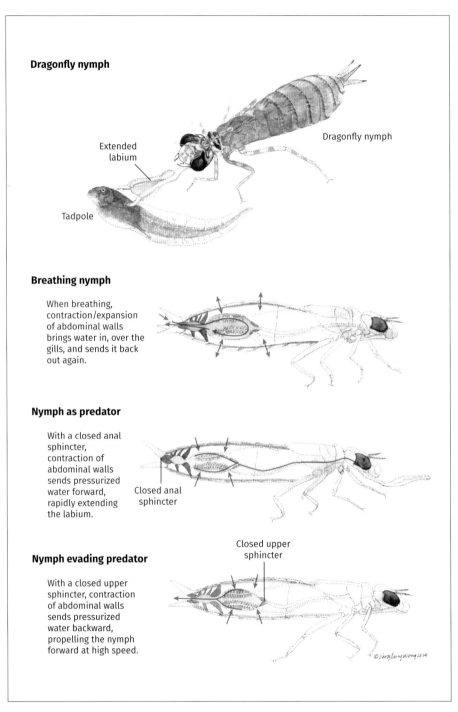

Dragonfly nymph

Extended labium

Tadpole

Dragonfly nymph

Breathing nymph

When breathing, contraction/expansion of abdominal walls brings water in, over the gills, and sends it back out again.

Nymph as predator

With a closed anal sphincter, contraction of abdominal walls sends pressurized water forward, rapidly extending the labium.

Closed anal sphincter

Nymph evading predator

With a closed upper sphincter, contraction of abdominal walls sends pressurized water backward, propelling the nymph forward at high speed.

Closed upper sphincter

Dragonfly nymphs breathe by drawing water through the anus into the abdomen and over internal gills. The closing of sphincters located in the abdomen pressurizes this water. The strategic release of this pressure provides nymphs with the power to thrust a hinged underlip for nabbing prey or to propel themselves away from predators at lightning speeds of 19.5 inches per second. Illustration by Vera Ming Wong.

riots in bloodshed and carnage. . . . Could we understand the language of insects, what tales of horror would be revealed! What traditions, sagas, fables, and myths must adorn the annals of animal life regarding this Dragon among insects!"

Dragonfly nymphs have earned their reputation as killers largely due to the speed with which they nab prey. They are equipped with two sphincters in the abdomen, anatomical structures that are also involved in breathing. To obtain oxygen, nymphs close the upper sphincter and draw water through the anal sphincter to circulate it over internal gills located in the abdomen. The water is then expelled back through the anus. By closing both the anal or upper sphincters, the nymphs can briefly pressurize this water, using the force to either propel themselves away from predators by opening the anal sphincter or to thrust their underlip, known as a labium, to nab victims by opening the upper. Both actions are performed at lightning speeds. Wood describes how a larva of the whirligig beetle in his aquarium was snatched by a dragonfly nymph with such speed "that it looked very much as if the larva had intentionally darted into its destroyer's mouth."

The swiftness of such surprise attacks is made possible by a mouthpart that rivals any instrument from the Spanish Inquisition. Not only is the labium long—it measures about one-third the length of a nymph's body—but it is also topped by a pair of clawlike lobes. Hinged for easy deployment, this underlip can lunge to nab prey and then retract long before victims are able to even react, much less escape. Aquatic entomologist James Needham describes them as a "combination of hands, carving tools, and serving table" rolled into one.

The larvae of other insects are not their only prey. Teale recalls a nymph-infested pond that had been stocked with fifty thousand young fish. By the time the dragonflies were sated, only one thousand fish remained. The survival odds for succulent young tadpoles are not much better. One of the pools in Smith's study area started out with one hundred tadpoles. Their numbers were quickly reduced to a mere handful.

How were the gravid female frogs on North Government Island tipped off to the potentially menacing presence of dragonfly nymphs? Scientists have long known that frogs and tadpoles may be able to chemically "read" the waters they enter. Laboratory tests conducted in the 1970s and 1980s have shown that tadpoles swimming in common pools of water, but separated by dividers, are able to sense kin, even distinguishing full siblings from half siblings.

This heightened sensitivity to chemicals also appears to help them to relay and intercept molecular warnings about enemy activities, a life-saving skill for an animal that science writer Susan Milius describes as the "universal snack food of aquatic life." In 1999 researchers at Yale University found that stressed tadpoles send at least one chemical signal that sounds an alarm to other tadpoles and frogs. When laboratory tadpoles were stalked by a decoy heron, for example, they discharged foul-smelling ammonium, a major component of their metabolic waste. Tadpoles that shared the ammonium-laced water in an adjoining tank responded with a series of defensive behaviors even though the decoy was shown only to the tadpoles in the first tank.

Life in the midzone pools, although free of dragonfly nymphs and relatively sheltered from storms, was no picnic either. Armed with rows of teeth both above and below the exterior of their mouths, hungry tadpoles can quickly scour pools clean of algae. The scarcity of food stunts the growth of midzone tadpoles, and in the North's short growing

THE CHORUS FROGS OF ISLE ROYALE

season, size is critical. Bigger tadpoles not only dominate smaller ones, but they are more likely to be successful breeders and to pass on their genes. Indeed, although fewer tadpoles escaped death in the nymph-patrolled upper pools, the survivors had greater access to food and were more robust. Consequently, they had an 80 percent greater chance of surviving to reproductive adulthood.

In time, Smith began to notice other differences as well. Tadpoles in the upper pools developed a higher and more muscular tail fin. They also tended to be more sedentary than tadpoles in nymph-free pools. Subsequent research demonstrated that upper-zone tadpoles exhibited these differences within ten days of hatching. Smith suspects that the tadpoles lie low to escape the notice of the nymphs, and when necessary, rely on their powerful tails to help them evade the nymphs' predatory lunges.

Were evolutionary pressures creating a new species or subspecies of chorus frog in the forest-edge pools, Smith wondered, or were all the island's chorus frogs capable of developing similar physical and behavioral characteristics when exposed to nymphs? To find the answer to that question, Smith and his team conducted an experiment to test the reaction of tadpoles to the presence of dragonfly nymphs. They began by removing a sampling of tadpoles from the midzone pools and transporting them back to camp, where they were photographed—no easy feat. Using a turkey baster underwater to gently blow the tadpoles onto a fine-mesh screen, the researchers were able to orient the animals' heads to the left before snapping baseline mug shots. Subsequent photographs were taken in the same position, a technique that allowed researchers to identify individual tadpoles by their distinctive tail spots.

The tadpole numbers were then divided into two groups and released into a pair of experimental pools. One contained caged dragonfly nymphs that were fed meals of tadpoles every few days; into the other—the control pool—the scientists inserted an empty cage. Ten days after the tadpoles' release, the team rephotographed each animal to record any changes.

Smith's team returned to the Williams College laboratory with hundreds of photos and notes on behavior. The negatives were digitized by painstakingly scribing their proportions on a computer screen so that the information could be transferred into a computerized database. The results? The researchers found that tadpoles in the pool with the empty cage did not alter their behavior or physical shape. But the biological and behavioral dynamics of pools containing caged dragonfly nymphs were dramatically different. Within hours of the introduction of the caged nymphs, the tadpoles became noticeably less active. By day ten, they had developed the exaggerated tail fins of their upper-zone cousins. Smith was stunned by the frogs' degree of plasticity, that is, their ability to quickly change their behavior and shape in response to new environmental factors.

Studies published in a 2001 issue of *Ecology* by Rick Relyea of the University of Pittsburgh showed that these changes persisted into adulthood for wood frog tadpoles raised with dragonfly larvae. The adult frogs sported hind legs that were longer and stouter than those reared in larvae-free waters.

The chorus frogs of North Government Island are just one in a series of recent revelations from the field that scientists are just beginning to fully grasp—that evolutionary change does not move only at glacial speeds but can also occur in the proverbial wink of an eye, not just for microbes but also for more complex organisms. Life is more fluid than fixed. What better example than the chorus frog, whose life swings in the balance of matter and water.

In his Pulitzer Prize–winning book *The Beak of the Finch,* Jonathan Weiner writes about finches in the Galapagos Islands whose beaks and other physical characteristics respond dramatically to changes in their environments in the span of a couple generations. "The original meaning of the word evolution—the unrolling of a scroll—suggested a metamorphosis, as of moths or beetles or butterflies," Weiner writes. "But the insects' metamorphosis has a conclusion, a finished adult form. The Darwinian view of evolution shows that the unrolling scroll is always being written, inscribed as it unrolls. The letters are composed by the hand of the moment, by the circumstances of the day itself. We are not completed as we stand, this is not our final stage. There can be no finished form for us or for anything else alive, any thing that travels from generation to generation." ❧

SUGGESTIONS FOR FURTHER READING

Kiesecker, Joseph M. "Identification of a Disturbance Signal in Larval Red-Legged Frogs, *(Rana aurora)." Animal Behaviour* 57, no. 6 (1999): 1295–1300.

Milius, Susan. "Tadpole Science Gets Its Legs . . . and Reveals Complex Lives for the Swimming Squiggles." *Science News* 160 (2002): 26–28.

Packard, Alpheus S., Jr. *Our Common Insects.* Boston: Estes and Lauriat, 1873.

Relyea, Rick A. "The Relationship between Predation Risk and Antipredator Responses in Larval Anurans." *Ecology* 82 (2001): 541–54.

Smith, David C. "Factors Controlling Tadpole Populations of the Chorus Frog *(Pseudacris triseriata)* on Isle Royale, Michigan." *Ecology* 64, no. 3 (1983): 501–10.

Storey, Kenneth B., and Janet M. Storey. "Lifestyles of the Cold and Frozen." *The Sciences,* May/June 1999: 33–37.

———. "Persistence of Freeze Tolerance in Terrestrially Hibernating Frogs after Spring Emergence." *Copeia* 3 (1987): 720–26.

Teale, Edwin Way. *The Strange Lives of Familiar Insects.* New York: Dodd, Mead and Company, 1962.

Tillyard, Robin J. *The Biology of Dragonflies.* Cambridge: Cambridge University Press, 1917.

Van Buskirk, Josh, and David C. Smith. "Between the Devil and the Deep Blue Lake." *Natural History* 102, no. 4 (1993): 38–41.

Wood, John G. *World of Little Wonders or, Insects at Home.* New York: Hurst and Co., Publishers, 1871.

Moose and Wolves
on Isle Royale

Dᴀᴠɪᴅ Sᴍɪᴛʜ ʜᴀs ʙᴇᴇɴ sᴛᴜᴅʏɪɴɢ ᴛʜᴇ ᴄʜᴏʀᴜs ꜰʀᴏɢs ᴏꜰ Nᴏʀᴛʜ Gᴏᴠᴇʀɴᴍᴇɴᴛ Island since 1978. In the annals of scientific field research, his studies are a rarity—both for the subject he has chosen and for the duration of his data collection.

In 1989, ecologist David Tilman issued a review of 749 papers that had been published over a ten-year period in the prestigious journal *Ecology*. Field studies that lasted at least five years accounted for only 1.7 percent of the total. In a 1986 study on the same subject, biologist Patrick J. Weatherhead published a review of 308 papers in major ecology, evolution, and animal-behavior journals and found that the mean duration of these studies was 2.5 years, the average length of a research grant or the research phase of a graduate degree. The result: scientific research that is guided by the artificial constraints of academic careers or government funding can result in a skewed picture of the real-world workings of ecosystems. All too often, we get little more than snapshot views of ecological processes, a brevity that can be dangerously misleading when drawing conclusions—or making management decisions—about organisms, particularly those whose populations naturally ride sine curves of boom-and-bust cycles, such as amphibians.

One exception to this has been the highly publicized research on the relationship between moose and wolves on Isle Royale that was begun in 1958. "Few studies on animal populations have continued for so long, have produced so much useful information, or have achieved such notoriety," says R. Gerald Wright in *Science and Ecosystem Management in the National Parks.* (Studies of Isle Royale moose are even shedding light on the incidence of osteoarthritis in humans, which afflicts some twenty-seven million Americans. Scientists have linked osteoarthritis in adult Isle Royale moose with nutritional deficits in the animals' early life. Fledgling research shows that nutritional deficits in childhood may similarly contribute to the incidence of the disease in adult humans.)

Although focused on moose and wolves, the Isle Royale research provides a cautionary tale about the perils of short-termism that is applicable to all species. Over the decades, Wright observes, Isle Royale researchers have learned that "events in nature should always be interpreted cautiously and that any conclusion is subject to change." Indeed, he adds, "there have been many instances in which interpretations about a phenomenon made during one period were later altered."

Part of the reason why the moose-wolf research has continued while other studies have long since disappeared into the pages of science journals is the personal and professional dedication to the project of several generations of wildlife researchers and park managers. Writing in 1996, Wright observed that the wolf-moose research consumed 40 percent of the park's $100,000 budget and took up most of the time of the park's natural resource personnel.

But the longevity of the study is also due, in no small part, to the fascination that these animals hold for the public, what conservation biologists refer to as charismatic species or so-called glamour mammals. "The first question visitors ask park staff when they embark on the island usually concerns the status of the wolves," Wright points out. "The wolf has essentially molded the visitors' perception of Isle Royale and is a major attraction. . . . A certain irony, however, pervades the current fascination with the wolf. Before the wolves arrived, moose symbolized the island and were the primary concern of management. Isle Royale National Park was often described as America's greatest moose refuge, and visitors often came just to photograph them."

Too often overlooked by the public, researchers, and natural resource managers alike are the millions of interactions among smaller, less photogenic species that make life possible for thrilling animals such as moose and wolves. As Jon Luoma points out in *The Hidden Forest,* "Not a single forest ecosystem exists where biologists have managed even to catalog the full panoply of organisms, particularly the thousands of types of insects, bacteria, and other tiny organisms, much less document the complex relationships among and between them."

The omission is all the more troubling when the health of forests fail, say, or some of these inconspicuous species run into trouble, as frogs have in many parts of the world. According to a review of ecology journals in the 1980s, amphibians were the least-studied vertebrates. Less than 5 percent of major research articles were devoted to their study (with frogs having to share the dimmed spotlight with other amphibians, such as salamanders). Researchers do not even have dependable long-term population counts

Research on the relationship between moose and wolves on Isle Royale, which began in 1958, is one of the few long-term scientific studies on animal populations in the world. Photograph by Rolf O. Peterson.

much less detailed information about their behavior in the wild or their role in aquatic and terrestrial ecosystems.

For park managers and other natural resource specialists who have had ecosystem management added to their job descriptions, such glaring knowledge gaps present difficult challenges. From the founding of the National Park Service in 1916, says Ervin H. Zube in *Science and Ecosystem Management in the National Parks,* the "dominant policy emphasis was on tourism and scenery management." Increasingly, managers now are being asked to manage parks along ecological lines—with little information. "Relatively little research has been directed to basic understanding of ecosystem function and relationships among ecosystems," Zube says. "Research has too frequently been directed to fighting management brush fires and not to understanding broader ecosystem and landscape issues."

"The central dilemma of national park management has long been the question of exactly what in a park should be preserved," writes Richard West Sellars in *Preserving Nature in the National Parks: A History.* "Is it the scenery—the resplendent landscapes of forests, streams, wildflowers, and majestic mammals? Or is it the integrity of each park's entire natural system, including not just the biological and scenic superstars, but also the vast array of less compelling species, such as grasses, lichens, and mice?"

Or perhaps the central dilemma, more accurately, is this: we need to understand that resplendent landscapes cannot be separated from intact ecosystems, that the behind-the-scenes work of the so-called lowlier creatures—tadpoles, fungi, nematodes, and beetles—makes possible the howl of a wolf in the night. ❧

SUGGESTIONS FOR FURTHER READING

Belluck, Pam. "Moose Offer Trail of Clues on Arthritis." *New York Times,* 16 August 2010.

Blaustein, Andrew R., David B. Wake, and Wayne P. Sousa. "Amphibian Declines: Judging Stability, Persistence, and Susceptibility of Populations to Local and Global Extinctions." *Conservation Biology* 8 (1994): 60–71.

Carpenter, Stephen R. "Microcosm Experiments Have Limited Relevance for Community and Ecosystem Ecology." *Ecology* 77, no. 3 (1996): 677–80.

Likens, Gene E., ed. *Long-Term Studies in Ecology.* New York: Springer-Verlag, 1989.

Luoma, Jon. R. *The Hidden Forest: The Biography of an Ecosystem.* New York: Henry Holt and Company, 1999.

May, Robert M. "The Effects of Spatial Scale on Ecological Questions and Answers." In *Large-scale Ecology and Conservation Biology,* ed. Peter J. Edwards, Robert M. May, and Nigel R. Webb. Oxford: Blackwell Scientific Publications, 1994.

Sellars, Richard West. *Preserving Nature in the National Parks: A History.* New Haven, Conn.: Yale University Press, 1997.

Wright, R. Gerald. "Wolf and Moose Populations in Isle Royale National Park." In *Science and Ecosystem Management in the National Parks,* ed. William L. Halvorson and Gary E. Davis. Tucson: University of Arizona Press, 1996.

Gulls in the Great Lakes

[Gulls] fill the valuable office of scavengers of the sea. . . .
[They] do for the water what the turkey buzzard does for the land—
rid it of enormous quantities of refuse. When one watches hundreds of gulls
following the garbage scows out of New York harbour, or sailing
in the wake of an ocean liner a thousand miles or more away from land,
to pick up the refuse thrown overboard from the ship's kitchen,
one realises the excellence of Dame Nature's housecleaning.

Neltje Blanchan, *Birds Every Child Should Know* (1907)

IN March 1998, officials of Outboard Marine Corporation (OMC), manufacturers of boat motors in Waukegan, Illinois, declared war on their neighbors. Droppings from thousands of nesting seagulls had so whitewashed their harbor-front facilities
on Lake Michigan that employees feared for their health. OMC workers were not alone.
The previous summer the county health department had closed a nearby beach a record
nineteen times, blaming the seagull guano for dangerously high bacteria counts in nearshore waters.

But before they hauled out the artillery, company leaders consulted the U.S. Fish
and Wildlife Service (USFWS), the federal agency charged with administering an international treaty that protects gulls and other migratory birds. OMC officials obtained a
permit to startle the birds using a propane-powered cannon that was timed to emit
several blasts each minute from sunrise to sunset. Plans called for keeping up the audio
harassment until the end of gull nesting season in June. But by early May officials
convened to reassess their strategy. It was not working. The gulls had become acclimated to the noise. A second permit was issued, this time authorizing sharpshooters
to periodically cull a few birds from the flock in an attempt to get the others to associate real danger with the cannon shots. Despite protests by animal-rights advocates,
some of whom offered to dive-bomb the nesting colonies using motorized paragliders,
officials kept up the negative reinforcement. It worked. The gulls avoided setting up
housekeeping on company property—at least for the duration of the 1998 nesting season.

Guns, cannons, and dive-bombers. From the Atlantic Coast and the harbors of the
Great Lakes to the islands of the Pacific Northwest, humans have employed every scare
tactic in the book to control populations of nuisance gulls. And where harassment has
not worked, they have tried to stall the growth of runaway gull populations by draping
monofilament over nesting areas, setting marauding herds of pigs loose in colonies with
incubating eggs and hatchlings, replacing eggs with wooden decoys, and, in extreme
cases, bulldozing problem colonies.

Lost in all the antigull fervor is the fact that humans have not always regarded gulls
as rats with wings. In Salt Lake City, Utah, for example, a monument in Temple Square
commemorates the hungry California gulls that saved the crops of Mormon settlers
from a plague of katydids in 1848.

Gulls once were beneficial to other commercial enterprises as well. During the nineteenth century, hunters along the Atlantic Coast raided nests for eggs and killed adult
birds to extract their feathers for the lucrative millinery trade. Overcollecting, however,
along with the destruction of nesting habitat through the building of cities and plowing
of farmland finally took their toll. Between the 1840s and the 1920s, crashes occurred
even in populations of ring-billed gulls, once so abundant that the great naturalist John
James Audubon dubbed them the Common American Gull. According to ornithologist
John P. Ryder, by the early 1900s, breeding ring-bills had been largely eradicated in the
East and the Great Lakes region and banished to the remote prairie interiors of Canada
and the United States.

Before human persecution and habitat reduction, gulls thrived along the nation's
shorelines. But populations that today have reached the proportions of Old Testament
plagues are a relatively recent development. Like other animal species that have made a
comeback from near decimation, such as white-tailed deer and the Canada goose, the
recovery of gulls is two parts success story to one part nightmare.

Take ring-bills, the species most often to blame for causing nuisance or hazardous conditions for humans. By 1990 their once-depleted population had rebounded in North America to an estimated three to four million individuals. Nowhere is this resurgence more noticeable—and problematic—than in the Great Lakes, where they have become the most common gull species. Ever since three hundred nesting pairs were discovered in 1926 near the Straits of Mackinac in Lake Huron, ring-bills have been on the increase. Their numbers were held in check until the late 1960s. Then suddenly ring-bills underwent a dramatic population spike. There is evidence to suggest that they may be experiencing a second population boom—with no peak in sight. Studies conducted by Francesca Cuthbert and Joan McKearnan at the University of Minnesota reviewed gull census data from 1977, 1989–91, and 1997–99. The surveys showed that ring-billed gull populations exploded throughout the Great Lakes during that time period. In 1977 researchers tallied 102,539 breeding pairs of ring-bills. In the 1989–91 survey the number had more than doubled to 283,970 breeding pairs. By 1997–99 the population had soared to 308,856.

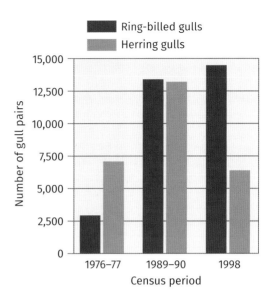

Census data collected by researchers at the University of Minnesota, St. Paul, show that, as in the rest of the Great Lakes, Lake Superior's ring-billed gull population increased dramatically in the last decades of the twentieth century.

The populations of other gull species are up too, although their growth curves are far more modest. The numbers of herring gulls climbed from 29,406 breeding pairs in 1977 to 44,823 in 1989–91. (At the time of the 1997–99 survey, the population had dropped slightly to 35,199 pairs.) And nesting pairs of species that historically have rarely bred in the Great Lakes, such as great black-backed gulls, also are slowly becoming more common.

Why the surge in gull populations? It is attributable, ornithologists say, to a variety of factors. The Migratory Bird Protection Treaty, signed by the United States and Canada in 1916, led to a series of federal, provincial, and state safeguards for gulls and other birds. As the Great Lakes forest gave way to the plow, the exposed soils of farm fields offered gulls such as ring-bills a supplementary buffet of earthworms and insects. Record low water levels in the Great Lakes in the 1930s and then again in the 1960s and in 2006 exposed new islands of bare rock, thereby increasing nesting habitat. Stepped-up industrial activity also gave adaptable species such as ring-bills a wide range of artificial nest sites, from dredge-spoil piles and rubble heaps to slag dumps.

And just as the birds expanded their nurseries and were faced with scores of hungry nestlings to feed, they found a greater availability of food, directly or indirectly supplied by humans. In 1922, rainbow smelt were deliberately introduced into the Great Lakes and soon became a staple for gulls. By the mid-twentieth century, populations of alewives, invaders from the Atlantic Ocean that gained entry to the Great Lakes after the Welland Canal was constructed in 1824, also began to build. Alewife numbers initially were kept in check by the Great Lakes' top predator—lake trout. In surveys conducted between 1922 and 1955, alewife numbers were so low that researchers found no traces of the fish in the stomach contents of the gulls they sampled. But by midcentury, yet another Atlantic intruder—the sea lamprey—had taken its toll on lake trout. Under the added pressure of overfishing, the top piscatorial predator became virtually extinct in the lower Great

Alewives invaded the Great Lakes from the Atlantic Ocean after the Welland Canal was constructed in 1824. The increase in fish forage fueled an explosion in gull populations. U.S. Fish and Wildlife Service Digital Library.

Lakes. Not surprisingly, by 1965 alewife numbers had exploded, so much so that they comprised 80 percent of the herring gulls' diet and 50 to 60 percent of the diet of ring-billed gulls.

Not only were alewives more numerous, but they were also more readily available to gulls at a time when they were raising young. The massive spring die-offs of alewives, which made national headlines when they fouled beaches in the lower lakes, were a bonanza to adult gulls with chicks to feed. Fledglings could easily secure their own food by diving for live fish in shallow water around shoals where alewives congregated during the summer. The ready availability of alewives is widely credited for the surge in gull numbers between 1960 and 1965.

Today, another major contributing factor to the continuing rise in gull numbers is the careless disposal of waste. In a 1977 survey of herring gulls on North Shore islands conducted by the U.S. Bureau of Land Management, scientist Don Goodermote concluded that the "availability of food [and not optimal nesting habitat] may well be the limiting factors for North Shore birds." But with the explosion of the shore's human population has come a feast of new gustatory offerings for birds such as gulls. According to Duluth ornithologist Laura Erickson, as soon as ring-billed gulls became accustomed to the presence of people, they learned to exploit the contents of dumps and dumpsters as well as feed on the handouts of picnickers, the scraps left around fast-food restaurants, and the fish remains from commercial fisheries. Although the warier of the two gulls that nest in the Lake Superior region, the herring gull is an equally successful exploiter of the smorgasbord of human refuse. The difference is that "they take their garbage at the dump," Erickson says, laughing. "They won't take the french fries while they're still hot." So acclimated are these gulls to humans that bird guides now list parking lots, parks, and restaurants as the habitats in which they can be found.

The year-round availability of food is perhaps no more important than in winter, when snow and ice cover limit gulls' access to natural forage such as fish and insects. The connection between human garbage and gull diets has been little studied in the north country. Nonetheless, many ornithologists believe that open landfills have dramatically boosted survival rates, especially for juvenile birds that otherwise would have starved during stressful times of the year and for the increasing numbers of gulls that now overwinter in the region instead of migrating to southerly climes.

Yet it is more than just the ability to make behavioral adjustments that enables gulls to eat hot dogs one day and hellgrammites the next. Woodcocks, for example, have long, slender bills that are specially adapted to probing the ground for worms, insect larvae, and slugs. Confine them to a parking lot with an open dumpster and they would starve. Gulls, on the other hand, are not picky eaters but capable omnivores. Their success is in no small part due to their physiology, which allows them to partake of a wide variety of foods. They have long, sharp beaks that can root out earthworms and pierce the tough, scaly skin of a lake trout as easily as they can break apart stale hamburger buns. They also have wide mouths that accommodate either bite-sized pieces or big chunks. Their intestines are intermediate in length, allowing them to metabolize a wide variety of both plant and animal materials. Agility and adaptability, say gull researchers Hans Blokpoel and Gaston Tessier, are key to a gull's diverse diet. "It can plunge dive for fish," they point out, "hawk for insects in the air, follow the plow looking for earthworms and grubs, scrounge french fries at fast food outlets, hunt for voles in fields, and forage at garbage dumps among dump trucks and bulldozers."

The Impact of Exploding Gull Populations on Other Species and Ecosystems

Although human castoffs undoubtedly have given gulls a survival advantage, these vigilant opportunists still turn to natural options whenever possible. And therein lies another, more serious problem. Their inflated numbers, Erickson says, may be placing heavy burdens on other organisms, especially fellow birds, as well as damaging some vulnerable ecological systems beyond their ability to rebound. "Predators have always been part of nature," Erickson points out, "but predator-prey relationships get kind of skewed when people enter the equation by subsidizing the predator," such as disposing waste in open landfills.

To understand the ways in which the increase in gull populations has reshuffled the deck of participants in the North Shore's ecology, you need only examine changes in the composition of the shorebird community. For example, herring gulls historically have nested on islands along the

Generalists possess physical features and behaviors that allow them to exploit a wide variety of foods. The herring gull, a successful omnivore, has a long, sharp beak and wide mouth that enable it to snatch both herring out of the lake and french fries from a dumpster. By contrast, specialists, such as the American woodcock *(below)*, sport bills that are specially adapted to probing the ground for worms, insect larvae, and slugs. U.S. Fish and Wildlife Service Digital Library.

Minnesota North Shore, but ring-billed gulls are relative newcomers to the region's ecology. In his 1932 book *The Birds of Minnesota,* Thomas S. Roberts noted that the ring-billed gull "does not nest in the state so far as known." So rare were nesting ring-bills, Roberts pooh-poohed the sighting in 1905 of a pair near the Cross River by an Aubudon Society ornithologist, saying "The observation is open to doubt," he wrote, "made as it was from a passing steamer."

According to Thomas E. Davis and Gerald J. Niemi of the University of Minnesota Duluth, the first verified record of nesting ring-bills in the Duluth-Superior harbor was made in 1958. Only one nest was reported. Whether the absence of breeding ring-bills in Lake Superior before this time was due to the human-caused decimation of their numbers or the fact that the birds did not breed naturally in the region remains an unanswered question. What scientists do know is that this single nest would herald the beginning of a dramatic shift in Superior's gull populations. Breeding ring-bills began to appear in unprecedented numbers during the early 1970s, when record-high lake levels in Lakes Michigan, Huron, and Superior wiped out nesting habitat throughout the lower Great Lakes. These extreme conditions prompted the birds to disperse into new breeding territories, including the western end of Lake Superior. Here the birds found refuge on

Gulls can take advantage of natural nesting sites as well as artificial ones such as islands of spoils at dredge-disposal facilities. An increase in such human-made nesting habitats contributes to the rise in gull numbers. Photograph by U.S. Fish and Wildlife Service.

human-made islands that were created by harbor dredging earlier in the century. Unlike rocky outcrops in the open lake, these islands provided dependable nurseries since they were relatively unaffected by rising and falling waters.

In 1973 about thirty ring-bill nests were discovered in the Duluth harbor along the Minnesota shoreline. Although these nests ultimately were abandoned, the gulls were not discouraged by their lack of reproductive success. In 1974 an estimated one thousand individuals were reported nesting on a spit of land in the harbor that formerly was used as a coal-handling site. According to a survey by the USFWS, by 1977 the population of nesting ring-billed gulls in the Minnesota portion of Lake Superior had soared to 984 breeding pairs, most of them clustered in three colonies in the Duluth-Superior harbor. By 1998, according to the data collected by Cuthbert and McKearnan, the majority of nesting sites for ring-billed gulls along the Minnesota shore of Lake Superior occurred in four main areas in the Duluth-Superior harbor.

It is unlikely that ring-billed gulls will disappear along the North Shore as suddenly as they appeared. For one thing, ring-bills are faithful to their natal grounds, returning as adults to breed in or near the colonies in which they were born. In addition, the human population continues to grow along the North Shore. Unless people make dramatic changes in the disposal of food waste, the birds will be assured of dependable sustenance well into the future.

And there is still lots of vacant nursery space along the North Shore's rocky island and mainland shores. When natural space is in short supply, ring-bills have demonstrated a remarkable resourcefulness. The shortage of optimum island habitat in the lower Great Lakes, for example, has prompted ring-bills to seek out a variety of other natural and human-made habitats on the mainland. McKearnan points out that in 1999 in Milwaukee, Wisconsin, surveyors tallied some six hundred ring-bill nests on twenty downtown rooftops along the Milwaukee River about one-half mile from Lake Michigan.

What, then, will halt the growth of runaway ring-bill populations if humans do not intervene or dramatically change their waste-disposal practices? Disease, pollution, and predation, even for prolonged intervals, would make only a small dent in ring-bill populations, say Blokpoel and Tessier, and would "not, in the long term, limit gull numbers on the entire breeding range. In sum, we do not yet know the limiting factors of the Great Lakes Ring-bill population."

The ecological impact of new and growing gull populations has been little studied on Lake Superior. To date, much of the evidence linking gulls to environmental harm is purely conjectural; yet enough anecdotal information has been amassed to warrant concern, particularly about the effects of gulls on other birds. In an article for *Audubon* magazine in 1965, Chief Officer J. P. Perkins, who crisscrossed the Great Lakes aboard commercial shipping vessels, recorded his observations about bird migrations in the region. A consummate bird-watcher, Perkins described how gulls expertly exploited migrating flocks:

> After cruising the Great Lakes for 31 years, I am convinced that gulls are a formidable hazard to migrating birds. Few people realize how predaceous herring and ring-billed gulls are. Gulls seldom snatch birds from the air, but I've seen a flock of them annihilate a small flock of sparrows.

The usual procedure is to chase the birds until the victims are exhausted and fall into the water. The gulls then swoop down on the hapless prey.

Small birds put on some flying acrobatics to avoid this fate, but the gulls, with their great endurance and numbers, usually win. I've seen gulls destroy various species from small warblers and sparrows to saw-whet owls, least and American bitterns, and coots. After these larger birds have fallen or been knocked into the water, the gulls gather around and tear them apart. . . .

Sometimes the gulls herd birds like seagoing cowboys. Once daylight arrives and smaller birds are aboard ship, the gulls tend to ride herd on them. If a wandering bird tries to get back to the ship, the gulls try to keep it away and tire it out. Gulls inhabit the Great Lakes in such numbers that quite a sizable toll of land birds must be taken.

Like Perkins, Erickson has witnessed gulls snatch migrating songbirds right out of the air. Listeners to her Duluth public radio program on birds have called to complain of gulls pushing the nests of other birds out of trees and then eating the eggs that have fallen to the ground. Already gull predation on eggs and chicks has been implicated in the demise of state and federally endangered populations of piping plovers on the Duluth waterfront. In the early 1970s researchers identified six to eight pairs of plovers that nested out in the open on dredge-spoil islands in the harbor. By 1985 the number of nesting pairs hovered between one and three; eventually the population disappeared.

Ring-bills also share breeding grounds with common terns. Because the gulls arrive earlier in the spring than terns and have their pick of the best nesting habitats, concerns are mounting that the gulls eventually may displace terns in the region. The competition is particularly worrisome since 63 percent of the entire breeding population of common terns in Lake Superior seek out nest sites in the Duluth-Superior harbor.

Populations of birds along the Superior shoreline are not the only ones at risk from gull predation and competition. Many longtime North Shore observers have noted a greater presence of resident gulls—and nesting gulls—on the region's inland lakes, which they suggest coincides with an increase in human activities in these areas. When gulls are lured to inland shorelines by the picnic leftovers of campers and fish remains that are dumped on land or in shallow water by recreational anglers, they may also prey on the eggs and chicks of loons and other vulnerable species.

There is also the unanswered question of habitat destruction as gulls expand their colonial nesting areas. For example, the North Shore's rocky shores and ledges of islands and mainland cliffs shelter many of the region's rare species of plants. According to research conducted on Knife Island in Lake Superior (a longtime herring-gull breeding colony) as well as on other islands throughout the lower Great Lakes, gulls destroy native vegetation and dramatically alter the soil chemistry of their nest sites.

Ring-bill colonies are especially problematic since their nurseries are far more dense than those of most other seabirds. Their nests, which measure up to two feet across, have attained documented densities in the Great Lakes of up to 5,170 nests per acre—about 1 nest every two feet. In most instances, the trampling alone of so many birds in such close quarters was enough to wreak havoc on existing vegetation. But researchers also documented the elimination of some woody plants through overfertilization from bird droppings. Especially damaging, they discovered, were the defensive behaviors of adult birds. In agitated territorial displays that accompanied boundary clashes, the birds used their bills to peck and uproot surrounding plants, even jabbing the soil when vegetation was not available. In studies of herring gulls on islands off the coast of Scotland, researchers found that on average there were 2.5 such clashes per square foot during the April–July nesting season.

Not only do gulls damage or eliminate existing island vegetation, but they also serve as inadvertent conduits for the transport of alien vegetation to isolated habitats. First, they clear the land by shading, trampling, and plucking out existing vegetation. As nests, dead chicks, and unhatched eggs decompose, soils build up in what was once rocky habitat with little or no substrate for plant growth. Foraging in garbage dumps and agricultural fields on the mainland, the birds return to their island nests with the still-viable seeds of alien weed species in their crops and regurgitated pellets. It is not long before weeds, deposited and fertilized by their unwitting hosts, take over the islands and completely transform their botanical regimes. Researchers who surveyed Superior's Knife Island in the late 1960s observed that generations of nesting herring gulls had decimated the native vegetation. The island, they wrote, largely supported only "weedy species from the mainland. There are with a few exceptions no species capable of surviving the impact of the Herring Gull activity on this small island."

Some of these effects persisted long after the islands were abandoned as nest sites. On Barrier Island in Lake Huron, researchers found that even after four years without nesting gulls the levels of phosphorus on the island remained about ten times higher than normal. And although native grasses had returned to the island, the species composition was not the same as that of a control-plot grassland that was untouched by nesting gulls.

Ring-bills compound this damage through their restless breeding patterns. Unlike most other seabird species, such as herring gulls, which colonize a site and return to it season after season, ring-bills are known to periodically abandon nest sites, moving on to adjacent habitats for a few seasons before returning to recolonize old ones. As a result, they have the potential to alter isolated habitats over a much wider area than other nesting gulls. Studies conducted on Lake Michigan's South Manitou Island in the early 1970s, for example, showed that trampling and overfertilization by bird droppings had killed woody vegetation in a nesting area used by ring-bill gulls in the 1960s. In response, the gulls relocated their breeding ground to a nearby stretch of shoreline where a live assemblage of low-growing shrubs such as creeping juniper and bearberry provided shade and cover for vulnerable chicks. By 1974, after only a few seasons of nesting activity, much of the dune and beach vegetation in the new site also had begun to die.

Especially troubling is that the destructive effects of runaway gull populations may be magnified by other species that for similar reasons have learned to flourish in the presence of humans. Like gulls, crow populations in the Duluth area, Erickson observes, also have skyrocketed. And with the crows have come greater numbers of merlins. Why? Merlins do not build their own nests but instead occupy those constructed by other birds. The abandoned crows' nests that festoon the city's trees have given them their pick of vacant real estate. Efficient aerial predators, the merlins are perfectly positioned to take advantage of the songbirds that pass through the city as they follow the Superior shoreline on their annual migrations.

With unlimited opportunities for food and nesting, Erickson foresees little that will curb gull populations in the future. "There's this thought that everything in nature just balances out in the long run," she says. "That's not true. The animals that are becoming the most abundant now are ones that have learned to exploit people in one way or another—crows, gulls, raccoons, coyotes, foxes. They tend to be opportunistic omnivores, the kind that are all now found in the congested suburbs of Chicago." ❧

SUGGESTIONS FOR FURTHER READING

Bernard, Jessie M., Donald W. Davidson, and Rudy G. Koch. "Ecology and Floristics of Knife Island, a Gull Rookery on Lake Superior." *Journal of the Minnesota Academy of Science* 37 (1971): 101–3.

Blokpoel, Hans, and William C. Scharf. "The Ring-Billed Gull in the Great Lakes of North America." *Acta XX Congressus Internationalis Ornithologici* 4 (1991): 2372–77.

Blokpoel, Hans, and Gaston D. Tessier. "The Ring-Billed Gull in Ontario: A Review of a New Problem Species." Occasional Paper, no. 57. Ottawa, Ont.: Canadian Wildlife Service, 1986.

Bukro, Casey. "Pilot Offers to Frighten, Not Shoot, Gulls." *Chicago Tribune,* 12 May 1998.

———. "Waukegan Firm Shoots to Teach Gulls a Lesson." *Chicago Tribune,* 7 May 1998.

———. "Waukegan Gulls Flee Following Shooting." *Chicago Tribune,* 8 May 1998.

Cuthbert, Francesca J., and Joan McKearnan. "U.S. Great Lakes Gull Survey: 1998 Progress Report." University of Minnesota, Department of Fisheries and Wildlife. St. Paul: University of Minnesota, 1998.

———. 1999. Unpublished data.

Davis, Thomas E., and Gerald J. Niemi. "Larid Breeding Populations in the Western Tip of Lake Superior." *The Loon* 52 (1980): 3–14.

Goodermote, Don. "Herring Gull Nest Counts on the North Shore of Lake Superior." *The Loon* 52 (1980): 15–17.

Green, Janet C., and Robert B. Janssen. *Minnesota Birds: Where, When, and How Many.* Minneapolis: University of Minnesota Press, 1975.

Hogg, Edward H., and John K. Morton. "The Effects of Nesting Gulls on the Vegetation and Soil of Islands in the Great Lakes." *Canadian Journal of Botany* 61 (1983): 3240–54.

Ludwig, James Pinson. *Herring and Ring-Billed Gull Populations of the Great Lakes 1960–1965.* Great Lakes Research Division. Publication no. 15. Ann Arbor: University of Michigan, 1966.

———. "Recent Changes in the Ring-Billed Gull Population and Biology in the Laurentian Great Lakes." *The Auk:* 91 (1974): 575–94.

Perkins, J. P. "A Ship's Officer Finds 17 Flyways over the Great Lakes." Part II. *Audubon* 67, no. 1 (1965): 42–45.

Roberts, Thomas S. *The Birds of Minnesota.* Minneapolis: University of Minnesota Press, 1932.

Ryder, John P. "Ring-Billed Gull." No. 33, in *The Birds of North America,* ed. Alan Poole, Peter Stettenheim, and Frank Gill. Philadelphia: Academy of Natural Sciences; Washington, D.C.: American Ornithologists' Union, 1993.

Shugart, Gary W. "Effects of Ring-Billed Gull Nesting on Vegetation." *The Jack-Pine Warbler* 54, no. 2 (1976): 50–53.

Sobey, Douglas G., and John B. Kenworthy. "The Relationship between Herring Gulls and the Vegetation of Their Breeding Colonies." *Journal of Ecology* 67, no. 2 (1979): 469–96.

Southern, William E. "Comparative Distribution and Orientation of North American Gulls." In *Behavior of Marine Animals,* ed. Joanna Burger, Bori L. Olla, and Howard E. Winn. New York: Plenum Press, 1980.

Southern, William E., William L. Jarvis, and L. Brewick. "Food Habits and Foraging Ecology of Great Lakes Region Ring-Billed Gulls." In *Proceedings of the Fish-Eating Birds of the Great Lakes and Environmental Contaminants Symposium,* Hull, Quebec, 1976.

The Return of
Lake Trout
to Superior

At 1.1 billion years old, the shoreline basalt of Lake Superior is among the world's oldest rock. Indeed, these exposed outcrops are part of Earth's basement, the very foundation on which the planet is built. Remarkably, however, the shoreline embraces waters that are only some ten thousand years old. It is the equivalent, in human terms, of a Han dynasty cup that has been filled with freshly brewed tea.

By geological reckoning, Lake Superior is just beginning the long evolutionary journey of its life. For two million years, glaciers bore down from the north, driving plants and animals into what are known as glacial refugia on the margins of the ice sheet. The resident organisms that we see today once were biological migrants chasing a habitat mosaic that slowly shifted back and forth across the landscape as the ice sheets came and went.

When the Wisconsin ice sheet mounted its final retreat to the northern edge of the present-day lake, it left behind a grand and largely empty aquatic mansion. Refugees from as far away as present-day Alaska, the Mississippi River basin, and the Atlantic Ocean drifted or swam through the connective arteries of glacial rivers and lakes to colonize the newly ice-free lake. Animals ranging from tiny diaphanous opossum shrimp to sturgeons the size of small whales slowly populated the lake's cavernous spaces and secreted corners. Freed from the weight of the ice sheets, the land began to rise, slowly cutting off the channels that funneled water into Lake Superior from faraway rivers and lakes. Until about 2,200 years ago the only remaining ingress for waters outside the Lake Superior watershed was through the St. Marys River near what is now Sault Ste. Marie. But as the land continued to rebound, a bedrock barrier rose up between Lakes Superior and Huron, permitting only an outflow of water from Superior. The lake's biota had finally found a stable home.

Some organisms were especially enterprising, quickly developing the ability to occupy the lake's numerously varied chambers. Among the most diversified and universally distributed group of animals was the lake trout. This 2.5 million-year-old species of fish survived the restless comings and goings of the glaciers by hiding out in the rivers and small lakes that lay both to the south and the northeast of the ice sheet. Once they became established in Superior, lake trout set up housekeeping, metaphorically speaking, in niches as different from each other as the basement, attic, and garage of a house.

As fisheries biologist Stanford H. Smith points out, the lake trout was "the only predator that occupied the entire lake from shore to shore, surface to bottom." They could be found holding in large bays or around islands or frequenting reefs just offshore. Their diet included terrestrial insects that had been blown onto the surface of the water, and sculpins and burbot in depths of more than 900 feet, where hydrostatic pressure is extreme.

Biologists have recognized three phenotypes of lake trout in Lake Superior, that is, groups of lake trout that share certain physical features, some of which are conditioned by the habitats in which they live and others by genetic makeup. The siscowet generally occupies depths of 250 feet or more. This habitat comprises 77 percent of the lake's waters, a commodious territory that has enabled siscowets to outnumber lean lake trout by twenty to one.

Siscowets are distinguished by their oily flesh. Some studies have shown that they can devote nearly 50 percent of their body weight to fat. In fact, so packed is the fish's flesh and viscera with fat that the Ojibwe name *siscowet* is literally translated as "cooks itself."

The animal's high fat content is not, as one might assume, to provide insulation from deep-lake temperatures, which rarely stray above a constant 39 degrees Fahrenheit. Scientists studying marine fishes have found that fat serves as a critical mechanism in buoyancy control for animals that must negotiate extremes in hydrostatic pressure as they migrate from the ocean's depths to shallower waters. Researchers suspect that it may play a similar role for siscowets and their cousins the humper lake trout. By day siscowets are routinely caught in depths of more than 328 feet. But analyses of their stomach contents—a collection of flotsam that includes everything from terrestrial insects, bats, birds, pine needles, grain, wood, fish guts, and net twine—suggest that they spend considerable time grazing at the lake's surface under the cover of night. Indeed, scientists have documented dramatic depth changes of about 500 feet over the course of just a few hours.

Great Lakes biologists also distinguish lake trout on the basis of their physical features. Squat and stout with a small head and blunt nose, the siscowet has "an altogether dumpy appearance," write fisheries experts Andrew H. Lawrie and Jerold Rahrer in an unusual lapse of anthropomorphizing.

Not so the lean lake trout. With its tapered body and

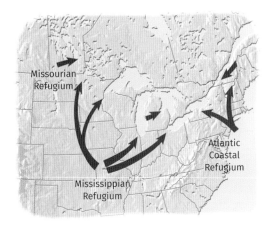

The resident organisms in the Great Lakes today were once biological refugees from places as far away as present-day Alaska, the Mississippi River basin, and the Atlantic Ocean.

Lake Superior's habitats support three phenotypes, or strains, of lake trout. Lean lake trout most often are found in depths of less than three hundred feet; siscowets live deep in the lake by day and surface at night; and humpers frequent isolated offshore reefs and the pinnacles of great underwater mountains. Illustration by Vera Ming Wong.

snout, the lean lake trout is "more salmon-like in appearance," Lawrie and Rahrer observe. Possessing an average of 9.4 percent body fat, leans also have a wider taste appeal. Their marketability and their greater accessibility (lean lake trout are found most often in depths of less than 300 feet) combine to make them the strain most prized by recreational and commercial fishermen.

A third type of lake trout, a lean trout known as a humper, is distinguished by its preference for spawning on isolated offshore reefs and shoals on the pinnacles of great underwater mountains whose steep slopes drop into waters hundreds of feet deep. Unlike siscowets and leans, humpers have not been officially accorded subspecies status. That is because many scientists theorize that humpers are descendants of once wide-roaming populations of leans that colonized underwater peaks in the open lake, particularly around Isle Royale and Caribou Island in Superior's southeastern waters. Nonetheless, humpers exhibit distinctive physical characteristics. Their body fat content, for example, ranges halfway between siscowets and leans. Their appearance also singles them out. The backs of humpers have a more bronzy-green hue, and their fins are more brightly colored than those of leans. Their most notable feature, however, is a thin abdominal wall, for which they have earned the nickname "paperbellies."

So goes the simple version of the lake trout's story. Before the collapse of the fishery in the mid-twentieth century, the narrative of the fishes' evolutionary journey was far more complex. Not only had siscowets and leans begun to differ from one another, but a variety of observers, from the region's Ojibwe, fur traders, and missionaries to commercial fishermen and fisheries biologists, also noted that considerable variations existed within these phenotypes. Based on differences in skin and flesh color, for example, fishermen historically recognized two varieties of siscowets. Of the three phenotypes of Superior lake trout, however, lean lake trout appeared to have traveled the faster evolutionary track judging by their diversity in behavior, habitat preferences, and physical characteristics (including weight, shape, fat content, and flesh, fin, and skin color). Such evolutionary refinements became especially noticeable to fishermen during spawning season. Some groups of lean trout, for example, shared spawning grounds but avoided interbreeding by spawning at staggered intervals. Others divvied up breeding territories so that some strains frequented the lee shores of islands while others favored windward locations. Because of this relative reproductive isolation, many lean lake trout populations (known variously as stocks, breeds, strains, or morphotypes) developed and maintained such distinctive features that fishermen used these physical cues to identify them. A list of the informal names given to the fishes by the people most intimately acquainted with them reflects the biological flowering that had begun to unfold in the lake. To the fishermen, who reportedly knew lake trout so well that they could distinguish lean breeds on the basis of taste alone, the fish were variously known as blacks, redfins, yellowfins, grays, salmon-trout, red trout, moss trout, sand trout, and racers. Indeed, the detailed knowledge that fishermen possessed was a matter of considerable amazement among early visitors to the Lake Superior region. In a 1926 article for *American Game* magazine, Frank M. Warren observed that the flesh color of the various breeds found on the reefs around Isle Royale ranged from "'white' to amber and salmon." But, he adds, "only the older and more experienced commercial fishermen can tell the light from the darker fleshed ones before dressing them."

According to researchers, the unique environmental conditions of each strain's habitat—from temperature regimes and water-circulation patterns to the availability of food—accounted for these marked differences. As a result, writer Steve Grooms points out, fish stocks can "share genetic traits that differ in subtle but crucial ways from those of other individuals of their species. Evolution 'fine-tunes' such a group of fish to make optimal use of the survival and reproductive opportunities offered by a specific micro-environment, a region that could be as small as a particular creek or reef. The lake trout is a highly 'plastic' [i.e., evolvable] fish, displaying remarkable morphological variety."

This specialization, particularly among leans, occurred throughout the Great Lakes, says Canadian researcher John Goodier. But Superior laid claim to the greatest lake trout diversity. In the 1970s Goodier interviewed commercial fishermen in Ontario and reviewed such historical documents as government reports, fur-trade records, and explorers' accounts. In Canadian waters alone, he identified more than two hundred different spawning grounds, including six rivers that were used by lean lake trout during fall spawning runs.

In time, these stocks had the potential of evolving into new species. Several events, however, conspired to turn back the evolutionary clock. As Grooms observes, the story of lake trout in the Great Lakes is a "nightmare case study of what happens when humans mess around with Mother Nature."

The History of Lake Trout Fishing

When commercial fishing began in earnest in the 1840s, the fishing economy targeted whitefish, not trout. Lake Superior catches were feeding not only workers in the region's

When populations of targeted species such as whitefish and lake trout were depleted, commercial fishermen switched to alternatives such as herring. Photograph by Gallagher Studio, circa 1940. Courtesy of the Minnesota Historical Society.

In the early years, commercial fishermen targeted large, mature fish, shown here in an Isle Royale catch in the 1920s. Since females do not reach reproductive maturity until age eight or nine, this practice culled spawning adults from the population. Photograph by Paul Gaylord. Courtesy of the Kathryn A. Martin Library, University of Minnesota Duluth, Archives and Special Collections.

burgeoning lumber and mining camps but also inhabitants of the Midwest's booming cities. Over time, technological improvements increased efficiency. Handmade cotton twine, for example, gave way to machine-made nylon, which was more durable, water resistant, and nearly invisible to fish. Diesel engines replaced oars and hand-cranked net haulers. New net designs effectively trapped more fish.

It did not take long for problems to surface. In an 1872 survey of the Great Lakes that included Lake Superior, J. S. Milner of the U.S. Commission of Fish and Fisheries reported serious depletions in whitefish populations and urged that protective legislation and hatchery stocking be instituted. The construction of a hatchery in Duluth in the late nineteenth century, however, did not reverse the downward trajectory. As whitefish dwindled, recreational anglers and commercial outfits switched their focus to lake trout. Because they targeted trout on their spawning grounds where the fishes gathered seasonally in huge numbers, fishermen had an inflated notion of their abundance. In a 1921 article for *National Geographic Magazine,* George Shiras describes how the trout population in one spawning ground was decimated:

In 1872, when 12 years old, I had an early introduction to lake trout. A report was brought to Marquette [Michigan,] by a lumber-laden schooner, becalmed for a while in the vicinity of Stannard Rock, a sandstone reef lying a few feet below the surface, some forty-five miles northwest of the town, that the waters about the reef were surrounded by immense schools of lake trout. It was said that the fish could be hauled aboard the schooner by simply casting a trolling spoon overboard, when there was such a rush by the fish that one could imagine it was a contest to see which one might be caught first.

An enterprising captain of a local excursion steamer thereupon advertised an expedition to this vicinity, and about seventy-five persons, including women and children, departed under bright skies and unruffled waters. At noon the dangerous reef was approached cautiously, and the steamer anchored in about thirty feet of water.

Soon ten boats were lowered over the side, each expectant fisherman having a trolling line, as many as four or five lines trailing behind each boat. In a few minutes there was a rush of eager fish, and to my youthful mind there never were such scenes of excitement, for as the boats circled about the reef the long lines were diverted at various angles by the larger fish, becoming entangled, while the continued flopping of those captured caused the women and children to shriek in triumph or dismay. . . .

Halvorsen and Froyset Two Harbors Minnesota
Ed. Running. Two Harbors Minnesota
Aarak Brothers And Egeland Two Harbors Minnesota
John Degerstedt. Two Harbors Minnesota
Johnson and Overby. Two Harbors Minnesota
Johnson And Hendrickson Two Harbors Minnesota
Jacobson and Sons Two Harbors Minnesota
Johnson and Carr Two Harbors Minnesota
Jensen and Edisen Two Harbors Minnesota
Jensen Brothers Two Harbors Minnesota
Fenstad Brothers Little Maria Minnesota
Andrew Frederikson Two Harbors Minnesota
Nick Nelson. Two Harbors Minnesota
Julian Jacobson And Company. Two Harbors Minnesota
Erick Johnson Two Harbors Minnesota
Mattson and Goss Two Harbors Minnesota.
Sjoquist and Carlson Brothers Two Harbors Minnesota
Andrew Pederson Two Harbors Minnesota
Sam Johnson and Son Two Harbors Minnesota
S.M. Jenson. Silver Creek Minnesota
John Running Two Harbors Minnesota
C.A. Nelson and Son Two Harbors Minnesota
Slotness and Co. Two Harbors Minnesota
Peterson Bros Two Harbors Minnesota
Lind Brothers, Jr. Two Harbors Minnesota
Ole Wick Two Harbors Minnesota
S. Myrvold and Co. Two Harbors Minnesota
Larson and Lind Two Harbors Minnesota
Salted 100 lbs.
Salted 50 lbs.
Fresh. 100 lbs.
Roleff.

In the heyday of commercial fishing along Minnesota's North Shore, abundant fish harvests were commonplace. By the 1960s, overfishing and the invasion of nonnative sea lampreys contributed to the population collapse of such commercially valuable species as lake trout. Photograph by William F. Roleff, circa 1916. Courtesy of the Minnesota Historical Society.

> In less than three hours a thousand fish were taken, averaging ten pounds, and then this riot of destruction came to an end, for it finally became apparent how difficult it would be to give away five tons of trout among the friends and neighbors of the participants.
>
> The results of this expedition soon reached the ears of the local fishermen, and for several succeeding seasons immense catches were made. Now a towering lighthouse surmounts this rock as a warning to the mariner and a fitting monument to the myriad of fish that have long since passed away.

By 1939 nearly one million pounds of Lake Superior fish were exported each year. Biologists now suspect that by the 1930s and 1940s lake-wide trout populations were in serious decline even though commercial catch records of the period registered no decrease in the trout harvest. The reason? Operators maintained their yields by making the rounds of the various subpopulations of trout, cropping one stock before moving on to the haunts of another. "In this way," write Rahrer and Lawrie, "stock after stock was depleted while conventional yield statistics gave an impression of relative stability."

Catch numbers also did not take into account the intensification of fishing effort that occurred over time or the fact that fishermen switched to other species of fish when the numbers of their targeted species were exhausted. According to University of Wisconsin Sea Grant's *The Fisheries of the Great Lakes,* "As fish became scarcer, instead of easing up, Great Lakes fishermen only tried harder, employing more boats, working longer hours, going further afield. Whitefish stocks in Lake Michigan, for example, were already in decline by the 1860s. The scarcity had become a common complaint among fishermen by the 1870s, yet the harvest remained high—totaling some 12 million pounds in 1879—as fishermen increased their fishing effort, shifted to new fishing grounds and used more efficient gear, including smaller meshes and finer twine in the gill nets. The whitefish catch dropped abruptly after that, though the total catch of Great Lakes fish remained stable until 1892 because of increases in the harvest of other species."

Especially damaging to lake trout populations was a form of fishing known as "fishing up." Lake trout are not a particularly fecund species of fish. In Superior most females do not reach sexual maturity until age eight or nine. The species made up for a relatively low rate of reproduction, however, by living to grand old ages. Before the advent of large-scale fishing on the lake, scientists estimate that about one-half of the lake's biomass was composed of large, mature fish weighing more than ten pounds and measuring up to three feet long. Fish older than fifty years of age and weighing in at more than twenty pounds were not uncommon. As commercial operations systematically exploited one locale after another, they removed the largest fish, a practice that culled spawning adults from the population. Reproduction was further suppressed when fishermen reaped larger numbers of younger age classes to make up for yield reductions in large, older fish.

Had fisheries managers known some of the most fundamental aspects of the lake trout's life history, many crises possibly could have been averted. But scientists did not begin to thoroughly study the biology of this linchpin species until the mid-1940s. By then, a far more efficient—and devastating—predator had burst on the scene: the sea lamprey.

The lamprey is thought to be a marine relict that adapted to freshwater following the last glacial retreat. How it gained access to the Great Lakes remains a matter of debate. Some scientists believe that it was indigenous to waters below Niagara Falls, while others maintain that it entered the Great Lakes system only after the construction of the Erie Canal in 1823.

The stuff of made-for-TV horror movies, sea lampreys attach themselves to their prey using a primitive mouth shaped like a suction cup and equipped with rows of file-like teeth. After rasping through the fish's tough outer skin, the lamprey digs into its victim's tender tissues and then slowly drains it of blood and other vital fluids. Within a mere twenty months, a single lamprey can kill forty-five to fifty pounds of fish.

The lamprey cut a methodical and deadly path through the Great Lakes, dealing what most biologists consider the decisive blow to already-declining lake trout populations in the Great Lakes. Some researchers contend that the lamprey invasion would not have been so ruinous had commercial fishing operations not depleted the number of jumbo-sized fish. Indeed, the lamprey executed its own variation of the fishing-up technique on an already vulnerable fish population. Lampreys prefer large host fish, which can better survive their parasitism, albeit in a weakened state. In the absence of

The rasping, file-like teeth of the sea lamprey make it a formidable predator of fish in Lake Superior. Photograph by T. Lawrence, Great Lakes Fishery Commission / NOAA Great Lakes Environmental Research Laboratory.

these mature fish, the sea lamprey pursued a younger, more vulnerable segment of the population—trout as young as four years of age. By killing juvenile fish before they could reach a spawning age, the lamprey virtually shut down lake trout reproduction within a few short years. Remarkably adaptable, the lamprey not only accommodated itself to smaller lake trout when big fish were no longer available, but it even switched species, parasitizing large whitefish and herring. With the exception of fish populations in two locations in eastern Lake Huron, by the late 1940s lake trout stocks in the lower lakes were largely extirpated. Gone was the vast storehouse of genetic diversity, built up over thousands of years of evolutionary tinkering.

The trout fishery lasted longer on Superior than on the lower lakes for a variety of reasons. Not only is it the largest of the Great Lakes, Superior also was the last lake to be invaded by the sea lamprey. And within its remote pockets of water, particularly on the isolated reefs around Isle Royale in the north and Stannard Rock in the southern reaches of the lake, many stocks persisted out of reach of the lamprey's deadly grip or the fisherman's net. (Historically, these were the sites of greatest lake-trout abundance in U.S. waters, a biological reputation they have maintained to this day.)

In time, however, Superior's trout fishery too would succumb. A shorthand of yield statistics tells the disturbing story. In the late 1940s, the lake trout catch averaged about 4.5 million pounds per year. By 1960 in the wake of the lamprey's rampage, yields had plummeted to 500,000 pounds, a drop of nearly 90 percent. The commercial losses were staggering since the lake trout catch assumed an economic value far out of proportion to actual yields. During the years 1945 to 1949, lake trout accounted for only 14 to 20 percent of the U.S. catch. However, they comprised 51 to 60 percent of the total value.

Though severely depleted, remnant stocks persisted, even in coastal reaches where

Sea lampreys use their sharp teeth to cut through a fish's tough outer skin, dig into its tender tissues, and slowly drain it of blood and other vital fluids. Within a mere twenty months, a single lamprey can kill forty-five to fifty pounds of fish. Photograph by U.S. Fish and Wildlife Service.

By 1960 lake trout populations had crashed in Lake Superior. The losses to commercial fisheries were staggering since the economic value of the lake-trout catch was far out of proportion to actual yields.

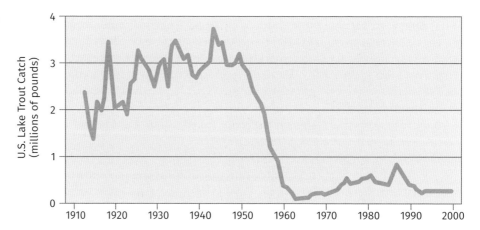

lamprey predation was heaviest. In 1955 the international Great Lakes Fishery Commission (GLFC) was formed to save these strains in Superior before they too underwent the mass extinctions of the lower lakes. Chief among the commission's goals was to establish effective controls of lamprey populations. Researchers discovered that adult lampreys migrated up some of Superior's tributaries to spawn. Early efforts using mechanical controls, such as weirs set up at the mouths of streams, proved largely ineffective in curbing the lamprey population. In 1958, following additional research into lamprey life history, scientists began treating Superior streams that served as nurseries for sea lamprey with the lampricide 3-trifluoromethyl-4-nitrophenol (TFM), a chemical that kills lamprey larvae before they have a chance to develop into juveniles and head into the lake. In 1960, the lamprey population in Lake Superior was estimated at eight hundred thousand. By 1999, an integrated-pest-management strategy—including applications of TFM, the construction of stream barriers and traps, and the release of sterile males—cut the number to two hundred thousand.

Still, lamprey control remains controversial. Although an effective method for controlling lampreys, TFM is not without its drawbacks. For one thing, chemical treatment is expensive. In 2005, the GLFC spent $15 million on sea lamprey control in the Great Lakes, more than $10 million of which was earmarked for stream treatments and assessment and monitoring activities. Twenty percent of the field budget, some $2 million, was devoted to lamprey control in Lake Superior tributaries alone.

TFM applications are not only pricey but also controversial because of their potential to kill stream invertebrates and native lampreys. The Lake Superior watershed hosts three species of endemic lampreys: the American brook lamprey, northern brook lamprey, and silver lamprey. (The only two Minnesota tributaries that host native lampreys are the St. Louis and Nemadji Rivers.) Of the three, only the silver lamprey is parasitic. In their larval stage, these fishes serve as biovacuums, straining prodigious amounts of tiny plants, animals, and organic matter from the water. According to Peter Sorensen, professor of fisheries and wildlife conservation at the University of Minnesota, "They clearly have a valuable role in biodiversity and the ecosystem. These ancient fish aren't well understood but they are incredibly efficient filter feeders, an extremely important link between the benthos and fish communities, and sensitive to environmental degradation."

The populations of all three species have seriously declined since the beginning of TFM use in Superior's tributaries. The most sensitive species, the silver lamprey, has registered the greatest losses, largely because it spawns in the lower reaches of streams where TFM is most commonly applied. In 1959, for example, scientists captured 4,278 silver lampreys from thirty-five riverine monitoring sites. Sampling conducted between 1973 and 1977 turned up only 91 animals.

Although TFM remains the most effective means of sea lamprey control to date, the GLFC has allocated a portion of its budget to alternative controls, such as installing mechanical barriers and releasing sterile males. The commission has also invested in research to find additional biologically benign measures to curb sea lamprey populations. Among the most promising is the discovery of a sex pheromone known as 3-keto petromyzonol sulfate (3KPZS), which male lampreys exude to signal their reproductive readiness and the location of the stream-bed nests they have constructed. So potent is this sex attractant that it can be picked up by females over long distances. Scientists are looking at ways in which to use this pheromone as the basis for a more targeted and environmentally benign form of lamprey control, such as releasing sterile males with enhanced pheromone-pumping capabilities or applying 3KPZS in streams to lure gravid females into traps.

While they instituted lamprey control, officials also stepped up an extensive stocking program of hatchery-raised fish, which began in 1952. According to the U.S. Fish and Wildlife Service, by 2002 an estimated 173.4 million lake trout had been stocked in the lower Great Lakes. An additional 100 million fish were planted in Lake Superior. The cost of raising and releasing one yearling trout ranges between $0.35 and $0.50, adding an estimated $95.7 to $136.7 million to the trout-rehabilitation budget. This figure does not include millions more spent on monitoring and managing fish populations as well as funding scientific research. Tabulating accurate figures of the total cost of lake trout restoration is also complicated by the number of parties involved in the effort, including numerous Indian tribes, eight states, the Province of Ontario, the federal governments of the United States and Canada, and the binational GLFC. This much, however, can be said: staving off extinction does not come cheap. "The effort to restore self-sustaining, wild lake trout populations," Grooms writes, "is one of the longest running and most expensive programs in the history of fisheries management."

The investment has paid handsome dividends in Lake Superior. Today, populations of naturally reproducing lake trout have rebounded dramatically. In 1982 netting surveys conducted by the Minnesota Department of Natural Resources (DNR), only 9 percent of the lake trout sampled were fish that had been born in Superior. By 1998 the percentage of wild fish in survey catches rose to 25 percent. In fact, Superior lake trout are reproducing so well on their own that for the first time in twenty-five years U.S. and Canadian agencies are cutting back dramatically on the number of fish they stock. In some places, lake trout stocking has ceased altogether. Beginning in 2003, the Minnesota DNR suspended its stocking of lake trout in Minnesota waters from the Cascade River near Grand Marais to the Canadian border. The numbers of wild trout have held strong, causing managers to extend the suspension from the Encampment River to the border.

The lower Great Lakes have not been so lucky. True, stocked lake trout have survived to fishable size, but the fishery is not self-sustaining. Like patients who cannot make their own blood and consequently must rely on intravenous transfusions to survive, "without continued massive stocking," Grooms observes, "lake trout would disappear from four of the Great Lakes for the second time this century."

Although many stocked fish survive to spawning age in the lower Great Lakes, few are able to successfully reproduce. One of the factors implicated in reproductive failure may be inappropriate stocking procedures that do not take into account the reproductive biology of trout. In the wild, young lake trout home on their natal grounds, returning to them as adults to spawn. The mechanisms of homing are not perfectly understood. One hypothesis suggests that young fish imprint on a whole complex of smells that may include the odor of the rocks and resident plants and animals within their nursery habitats as well as the odor of mucous and feces that are shed by the fish themselves and remain lodged within the crevices where they hatch and develop. Scientists who have conducted research on lake trout stocks in Lake Michigan suggest that "each discrete stock might have its own distinctive olfactory signature by which its natal waters are recognized by the fish when they return years later as homing adults."

The evolutionary advantages to homing in a lake the size of Superior seem clear: homing keeps trout from having to reinvent the reproductive wheel each season. Devotion to sites that have paid dividends in the past boosts the odds of reproductive success in the future and keeps fish from expending precious energy in wandering a vast lake in search of new, and potentially risky, spawning grounds. In time, such a strategy also helps to cut down on needless competition for scarce resources by distributing populations around localized nodes of plenty in the lake.

This piece in the biological puzzle was missing in most of the fish management practices of the past. The result: little, if any, natural reproduction. Stocked fish were observed, for example, spawning at such doomed sites as boat ramps where they were released as youngsters into the lake from hatchery trucks. In other cases, fish were retained in hatcheries until they reached a certain size in the belief that bigger fish would have a better chance of survival. As adults, these fish were literally adrift, lacking any compass to guide them through the reproductive process.

Biologists also cite the "stock concept" as another potential reason for the reproductive failure of lake trout in the lower lakes. The stock concept was formulated in the 1970s by biologists studying Pacific West Coast salmon. Like keys cut for individual locks, salmon strains possess specific morphological, behavioral, and genetic characteristics that vary depending on the rivers in which they spawn. Adapting the stock concept to the Great Lakes, scientists argued that the lake trout's reproductive failure in the lower lakes was due to the planting of stocks that were ill adapted to the habitats in which they were released. The problem was this: fisheries biologists had little genetic capital to draw upon since, for all practical purposes, lake trout were extirpated from the lower lakes. (In Lake Michigan, for example, biologists turned to fish stocks from as far away as Green Lake, Wisconsin, and Lewis Lake, Wyoming, where Lake Michigan lake trout were transplanted in the early twentieth century. Scientists have expressed concerns, however, that even though these stocks originated in Lake Michigan, they now may be too far removed genetically from their parent stocks to become established in the lake.)

The only stock widely available for reintroduction was the Marquette strain, a stock of lean lake trout taken from Lake Superior around Marquette, Michigan. But transplanting Lake Superior stocks into foreign waters, some biologists argue, is the fisheries equivalent of trying to grow varieties of wheat that are adapted to the cool climate of

Saskatchewan's prairies in the fields of Kansas. "Precious genetic coding was lost when the lake trout was eliminated from the lower lakes. . . . Thus the fate of the rehabilitation program," Grooms writes, "depends upon stocking the lower lakes with fish whose genes evolved in icy, deep Lake Superior. Maybe that will be no special problem. Yet only one lake has seen strong natural reproduction, and it is the only lake planted with fish stocks that originally developed in it."

The lack of recruitment among lake trout in the lower lakes can also be blamed on the major changes in the lakes' biota that have occurred in the past one hundred years. Minnesota fisheries biologist Thomas F. Waters points out that the forage system, which included lake trout as the top predators and microscopic zooplankton on the lowest rung, was the "most stable and productive predator-prey system of the Great Lakes." Young trout consumed a wide array of invertebrates that could be found both on the lake bottom and in its open waters. Once they matured, trout feasted on schools of fish that were, like themselves, adapted to specific environments within the lake. Prey such as bloaters, a deepwater cisco, for example, inhabited the lake's depths, while other cisco species occupied pelagic zones. The fact that this stable matrix of relationships was not broken until relatively recent times, despite more than a century of serious assaults, attests to its extraordinary strength and durability. Lake trout were able to maintain their hold as top predator in large part because they shared the lake with few other ecologically similar species. Today, however, lake trout must compete for the position of top predator with numerous introduced species of salmon and trout. Furthermore, the natural dynamics between predator and prey have changed dramatically, particularly in the lower lakes, where the forage base now includes such successful alien invaders as alewives.

The Future for Superior Lake Trout

Lake Superior maintains the most intact predator-prey structure in the entire Great Lakes. Not surprisingly, it is the only Great Lake with successful wild trout reproduction. Even in the remote reaches of Superior where native stocks have survived, however, scientists estimate that restoration requires a minimum of thirty years. "The stunning success with wild lake trout in Superior teaches—or should teach—the importance of patience in fish and wildlife restoration," observes *Audubon* writer Ted Williams. "If you bust up a complicated machine that took nature 10,000 years to build, you don't slap it back together before lunch."

Given sufficient protection from lampreys and overfishing, nature has exhibited an extraordinary resilience. Studies of lake trout populations at Stannard Rock, for example, showed that fish recovered from the historic drop in their numbers by compensating with greater reproductive efficiency: they matured at an earlier age and developed greater fertility.

Siscowets have made such a dramatic comeback that biologists have considered using the strain as the basis for a whole new commercial fishery. The fish's fat has been shown to be rich in two kinds of omega-3 fatty acids, a popular dietary supplement. Siscowets also show promise as a high-quality fish meal for both human and animal consumption.

But the animals' close brush with mass extinction should teach us not to take advantage of such flexibility but to manage recovering fish populations with greater wisdom and caution. Amazingly, the hard lessons of history have yet to fully sink in. Take the example of Gull Island Shoal in western Lake Superior. Following the crash of lake trout populations in the Apostle Islands, state officials outlawed commercial fishing in Wisconsin waters from 1962 to 1970. The prohibition helped to increase trout numbers, particularly around the complex of reefs that composed Gull Island Shoal, the only remaining spawning ground in the Apostle Islands that continued to produce native fish. So healthy was the lake trout population that it supplied 90 percent of the eggs used to restock lake trout in other parts of the Wisconsin waters in Lake Superior.

The recovery of trout in the region led to an easing of fishing restrictions in the early 1970s. But by 1974 the spawning population had declined by 50 percent, largely due to pressure from tribal commercial fisheries and recreational anglers. Overharvesting had once again prompted fishing closures. To permanently protect these stocks, in 1975 the State of Wisconsin set aside the 173,000-acre Gull Island Shoal as a fish refuge. Bans on fishing within the refuge have helped to boost fish numbers such that the Gull Island Shoal now is a major exporter of lake trout to the surrounding waters. It has also retained its status as the region's main nursery for native fish.

But the refuge protects fish only on their spawning grounds, not throughout their entire home ranges. Trout that wander outside the refuge are subject to intense pressure once again from both recreational anglers and Native American commercial operations, particularly in waters west and south of the refuge.

In the lower Great Lakes, humans have worked changes of such glacial proportions that they have taken the lake trout back to their evolutionary beginnings in these inland waters. Like the ecological havens that once existed on the margins of the ice sheet, Lake Superior now serves as a kind of modern-day refugium from which the lake trout might disperse and, given lots of luck and enough time, once again reign supreme throughout its former range. The future of the lake trout, however, hangs in the balance. The question is, How willing are we to learn from our mistakes—as well as our successes? ❧

SUGGESTIONS FOR FURTHER READING

Bailey, Reeve M., and Gerald R. Smith. "Origin and Geography of the Fish Fauna of the Laurentian Great Lakes Basin." *Canadian Journal of Fisheries and Aquatic Sciences* 38 (1981): 1539–61.

Bogue, Margaret Beattie. *Fishing the Great Lakes: An Environmental History, 1783–1933.* Madison: University of Wisconsin Press, 2000.

Bronte, Charles R., Mark P. Ebener, Donald R. Schreiner, David S. DeVault, Michael M. Petzold, Douglas A. Jensen, Carl Richards, and Steven J. Lozano. "Fish Community Change in Lake Superior, 1970–2000." *Canadian Journal of Fisheries and Aquatic Sciences* 60 (2003): 1552–74.

Brown, Edward H., Jr., Gary W. Eck, and Neal R. Foster. "Historical Evidence for Discrete Stocks of Lake Trout *(Salvelinus namaycush)* in Lake Michigan." *Canadian Journal of Fisheries and Aquatic Sciences* 38 (1981): 1747–58.

Christie, W. Jack. "Changes in the Fish Species Composition of the Great Lakes." *Journal of the Fisheries Research Board of Canada* 31 (1974): 827–54.

Cunningham, Aimee. "Whiff Weapon." *Science News* 168 (2005): 308–9.

Dann, Shari L., and Brandon C. Schroeder. *The Life of the Lakes: A Guide to the Great Lakes Fishery.* Ann Arbor: Michigan Sea Grant College Program, 2003.

Dehring, Terrence R., Anne F. Brown, Charles H. Daugherty, and Stevan R. Phelps. "Survey of the Genetic Variation among Eastern Lake Superior Lake Trout *(Salvelinus namaycush).*" *Canadian Journal of Fisheries and Aquatic Sciences* 38 (1981): 1738–46.

Edsall, Thomas A., and Gregory W. Kennedy. "Availability of Lake Trout Reproductive Habitat in the Great Lakes." *Journal of Great Lakes Research* 21 (Supplement 1) (1995): 290–301.

Goodier, John L. "Native Lake Trout *(Salvelinus namaycush)* Stocks in the Canadian Waters of Lake Superior prior to 1955." *Canadian Journal of Fisheries and Aquatic Sciences* 38 (1981): 1724–37.

Grooms, Steve. "The Enigma of Lake Trout." *Trout,* Spring 1992: 21–49.

Gunderson, Jeff. "Stout Trout Eyed for Market." *Seiche,* September 2009: 5–8.

Hrabik, Thomas R., Olaf P. Jensen, Steven J. D. Martell, Carl J. Walters, and James F. Kitchell. "Diel Vertical Migration in the Lake Superior Pelagic Community. 1. Changes in Vertical Migration of Coregonids in Response to Varying Predation Risk." *Canadian Journal of Fisheries and Aquatic Sciences* 63 (2006): 2286–95.

Kaups, Matti. "North Shore Commercial Fishing, 1849–1870." *Minnesota History,* Summer 1978: 43–58.

Kozhov, Mikhail. *Lake Baikal and Its Life.* The Hague: W. Junk, 1963.

Krueger, Charles C., and Peter E. Ihssen. "Review of Genetics of Lake Trout in the Great Lakes: History, Molecular Genetics, Physiology, Strain Comparisons, and Restoration Management." *Journal of Great Lakes Research* 21 (Supplement 1) (1995): 348–63.

Lawrie, Andrew H. "The Fish Community of Lake Superior." *Journal of Great Lakes Research* 4, no. 3–4 (1978): 513–49.

Lawrie, Andrew H., and Jerold F. Rahrer. *Lake Superior: A Case History of the Lake and Its Fisheries.* Technical Report No. 19. Ann Arbor, Mich.: Great Lakes Fishery Commission, 1973.

———. "Lake Superior: Effects of Exploitation and Introductions on the Salmonid Community." *Journal of the Fisheries Research Board of Canada* 29 (1972): 765–76.

Li, Weiming, Alexander P. Scott, Michael J. Siefkes, Honggao Yan, Qin Liu, Sang-Seon Yun, and Douglas A. Gage. "Bile Acid Secreted by Male Sea Lamprey That Acts as a Sex Pheromone." *Science* 296, no. 5565 (2002): 138–41.

Loftus, Kenneth H. "Studies on River-Spawning Populations of Lake Trout in Eastern Lake Superior." *Transactions of the American Fisheries Society* 87 (1958): 259–77.

Marsden, J. Ellen, John M. Casselman, Thomas A. Edsall, Robert F. Elliott, John D. Fitzsimons, William H. Horns, Bruce A. Manny, Scott C. McAughey, Peter G. Sly, and Bruce L. Swanson. "Lake Trout Spawning Habitat in the Great Lakes—A Review of Current Knowledge." *Journal of Great Lakes Research* 21 (Supplement 1) (1995): 487–97.

Moen, Sharon. "Genetic Evidence Confirms 'There's No Place Like Home.'" *Seiche,* August 2001: 1–4.

———. "Lake Superior's Native Lampreys." *Seiche,* March 2002: 4–5.

———. "Romancing the Sea Lamprey (Love Potion Number 3KPZS)." *Seiche,* March 2002: 4–5.

———. "Siskowet Trout: A Plague of Riches." *Seiche,* December 2002: 1–4.

Ryder, Richard A., Stephen R. Kerr, William W. Taylor, and Peter A. Larkin. "Community Consequences of Fish Stock Diversity." *Canadian Journal of Fisheries and Aquatic Sciences* 38 (1981): 1856–66.

Schram, Stephen T., James H. Selgeby, Charles R. Bronte, and Bruce L. Swanson. "Population Recovery and Natural Recruitment of Lake Trout at Gull Island Shoal, Lake Superior, 1964–1992." *Journal of Great Lakes Research* 21 (Supplement 1) (1995): 225–32.

Schuldt, Richard J., and R. Goold. "Changes in the Distribution of Native Lampreys in Lake Superior Tributaries in Response to Sea Lamprey *(Petromyzon marinus)* Control, 1953–77." *Canadian Journal of Fisheries and Aquatic Sciences* 37 (1980): 1872–85.

Shiras, George, III. "The Wild Life of Lake Superior, Past and Present." *National Geographic Magazine* 40, no. 2 (1921): 113–204.

Smith, Stanford H. "Factors of Ecological Succession in Oligotrophic Fish Communities of the Laurentian Great Lakes." *Journal of the Fisheries Research Board of Canada* 29 (1972): 717–30.

————. "Species Succession and Fishery Exploitation in the Great Lakes." *Journal of the Fisheries Research Board of Canada* 25 (1968): 667–93.

Todd, Thomas N. "Allelic Variability in Species and Stocks of Lake Superior Ciscoes (Coregoninae)." *Canadian Journal of Fisheries and Aquatic Sciences* 38 (1981): 1808–13.

University of Wisconsin Sea Grant Institute. *The Fisheries of the Great Lakes.* Madison: University of Wisconsin Sea Grant Institute, 1986.

Warren, Frank M. "The Wildlife of Isle Royale." *American Game* 15, no. 1 (1926): 15–17.

Waters, Thomas F. *The Superior North Shore.* Minneapolis: University of Minnesota Press, 1987.

Williams, Ted. "Lessons from the Lakes." *Fly Rod & Reel,* January/February 1997: 21–26.

INTERNET RESOURCES

Great Lakes Fishery Commission, www.glfc.org

EPILOGUE

The Wild Card
of Climate Change

O N July 1, 1993, shortly before 3 p.m. local time, a remote outpost in central Greenland suddenly exploded in a chorus of cheering. For the small group of jubilant scientists gathered that day, five years of hard labor paid off when a massive drill used to bore into the ice struck bedrock nearly two miles beneath their feet. With the clunk of metal against rock, the team made scientific history. They had reached the bottom of the deepest glacier in the Northern Hemisphere, extracting from its interior a core of ice 10,018 feet long.

Their success came none too soon. Scientists working on the Greenland Ice Sheet Project Two, along with their colleagues on glaciers around the world, are in a hot race—literally—to gather samples from ice that is melting out from under them at an alarming rate. As writer Sid Perkins noted in a 2003 article for *Science News,* "Glaciers have become the geological equivalent of endangered species." Himalayan glaciers, for example, which supply water for the people of south Asia—a quarter of Earth's population—have lost an estimated 174 metric tons—nearly 46,000 gallons—of water annually between 2003 and 2009 alone, writes Javaid Laghari in a 2013 issue of the prestigious journal *Nature.* (In a one-two punch, the deposition of soot from pollution from India and China is exacerbating the effects of climate change on the melting of the ice.) Glaciers throughout the Andes Mountains are disappearing so rapidly that experts foresee water shortages in cities like Lima and La Paz in the not-too-distant future. In 2013 researchers reported that margins along the Quelccaya ice cap in Peru, which took sixteen hundred years to form, had melted back in just the past twenty-five years. Sea ice too is dwindling. In fall 2013 the National Snow and Ice Data Center declared 2013 to be the sixth lowest for ice cover in the Arctic sea since satellite monitoring began in 1979. The dwindling ice appears to signal a long-term trend since the seven lowest minimum extents for ice cover have all occurred since 2007.

Much of the world's ice is melting because the surface of Earth is rapidly warming. The opening decade of the twenty-first century was the warmest since accurate temperature records began to be kept in 1880. As of this writing, the year 2014 had bested the year 2010 for the dubious distinction of being the hottest on record. New records continue to be set before the ink has dried on the page.

Climatologists are quick to point out that climate change is nothing new. Indeed, Earth's temperatures have naturally fluctuated—sometimes abruptly—throughout geologic time. And glaciers, like fluid amoebas on the land, have responded to these trends, waxing when Earth's temperatures dip, and waning when they rise.

What has snagged the attention of scientists is the magnitude and rate of the current warming. According to the National Aeronautics and Space Administration (NASA), from 1880 to 2010, the average surface temperature of Earth has risen about 1.1 to 1.6 degrees Fahrenheit. Moreover, the *rate* of the temperature increase has nearly doubled in the past fifty years. By the end of the twenty-first century, the "best estimate projections" from the Intergovernmental Panel on Climate Change (IPCC), a consortium of some twenty-five hundred climate experts from a hundred nations, describe a likely warming of the average global surface temperature that ranges from 2.7 degrees Fahrenheit to more than 3.6 degrees Fahrenheit, depending on the model used.

Equally troubling is the fast pace of the temperature increase. Even turning up the planet's thermostat by only 1.8 degree Fahrenheit over the next century will exceed the pace of any warming that has occurred over the past ten thousand years.

Many of the international climate accords have officially drawn the line in the sand at a warming of 3.6 degrees Fahrenheit. To cross this threshold would cause planetary chaos and catastrophe. But for climate policy experts such as Bill McKibben, this temperature ceiling already is far too high. Even current levels of warming, he points out, have "caused far more damage than most scientists expected. A third of the summer sea ice in the Arctic is gone, the oceans are 30 percent more acidic, and since warm air holds more water than cold, the atmosphere over the oceans is a shocking 5 percent wetter, loading the dice for devastating floods."

Largely to blame for this extreme warming trend is the buildup of heat-trapping carbon dioxide (CO_2) in the atmosphere. According to climatologists, today's carbon dioxide levels are higher than they have been for the past 800,000 years, the interval of time that scientists have been able to verify in the climate record. They were able to make this determination by studying what they call proxy records, such as deep-sea sediments, the growth rings of corals and trees (some lay down thicker growth rings during warm periods than cold ones), and the chemicals in the shells of marine fossils known as foraminifera. Among the oldest temperature logs are those locked deep within the ice. The great core of ice that was extracted by the Greenland team contained some 110,000 years of Earth's climatic history. Glaciologist Paul Mayewski, the team's leader, dubbed it the ice chronicles, for it contained "stories of the ancient climate that we can read in modern times," as he observed in his 2002 book of the same name. (Scientists have since topped the ice core record with a cylinder of ice extracted from Antarctica that dates back 800,000 years.)

Because of their great size, the ice sheets of Greenland and Antarctica provide a particularly extensive record of past climatic events. This record is preserved because of the way in which glaciers form. Glaciers take shape in places with abundant precipitation and cold temperatures. The combination allows snowfields to survive the summer and over time become buried under successive new layers of snow.

Snow grains trap tiny pockets of air. The weight of accumulating snowfalls compresses underlying layers into ice, turning the air pockets into bubbles that contain concentrations of telling gases. Of special interest to scientists are two oxygen molecules with different molecular weights: ^{16}O and ^{18}O. By measuring the ratio of these two isotopes, they can deduce the temperatures that existed when the snow was deposited. The colder the temperature, the lower the ratio of ^{18}O to ^{16}O.

The bubbles of ancient air also contain information on past concentrations of carbon dioxide, the levels of which are tightly correlated with global temperatures. Carbon dioxide is part of a group of gases, including water vapor and methane, that occur naturally in the atmosphere and create a phenomenon known as the greenhouse effect. Greenhouse gases are transparent to sunlight, allowing the shortwave radiation of the sun to penetrate Earth's atmosphere. The surface of Earth, everything from sun-warmed farm fields to sidewalks, reflects energy back into the atmosphere in the form of long wavelength radiant energy known as terrestrial infrared radiation. Greenhouse gases, which are opaque to infrared radiation, absorb some of this energy and radiate it back to Earth instead of into space, a mechanism that keeps Earth at a relatively balmy 59 degrees Fahrenheit. Without it, the average temperature of the planet would hover at a frigid, uninhabitable −4 degrees Fahrenheit.

The climatic record shows that during cold periods of glaciation, carbon dioxide levels were low, about 180 parts per million (ppm), and rose to 280 ppm during warmer interglacial times. Natural events, such as major volcanic eruptions and megalithic meteors slamming into Earth, produced wild fluctuations in global temperatures and precipitation. But even such extreme events could not swing the planetary pendulum beyond this established range of variation.

Over the past eight thousand years, the period of human civilization, carbon dioxide levels have largely hovered in the 280 ppm range. Since the advent of the Industrial Revolution two centuries ago, however, scientists have measured a 41 percent increase in this greenhouse gas—with no end in sight. Today, carbon dioxide levels continue to rise by 20 ppm each decade. In 2012, concentrations of atmospheric carbon dioxide reached record highs, briefly topping 400 ppm over the Arctic and Mauna Loa, the monitoring station that is run by the National Oceanic and Atmospheric Administration in Hawaii. In May 2013, the Mauna Loa researchers confirmed fears that these high levels were becoming the new normal when instruments recorded 400.03 ppm of carbon dioxide as an average daily level and not just a blip on the screen. And carbon dioxide is not the only culprit. During the past two hundred years, levels of another greenhouse gas, methane (which has twenty times the heat-trapping capacity of carbon dioxide in the atmosphere), have increased 150 percent.

On February 2, 2007, the IPCC declared with more than 90 percent certainty that humans have caused this dramatic warming in the past half century. The date "will be remembered," stated Achim Steiner of the United Nation's Environment Programme, "as the date when uncertainty was removed as to whether humans had anything to do with climate change on this planet. The evidence is on the table." In its Fifth Assessment Report, published in 2013, the IPCC was even more emphatic. "It is extremely likely [95 to 100 percent certain] that human influence has been the dominant cause of the observed warming since the mid-20th century," the authors concluded.

Although human-caused climate change first appeared on the national science agenda in the 1950s, the potential for human alteration of the atmosphere was predicted with uncanny accuracy as industrialization exploded across Europe in the nineteenth century. As early as 1827 the French mathematician Jean Baptiste Joseph Fourier suggested that a doubling in the release of carbon dioxide from the burning of fossil fuels could raise global temperatures by 7.2 to 10.8 degrees Fahrenheit. Nearly seventy years later, independent forecasts by the Swedish chemist Svante Arrhenius and the American geologist T. C. Chamberlain identified that carbon dioxide emissions from industrial activity could cause global warming. In 1979 a report from the National Academy of Sciences gave wider currency to such suspicions, arguing that the global climate would likely warm with the continued pumping of carbon dioxide into the atmosphere.

The theory of global warming is widely accepted in scientific circles today, and its predicted outcomes, though still crudely calculated, are sobering. Yet the citizens of many industrialized nations and their governments still fail to grasp the urgency of curbing greenhouse gases. Indeed, climatic warming is often welcomed, particularly by many residents of northern climes. In September 2003, for example, when Russian leader Vladimir Putin announced his country's refusal to endorse the 1997 Kyoto Protocol, an international treaty obliging industrialized nations to reduce emissions of greenhouse gases, he echoed the sentiments of many of his citizens who dismiss the scientists' alarm. "Here in Russia," Putin explained, "you can often hear people say, sometimes jokingly, sometimes seriously, that Russia is a northern country, so if it warms up 2 or 3 degrees it's not terrible. It might even be good; we'd spend less money on fur coats and other warm things."

But scientists find few silver linings in the cloud of global climate change. In June 1988 experts from forty-six nations at the World Conference on the Changing Atmosphere issued this dire warning: "Humanity is conducting an unintended, uncontrolled, globally pervasive experiment whose ultimate consequences could be second only to a global nuclear war."

A warming of only a few degrees, they caution, may appear insignificant, but it can have dramatic consequences. A rise of 9 to 11 degrees Fahrenheit in average global temperatures over a period of about ten thousand years was enough to melt the Wisconsin glacier, which during its maximum extent roughly twenty thousand years ago, covered regions as far south as southern Indiana and Ohio with a cap of ice some two miles deep. The rise in the global thermostat was enough to drive spruce trees out of Kansas and into Canada and contribute to the extinction of such megafauna as woolly mammoths, giant beavers, and saber-toothed cats. "Put in these terms," say the authors of a 2003 report from the Union of Concerned Scientists (UCS), the magnitude of warming "in less than 100 years should ring bells of alarm."

To help predict the impact of global warming on Earth, scientists have developed what are known as general circulation models (GCMs), a series of complex, but nonetheless imperfect, mathematical models of the physical forces that govern Earth and its atmosphere. GCMs allow climatologists to simulate links among a large number of climatic variables, including cloud behavior, the globe's hydrological cycle, and the ways in which oceans absorb, release, and redistribute atmospheric heat around the planet's surface. To test their validity, scientists check the results of models against data collected from the real world.

GCMs are works in progress that are continually updated as the computational power of computers is increased or as scientists discover new climatic variables. Prior to the mid-1990s, for example, most GCMs did not include the effects of aerosols (suspended airborne particles from sources as diverse as sea spray, auto exhaust, and windblown dust), which scientists now believe help to depress rising global temperatures by reflecting some of the sun's energy back into space. In 2003, NASA scientists revealed that atmospheric soot—from burning diesel fuels and such biofuels as animal dung—reduced the ability of snow and ice to reflect sunlight. Indeed, soot was twice as effective as carbon dioxide in changing the surface temperatures in the Arctic and Northern Hemisphere.

Perhaps most worrisome, however, is the revelation that the warming of Earth's surface may not proceed at a gradual, measured pace that will allow human societies to make adjustments in an orderly fashion. Recent melting of the Greenland ice cap (a great crust of ice as large as the United States east of the Mississippi River) has raised the question among those studying global climate of whether a flood of freshwater into the North Atlantic could shut down what is known as the thermohaline circulation, or great ocean conveyor belt. Already, scientists have reported a freshening of North Atlantic and Arctic waters along with an increase in the saltiness of waters in tropical Atlantic reaches.

The conveyor, whose current strength is equivalent to that of 150 Amazon Rivers, begins in the South Atlantic, gathering warmth as it moves through the Caribbean and around the tip of Florida, where it becomes the Gulf Stream. The flow continues up the eastern coast of the United States to Cape Hatteras, North Carolina, before charting a course for the North Atlantic. As the conveyor reaches the North Atlantic, its waters discharge their great payload of tropical heat, which "helps make northern Europe warm enough that roses can be grown at latitudes that elsewhere support polar bears," writes Andrew Revkin of the *New York Times*.

Scientists observe that warming global temperatures could shut down this ocean conveyor by disrupting the physical phenomena that drive it. As the current loses heat, its waters become cooler, and therefore denser, causing them to slip beneath the warmer, more buoyant surface waters. Also, on its journey north, the current loses water to evaporation along the way, which makes its waters saltier. Saltwater is denser than freshwater, a physical factor that also contributes to the sinking of the conveyor's waters. Between Iceland and Norway, this strong downward flow pushes the conveyor's waters south along the ocean floor and into the Pacific Ocean where its waters eventually warm and rise to the surface to begin the process all over again.

In October 2003, a pair of energy consultants produced a twenty-two-page report for the Pentagon titled "Imagining the Unthinkable." The report warns of the potential for such a disruption in the ocean's circulation, which almost certainly would set off a cooling trend in Europe and quite possibly could cause prolonged droughts in some regions of the world. The authors based their report on recent studies of ocean sediments and fossils that have identified how just such a shutdown occurred when the freshwaters of Glacial Lake Agassiz surged into the Atlantic Ocean via the Great Lakes and St. Lawrence River system. For thousands of years, Lake Agassiz, a vast inland lake that once submerged portions of present-day Minnesota, North Dakota, and much of southern Canada, drained south through what is now the Mississippi River. But the receding of the last glacier—the Wisconsin ice sheet—opened other outlets to the north and east. Beginning about 10,900 years ago, the breaking of a series of ice dams allowed water from Agassiz to pour into Lake Nipigon, which in turn overflowed through a series of channels into Lake Superior. The size of these deluges beggars the imagination. In one period alone, 1,180 cubic miles of water—the equivalent of more than half the volume of Lake Michigan—poured into Lake Superior. The deluge, which lasted nearly two years, may have temporarily raised Superior's lake levels by 164 feet.

The floodwaters are believed to have eroded down to bedrock a 131-foot-tall barrier of glacial drift that lay across Whitefish Bay on the southeastern end of Superior, thereby opening what is today the modern outlet for the lake through the rapids of Sault Ste. Marie. Eventually they poured into the Atlantic Ocean via the St. Lawrence River. So great was the sudden influx of freshwater that it led to a change in the density of seawater, which halted traffic on the conveyor's great highway of ocean currents that move heat from equatorial regions to higher latitudes. The disruption plummeted large portions of the Northern Hemisphere into a cold spell known as the Younger Dryas, creating conditions so frigid that the glacier began to reinvade portions of its former territory, including the Lake Superior basin. (See also "Searching High and Low: The Science of Lake Superior Exploration.")

The GCMs cited most frequently in projections of climate changes in the Great Lakes region include what are known as the Canadian Coupled-Climate Model (CGCM) and the Hadley Centre Climate Model (HadCM). Among the conclusions to emerge from such modeling is the prediction that warming will not proceed at uniform rates around the world. The middle to higher latitudes are expected to experience more dramatic warming than regions near the equator. According to a 2003 report of the UCS and Ecological Society of America, which also incorporates projections based on the Midwest-Focused Parallel Climate Model (PCM) of the U.S. National Center for Atmospheric Research, the increase in the average global temperature of 2 to 11 degrees Fahrenheit over the next century could be as much as 50 percent higher in the northern United States and southern Canada.

Temperature trends in Minnesota appear to bear this prediction out. In 1960 Donald Baker, a University of Minnesota climatologist, published a study examining temperature data from nineteen stations in Minnesota between 1900 and 1958. He found that average winter temperatures rose by 2.8 degrees Fahrenheit, and average summer temperatures increased by 1.6 degrees Fahrenheit.

In 1989 Baker and his colleagues analyzed records from Fort Snelling. Begun in 1820, this data collection is regarded as the most authoritative of any west of the Mississippi. The team found that between 1820 and 1987 the average annual temperature rose 2.9 degrees Fahrenheit. The results were startling: Minnesota temperatures had risen nearly three times the global average. After factoring in such historical data with results from the most current GCMs, scientists estimate that by 2050 average temperatures in Minnesota will rise 2 to 6 degrees Fahrenheit, and by 2100, 5 to 10 degrees Fahrenheit, with some of the most dramatic changes taking place in the northern part of the state.

Minnesota ecosystems are sensitive to even subtle shifts in temperature, much less dramatic leaps. That is because Minnesota lies at an ecological crossroads between three great rivers of air known as jet streams. These jet streams bring cold, dry air from the Arctic, warm, humid air from the Gulf of Mexico, and drier air from the Pacific Ocean. The boundaries among these air masses reflect fairly abrupt ecotonal provinces on the ground, including the prairie to the west, the deciduous forest in the center and south, and the boreal forest to the north.

Relatively minute differences keep one ecosystem from overrunning the other. University of Minnesota ecologist John Tester points out that the separation between the tallgrass prairie of western Minnesota and the tall-pine woodlands of Itasca State Park is a narrow climatic boundary of less than 4 degrees Fahrenheit in temperature and six inches in moisture. Today the difference of a mere 3.6 degrees Fahrenheit in mean summer temperatures divides the oak and maple forests around the Twin Cities from the spruce-fir forests of Duluth.

Such narrow differences mean that changes to the region's ecology will likely be dramatic and occur rapidly. "Move the climate of Minneapolis to Duluth and you'll get Minneapolis vegetation. And that could happen in 50 years," observes University of Minnesota forest ecologist John Pastor.

But whether or not Minneapolis habitats are able to shift their range northward by hundreds of miles within a few decades remains an open question. Paleobotanists point out that the average speed with which plants recolonized land vacated by the Wisconsin ice sheet was 12 to 25 miles per century. Even the most rapid colonizers advanced north at a rate of only 63 to 94 miles per century. The current warming of the planet would require taxa to shift ranges at speeds of 188 to 314 miles per century. Even if trees were able to make the rapid transition, many other contributing members of a healthy forest

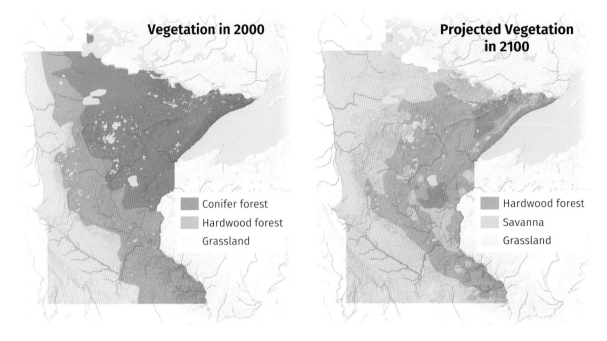

Narrow climatic differences in temperature and precipitation separate Minnesota's three major biomes. Under climate change predictions, their current positions are likely to shift rapidly and dramatically.

community—herbaceous plants, fungi, insects, and soil microbes—would likely lag behind.

Not only will taxa be required to establish new homes at speeds that are unprecedented in the biogeographical record, but they will also need to surmount obstacles that did not exist in the wake of the last glacial retreat. "It seems unlikely that all of the migrating species that survived the Ice Age would be able to safely reach refuges after migrating across freeways, agricultural zones, industrial parks, military bases and cities of the 21st century," say the authors of the UCS report. Adding yet another degree of difficulty is the fact that native North American species must now contend with competition from a whole range of imported plants and animals.

Despite the uncertainties, scientists have made several educated guesses as to the future of Minnesota's landscape. Here is what they say about the North Shore watershed.

Headwaters

If you were to draw a picture of the quintessential northern lake, it would have a shoreline ringed by the ragged tree tops of cedar, spruce, and fir and the branches of a few giant old pines laddered against the sky. You might add a bull moose blinking to attention from a marshy cove, his antlers tangled with clumps of pondweeds. And if you could dub an audio component, you might choose the song of the white-throated sparrow, oh, sweet Canada, Canada, Canada, calling from a dark thicket of forest.

As carbon dioxide levels and Earth's temperatures continue to rise, this scene will undoubtedly change. Although climate models differ on their prognostications regarding changes in precipitation levels, they agree that temperatures will grow warmer, so much so that forest ecologists predict that within thirty to sixty years—within the lifetimes of many of the readers of this book—boreal tree species will begin to decline. The southern limit of the boreal forest, which today reaches into the headwaters region of the Lake Superior watershed, is expected to retreat north by as much as 620 miles.

According to University of Minnesota forest ecologists Lee Frelich and Peter Reich, natural disturbance regimes, such as the frequency of fire and windthrow, will be altered. These changes will be coupled with outbreaks of native and exotic pests and diseases, and increasing populations of deer and European earthworms. Together, they will hasten the changing of the guard from boreal vegetation to species that are better adapted to a regional climate that will resemble that of present-day Mason City, Iowa, by the end of this century. The changed climate and natural disturbance regimes will interact with such localized environmental factors as soil type to produce a new forest mosaic. A model developed by forest ecologist Pastor, for example, shows that in northern Minnesota, on moister clayey soils, the dim boreal forest with its mossy feet and head of pointed firs and spruce will likely give way to hardwood forest uplands dominated by sugar maple and yellow birch. On sandy soils that currently support mixed hardwood and conifer forests, a stunted forest of pine and oak will take hold. Grassland, oak savanna, and brush are likely to succeed today's northern sand-plain and outwash forest of jack pine and red pine.

Changed too will be the suite of animals that generations of Minnesotans have come to identify with life in the north country. Especially vulnerable are animals, such as moose, whose populations in northwestern and northeastern Minnesota already skirt the southern limits of their range. Herds in both corners of the state are imperiled. Moose numbers in northwestern Minnesota crashed from a high of 4,700 in the mid-1980s to less than 100 individuals in 2007. Despite the Minnesota DNR's suspension of hunting licenses and efforts to enhance habitat, moose populations continued on a downward spiral such that by 2013 the animals were rare on the landscape.

For a time, it appeared as if moose herds were able to hold their own in the northeastern part of the state. Wildlife experts believed that the region's longer winters, colder temperatures, deeper snows, and other critical habitat differences gave Arrowhead moose a survival advantage. Warmer and drier, the northwestern corner of the state is dominated by patchy forests of aspen parkland interspersed with open blocks of agricultural fields. With its abundance of evergreen forests, lakes, rivers, and beaver ponds, northeastern Minnesota, on the other hand, was thought to offer moose a better chance to escape warm temperatures. Historically, the Arrowhead's severe winters also have kept white-tailed deer in check. They serve as the primary vectors of the brainworm parasite that often is fatal to moose. Despite these apparent advantages, moose numbers in northeastern Minnesota too are in steep decline. Between 2008 and 2013, the population dwindled to 2,700 animals, a 65 percent drop. Between 2012 and 2013 alone, the herd decreased by about a third.

Thirty-two percent of adult moose mortality has been attributed to health-related factors and another 33 percent to unknown factors, some of which probably include health issues, says Mike Schrage, a wildlife biologist with the Fond du Lac Resource

Parasites, including brainworms and winter ticks, have been implicated in the steep decline of moose in northeastern Minnesota. Warming temperatures could further compromise the ability of moose to withstand these chronic health challenges, leading to predictions that they will disappear from the state by midcentury. Photograph by Sandy Updyke.

Management Division, which has partnered with the DNR and other state, tribal, and federal agencies in moose-mortality research. Analyses of animal remains have implicated a variety of parasites, including brainworm, winter ticks, and possibly liver flukes, all of which can either kill a moose outright or weaken it such that the animal becomes susceptible to other health issues or to predators. Research has shown that Arrowhead moose have also been exposed to a variety of new diseases, including Lyme disease, West Nile virus, eastern equine encephalitis, and malignant catarrhal fever. Whether these diseases are affecting moose health is currently unknown.

A warming trend in the north country due to climate change could further compromise the ability of moose to defend against chronic health challenges, tipping the balance against their persistence in northern Minnesota. Although Schrage points out that heat stress has not been shown to kill moose outright, it may play a role in moose mortality. One laboratory study indicated that when moose are in their summer coats, they exhibit signs of heat stress when temperatures rise above 57 degrees Fahrenheit. When sporting their dark, heavy winter coats, heat stress sets in at 23 degrees Fahrenheit. "Extended warm temperatures in summer and winter may cause moose to reduce feeding and ultimately make them more susceptible to other health-related issues," Schrage says. "Certainly moose are not and never were a southern species and northern Minnesota has always been about the southern extent of their mid-continent range. We might logically expect moose to struggle more and more as the climate warms and the forest changes, and clearly our moose herd is declining rapidly. What is still unclear is whether this is due to increased heat stress on the animal or because the environment has become more favorable to diseases, parasites, and their vectors (deer, mosquitoes, snails) which in turn impact moose. Or maybe all of the above." An added blow to the viability of the moose herd in Minnesota's north country, Schrage adds, may be the abundant populations of wolves and bears, which may be able to more easily prey on animals weakened by disease and heat stress. Based on population trends, researchers say that moose are likely to disappear from the region by midcentury.

Gone too under warmer conditions will be populations of lynx, pine marten, and fisher. And depending on their individual characteristics, many inland lakes, particularly smaller lakes, are expected to undergo physical, biological, and chemical changes that, among other effects, will eliminate lake trout from their reaches. For example, warmer air temperatures will reduce the duration of winter ice cover. As more water evaporates, lake levels plummet, allowing sunlight to warm the cold, deep habitats in which lake trout live. And as the water temperature grows warmer, oxygen levels drop, eliminating the biochemical conditions on which these fish depend. These same changes in lake ecosystems may also accelerate the accumulation of mercury and other toxic pollutants in aquatic food webs, raising contamination levels in organisms such as fish.

Especially hard hit, however, will be birds that travel across whole continents to breed in the headwaters region. The three-million-acre Superior National Forest forms part of a sickle-shaped swath of prime bird-breeding territory that dips down from the prairie provinces of Canada, stretches across the northern Great Lakes, and then veers up through New England and the Maritime Provinces. Ornithologists have tallied an average of sixty-one to sixty-seven species per surveyed route, making these regions a hot spot for breeding birds, richer than any region north of Mexico. (See also "As Good As It Gets: Bird Diversity on the North Shore.")

With the loss of the boreal component of Minnesota's forests would come the disappearance of an estimated thirty-six species of birds that depend on conifers for food and cover, including pine siskins, boreal chickadees, dark-eyed juncos, evening grosbeaks, crossbills, and a host of vireos and warblers.

The transitional conditions caused by global warming are expected to threaten even migratory birds that do not face such wholesale habitat loss. To reach their breeding grounds, migrants must contend with a host of natural and human-caused obstacles, such as cell phone towers, sudden ice storms, and the ongoing diminishing and fragmentation of habitat in which to rest and refuel along the way.

By breaking the synchronous chain that binds birds, their insect prey, and the spring leafout, global warming may prove to be one of the most lethal obstacles of all. Many birds use the lengthening days of spring, rather than changes in temperature, to cue their departure to northern locales. Their food

Evening grosbeaks are among the thirty-six species of migratory birds predicted to disappear as Minnesota's forests lose their boreal component in a changing climate. U.S. Fish and Wildlife Service Digital Library.

As habitats shift under climate change, cardinals are expected to continue to extend their range northward. U.S. Fish and Wildlife Service Digital Library.

Changes in several annual patterns, such as the length of the growing season and the timing of fall and spring frosts, indicate that global warming already has had pronounced effects on the seasonal clock.

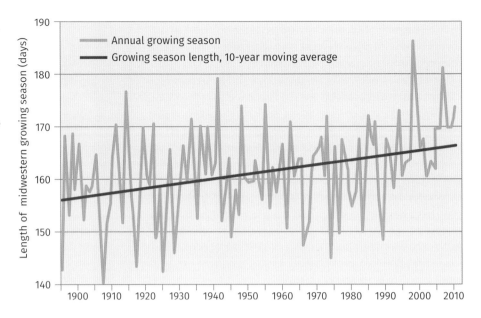

supply, however, responds to a more complex set of environmental signals. The leafout of spring vegetation—and the emergence of insects that accompany the flush of new foliage—are influenced not simply by an increase in daylight hours but also by rising temperatures.

Evidence suggests that warmer conditions will trigger increasingly earlier spring life-cycle events for a host of organisms including plants and insects, potentially leaving some migratory birds with a critical shortage of food just when they need it most—along physically taxing migratory routes or on their breeding grounds, where adults have hungry nestlings to feed.

A 2003 study published in the prestigious journal *Nature* showed that for a wide range of plants and animals, the seasonal clock has already advanced. A team of scientists analyzed sixty-one long-term studies that tracked life-cycle changes from locations around the world over time periods ranging from 16 to 132 years. The studies focused on such phenomena as frog breeding, bird nesting, first flowering, the bud bursts of trees, and the arrival of migrating animals such as birds and butterflies. Of the 677 species examined, 62 percent showed marked acceleration of these springtime events.

Unfortunately, not all organisms are marching in lockstep at the same quickened pace. Ecologists point out that climates are warming more rapidly in northern latitudes. This mismatch of seasonal rhythms could be especially catastrophic for migrating birds. Long-distance migrants, such as scarlet tanagers, warblers, thrushes, and flycatchers, need to bulk up on temperature-activated caterpillars in southerly locales to fuel each leg of their cross-continental journey north. The accelerated rate of warming at higher latitudes could mean that many of these bird species will arrive on their northern breeding grounds long after the seasonal flush of insects has peaked. Missing the banquet could be especially hard on birds that depend on particular insect species for food and cannot readily shift to other prey.

If the results of recent studies of European bird communities hold true for those in North America, such lag times could further skew the odds in favor of stay-at-home birds or birds that migrate shorter distances. In a study of European birds published in the 2003 issue of *Conservation Biology,* German researchers Nicole Lemoine and Katrin Bohning-Gaese found that warmer winter temperatures coincided with a reduction in the number of long-distance migratory bird species and an increase in the number of resident species and short-distance migrants, a trend that they predict will be exacerbated if the differential rise in winter temperatures in northern latitudes continues.

Should a similar pattern unfold in North America, resident birds in the headwaters of the North Shore watershed, for example, may be able to take advantage of the earlier emergence of insects and increase the odds of success in raising their broods. By the time long-distance migrants finally arrive on their northern breeding sites, dwindling food supplies could reduce their nesting success, putting them at a competitive disadvantage with resident birds. "You have to change what you eat, or rely on fewer things to eat, or travel farther to eat, all of which have costs," says Pennsylvania State University paleoclimatologist Richard P. Alley.

Warmer temperatures have also allowed the expansion of ranges for mammals that prey on ground-nesting songbirds. Opossums have now made their way into central and northern Minnesota. Raccoons have expanded their ranges into Canada and are making their way along the North Shore at an estimated ten miles per year. Indeed, so unusual was a raccoon sighting along the Sawbill Trail in November 2003 that Frank Hansen noted it in the column he regularly wrote for the *Cook County News-Herald.* Milder winters may allow populations of these and other predatory mammals, such as skunks, to grow, thereby increasing the pressure on ground nesters.

Putting a price tag on the loss of the north country's songbirds is difficult. The 2003 report from the UCS points out that wildlife watching—primarily bird watching—is a $3.5 billion industry in northern Michigan, Minnesota, and Wisconsin combined. Even trickier to tally is the price tag for the lost ecological services that these birds currently provide free of charge, such as pollination, seed dispersal, and insect control.

But perhaps hardest of all is to calculate the loss of these species in the arithmetic of the human heart. The call of the white-throated sparrow, the sight of a moose cow and calf knee-deep in a lake, and a raven calling from the top of a black spruce are among the essential ingredients that make up the borderlands' singularly haunting character. The authors of the UCS report observe that a shift from a boreal spruce-fir to hardwood forest in the northern Great Lakes with its attendant shift in animal species "could occur in the lifetime of current residents" and be "jarring to many residents' sense of place. . . . [But] while a 'sense of place' is felt strongly by many people, it is hard to assign it a dollar value equivalent to dollar values for changes in harvestable timber."

Highlands

Of all the scenarios developed by today's climate models, forecasting precipitation levels and patterns is fraught with the greatest uncertainty. But based on the available evidence, scientists predict that fifty years from now when nature buffs in Minnesota compare the annual precipitation levels captured in their backyard rain gauges with those recorded today, they probably will not see an appreciable difference. Under several

The advanced onset of spring under global warming could favor resident birds such as blue jays, which can exploit the earlier emergence of insects for raising their broods, unlike long-distance migrants such as hermit thrushes (below), which arrive later on their northern breeding sites. U.S. Fish and Wildlife Service Digital Library.

warming scenarios, winter rain and snowfall are expected to increase by 15 to 40 percent, while summer totals are expected to decrease by up to 20 percent. With the ups and downs of seasonal precipitation totals more or less canceling each other out, then why is it that climate experts predict a drying trend for the state, so much so that the Land of Ten Thousand Lakes may be forced to seriously recalculate the number of its aquatic endowments?

For one thing, precipitation gains are likely to be offset by higher temperatures, which increase evapotranspiration (the amount of water returned to the atmosphere from soil and surface water evaporation and the transpiration of plants). While predictions call for increases in soil moisture in winter, the overall warmer conditions will cause soil moisture in the Great Lakes region to drop by 30 percent in summer and fall. Such changes would have significant negative repercussions for ecosystems that rely on recharge of groundwater during the summer months. Water levels are expected to decline in groundwater tables, lakes, streams, and wetlands. Many will simply dry up as bouts of drought grow more frequent.

Patterns of precipitation are expected to shift drastically as well. And when it comes to replenishing the moisture in soils and groundwater, the timing of precipitation is just as important as quantity of snow or rain. In winter, the precipitation that falls to Earth is largely locked away as ice and snow. During the spring thaw, the still-frozen ground sends snow-melt and rain coursing over the surface, preventing them from penetrating deeply into the soil and percolating into ground-water tables.

Soil moisture and groundwater will not be replenished during the summer months. Not only will the state receive less summer precipitation overall, but the rain will also tend to fall in large storm events. Instead of allowing rain to seep deeply into the ground as it does during gentler rainfalls, these deluges increase the amount of water that is shed from the land.

According to atmospheric scientists Donald J. Wuebbles and Katharine Hayhoe, the incidence of heavy rainfall events is likely to increase. For example, they point to paleoflood records from the upper Mississippi and lower Colorado Rivers that show how seemingly minor climatic changes in the past—a warming of 1.8 to 3.6 degrees Fahrenheit and a 10 to 20 percent increase in precipitation—have stepped up the size and frequency of major floods. According to the researchers, the frequency of heavy rainfalls in the Midwest already has doubled since the early part of the twentieth century. Some models predict that the incidence of heavy rainstorms may double again by the end of the century.

All of these trends are troubling news for the populations of frogs and salamanders that make their home in the North Shore Highlands. Many of them depend on seasonal water pockets, known as ephemeral pools, for breeding and rearing waters. In the past, these species have been able to weather the prolonged dry spells that cause small pools to evaporate by finding nearby refuge in deeper, more expansive wetlands. But under global warming scenarios, even these larger wetland complexes are expected to greatly diminish. According to Lucinda Johnson, a wetlands ecologist at the University of Minnesota Duluth, shallower waters are likely to cause a host of problems for amphibians, including greater exposure to damaging ultraviolet radiation, which typically penetrates the top two to eight inches of the water.

For many species of amphibians, however, scouting out new pools may not be a viable option. Biologists Andrew Blaustein, David Wake, and Wayne Sousa note that a large number of amphibian species are intensely philopatric; that is, they are faithful in the extreme to their home territories, which in one species is limited to as little as eleven square feet, about the size of a dining room table. With good reason. To carry out day-to-day functions, amphibians rely on moderate temperatures and, just as important, sufficient moisture since the lack of protective covering such as hair and feathers leaves them vulnerable to desiccation. Such optimal circumstances are not only limited, however, but also relatively isolated across the landscape. In general, the authors observe, amphibians are consigned to "patches of suitable habitat surrounded by conditions that are relatively harsh to them." Those that venture out of the boundaries of their hospitable home turf face the life-threatening problem of evaporative water loss. Therefore, the biologists point out, "Movements may occur only during a narrow range of environmental conditions and are often limited to relatively short periods during the annual activity period." (See also "The Secret Life of Salamanders.")

As conditions become drier and as the landscape is increasingly sliced and diced by logging, residential development, and roads, amphibian refugia will become more isolated, leaving the less-mobile animals high and dry and the more active ones vulnerable to predators and desiccation as they roam over ever-greater distances to find reproductive waters. The net effect of the combined forces of climate change and new development may be an increase in the localized extinctions of amphibian populations across the landscape.

Furthermore, as the increased magnitude of spring floodwaters and heavy summer downpours create surging stormwater flows, biologists fear that many of the safe havens that do persist will be flushed of tiny organisms, such as tadpoles, that rely on quiet pools and backwaters.

The combination of floodwater surges and frequent drought is also expected to pose chemical hazards to sensitive animals such as amphibians as well. During winter, acid rain chemicals, for example, are largely locked up in the snow. Early spring warm-ups can cause rapid meltdowns, sending acidic waters pouring over the frozen landscape and directly into streams and lakes. Strong acid pulses are especially damaging to these aquatic systems in spring since many surface waters are at seasonal lows and cannot adequately dilute inputs of concentrated pollutants. During droughts, these chemicals accumulate in the soil, where they become washed into surface waters during intense but infrequent rainstorms. Ironically, limnologist David Schindler says, global warming could reintroduce the problems of acid rain just as many aquatic ecosystems are beginning to stabilize or recover from acid rain pollution.

But for some species, the climate-warming cloud could have a silver lining. Take, for example, the highlands' maple forests. Studies have shown that when maple trees are grown under enhanced levels of carbon dioxide in laboratory chambers, they not only increase their growth rate but also become more tolerant of shade. Such adaptations could give maples a distinct successional advantage in the forests of the future.

How much of an advantage, however, remains an unanswered question. Climate researchers caution that the results of experiments conducted in controlled laboratory settings are a far cry from replicating conditions in the real world where wild cards—such as the rise in concentrations of ozone, which is poisonous to trees and may offset carbon dioxide–enhanced productivity, and the migration of new insect pests—make forest forecasting an iffy proposition. For example, as temperatures warm, the gypsy moth, an alien invader from Europe, is expected to dramatically expand its range. In 2005 a breeding population of this voracious defoliator of hardwood forests was found in the heart of the Superior National Forest near Tower. Trap surveys registered an exponential increase in this pest along the North Shore as well. Monitors from the Minnesota Department of Agriculture captured more than seven hundred individuals, a dramatic jump from previous years of only twenty-five to thirty moths. By 2009 gypsy moths had been documented throughout the Arrowhead region. Milder winters will only increase the odds of survival for these and other pests.

Modeling the impacts of gypsy moths much less predicting how the deck of existing pests and diseases in northern forests will be reshuffled under global warming scenarios have a level of ecological complexity that is impossible to track. "Insects, disease and their vectors (e.g., other insects, fungi, and bacteria) have their own, often complicated life cycles which depend on weather and climatic events in a manner different from that of the host trees," say forest ecologists Allen M. Solomon and Darrell C. West.

Studies that examine the relationship between the caterpillars of butterflies and moths and their host plants show just how complicated these relationships might be. Take, for example, the nutritional instability that elevated levels of carbon dioxide could introduce to some forests. Like maple, other plant species have responded to the fertilizing stimulus of elevated carbon dioxide levels by stepping up their growth. But instead of offering insects a vegetational feast, these plants are the equivalent of a junk-food warehouse. Many plants grown under enhanced carbon dioxide levels tended to produce leaves that were high in starch and toxic compounds known as phenolics but low in nitrogen, a correlation of protein levels. Low nitrogen levels forced caterpillars to eat more in order to obtain the nutrients they needed to tide them through pupation. In some cases, they went on veritable food rampages—with no benefit. Not only was their growth slowed by 10 percent, but the insects made the transition from caterpillars to adult moths and butterflies much more slowly. And when the researchers compared the insects fed on carbon dioxide–boosted plants with those that grazed on regular leaves, they found that the former were far smaller as adults. One plant species proved the exception to the eat-more, weigh-less rule: sugar maples. Caterpillars fed carbon dioxide–enhanced sugar maple leaves obtained their nutrition from the same quantity of food as those that were fed on regular leaves. In the forest, would insects evolve responses to capitalize on such nutritional advantages, shifting their foraging to maple leaves and thus putting more pressure on maples to the exclusion of other tree species?

Experts add that the changed chemical composition of leaves could affect aquatic organisms as well as terrestrial ones. Stream ecosystems rely on the energy and nutrients that are produced when microbes break down the leafy material that enters the water from surrounding forests. The skewing of the chemical ratio between starches and proteins could slow the pace of this microbial decomposition as well as shortchange the aquatic insects that feed on leaves, impairing their growth and survival just as it does for insects that live on land.

Such changes are sure to have repercussions throughout the food web. And combined with other stressors, some species may be pushed over the edge into extinction. Take, for example, brook trout. Lower stream flows mean that even in relatively mild winters, freeze-up can lock waterways in ice from top to bottom, wiping out young fish along with frogs, turtles, and stream invertebrates. Low water can also cause sandbars to form at the mouth of rivers, blocking the migration of coaster brook trout into Lake Superior tributaries.

Brook trout are vulnerable on other fronts as well. Many North Shore streams emerge from warm-water wetlands and lakes. Dense streamside stands of conifers—white and black spruce and cedar—are credited with creating the cool pockets of water in which brook trout survive. Some biologists hypothesize that the loss of these species due to human-caused catastrophic wildfires and streamside logging may help to account for the disappearance of brook trout from the Temperance River, where, according to historical accounts, they once were plentiful. As global warming increases air temperatures and forces conifers to retreat to the north, stream temperatures will continue to rise. A climate model developed by Bill Herb of the University of Minnesota's St. Anthony Falls Lab and his colleagues at the University of Minnesota's Natural Resources Research Institute predicts a warming of 3 degrees Fahrenheit in North Shore streams. Because of their extensive lengths of unshaded banks and sluggish backwater pockets that retain heat, the largest waterways are expected to be the most affected by climbing temperatures. Chemical shifts in the fishes' food base would only compound the effects of thermal stress, threatening, among others, the endangered coaster brook trout that use these rivers for spawning and rearing of young.

Nearshore

Since the retreat of the last ice sheet some nine thousand years ago, many plant species and communities whose main ranges are more commonly found much farther to the north have been able to persist along Lake Superior's shoreline, thanks in large part to the lake's cool, moist influence. On Shovel Point in Tettegouche State Park, for example, a miniature boreal forest of spruce and fir, its trees stunted and crabbed by cold onshore winds and freezing wave spray, has taken hold. And for more than a century, botanists have combed the Minnesota coastline for rare arctic-alpine disjuncts, so called because they are separated more than six hundred miles from their main range in the Arctic tundra or in the cordilleran, or mountainous, regions of western North America. (See also "Hay Pickers and Grass Gatherers: Botanical Exploration along the Lakeshore.")

These arctic and boreal plant holdovers survived a warming of the climate by some

9 to 11 degrees Fahrenheit over the past twenty thousand years as the world slowly emerged from the depths of the Ice Age. But they may not survive the warming of 5 to 10 degrees Fahrenheit that is expected to occur in Minnesota by the end of this century.

According to conservative estimates, by 2090 Superior's lake levels could drop by as much as 1.38 feet, exposing great swaths of new lakeshore. Even if disjunct populations could migrate quickly enough to colonize this new real estate, would conditions be suitable for their long-term survival? The duration and extent of ice cover on the lake are expected to decline substantially. Without the scouring effect of a phenomenon known as ice push—a kind of blunt mowing action that keeps most vegetation from gaining a foothold along much of the lake edge—would the shoreline become colonized by plants that crowd or shade out sun-loving disjuncts? And as lake temperatures rise, will the shore edge become too warm to support arctic and boreal vegetation? Observation of historic and current locations of disjuncts shows they drop out along portions of the shoreline where the midsummer surface temperatures of adjoining waters exceed 59 degrees Fahrenheit. Water temperature is suspected to play a role in this pattern.

Some experts, including the authors of the 2003 UCS report, do not hold out much hope for the long-term survival of the arctic-boreal remnants in Minnesota's Lake Superior watershed. Even the capacity of the "lake effect" to provide a protective buffer against warming along Superior's shoreline remains a question. Instead of butterwort and black spruce or mixed forests of birch, spruce, and fir, parts of the lake terrace and shore will likely host more southerly species of trees such as sugar maple and oaks. Some scientists even hypothesize that savanna or open grasslands could replace today's forests.

While longtime North Shore lovers may mourn the change of its northland character, experts predict that it will not lack for people eager to settle along its shores. As lake levels and water quality decline in lakes around the state and national water disputes become more contentious, especially in the southern and western regions of the country, people will increasingly be drawn to the Great Lakes, which, though greatly diminished, will still contain comparatively large volumes of water. Indeed, according to the 2000 report *Preparing for Climate Change: The Potential Consequences of Climate Variability and Change,* "Climate change will no doubt worsen the effects of [the current trend of rural sprawl]."

The report notes that small towns and rural counties may not have developed adequate land-use rules or possess the funds to create new infrastructure that will protect fragile ecosystems. Take, for example, the problems sewage treatment has caused for the residents of Grand Traverse County, a burgeoning retirement and vacation destination on the shores of Lake Michigan. The failure of on-site septic systems to stem the growing pollution of local waters has prompted county officials to institute more centralized methods of sewage treatment. But, say the authors of *Preparing for Climate Change,* "the sudden need to expand sewerage service to many new residents and visitors has taxed municipal budgets. The problem will probably continue for the foreseeable future."

Similar cost-related barriers have plagued the Cook County communities of Schroeder, Tofte, and Lutsen as they seek to resolve water pollution problems by building shared waste-treatment facilities to serve new and existing development along the Lake Superior shore and Highway 61. The cost of such infrastructure has outstripped the financial resources of these tiny communities, and many local property owners have opposed the levying of tax assessments that would relieve some of the financial burden.

But the momentum to develop effective and affordable solutions to wastewater woes is not likely to slow in the future. Yet, without careful long-term planning, these technological fixes can have dramatic effects on land-use patterns. According to current Cook County zoning ordinances, for example, properties fronting Lake Superior must be a minimum of two hundred feet in width. Not only does this regulation allow adequate space for the operation of traditional septic systems, but it also ensures that these systems are located at a safe distance from drinking-water wells. By default, such rules have limited the density of development along much of Superior's shoreline. However, the increased use of shared waste-treatment systems—whether conventional or new field-tested methods such as constructed wetlands—could accommodate far more concentrated development. Coupled with increases in demand—and prices—for real estate along the shore, owners of large properties will find even greater incentives for subdividing their lots, particularly under the pressure of rising taxes. Even if the design of these new projects employs such land-saving strategies as cluster development, in which open space is preserved by concentrating development, the North Shore coastline in Lake and Cook Counties could end up looking much more like the shoreline along the outskirts of Duluth, a change that would further fragment coastal habitats as well as impart a suburban feel to what was once a heavily forested rural landscape. (See also "Where Has All the Sewage Gone? Development and Water Quality.")

Lake Superior

Water has long been the life blood of communities around the Great Lakes, so it is not surprising to find that the physical, chemical, and biological characteristics of lake water have been tracked with greater scrupulousness than almost any other feature of the Great Lakes watershed. As early as the mid-seventeenth century, travelers posted themselves at the mouths of streams feeding into the Great Lakes, recording the fluctuations of water levels in their journals. In 1860 officials began to gauge Great Lakes' water levels with greater regularity, amassing the longest, uninterrupted record of hydrological data in North America. Continuous records of ice phenologies—a seasonal ledger of such things as the timing of freeze-up and ice-out and the extent of ice cover and its thickness—go back as far as 1823. And beginning in the nineteenth century, city sanitarians routinely kept notes on a variety of conditions concerning municipal water supplies, including the temperature of lake water at the mouth of intake pipes.

The result of such record keeping is a long and fairly precise accounting of seasonal variations over the past 150 years. Many of them, such as ice cover and water temperature, are considered sensitive indicators of climate change. Analyses of the numbers have revealed some ominous trends. Take ice cover, for example. A study of the winter navigation season on Lake Superior near Bayfield, Wisconsin, between 1856 and 2007 shows that the formation and duration of ice on the lake has declined substantially. According to the research, the period of ice-free navigation has been extended by sixteen days in early winter and has reopened seventeen days earlier at winter's end, with the most dramatic lengthening of the season occurring since 1975. In 2012, a team of scientists in the *Journal of Climate* published research showing that ice cover fell 79 percent on Lake Superior between the years 1973 and 2010. "At the current rate of decline of ice cover," write Jay Austin and Steven Colman of the Large Lakes Observatory (LLO), University of Minnesota Duluth, "Lake Superior will be ice-free in a typical winter" by 2040.

The diminished ice cover has already tripped a cascade of changes in Superior. For example, the winter ice cover serves as a kind of heat shield, reflecting the warming rays of the sun back into space. With the loss of ice cover has come a warming of Superior's waters. In 2007 Austin and Colman published startling news showing that the average temperature of Superior's surface waters in summer had warmed up more than 4 degrees Fahrenheit from 1979 to 2006, twice as fast as the warming of the air.

Because so many of Superior's biological, chemical, and physical functions are driven by air and water temperatures, such increases have disrupted even a system as large and seemingly immutable as Lake Superior. The temperature increase, the researchers say, has already reset the clock for spring turnover two weeks earlier to mid-June rather than July, thus extending the seasonal stratification of the lake's waters. Lengthening the duration of this thermal barrier further narrows the critical window of opportunity for food to drift to the lake bottom to feed bread-and-butter species such as amphipods. (See also "Amphipods and Diatoms: The Big Lake's Bread and Butter.")

The potential for significant changes in the food web have already been noted by researchers, including in copepod populations, which comprise the largest component of biomass in summer zooplankton communities. Robert Megard, a researcher at the University of Minnesota, Twin Cities, published research in 2009 demonstrating that rising water temperature trends in the lake have resulted in a doubling of the abundance of copepods but a decrease in their average size, favoring smaller species over larger ones. Changes in such an important part of the ecosystem's prey base would be expected to have ripple effects in the structure of the community of predators that feed on zooplankton.

The diminished ice cover is also expected to exacerbate another downside of climate change. As air temperatures rise, water levels are expected to drop. That is because warmer temperatures increase evaporation. The loss is compounded, however, by the fact that Superior's surface waters are warming more quickly than the air above them. The temperature difference is creating unstable air masses, which result in higher winds over the lake. In a 2008 article in *Nature Geoscience,* Ankur Desai and his colleagues published data showing that since the 1980s, average wind speeds over Superior have increased by 5 percent. "Less ice in the winter means stronger winds in the summer," Desai and colleagues observe. In addition to increasing evaporation, greater wind speeds have also stepped up the velocity of lake currents, which perform such vital tasks as distributing nutrients throughout the lake.

The increased evaporation combined with a projected 20 percent decline in the amount of runoff the lake receives from surrounding rivers and streams will cause a lowering of lake levels. By 2030 Superior is expected to drop by about nine inches. By 2090, when predictions call for another doubling of carbon dioxide levels, Superior's waters could decline by about sixteen inches. The extreme lows of the past, say LLO researchers, could become the norm for the future.

Plummeting water levels could have disastrous consequences for many ecological functions. Coastal wetlands, including the Kakagon Sloughs on Superior's southern shore, one of the most pristine wetlands in the entire Great Lakes, could be cut off from the periodic flooding that flushes revitalizing oxygen and nutrients into backwater reaches. Without the protective crust of ice and snow, erosion-prone stretches of the shoreline could be severely damaged by violent winter storm waves. And under low water levels, Superior's most productive lake habitat—a narrow band of relatively

Normal Winter Maximum Ice Cover
(March 20–30)

Open water   Open pack
Close pack   Consolidated
              ice pack

Measurements of typical winter ice cover, as of 1969, showed large portions of Lake Superior covered in ice. Research shows that ice cover has fallen by 79 percent between the years 1973 and 2005 *(below)*.

shallow nearshore waters that hugs the coastline—could be greatly diminished. The greater storm turbulence could churn up the bottom of shallow bays, disrupting spawning habitat for such species as whitefish. Shallower waters would also allow pollutant-laced sands and muds to be more deeply and frequently stirred up by storm waves.

The effects of global warming are expected to be far more severe downstream of Lake Superior. Although the science is far from conclusive, experts say that by 2030 Lakes Michigan and Huron could experience large declines in water levels. By 2090 drops in excess of three feet are predicted. These declines do not sound excessive until you consider that a loss of one inch equals 9.5 trillion gallons of water, enough to supply a city the size of Cleveland for eighty-eight years.

Because of these losses, downstream interests are likely to pressure officials from the United States and Canada to renegotiate binational water agreements. To mitigate some of the effects of global warming on the lower lakes, political and industrial lobbies undoubtedly will demand a greater drawdown of Lake Superior's waters through the lock-and-dam structures at Sault Ste. Marie. Their economic—as well as ecological—arguments will be compelling. For example, under some climate scenarios the reduced outflow of Superior combined with reductions throughout the Great Lakes could result in a 40 percent decrease in the outflow of the St.

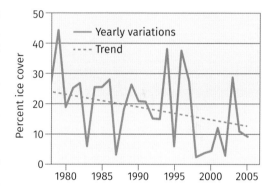

Lawrence River. Without the great discharge of freshwater that pulses through the river, saltwater will penetrate farther upstream into the St. Lawrence, with catastrophic consequences to the delicate freshwater-saltwater balance that is critical to sustaining life in the river and the Gulf of St. Lawrence.

The need for water is just as urgent in the municipal and industrial sectors. A 3.28-foot drop in Lake Michigan's water levels, for instance, will cripple the operation of the Chicago Diversion, a waterway that relies on the siphoning of Michigan's waters for sanitary as well as inland irrigation and navigation purposes. Politically powerful Illinois legislators will undoubtedly reopen their campaign for increasing flows through the diversion, as they did during the drought of 1988 when they petitioned the U.S. Army Corps of Engineers to open the diversion's floodgates in order to alleviate navigation problems on the Mississippi River.

As the bills incurred by global warming come due in other sectors of the economy, commercial interests will only step up the political pressure for regulatory water-flow change. According to the authors of *Preparing for Climate Change,* Great Lakes shipping, a $3 billion industry that employs sixty thousand people, will be adversely affected by lower water levels. The Lake Carriers' Association estimates, for example, that for every drop of one inch in lake levels, carriers must leave 270 tons—a whopping 540,000 pounds—of cargo on the docks.

To increase clearance for the great cargo carriers, commercial harbors and connecting channels must be deepened through dredging of the lake bottom. Such activities will stir up heavy metals and toxic contaminants in the sediment, an activity

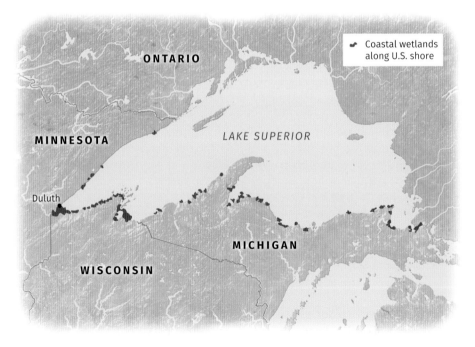

The lower water levels predicted in some climate-change scenarios will cut off Lake Superior's coastal wetlands from the periodic flooding that flushes revitalizing oxygen and nutrients into backwater reaches.

that is particularly worrisome in the Duluth-Superior harbor, one of the forty-three heavily polluted Areas of Concern that have been identified by the International Joint Commission in the Great Lakes. Compensatory measures also come with a hefty price tag. According to the Wisconsin Sea Grant, dredging costs alone could top $100 million annually, nearly doubling the $52 million that the U.S. Army Corps of Engineers devoted in 2002 to all maintenance and operation activities in the entire Great Lakes. Compounding costs is the fact that contaminated sediments must be sequestered in special waste-disposal sites known as confined disposal facilities (CDFs). Many of the dozens of CDFs that have been built around the Great Lakes by the federal government since the 1970s are filled to capacity. Constructing new ones is growing costlier at a time when public dollars for big infrastructure projects are dwindling.

Commercial shipping infrastructure would require rehabilitation as well. Take, for example, the port of Duluth. A legal agreement that was signed by the United States and Canada in 1914 stipulated that Superior's water levels must be maintained at levels sufficient to protect navigation and shoreline interests from excessive drawdowns by hydropower operators. According to the authors of "Lake Superior's High Water Levels: A Delicate Balance," "Most of the Duluth harbor boat slips, grain elevators, and ore and shipping docks were all designed or modified in this century when Lake Superior's water level was operated between 600.5 and 602.0 feet. A deviation from this range, that is, if the level exceeds 602.0 or falls below 600.5 causes structural damage and navigation problems."

A drop of four to eight feet in Great Lakes water levels, say the authors of the 2000 *Preparing for Climate Change,* will incur expenses for lakeside cities as well as the federal taxpayer. Experts put a price tag of $132 million to $228 million on the cost of extending water intake pipes, stormwater outfalls, and docks exposed by low water levels. This prodigious sum does not factor in expenditures needed to shore up or rebuild breakwaters that are weakened by exposure to the air. Breakwaters throughout the Great Lakes routinely feature concrete caps that are supported by underwater wooden cribs, or boxes, filled with rocks. Others rest on timber piles driven into the lake bed. Thanks to the cold temperatures and low microbial activity of Great Lakes' waters, these submerged wooden structures can last indefinitely. But when they are exposed to the air for prolonged periods, as is expected by lowered water levels in the Great Lakes, the wooden supports can rapidly deteriorate, leaving many breakwaters vulnerable to catastrophic failure during major storm events.

Warmer temperatures could usher in biological changes whose consequences are just as far-reaching as those caused by lowered water levels. For example, in the future, municipalities on Lake Superior could be adding the costs of combating zebra mussels to their infrastructure budgets. Unlike the lower lakes, Superior's cold waters and low calcium levels have held in check wholesale infestations by zebra mussels. But the warmer waters expected under a doubling of carbon dioxide could create a more hospitable habitat for zebra mussels in Superior's bays and nearshore zones, as they already have in the Duluth-Superior harbor. In fall 2000, Dan Kelner, a mussel expert with the DNR, dove to the bottom of the harbor to survey the mussel population. Kelner discovered densities of zebra mussels that rivaled those of long-infested sites in the Ohio and Mississippi Rivers. He blamed the warm summers and mild winters in the northland over the prior three years for creating the conditions that allowed the nuisance mussels to flourish.

Not only do zebra mussels increase costs for maintenance of such infrastructure as water intake pipes, but they have also been implicated in disastrous disruptions of the food web in the lower lakes. Martin Auer of Michigan Technological University's Department of Civil and Environmental Engineering observes that "zebra mussels are like shop-vacs and are sucking up all the *Diporeia* food." The shrimplike *Diporeia,* he says, appears to be "headed for an early retirement" in the lower Great Lakes. Since the mid-1990s, Auer points out, "they've all but vanished in lakes Erie and Ontario, and they are going fast in Huron and Michigan. Soon, all that will be left is a *Diporeia* museum, and it'll be right in the middle of Lake Superior." Since zebra mussels also accumulate higher rates of toxins in their tissues, they could become a major source of contamination for the animals that eat them, including fish and diving ducks.

Among the greatest concerns, however, is a decrease in fall and spring turnover rates. During the summer, a layer of warm, buoyant water serves as a kind of lid that isolates bottom waters from oxygen in the atmosphere. But just as the bottom begins to run low on oxygen, seasonal air temperatures start to drop, cooling the surface waters of the lake. When water temperatures reach 39 degrees Fahrenheit, the maximum density of water, they begin to sink, breaking down summer's stratification of water in what is known as the fall turnover. This process, which is repeated in the spring, mixes lake waters from top to bottom, replenishing nutrients and oxygen supplies even in the deepest portions of the lake. The mixing allows such bread-and-butter organisms as the tiny shrimplike *Diporeia* to flourish on the lake bottom. (See also "Amphipods and Diatoms: The Big Lake's Bread and Butter.")

Scientists predict that warmer temperatures could alter this life-giving dynamic, causing the summer stratification of lake waters to occur earlier and last longer on all the Great Lakes, including Superior. Prolonged stratification could starve bottom-dwelling organisms of critical supplies of oxygen.

The results of several climate models show that under a doubling of carbon dioxide, such a scenario is likely to become the norm. Instead of experiencing a fall and spring mixing, in some years even the deepest of the Great Lakes—Superior, Michigan, and Huron—would likely experience only a single, short period of complete mixing in late winter.

Not only will the warming of lake waters disrupt the life cycles of organisms on the lake bottom, but it also will change the balance of plants and animals that live in the water column. The Great Lakes food web depends on the spring bloom of highly nutritious algae known as diatoms. The timing of the bloom is calibrated to several seasonal events including the onset of spring runoff and the spring turnover of the lake waters. According to the UCS report, "Earlier ice-out (thaw of lake ice) and spring runoff will shift the timing of the spring algal bloom, and earlier and longer periods of summer stratification tend to shift dominance in the algal community during the growing season from diatoms to inedible blue-green algae. If climate change causes inedible nuisance species to dominate algal productivity, or if the timing of algal production is out of synch with the food demands of fish, then all upper levels of the food chain, particularly fish, will suffer." Shifts in forage will only add to the survival woes of such cold-water species as lake trout, which will experience greater thermal stress from warming water temperatures.

Lake Superior's response to near-record-high temperatures in 1998 forecast what may become routine conditions under a doubling of carbon dioxide in the future. In the summer of that year, recalls Elise Ralph, a physical oceanographer at the LLO, stories of people swimming in Lake Superior made the front page of newspapers all around the lake. After sampling the temperature of surface waters from the western end of the lake, she discovered the reason why. Readings averaged 10 degrees Fahrenheit higher than the year before.

The higher water temperatures meant that Superior's waters stored more heat prior to winter that year and consequently took longer to cool down. When Ralph ventured out on the lake in late December, Superior, which normally begins the fall turnover in October, still had not turned over. Scientists also noted a dramatic increase in zebra mussels in the Duluth-Superior harbor.

And as the UCS report points out, rising temperatures and less oxygen could increase the levels of such problem nutrients as phosphorus in the water. Reductions in oxygen supplies could also make heavy metals, including mercury, more available to aquatic organisms. That is because oxygen molecules bind with the metals to form insoluble compounds. These compounds sink to the bottom of the lake, where they lie dormant in the sediments. In the absence of oxygen, however, they revert to a soluble state and enter the food web. These metals can become highly concentrated in the tissues of organisms that are high on the food chain, such as fish.

If the effects of global warming prove to be far more extreme than those predicted by current general circulation models, says former LLO director Thomas Johnson, Superior's water levels could drop below the St. Marys River outfall, causing Superior to become a closed basin. Over many hundreds of years, the lake's chemistry would change radically due to the dissolved minerals that are swept into the lake by water flowing over rocks and soils. When evaporation exceeds the amount of water flowing in, minerals become trapped and concentrated in the lake, causing lake waters to become salty. With a decrease or cessation of the turnover process, the lake bottom would be deprived of oxygen, and all but the most primitive forms of life would die.

The future of Superior, he says, could be a saltwater lake ringed by a coastal forest of oak trees and maples.

Islands

Ever since Superior's islands emerged from the waters of the lake, they have hosted visitors from the mainland. At first, prevailing winds delivered tiny, airborne seeds that lodged in pockets of rock or fell on plains of gravel, blossoming into delicate, frilly flowers and huddled tangles of trees. Birds braved the winds and found safe landfall in the islands' forests. Frogs clung to rafts of floating debris that washed up on the island's rocky shores. During severe winters, larger animals, such as wolves and caribou, dispersed across great highways of ice. On the more roomy islands, such as the sprawling Isle Royale archipelago, the migrants took up permanent residence.

Despite the brisk traffic, not many species have been able to overcome the barriers presented by Superior's waters. Compared to the mainland, which supports more than forty species of mammals, the islands count only fifteen species. The number of terrestrial vascular plants is similarly decreased.

Global warming is likely to cause shifts in these plant and animal populations. For the islands that lie just off the mainland, the lower water levels predicted under global warming may boost the number of species that take up residence by exposing new swaths of shore edge. Shallower nearshore waters could also uncover rocky land bridges that connect islands with the mainland, removing the barriers to animals such as small mammals and the seeds they transport.

The exposure of new habitat and greater species exchange may not, however, always benefit Superior's native complement of plants and animals. Spiraling populations of nuisance species such as ring-billed gulls, for example, have been quick to colonize new offshore nesting sites. Not only does their nesting activity disturb native plant communities, but their feeding habits also inoculate isolated islands with the seeds of invasive exotic plants transported from the mainland. (See also "Nature or Nuisance? Gulls in the Great Lakes.")

For islands that lie far out in the lake, such as Isle Royale, global warming may sever bridges to the mainland by decreasing the frequency of winter freeze-overs. The ice serves as an arterial for the transport of new genetic material. Without fresh recruits, many animal populations could become dangerously inbred and die out, which is the likely fate of the famous wolf packs on Isle Royale.

But even if park managers violate their principal of "letting nature take its course" by introducing new wolves to the archipelago, the future of the animals' favored prey remains uncertain. Balsam fir, a staple of the moose diet on Isle Royale, is one of the tree species expected to shift its range farther to the north under global-warming scenarios.

Populations of other plant species are expected to decrease as well. As they do along the mainland shore, rare species of arctic-alpine disjuncts make their home in the crevices of exposed rock. As the temperature of the air and water warms and shore-edge ice formations diminish, many of these plants are likely to disappear. Also at risk is the especially rich profusion of lichens and mosses. The Susie Islands alone host more than four hundred species of lichen. It is unlikely that such species will be capable of surmounting Superior's great watery barriers and migrating to cooler climes as conditions grow warmer and drier. As a result, many plants may face localized extinctions.

Preparing for the Future

In 1896, writing what would be a chillingly prescient description of today's ecological predicament, the nineteenth-century Swedish chemist Svante Arrhenius warned, "We are evaporating our coal mines into the air."

If we are to stave off the worst of global warming's effects on our ecologies, economies, and health, we would do well to heed Arrhenius's warning—reduce the amount of heat-trapping gases that are emitted into the atmosphere, especially the carbon dioxide that is released from the burning of fossil fuels.

Experts advocate taking greater steps—both individually and collectively—toward energy efficiency and conservation as we concentrate on shifting our carbon-based economy away from coal, oil, and natural gas. While the challenge is enormous, it is not impossible. Nor does it entail a drastic downscaling in our standard of living. According to the U.S. Department of Energy, if every American homeowner simply replaced the incandescent lighting of his or her top five most frequently used fixtures with fluorescent bulbs, the energy savings would amount to some eight hundred billion kilowatt hours of electricity. The savings from this simple measure alone would be enough to shutter the doors of twenty-one power plants.

Unfortunately, even if we met a zero-emissions standard today, the global thermostat would continue to rise for decades to come due to existing levels of greenhouse gases in the atmosphere. (Once it has entered the atmosphere, carbon dioxide lingers for a period of roughly one hundred years.) If by 2060, for example, the world cut its greenhouse gas emissions in half, the concentrations of these gases in the atmosphere would still be double the 1990 levels. Reductions in our consumption of fossil fuels would not reduce the amount of greenhouse gases in the air but rather slow their rate of increase.

Some scientists argue that the best method for preserving our native complement of plants and animals is not only to minimize human impacts but also to conserve and restore their habitats. But given that many species may be forced to migrate to keep pace with climate change, some ecologists urge that we expand our land-conservation strategies to set aside the green highways that many species need as migratory corridors. Instead of simply seeking to save species or entire ecosystems, scientist James Trefil suggests that "it may be time . . . to focus on a different goal. We should concentrate on preserving the ability of plants and animals to respond to environmental changes."

According to recent research, such advice appears to make good ecological sense. In a 2002 study published in the *Proceedings of the National Academy of Sciences,* researchers from four American universities conducted an experiment in which they clear-cut and burned selected areas in order to create islands of habitat within eight stands of pine trees. Some of these islands were connected to one another by eighty-foot-wide corridors of vegetation. Others were completely isolated by the clearings. The scientists found that butterflies visited the connected patches more frequently than the isolated ones, preferring to travel the vegetated corridors rather than risk exposure in the surrounding clear-cuts. In addition, pollen and seeds were better dispersed among the connected islands, largely due to the greater number of animals that used the vegetated corridors.

In the face of rising global temperatures, such conservation corridors may save untold numbers of plants and animals from extinction. In a 2004 study published in *Nature,* a team of scientists led by Chris D. Thomas, an ecologist at the University of Leeds, England, examined mapping and distribution trends for 1,103 plant and animal species from locations in Mexico, Australia, South Africa, Europe, and the Amazon rain forest. After factoring in the most conservative estimate of global climate change— a temperature rise of 1.44 to 3.06 degrees Fahrenheit by 2050—the group found that 13 percent of the species studied would become extinct or their numbers diminished beyond hope of recovery. Without migratory corridors that would allow species to move to other, more suitable habitats, the prospect of extinction jumped to 31 percent.

Some conservation biologists, however, charge that even the best efforts at conserving habitat and migration corridors will be insufficient for many species to outrun the race of climate change. They argue for additional measures such as a far more active human intervention in dispersing species of concern, a new management philosophy known as "assisted migration." Proponents advocate developing scientifically based management policies, for example, for determining which species may benefit from being moved, their best relocation sites, and the social and ecological effects of resituating these organisms into new territories.

Preserve urban parks, gardens, and groves of trees as steppingstones to connect core habitats

Preserve large patches of native vegetation as core habitats

Plant native vegetation along roadsides and around buildings to aid the dispersal of pollen and seeds and facilitate animal movement

©VeraLingWong 2014

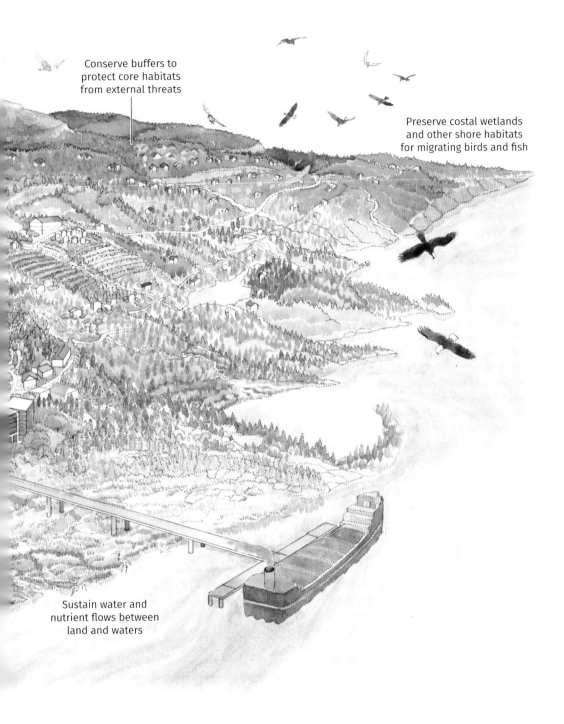

Conserve buffers to
protect core habitats
from external threats

Preserve costal wetlands
and other shore habitats
for migrating birds and fish

Sustain water and
nutrient flows between
land and waters

In the face of rising global temperatures, preserving native habitats and conservation corridors are
among the strategies that can allow plants and animals to migrate to more hospitable locations.
Illustration by Vera Ming Wong.

As the UCS report points out, the measures that we take to ensure the future of nonhuman lives will only enhance the quality of our own. "Fortunately, many of the actions that can be taken now to prevent the most damaging impacts of climate change can also provide immediate collateral benefits," the authors write, "such as cost savings, cleaner air and water, improved habitat and recreational opportunities, and enhanced quality of life in communities throughout the [Great Lakes] region."

In the beginning of this book, we posed the questions: "What will be the environmental legacy of our own era? How can we live in a manner that sustains this home so that all its inhabitants might flourish?"

The purpose of our book is to help readers nurture a deep sense of belonging to nature. We believe that this recognition of interdependence is essential to finding the motivation for preventing further injury to Earth and for healing old wounds, for sustaining and restoring the space that grants all of our fellow inhabitants the opportunity to adapt to change. This is, after all, the fundamental capacity that allowed life to first take hold and persist on our extraordinary planet some 3.8 billion years ago.

That we live in a time when this is still possible is a blessing beyond measure. ❧

SUGGESTIONS FOR FURTHER READING

Assel, Raymond, Kevin Cronk, and David Norton. "Recent Trends in Laurentian Great Lakes Ice Cover. *Climatic Change* 57 (2003): 185–204.

Austin, Jay A., and Steven M. Colman. "Lake Superior Summer Water Temperatures Are Increasing More Rapidly Than Regional Air Temperatures: A Positive Ice-Albedo Feedback." *Geophysical Research Letters* 34 (2007): L06604.

Blaustein, Andrew R., David B. Wake, and Wayne P. Sousa. "Amphibian Declines: Judging Stability, Persistence, and Susceptibility of Populations to Local and Global Extinctions." *Conservation Biology* 8 (1994): 60–71.

Broder, John M. "Past Decade Warmest on Record, NASA Data Shows." *New York Times,* 22 January 2010.

Brooks, Arthur S., and John C. Zastrow. "The Potential Influence of Climate Change on Offshore Primary Production in Lake Michigan." *Journal of Great Lakes Research* 28 (2002): 597–607.

Center for the Great Lakes. "*Effects of Global Warming on the Great Lakes: The Implications for Policies and Institutions.* Chicago: Center for the Great Lakes, 1988.

Cohen, Stewart J. "Impacts of CO_2-Induced Climatic Change on Water Resources in the Great Lakes Basin." *Climatic Change* 8 (1986): 135–53.

Committee on Abrupt Climate Change; Ocean Studies Board; Polar Research Board; Board on Atmospheric Sciences and Climate, Division on Earth and Life Studies; and National Research Council. *Abrupt Climate Change: Inevitable Surprises.* Washington, D.C.: National Academy Press, 2002.

Croley, Thomas E. "Laurentian Great Lakes Double-CO_2 Climate Change Hydrological Impacts." *Climatic Change* 17 (1990): 27–48.

Davis, Margaret B., and Ruth G. Shaw. "Range Shifts and Adaptive Responses to Quaternary Climate Change." *Science* 292 (2001): 673–79.

Desai, Ankur R., Jay A. Austin, Val Bennington, and Galen A. McKinley. "Stronger Winds over a Large Lake in Response to Weakening Air-to-Lake Temperature Gradient." *Nature Geoscience,* 15 November 2008.

Frelich, Lee E., and Peter B. Reich. "Will Environmental Changes Reinforce the Impact of Global Warming on the Prairie-Forest Border of Central North America?" *Frontiers in Ecology and the Environment* 8 (2009): 371–78.

Gillis, Justin. "Heat-Trapping Gas Passes Milestone, Raising Fears." *New York Times,* 10 May 2013.

"Global Warming: Potential Effects on BWCA Wilderness." BWCA *Wilderness News,* Autumn 1998: 7.

Hartmann, Holly C. "Climate Change Impacts on Laurentian Great Lakes Levels." *Climatic Change* 17 (1990): 49–68.

Hesman, Tina. "Greenhouse Gassed: Carbon Dioxide Spells Indigestion for Food Chains." *Science News* 157 (2000): 200–202.

Howk, Forrest. "Changes in Lake Superior Ice Cover at Bayfield, Wisconsin." *Journal of Great Lakes Research* 35 (2009): 159–62.

Hunter, Malcolm L., Jr. "Climate Change and Moving Species: Furthering the Debate on Assisted Colonization." *Conservation Biology* 21 (2007): 1356–58.

Intergovernmental Panel on Climate Change. "Summary for Policymakers." In *Climate Change 2013: The Physical Science Basis.* Contribution of Working Group I to the Fifth Assessment Report of the Intergovernmental Panel on Climate Change, ed. T. F. Stocker et al. Cambridge: Cambridge University Press, 2013.

Johnson, Stephanie L., and Heinz G. Stefan. "Indicators of Climate Warming in Minnesota: Lake Ice Covers and Snowmelt Runoff." *Climatic Change* 75 (2006): 421–53.

Keillor, Philip, and John Karl. "Living with the Lakes." *Littoral Drift,* March/April 2002: 1.

Kling, George W., Katherine Hayhoe, Lucinda B. Johnson, John J. Magnuson, Stephen Polasky, Scott K. Robinson, Brian J. Shuter, et al. *Confronting Climate Change in the Great Lakes Region: Impacts on Our Communities and Ecosystems.* Cambridge, Mass.: Union of Concerned Scientists; Washington, D.C.: Ecological Society of America, 2003.

Kronberg, Barbara I., Murray J. Watt, and Susan C. Polischuk. "Forest-Climate Interactions in the Quetico-Superior Ecotone (Northwest Ontario and Northern Minnesota)." *Environmental Monitoring and Assessment* 50 (1998): 173–87.

Laghari, Javaid R. "Melting Glaciers Bring Energy Uncertainty." *Nature* 502 (2013): 617–18.

Lemoine, Nicole, and Katrin Bohning-Gaese. "Potential Impact of Global Climate Change on Species Richness of Long-Distance Migrants." *Conservation Biology* 17 (2003): 577–86.

Lenters, John D. "Long-Term Trends in the Seasonal Cycle of Great Lakes Water Levels." *Journal of Great Lakes Research* 27 (2001): 342–53.

Lofgren, Brent M., Frank H. Quinn, Anne H. Clites, Raymond A. Assel, Anthony J. Eberhardt, and Carol L. Luukkonen. "Evaluation of Potential Impacts on Great Lakes Water Resources Based on Climate Scenarios of Two GCMs." *Journal of Great Lakes Research* 28 (2002): 537–54.

Magnuson, J. J., K. E. Webster, R. A. Assel, C. J. Bowser, P. J. Dillon, J. G. Eaton, H. E. Evans, et al. "Potential Effects of Climate Changes on Aquatic Systems: Laurentian Great Lakes and Precambrian Shield Region." In *Freshwater Ecosystems and Climate Change in North America: A Regional Assessment,* ed. Colbert E. Cushing. Chichester, Eng.: John Wiley and Sons, 1997.

Mandrak, Nicholas E. "Potential Invasion of the Great Lakes by Fish Species Associated with Climatic Warming." *Journal of Great Lakes Research* 15 (1989): 306–16.

Mayewski, Paul Andrew, and Frank White. *The Ice Chronicles: The Quest to Understand Global Climate Change.* Hanover, N.H.: University Press of New England, 2002.

McKibben, Bill. "Global Warming's Terrifying New Math." *Rolling Stone,* 2 August 2012.

McLachlan, Jason S., Jessica J. Hellmann, and Mark W. Schwartz. "A Framework for Debate of Assisted Migration in an Era of Climate Change." *Conservation Biology* 21 (2007): 297–302.

Megard, Robert O., Elise Ralph, and Michelle Marko. "Effects of Wind and Temperature on Lake Superior Copepods." *International Society of Limnology* 30 (2009): 801–8.

Meisner, J. Donald, John L. Goodier, Henry A. Regier, Brian J. Shuter, and W. Jack Christie. "An Assessment of the Effects of Climate Warming on Great Lakes Basin Fishes." *Journal of Great Lakes Research* 13 (1987): 340–52.

Moen, Sharon. "Climate and Lake Superior's Crunchy Creatures." *Seiche,* April 2009: 1–3.

———. "Mysteries of the Distilled Water Ice Bath." In *Superior Science: Stories of Lake Superior Research.* Duluth: Minnesota Sea Grant, 2004.

Mortsch, Linda D., Henry Hengeveld, Murray Lister, Brent Lofgren, Frank Quinn, Michel Slivitzky, and Lisa Wenger. "Climate Change Impacts on the Hydrology of the Great Lakes–St. Lawrence System." *Canadian Water Resources Journal* 25 (2000): 153–79.

Mortsch, Linda D., and Frank H. Quinn. "Climate Change Scenarios for Great Lakes Basin Ecosystem Studies." *Limnology and Oceanography* 41 (1996): 903–11.

Murphy, Helen T., Jeremy VanDerWal, and Jon Lovett-Doust. "Signatures of Range Expansion and Erosion in Eastern North American Trees." *Ecology Letters* 13 (2010): 1233–44.

Natural Resources Research Institute. "Warming Waters for Brook Trout?" NRRI *Now,* Spring/Summer 2013: 3. http://www.nrri.umn.edu/news/brook_trout.htm.

Ohio Sea Grant. *Global Change in the Great Lakes Scenarios.* Columbus: Ohio Sea Grant, 1991.

Overpeck, Jonathan T., Patrick J. Bartlein, and Thompson Webb III. "Potential Magnitude of Future Vegetation Change in Eastern North America: Comparisons with the Past." *Science* 25, no. 4 (1991): 692–95.

Pastor, John, and W. M. Post. "Response of Northern Forests to CO_2-Induced Climate Change." *Nature* 334 (1988): 55–58.

Perkins, Sid. "On Thinning Ice." *Science News* 164 (2003): 215–16.

Parmesan, Camille, and Gary Yohe. "A Globally Coherent Fingerprint of Climate Change Impacts across Natural Systems." *Nature* 421 (2003): 37–42.

Prasad, Anantha, Louis Iverson, Stephen S. Matthews, and Matt Peters. "Atlases of Tree and Bird Species Habitats for Current and Future Climates." *Ecological Restoration* 27 (2009): 260–63. http://www.nrs.fs.fed.us/atlas/.

Quinn, Frank H. "Likely Effects of Climate Changes on Water Levels in the Great Lakes." In *Proceedings of the First North American Conference on Preparing for Climate Change: A Cooperative Approach*, October 27–29, 1987, Washington, D.C.

Revkin, Andrew C. "An Icy Riddle as Big as Greenland." *New York Times,* 8 June 2004, D1–D4.

Robbins, Jim. "Moose Die-Off Alarms Scientists." *New York Times,* 14 October 2013.

Robertson, Tom. "DNR Likely to Cut Number of Moose Hunt Permits in Half." Minnesota Public Radio, 23 February 2011. http://minnesota.cbslocal.com/2011/02/23/minn-dnr-likely-to-cut-moose-hunt-permits-in-half/.

Root, Terry L., Jeff T. Price, Kimberly R. Hall, Stephen H. Schneider, Cynthia Rosenzweig, and J. Alan Pounds. "Fingerprints of Global Warming on Wild Plants and Animals." *Nature* 421 (2003): 57–60.

Rosenthal, Elizabeth, and Andrew C. Revkin. "Science Panel Calls Global Warming 'Unequivocal.'" *New York Times,* 3 February 2007.

Santos, Fernanda. "Inch by Inch, Great Lakes Shrink, and Cargo Carriers Face Losses." *New York Times,* 22 October 2007.

Sarewitz, Daniel, and Roger Pielke Jr. "Breaking the Global-Warming Gridlock." *Atlantic Monthly,* July 2000: 55–64.

Schneider, Stephen H., and Terry L. Root. "Climate Change." In *Status and Trends of the Nation's Biological Resources*, vol. 1, ed. Michael J. Mac, Paul A. Opler, Catherine E. Puckett Haecker, and Peter D. Doran. Reston, Va.: U.S. Department of the Interior, U.S. Geological Survey, 1998.

Smith, Doug. "The Mystery of the Disappearing Moose." *National Wildlife Magazine* 45 (2007): 46–50.

Smith, Joel B., and Dennis A. Tirpak, eds. *The Potential Effects of Global Climate Change on the United States.* New York: Hemisphere Publications, 1990.

Solomon, Allen M., and Darrell C. West. "Atmospheric Carbon Dioxide Change: Agent of Future Forest Growth or Decline?" In *Effects of Changes in Stratospheric Ozone and Global Climate*, ed. James G. Titus, 23–38. Washington, D.C.: United Nations Environment Programme/U.S. Environmental Protection Agency, 1986.

Sousounis, Peter J., and Jeanne M. Bisanz, eds. *Preparing for Climate Change: The Potential Consequences of Climate Variability and Change.* Ann Arbor: U.S. Global Change Research Program, University of Michigan, 2000.

Tewksbury, Joshua J., Douglas J. Levey, Nick M. Haddad, Sarah Sargent, John L. Orrock, Aimee Weldon, Brent J. Danielson, Jory Brinkerhoff, Ellen I. Damschen, and Patricia Townsend. "Corridors Affect Plants, Animals, and Their Interactions in Fragmented Landscapes." *Proceedings of the National Academy of Sciences* 99 (2002): 12923–26.

Thomas, Chris D., Alison Cameron, Rhys E. Green, Michel Bakkenes, Linda J. Beaumont, Yvonne C. Collingham, Barend F. N. Erasmus, et al. "Extinction Risk from Climate Change." *Nature* 427 (2004): 145–48.

Thompson, Lonnie G., Ellen Mosley-Thompson, Mary E. Davis, Victor S. Zagorodnov, Ian M. Howat, Vladimir N. Mikhalenko, and Pang-Nan Lin. "Annually Resolved Ice Core Records of Tropical Climate Variability over the Past ~1800 Years." *Science* 340, no. 6135 (2013): 945–50.

Trefil, James. "When Plants Migrate." *Smithsonian* 28 (1998): 22–24.

Union of Concerned Scientists. "Findings from *Confronting Climate Change in the Great Lakes Region:* Impacts on Minnesota Communities and Ecosystems." Cambridge, Mass.: Union of Concerned Scientists; Washington, D.C.: Ecological Society of America, 2003.

Vucetich, John A., and Rolf O. Peterson. *Ecological Studies of Wolves on Isle Royale 2009–2010.* Houghton: Michigan Technological University, 2010.

Wang, Jia, Xuezhi Bai, Haoguo Hu, Anne Clites, Marie Colton, and Brent Lofgren. "Temporal and Spatial Variability of Great Lakes Ice Cover, 1973–2010." *Journal of Climate* 25 (2012): 1318–29.

Weflen, Kathleen. "The Crossroads of Climate Change." *Minnesota Conservation Volunteer,* January/February 2001: 8–21.

Wuebbles, Donald J., and Katharine Hayhoe. "Climate Change Projections for the United States Midwest." *Mitigation and Adaptation Strategies for Global Change* 9 (2004): 335–63.

Wyckoff, Peter H., and Rachel Bower. "Response of the Prairie–Forest Border to Climate Change: Impacts of Increasing Drought May Be Mitigated by Increasing CO_2." *Journal of Ecology* 98 (2010): 197–208.

Zhuikov, Marie. "Great Salt Lake Superior?" *Seiche,* April 1999: 1–2.

———. "Major Zebra Mussel Infestation in Harbor Impacts Native Mussels, Boaters." *Seiche,* January 2001: 5–6.

INTERNET RESOURCES

Birds and climate change: U.S. Department of Agriculture, Forest Service, "Climate Change Atlas, http://www.nrs.fs.fed.us/atlas/

"Climate Policy: Cracking Abrupt Climate Change," www.geotimes.org/feb02/NN_AbruptCC .html

The Discovery of Global Warming, http://www.aip.org/history/climate/index.htm

Forests and climate change: Earth Observatory, "The Migrating Boreal Forest," www.earth observatory.nasa.gov/Study/BorealMigration/; U.S. Department of Agriculture, Forest Service, "Climate Change Atlas, http://www.nrs.fs.fed.us/atlas/

Fresh Energy, http://fresh-energy.org/

Great Lakes Information Network and Great Lakes Commission, www.great-lakes.net/

National Oceanic and Atmospheric Administration (NOAA), www.ngdc.noaa.gov/paleo/global warming/home.html

"Sudden Warming in the Past," www.geotimes.org/feb02/NN_lptm.html

United States–Canadian Great Lakes Atlas, www.epa.gov/glnpo/atlas/intro.html

U.S. Environmental Protection Agency, Global Warming Case Studies, Great Lakes and Upper Midwest, http://epa.gov/climatechange/impacts-adaptation/midwest.html#impactswater

U.S. National Assessment of Climate Variability and Change http://www.epa.gov/global-adaptation/ topics/national_assessment.html

Acknowledgments

We are grateful beyond measure to the many scientists, natural resource professionals, and North Shore residents who generously shared their research, experience, and personal insights with us. Without their extraordinary devotion to the Lake Superior region, this book would not have been possible.

First and foremost, we would like to express our heartfelt gratitude to the biologists who accompanied us into the field. We are in awe of your determination to carry out your research in the face of so many challenges, including the physical rigors and risks of fieldwork and the vagaries of project funding. May you and your organisms live long and prosper: Joan Edwards, JoAnn Hanowski, John Johnson, Lee Newman, Lynn Rogers, and David Smith.

Special thanks too to the people who invited us into their homes and offices and courageously forded the fast-running streams of our questions for hours at a time. Many of you followed up in-depth interviews with careful reviews of our manuscript: David Abazs, Rich Axler, Mary Balcer, Richard Buech, Ken Catania, Caven Clark, Tim Cochrane, Jeff Crosby, Robert Dunn, Laura Erickson, Karen Evens, Janet Green, LeRoy Halberg, Cindy Hale, Stephen Heard, George Host, Thomas Johnson, Tim Kennedy, George Kessler, Ben Korgen, Jon Kramer, Stephen Lozano, Colleen Matula, Barb McCarthy, Joan McKearnan, Sarah Miller, Gerald Niemi, Walt Okstad, Dave Olfelt, Phil Olfelt, John Pastor, Shawn Perich, Gordon Peters, Bill Peterson, James Petranka, David Radford, Lee Radzak, Elise Ralph, Carl Richards, Nigel Wattrus, Mark White, and Julia Wilder.

We would also like to express our appreciation to those who took time out of their busy schedules to respond to our inquiries with thoughtful and information-packed replies: John Bonde, John Brazner, Dave Cooper, David Divins, Michael Donahue, Roger Flood, Lawson Gerdes, Paul Gobster, John Green, Carol Hall, Karen Jensen, Tim Norman, Don Schreiner, David Schwab, Milt Stenlund, and Linda Zellmer.

Central to refining our thinking, clarifying our writing, and increasing the accuracy and relevance of the manuscript have been those who read and commented on large portions of the manuscript. Your generosity has been critical to the success of our enterprise: Warren Abrahamson, John Almendinger, Joel Baker, Tim Burton, Francesca Cuthbert, Thomas Edsall, Sara Green, Jeff Jeremiason, Lucinda Johnson, Clive Lipchin, Carolyn (Lindy) McBride, Robert Megard, Richard Ojakangas, Mike Schrage, Joseph Shorthouse, Eugene Stoermer, and Herb Wright. Our special thanks to Meredith Cornett. Using her prodigious literary talents and scientific expertise, she provided an exhaustive review of the manuscript, encouragement, and inspiration at a time when both were sorely lacking. Any errors that remain are entirely ours.

Our book would not have been possible without the help of staff from numerous museums, public libraries, and historical societies including the Minnesota Historical Society and the Isle Royale Natural History Association. We are enormously grateful for their too-little-heralded work.

Minnesota Sea Grant and Wisconsin Sea Grant graciously responded to requests for information and photographs on a variety of environmental and cultural topics.

We are deeply grateful to staff at the University of Minnesota Press who were essential to making our vision of this book a reality. Todd Orjala's steadfast support of this project for more than a decade is a testament both to his patience and his commitment to publishing regional material of the highest quality. Thanks to interns Ana Bichanich, Ashley Kes, and Shay Logan, who picked up the editorial load on the exhausting relay to the finish, and to Daniel Ochsner and Laura Westslund for their production prowess. Special thanks to Mary Keirstead, whose insights brought added accuracy and grace to our manuscript. And finally to Erik Anderson and Kristian Tvedten, who played Virgil to our Dante, leading us through the inferno of publication with enthusiasm, professionalism, and good guidance. We would not have finished the journey without you.

Our personal thanks follow.

From Adelheid Fischer

One of the biggest misconceptions about writing is that it is a solitary pursuit. In the crafting of our book, nothing could have been further from the truth.

This book is the product of numerous conversations, the best of which seem to have been somehow associated with either a fly rod, a glass of wine, or a pair of well-worn hiking boots. So, to my family and friends, this is a small down payment on a great debt. Beyond repayment, however, is my debt to the Montana fishing crowd. Thank you, thank you, thank you for those wild dinners, death marches to the Sphinx, and twilit summer nights on Varney Bridge after a day of writing: Ted Leeson, Betty Campbell, Greg and Cindy Leeson, Paul Hextell, Jimmy Schollmeyer, and George Hopper. Tatonka! May we have many more years to boot the braids for the big 'bows.

To Dr. Phillip Potter, from the kindness of a stranger the rarest of gifts: redemption.

To the people and organizations that have given me shelter and offered financial support: the Center for Arts Criticism; Mesa Refuge; the Graham Foundation for Advanced Studies in the Fine Arts; the incomparable Joan Drury and her generous gift of Norcroft; Jean Replinger of the Oberholtzer Foundation and the pleasure of her company on Mallard Island; and especially to Adrienne Waltking and Michael Harris, who funded the final leg of the book and offered many nights of dinner and dissertating under the stars.

And finally, to Dan Brouwer, my dancing partner in Carcass Canyon, and to Prasad "Rooz" Boradkar, who lives up to the meaning of his name—"a gracious gift"—to those of us lucky enough to know him. To my coauthor, Chel: Your life has changed my life. I walk the world with exponentially more gratitude, wonder, and creaturely joy.

And thanks most of all to my late husband, Paul Rothstein. For our first Christmas together, he tucked a pair of hiking boots under the tree. I've been walking ever since, discovering worlds more beautiful than I could ever have imagined. Wherever I go, Lizard, you are.

From Chel Anderson

Enduring gratitude to my parents and sister, who enjoyed and explored the natural world with me from the start and are unwavering in their love and support. I have also been enormously blessed with curious, thoughtful, and playful friends and colleagues too many to list—you know who you are. May delight, wonder, and connection with our earthly home always be yours! Our conversations and shared experiences and your own stories, observations, knowledge, and questions are an irreplaceable source of learning, inspiration, and joy to me.

To Heidi, my coauthor, our friendship has been and will remain this endeavor's greatest gift.

To my husband, John, my deep and abiding appreciation for being my exceptional companion in wild places and steadfast partner in so many aspects of daily life that sustain and fulfill me. ෴

Illustration Credits

Frontispiece Photographs

Page 1:   Photograph by Beau Liddell. Images by Beaulin.

Pages 8–9:   Photograph by Gary Alan Nelson.

Page 59:   Photograph copyright Stan Tekiela / NatureSmartImages.com.

Page 71:   Photograph by Per Foreby.

Page 79:   Photograph by Skip Moody / Dembinsky Photo Associates. Copyright 2014; all rights reserved.

Page 89:   Photograph copyright Patrick Clayton / Engbretson Underwater Photography.

Pages 110–11:   Photograph by Gary Alan Nelson.

Page 169:   Photograph copyright schizoform. Creative Commons BY 2.0.

Page 181:   Photograph copyright aecole2010. Creative Commons BY 2.0.

Page 195:   Photograph by Warren Uxley.

Page 203:   Courtesy of the Northwoods Research Center, Wildlife Research Institute, Ely, Minnesota.

Page 217:   Photograph by W. T. Helfrich.

Pages 224–25:   Photograph by Gary Alan Nelson.

Page 311:   Photograph by Sara Oyer.

Page 317:   Photograph by Rolf Hicker, HickerPhoto.com.

Page 333:   Photograph copyright Matthew Shipp. Creative Commons BY 2.0.

Page 349:   Photograph by Gary Alan Nelson.

Page 355:   Photograph copyright Stan Tekiela / NatureSmartImages.com.

Pages 362–63:   Photograph by John Gregor / ColdSnap Photography.

Page 419:   Courtesy of the James Ford Bell Library, University of Minnesota.

Page 425:   Courtesy of the U.S. Environmental Protection Agency Great Lakes National Program Office.

Page 459:   Courtesy of Mark Edlund, St. Croix Watershed Research Station.

Page 469:   Photograph by Bill McKibben.

Page 477:   Photograph by John Gregor / ColdSnap Photography.

Page 483:   Photograph by Greg Kretovic / www.michignanaturephotos.com.

Pages 488–89:   Photograph by Gary Alan Nelson.

Page 521:   Photograph copyright Stan Tekiela / NatureSmartImages.com.

Page 535:   Photograph by Rolf O. Peterson.

Page 539:   Photograph copyright Ray Dumas. Creative Commons BY 2.0.

Page 549:   Photograph copyright Paul Vecsei / Engbretson Underwater Photography.

Page 565:   Photograph by Gary Alan Nelson.

Sources for Illustrations

Pages 4, 5, 10:   Mark A. White and George E. Host, *Mapping Range of Natural Varia-tion Ecosystem Classes for the Northern Superior Highlands: Draft Map and Analytical Methods* (Duluth: University of Minnesota and Natural Resources Research Institute, 2000).

Pages 13, 46, 136–37, 140, 374: John R. Tester, *Minnesota's Natural Heritage: An Ecolog-ical Perspective* (Minneapolis: University of Minnesota Press, 1995).

Page 47:   Jim Manolis, *Project Summary: Results from the Minnesota Forest Spatial Analysis and Modeling Project* (St. Paul: Minnesota Forest Resources Council, 2003).

Page 60:   C. S. Robbins, D. Bystrak, and P. H. Geissler, *The Breeding Bird Survey: Its First Fifteen Years, 1965–1979.* U.S. Fish and Wildlife Service Resource Publication 157, Washington, D.C., 1986.

Pages 62, 67:   Janet Green, *Birds and Forests: A Management and Conservation Guide* (St. Paul: Minnesota Department of Natural Resources, 1995).

Page 72:   Minnesota Department of Natural Resources, "Special Forest Products Product Specifications, Units of Measure and Harvest Specifications," 2006.

Pages 118, 234, 366: Richard W. Ojakangas and Charles L. Matsch, *Minnesota's Geology* (Minneapolis: University of Minnesota Press, 1982), 107; Edmund C. Bray, *Billions of Years in Minnesota: The Geological Story of the State* (St. Paul: Science Museum of Minnesota, 1980).

Page 365 (below): John C. Green, *Geology on Display: Geology and Scenery of Minneso-ta's North Shore State Parks* (St. Paul: Minnesota Department of Natural Resources, 1996).

Page 252:   Edmund Jefferson Danziger Jr., *The Chippewas of Lake Superior* (Norman: University of Oklahoma Press, 1990).

Page 263 (below): *Sixteenth Annual Report of the Forestry Commissioner of Minne-sota for the Year 1910* (St. Paul: State of Minnesota, Office of Forestry Commissioner, 1911).

Page 266:   John Disturnell, *Sailing on the Great Lakes and Rivers of America* (Philadelphia, 1874).

Page 274:   Chilson D. Aldrich, *The Real Log Cabin* (New York: Macmillan Company, 1928).

Pages 326, 329: University of Minnesota Extension Service, Onsite Sewage Treatment Program, 2008, http://www.septic.umn.edu.

Page 338:   Minnesota Department of Natural Resources, Rare Species Guide, www.dnr.state.mn.us/rsg.

Page 356:   Jonathan Elphick, ed., *Atlas of Bird Migration: Tracing the Great Jour-neys of the World's Birds* (Richmond Hill, Ont.: Firefly Books, 2007).

Pages 365 (above), 434: Natural Resources Research Institute, University of Minnesota Duluth, 1998.

Page 372 (left): Justin G. Mychek-Londer and David B. Bunnell, "Gastric Evacuation Rate, Index of Fullness, and Daily Ration of Lake Michigan Slimy *(Cottus cognatus)* and Deepwater Sculpin *(Myoxocephalus thompsonii).*" *Journal of Great Lakes Research* 39 (2013): 327–35.

Pages 429, 430: Natural Resources Research Institute, University of Minnesota Duluth, 1999.

Page 438: Wallace S. Broecker et al., "The Chronology of the Last Deglaciation: Implications to the Cause of the Younger Dryas Event," *Paleoceanography* 3 (1988).

Page 450: Mark W. Harrington, *Surface Currents of the Great Lakes* (U.S. Department of Agriculture: Weather Bureau, 1895).

Page 455: W. Charles Kerfoot and George Lauster, "Paleolimnological Study of Copper Mining around Lake Superior: Artificial Varves from Portage Lake Provide a High Resolution Record," *Limnology and Oceanography* 39 (1994): 649–69.

Page 470: National Oceanographic and Atmospheric Administration, Great Lakes Environmental Research Lab, June 17, 2013.

Page 480: http://www.seagrant.umn.edu/newsletter/2005/12/from_the_great_lakes_to_the_gulf.html

Page 485: Ben J. Korgen, "Seiches," *American Scientist* 83, no. 4 (July–August 1995): 332.

Page 490: U.S. Geological Survey, Core Science Analytics and Synthesis–USGS–NPS Vegetation Characterization Program, "Isle Royale National Park Vegetation Map," June 2009.

Pages 502 (above), 504: Thomas Gale, *Isle Royale: A Photographic History* (Houghton, Mich.: Isle Royale Natural History Association, 1995).

Page 541: Francesca J. Cuthbert and Joan McKearnan, Department of Fish and Wildlife, University of Minnesota, 1998.

Page 551 (above): Nicholas E. Mandrak and E. J. Crossman. "Postglacial Dispersal of Freshwater Fishes into Ontario," *Canadian Journal of Zoology* 70, no. 11 (1992).

Page 551 (below): "A Mansion of Many Rooms: Lean and Siscowet Lake Trout and Their Habitat Depths," Minnesota Sea Grant, 2002.

Page 558: U.S. Geological Survey, Great Lakes Science Center, 2000.

Page 572: Kathleen Weflen, "The Crossroads of Climate Change," *Minnesota Conservation Volunteer,* January/February 2001.

Page 576: Kenneth E. Kunkel, David R. Easterling, Kenneth Hubbard, and Kelly Redmond, "Temporal Variations in Frost-Free Season in the United States: 1895–2000," *Geophysical Research Letters* 31, no. 3 (February 2004).

Page 585 (above): D. R. Rondy, "Great Lakes Ice Cover Winter 1962–63 and 1963–64," Basic Data Report 5–5, U.S. Army Corps of Engineers, Detroit, Mich.

Page 585 (below): Jay A. Austin and Steven Colman, "Lake Superior Summer Water Temperatures Are Increasing More Rapidly Than Regional Air Temperatures: A Positive Ice-Albedo Feedback," *Geophysical Research Letters* 34, no. 6 (March 2007), doi:10.1029/2006GL02902/.

Page 586: Minnesota Sea Grant , "Map of Coastal Wetlands in U.S. Portion of Lake Superior."

Index

Chel Anderson has lived and worked on Minnesota's North Shore since 1974. She has worked in the Superior National Forest, as a consulting ecologist and botanist in the private and public sectors, and with the Minnesota Biological Survey. She has written on forest ecology and natural area conservation, and she leads natural history classes and field trips for educational and conservation organizations. In 2001 she received the Conservation Award from the Minnesota Chapter of The Nature Conservancy for her "tireless efforts to inventory northeast Minnesota's plant communities, her work to support conservation action with sound science, and her extraordinary ability to inspire passion for wild things and wild places."

Adelheid Fischer is a writer who focuses on natural history, ecology, and environmental history. She has written hundreds of articles for many publications, including *Utne Reader, Orion, Conservation,* and *Arizona Highways.* She is the coauthor of *Valley of Grass: Tallgrass Prairie and Parkland of the Red River Region,* winner of a Minnesota Book Award for nature writing. She has received Critics' Travel grants from the Center for Arts Criticism, Twin Cities; a grant from the Graham Foundation for Advanced Studies in the Fine Arts; and the Ellen Meloy Desert Writers Award. A sixteen-year resident of Minnesota, she now lives at the foot of South Mountain in the Sonoran Desert of Arizona.